T0203752

Modeling and Inverse Problems in the Presence of Uncertainty

MONOGRAPHS AND RESEARCH NOTES IN MATHEMATICS

Series Editors

John A. Burns
Thomas J. Tucker
Miklos Bona
Michael Ruzhansky
Chi-Kwong Li

Published Titles

Iterative Optimization in Inverse Problems, Charles L. Byrne

Modeling and Inverse Problems in the Presence of Uncertainty, H. T. Banks, Shuhua Hu, and W. Clayton Thompson

Forthcoming Titles

Sinusoids: Theory and Technological Applications, Prem K. Kythe

Stochastic Cauchy Problems in Infinite Dimensions: Generalized and Regularized Solutions, Irina V. Melnikova and Alexei Filinkov

Signal Processing: A Mathematical Approach, Charles L. Byrne

Monomial Algebra, Second Edition, Rafael Villarreal

Groups, Designs, and Linear Algebra, Donald L. Kreher

Geometric Modeling and Mesh Generation from Scanned Images, Yongjie Zhang

Difference Equations: Theory, Applications and Advanced Topics, Third Edition, Ronald E. Mickens

Set Theoretical Aspects of Real Analysis, Alexander Kharazishvili

Method of Moments in Electromagnetics, Second Edition, Walton C. Gibson

The Separable Galois Theory of Commutative Rings, Second Edition, Andy R. Magid

Dictionary of Inequalities, Second Edition, Peter Bullen

Actions and Invariants of Algebraic Groups, Second Edition, Walter Ferrer Santos and Alvaro Rittatore

Practical Guide to Geometric Regulation for Distributed Parameter Systems, Eugenio Aulisa and David S. Gilliam

Analytical Methods for Kolmogorov Equations, Second Edition, Luca Lorenzi

Handbook of the Tutte Polynomial, Joanna Anthony Ellis-Monaghan and Iain Moffat

Blow-up Patterns for Higher-Order: Nonlinear Parabolic, Hyperbolic Dispersion and Schrödinger Equations, Victor A. Galaktionov, Enzo L. Mitidieri and Stanislav Pohozaev

Application of Fuzzy Logic to Social Choice Theory, John N. Mordeson, Davendar Malik and Terry D. Clark

Microlocal Analysis on R^n and on NonCompact Manifolds, Sandro Coriasco

Cremona Groups and Icosahedron, Ivan Cheltsov and Constantin Shramov

MONOGRAPHS AND RESEARCH NOTES IN MATHEMATICS

Modeling and Inverse Problems in the Presence of Uncertainty

H. T. Banks, Shuhua Hu, and W. Clayton Thompson

North Carolina State University
Raleigh, USA

CRC Press
Taylor & Francis Group
Boca Raton London New York

CRC Press is an imprint of the
Taylor & Francis Group, an **informa** business

A CHAPMAN & HALL BOOK

CRC Press
Taylor & Francis Group
6000 Broken Sound Parkway NW, Suite 300
Boca Raton, FL 33487-2742

First issued in paperback 2019

© 2014 by Taylor & Francis Group, LLC
CRC Press is an imprint of Taylor & Francis Group, an Informa business

No claim to original U.S. Government works

ISBN-13: 978-1-4822-0642-5 (hbk)
ISBN-13: 978-0-367-37875-2 (pbk)

Visit the Taylor & Francis Web site at
http://www.taylorandfrancis.com

and the CRC Press Web site at
http://www.crcpress.com

Contents

Preface

Writing a research monograph on a "hot topic" such as "uncertainty propagation" is a somewhat daunting undertaking. Nonetheless, we decided to collect our own views, supported by our own research efforts over the past 12–15 years on a number of aspects of this topic, and summarize these for the possible enlightenment they might provide (for us, our students and others). The research results discussed below are thus necessarily filled with a preponderance of references to our own research reports and papers. In numerous references below (given at the conclusion of each chapter), we refer to CRSC-TRXX-YY. This refers to early Technical Report versions of manuscripts which can be found on the Center for Research in Scientific Computation website at North Carolina State University where XX refers to the year, e.g., XX = 03 is 2003, XX = 99 is 1999, while the YY refers to the number of the report in that year. These can be found at and downloaded from http://www.ncsu.edu/crsc/reports.html where they are listed by year.

Our presentation here has an intended audience from the community of investigators in applied mathematics interested in deterministic and/or stochastic models and their interactions as well as scientists in biology, medicine, engineering and physics interested in basic modeling and inverse problems, uncertainty in modeling, propagation of uncertainty and statistical modeling.

We owe great thanks to our former and current students, postdocs and colleagues for their patience in enduring lectures, questions, feedback and some proofreading. Special thanks are due (in no particular order) to Zack Kenz, Keri Rehm, Dustin Kapraun, Jared Catenacci, Katie Link, Kris Rinnovatore, Kevin Flores, John Nardini, Karissa Cross and Laura Poag for careful reading of notes and suggested corrections/revisions on subsets of the material for this monograph. However, in a sincere attempt to give credit where it is due, each of the authors firmly insists that any errors in judgment, mathematical content, grammar or typos in the material presented in this monograph are entirely the responsibility of his/her two co-authors!!

We (especially young members of our research group) have been generously supported by research grants and fellowships from US federal funding agencies including AFSOR, DARPA, NIH, NSF, DED, and DOE. For this support and encouragement we are all most grateful.

H.T. Banks
Shuhua Hu
W. Clayton Thompson

Chapter 1

Introduction

The terms **uncertainty quantification** and **uncertainty propagation** have become so widely used as to almost have little meaning unless they are further explained. Here we focus primarily on two basic types of problems:

1. Modeling and inverse problems where one assumes that a precise mathematical model without modeling error is available. This is a standard assumption underlying a large segment of what is taught in many modern statistics courses with a frequentist philosophy. More precisely, a mathematical model is given by a dynamical system

$$\frac{d\boldsymbol{x}}{dt}(t) = \boldsymbol{g}(t, \boldsymbol{x}(t), \boldsymbol{q}) \tag{1.1}$$

$$\boldsymbol{x}(t_0) = \boldsymbol{x}_0 \tag{1.2}$$

with **observation process**

$$\boldsymbol{f}(t; \boldsymbol{\theta}) = \mathcal{C}\boldsymbol{x}(t; \boldsymbol{\theta}), \tag{1.3}$$

where $\boldsymbol{\theta} = (\boldsymbol{q}, \boldsymbol{x}_0)$. The mathematical model is an n-dimensional deterministic system and there is a corresponding "truth" parameter $\boldsymbol{\theta}_0 = (\boldsymbol{q}_0, \boldsymbol{x}_{00})$ so that in the presence of no measurement error the data can be described exactly by the deterministic system at $\boldsymbol{\theta}_0$. Thus, uncertainty is present entirely due to some **statistical model** of the form

$$\boldsymbol{Y}_j = \boldsymbol{f}(t_j; \boldsymbol{\theta}_0) + \boldsymbol{\mathcal{E}}_j, \quad j = 1, \ldots, N, \tag{1.4}$$

where $\boldsymbol{f}(t_j; \boldsymbol{\theta}) = \mathcal{C}\boldsymbol{x}(t_j; \boldsymbol{\theta})$, $j = 1, \ldots, N$, corresponds to the observed part of the solution of the mathematical model (1.1)–(1.2) at the jth covariate or observation time and $\boldsymbol{\mathcal{E}}_j$ is some type of (possibly state dependent) measurement error. For example, we consider errors that include those of the form $\boldsymbol{\mathcal{E}}_j = \boldsymbol{f}(t_j; \boldsymbol{\theta}_0)^\gamma \circ \tilde{\boldsymbol{\mathcal{E}}}_j$ where the operation $^\gamma\circ$ denotes component-wise exponentiation by γ followed by component-wise multiplication and $\gamma \geq 0$.

2. An alternate problem wherein the mathematical modeling *itself* is a major source of uncertainty and this uncertainty usually propagates in time. That is, the mathematical model has major uncertainties in its form and/or its parametrization and/or its initial/boundary data, and this uncertainty is propagated dynamically via some framework as yet to be determined.

Before we begin the inverse problem discussions, we give a brief but useful review of certain basic probability and statistical concepts. After the probability and statistics review we present a chapter summarizing both mathematical and statistical aspects of inverse problem methodology which includes ordinary, weighted and generalized least-squares formulations. We discuss asymptotic theories, bootstrapping and issues related to evaluation of the correctness of the assumed form of statistical models. We follow this with a discussion of methods for evaluating and comparing the validity of appropriateness of a collection of models for describing a given data set, including statistically based model selection and model comparison techniques.

In Chapter 5 we present a summary of recent results on the estimation of probability distributions when they are embedded in complex mathematical models and only aggregate (not individual) data are available. This is followed by a brief chapter on optimal design (what to measure? when and where to measure?) of experiments to be carried out in support of inverse problems for given models.

The last two chapters focus on the uncertainty in model formulation itself (the second item listed above as the focus of this monograph). In Chapter 7 we consider the general problem of evolution of probability density functions in time. This is done in the context of associated processes resulting from stochastic differential equations (SDE), which are driven by white noise, and those resulting from random differential equations (RDE), which are driven by colored noise. We also discuss their respective wide applications in a number of different fields including physics and biology. We also consider the general relationship between SDE and RDE and establish that there are classes of problems for which there is an equivalence between the solutions of the two formulations. This equivalence, which we term pointwise equivalence, is in the sense that the respective probability density functions are the same at each time t. We show, however, that the stochastic processes resulting from the SDE and its corresponding pointwise equivalent RDE are generally not the same in that they may have different covariance functions.

In a final chapter we consider questions related to the appropriateness of discrete versus continuum models in transitions from small numbers of individuals (particles, populations, molecules, etc.) to large numbers. These investigations are carried out in the context of continuous time Markov chain (CTMC) models and the Kurtz limit theorems for approximations for large number stochastic populations by ordinary differential equations for corresponding mean populations. Algorithms for simulating CTMC models and CTMC models with delays (discrete and random) are explained and simulations are presented for problems arising in specific applications.

The monograph contains illustrative examples throughout, many of them directly related to research projects carried out by our group at North Carolina State University over the past decade.

Chapter 2

Probability and Statistics Overview

The theory of probability and statistics is an essential mathematical tool in the formulation of inverse problems, in the development of subsequent analysis and approaches to statistical hypothesis testing and model selection criteria, and in the study of uncertainty propagation in dynamic systems. Our coverage of these fundamental and important topics is brief and limited in scope. Indeed, we provide in this section a few definitions and basic concepts in the theory of probability and statistics that are essential for the understanding of estimators, confidence intervals, model selection criteria and stochastic processes. For more information on the topics in this chapter, selected references are provided at the end of these as well as subsequent chapters.

2.1   Probability and Probability Space

The set of all possible outcomes in a statistical experiment is called the *sample space* and is denoted by Ω. Each element $\omega \in \Omega$ is called a *sample point*. A collection of outcomes in which we are interested is called an *event*; that is, an event is a subset of Ω. For example, consider the experiment of rolling a six-sided die. In this case, there are six possible outcomes, and the sample space can be represented as

$$\Omega = \{1, 2, 3, 4, 5, 6\}. \tag{2.1}$$

An event $\mathbb{A}$ might be defined as

$$\mathbb{A} = \{1, 5\}, \tag{2.2}$$

which consists of the outcomes 1 and 5. Note that we say the event $\mathbb{A}$ occurs if the outcome of the experiment is in the set $\mathbb{A}$. Consider another experiment of tossing a coin three times. A sample point in this experiment indicates the result of each toss; for example, HHT indicates that two heads and then a tail were observed. The sample space for this experiment has eight sample points; that is,

$$\Omega = \{HHH, HHT, HTH, THH, TTH, THT, HTT, TTT\}. \tag{2.3}$$

3

An event $\mathbb{A}$ might be defined as the set of outcomes for which the first toss is a head; that is,

$$\mathbb{A} = \{\mathrm{HHH, HHT, HTH, HTT}\}.$$

Thus, we see that the sample point could be either a numerical value or a character value. Based on the number of sample points contained in the sample space, the sample space can be either finite (as we illustrate above) or infinitely countable (e.g., the number of customers to arrive in a bank) or uncountable (e.g., the sample space for the lifetime of a bulb or the sample space for the reaction time to a certain stimulus).

Definition 2.1.1 *Let Ω be the given sample space. Then the σ-algebra $\mathcal{F}$ is a collection of subsets of Ω with the following properties:*

(i) *The empty set $\emptyset$ is an element of $\mathcal{F}$; that is, $\emptyset \in \mathcal{F}$.*

(ii) *(closed under complementation): If $\mathbb{A} \in \mathcal{F}$, then $\mathbb{A}^c \in \mathcal{F}$, where $\mathbb{A}^c$ denotes the complement of the event $\mathbb{A}$, which consists of all sample points in Ω that are not in $\mathbb{A}$.*

(iii) *(closed under countable unions): If $\mathbb{A}_1, \mathbb{A}_2, \mathbb{A}_3, \dots, \in \mathcal{F}$, then $\bigcup_j \mathbb{A}_j \in \mathcal{F}$.*

The pair $(\Omega, \mathcal{F})$ is called a measurable space. A subset $\mathbb{A}$ of Ω is said to be a measurable set (or event) if $\mathbb{A} \in \mathcal{F}$.

It is worth noting that a σ-algebra is also called a σ-field in the literature. In addition, one can define many different σ-algebras associated with the sample space Ω. The σ-algebra we will mainly consider is the smallest one that contains all of the open sets in the given sample space. In other words, it is the algebra generated by a topological space, whose definition is given as follows.

Definition 2.1.2 *A topology $\mathcal{T}$ on a set Ω is a collection of subsets of Ω having the following properties:*

- *$\emptyset \in \mathcal{T}$ and $\Omega \in \mathcal{T}$.*

- *(closed under finite intersection): If $\mathbb{U}_i \in \mathcal{T}$, $i = 1, 2, \dots, l$, with l being a positive integer, then $\bigcap_{i=1}^{l} \mathbb{U}_i \in \mathcal{T}$.*

- *(closed under arbitrary union): If $\{\mathbb{U}_\alpha\}$ is an arbitrary collection of members of $\mathcal{T}$ (finite, countable, uncountable), then $\bigcup_\alpha \mathbb{U}_\alpha \in \mathcal{T}$.*

The pair $(\Omega, \mathcal{T})$ is called a topological space. A subset $\mathbb{U}$ of Ω is said to be an open set if $\mathbb{U} \in \mathcal{T}$. A subset $\mathbb{A}$ of Ω is said to be closed if $\mathbb{A}^c$ is an open set.

A σ-algebra generated by all the open sets in a given sample space Ω is often called a *Borel algebra* or *Borel σ-algebra* (denoted by $\mathcal{B}(\Omega)$), and the sets in a Borel algebra are called *Borel sets*.

Associated with an event $\mathbb{A} \in \mathcal{F}$ is its probability Prob$\{\mathbb{A}\}$, which indicates the likelihood that event $\mathbb{A}$ occurs. For example, in a fair experiment of rolling a die, where one assumes that each possible sample point has probability $\frac{1}{6}$, then the event $\mathbb{A}$ as defined by (2.2) has probability Prob$\{\mathbb{A}\} = \frac{2}{6} = \frac{1}{3}$. The strict definition for the probability is given as follows.

Definition 2.1.3 *A probability Prob on the measurable space $(\Omega, \mathcal{F})$ is a set function Prob : $\mathcal{F} \to [0, 1]$ such that*

(i) Prob$\{\emptyset\} = 0$, Prob$\{\Omega\} = 1$.

(ii) (completely additive): If $\mathbb{A}_1, \mathbb{A}_2, \mathbb{A}_3, \ldots$, is a finite or an infinite sequence of disjoint subsets in $\mathcal{F}$, then Prob$\{\cup_j \mathbb{A}_j\} = \sum_j Prob\{\mathbb{A}_j\}$.

The triplet $(\Omega, \mathcal{F}, Prob)$ is called a probability space.

It is worth noting that a probability is also called a *probability measure* or a *probability function* (these names will be used interchangeably in this monograph), and a probability measure defined on a Borel algebra is called a *Borel probability measure*.

Using Definition 2.1.3 for probability, a number of immediate consequences can also be derived which have important applications. For example, the probability that an event will occur and that it will not occur always sum to 1. That is,
$$\text{Prob}\{\mathbb{A}\} + \text{Prob}\{\mathbb{A}^c\} = 1.$$
In addition, if $\mathbb{A}, \mathbb{B} \in \mathcal{F}$, then we have
$$\text{Prob}\{\mathbb{A}\} \le \text{Prob}\{\mathbb{B}\} \text{ if } \mathbb{A} \subset \mathbb{B}.$$
It can also be found that if $\mathbb{A}_j \in \mathcal{F}$, $j = 1, 2, \ldots$, then
$$\text{Prob}\left\{\cup_{j=1}^{\infty} \mathbb{A}_j\right\} \le \sum_{j=1}^{\infty} \text{Prob}\{\mathbb{A}_j\}.$$

If Prob$\{\mathbb{A}\} = 1$, then we say that the event $\mathbb{A}$ occurs "with probability 1" or "*almost surely* (a.s.)." A set $\mathbb{A} \in \mathcal{F}$ is called a *null set* if Prob$\{\mathbb{A}\} = 0$. In addition, a probability space $(\Omega, \mathcal{F}, \text{Prob})$ is said to be *complete* if for any two sets $\mathbb{A}$ and $\mathbb{B}$ the following condition holds: If $\mathbb{A} \subset \mathbb{B}$, $\mathbb{B} \in \mathcal{F}$ and Prob$\{\mathbb{B}\} = 0$, then $\mathbb{A} \in \mathcal{F}$. It is worth noting that any probability space can be extended into a complete probability space (e.g., see [15, p. 10] and the references therein). Hence, we will assume that all the probability spaces are complete in the remainder of this monograph.

Remark 2.1.1 *We remark that we can generalize a probability space to more general measure spaces. Specifically, a measure ν on the measurable space $(\Omega, \mathcal{F})$ is a set function $\nu : \mathcal{F} \to [0, \infty]$ such that $\nu(\emptyset) = 0$ and ν is completely additive (that is, the second property of probability in Definition 2.1.3 holds). The triplet $(\Omega, \mathcal{F}, \nu)$ is called a measure space. If $\nu(\Omega)$ is finite, then ν is said to be a finite measure. In particular, Prob is a normalized finite measure with $Prob(\Omega) = 1$. Hence, a probability possesses all the general properties of a finite measure.*

Remark 2.1.2 *Let $(\Omega, \mathcal{F}, \nu)$ be a measure space. If $\Omega = \bigcup_{j=1}^{\infty} \mathbb{A}_j$ and $\nu(\mathbb{A}_j)$ is finite for all j, then we say that ν is a σ-finite measure and $(\Omega, \mathcal{F}, \nu)$ is a σ-finite measure space. Another measure that we will consider in this monograph is the Lebesgue measure, which is defined on the measurable space $(\mathbb{R}^l, \mathcal{B}(\mathbb{R}^l))$ and is given by*

$$\nu((a_1, b_1) \times \cdots \times (a_l, b_l)) = \prod_{j=1}^{l} (b_j - a_j),$$

that is, the volume of the interval $(a_1, b_1) \times \cdots \times (a_l, b_l)$. We thus see that the Lebesgue measure of a countable set of points is zero (that is, a countable set of points is a null set with respect to the Lebesgue measure), and that the Lebesgue measure of a k-dimensional plane in $\mathbb{R}^l$ ($l > k$) is also zero. We refer the interested reader to some real analysis textbooks such as [11, 23] for more information on a measure as well as its properties.

2.1.1   Joint Probability

Instead of considering a single experiment, let us perform two experiments and consider their outcomes. For example, the two experiments may be two separate tosses of a single die or a single toss of two dice. The sample space in this case consists of 36 pairs (k, j), where $k, j = 1, 2, \ldots, 6$. Note that in a fair dice game, each sample point in the sample space has probability $\dfrac{1}{36}$. We now consider the probability of joint events, such as $\{k = 2, j = \text{odd}\}$. We begin by denoting the event of one experiment by $\mathbb{A}_k$, $k = 1, 2, \ldots, l$, and the event of the second experiment by $\mathbb{B}_j$, $j = 1, 2, \ldots, m$. The combined experiment has the joint events $(\mathbb{A}_k, \mathbb{B}_j)$, where $k = 1, 2, \ldots, l$ and $j = 1, 2, \ldots, m$.

The joint probability $Prob\{\mathbb{A}_k, \mathbb{B}_j\}$, also denoted by $Prob\{\mathbb{A}_k \cap \mathbb{B}_j\}$ (which will be occasionally used in this monograph for notational convenience) or $Prob\{\mathbb{A}_k \mathbb{B}_j\}$ in the literature, indicates the likelihood that the events $\mathbb{A}_k$ and $\mathbb{B}_j$ occur simultaneously. By Definition 2.1.3, a number of immediate consequences can also be derived for the joint probability. For example, $Prob\{\mathbb{A}_k, \mathbb{B}_j\}$ satisfies the condition $0 \leq Prob\{\mathbb{A}_k, \mathbb{B}_j\} \leq 1$. In addition, if $\mathbb{B}_j$ for $j = 1, 2, \ldots, m$ are mutually exclusive (i.e., $\mathbb{B}_i \cap \mathbb{B}_j = \emptyset, i \neq j$) such that $\bigcup_{j=1}^{m} \mathbb{B}_j = \Omega$, then

$$\sum_{j=1}^{m} \text{Prob}\{\mathbb{A}_k, \mathbb{B}_j\} = \text{Prob}\{\mathbb{A}_k\}. \tag{2.4}$$

Furthermore, if all the outcomes of the two experiments are mutually exclusive such that $\bigcup_{k=1}^{l} \mathbb{A}_k = \Omega$ and $\bigcup_{j=1}^{m} \mathbb{B}_j = \Omega$, then $\sum_{k=1}^{l} \sum_{j=1}^{m} \text{Prob}\{\mathbb{A}_k, \mathbb{B}_j\} = 1$. The generalization of the above concept to more than two experiments follows in a straightforward manner.

2.1.2 Conditional Probability

Next, we consider a joint event with probability $\text{Prob}\{\mathbb{A}, \mathbb{B}\}$. Assuming that event $\mathbb{A}$ has occurred and $\text{Prob}\{\mathbb{A}\} > 0$, we wish to determine the probability of the event $\mathbb{B}$. This is called the *conditional probability* of event $\mathbb{B}$ given the occurrence of event $\mathbb{A}$ and is given by

$$\text{Prob}\{\mathbb{B}|\mathbb{A}\} = \frac{\text{Prob}\{\mathbb{A}, \mathbb{B}\}}{\text{Prob}\{\mathbb{A}\}}. \tag{2.5}$$

Definition 2.1.4 *Two events, $\mathbb{A}$ and $\mathbb{B}$, are said to be statistically independent if and only if*

$$Prob\{\mathbb{A}, \mathbb{B}\} = Prob\{\mathbb{A}\}Prob\{\mathbb{B}\}. \tag{2.6}$$

Statistical independence is often simply called *independence*. By (2.5) and (2.6), we see that if $\mathbb{A}$ and $\mathbb{B}$ are independent, then

$$\text{Prob}\{\mathbb{B}|\mathbb{A}\} = \text{Prob}\{\mathbb{B}\}. \tag{2.7}$$

In addition, we observe that if (2.7) holds, then by (2.5) and Definition 2.1.4 we know that $\mathbb{A}$ and $\mathbb{B}$ are independent. Thus, (2.7) can also be used as a definition for the independence of two events.

Two very useful relationships for conditional probabilities can be given. If $\mathbb{A}_k$, $k = 1, 2, \ldots, l$, are mutually exclusive events such that $\bigcup_{k=1}^{l} \mathbb{A}_k = \Omega$ and $\mathbb{B}$ is an arbitrary event with $\text{Prob}\{\mathbb{B}\} > 0$, then by (2.4) and (2.5) we have

$$\text{Prob}\{\mathbb{B}\} = \sum_{j=1}^{l} \text{Prob}\{\mathbb{A}_j, \mathbb{B}\} = \sum_{j=1}^{l} \text{Prob}\{\mathbb{B}|\mathbb{A}_j\}\text{Prob}\{\mathbb{A}_j\}, \tag{2.8}$$

and

$$\text{Prob}\{\mathbb{A}_k|\mathbb{B}\} = \frac{\text{Prob}\{\mathbb{A}_k,\mathbb{B}\}}{\text{Prob}\{\mathbb{B}\}} = \frac{\text{Prob}\{\mathbb{B}|\mathbb{A}_k\}\text{Prob}\{\mathbb{A}_k\}}{\sum_{j=1}^{l}\text{Prob}\{\mathbb{B}|\mathbb{A}_j\}\text{Prob}\{\mathbb{A}_j\}}. \tag{2.9}$$

Equation (2.8) is often called the *law of total probability*. Equation (2.9) is known as Bayes' formula or Bayes' Theorem. Here $\text{Prob}\{\mathbb{A}_k\}$ is called a *prior* probability of event $\mathbb{A}_k$, $\text{Prob}\{\mathbb{B}|\mathbb{A}_k\}$ is called the likelihood of $\mathbb{B}$ given $\mathbb{A}_k$, and $\text{Prob}\{\mathbb{A}_k|\mathbb{B}\}$ is called a *posterior* probability of event $\mathbb{A}_k$ obtained by using the information gained from $\mathbb{B}$.

2.2   Random Variables and Their Associated Distribution Functions

In most applications of probability theory, we are not interested in the details associated with each sample point but rather in some numerical description of the outcome of an experiment. For example, in the experiment of tossing a coin three times, we might only be interested in the number of heads obtained in these three tosses. In the language of probability and statistics, the number of heads obtained in these three tosses is called a *random variable*. The values of this particular random variable corresponding to each sample point in (2.3) are given by

$$\text{HHH HHT HTH THH TTH THT HTT TTT}$$
$$32221110.$$

This implies that the range (i.e., the collection of all possible values) of this random variable is $\{0, 1, 2, 3\}$. The strict definition of a random variable is as follows.

Definition 2.2.1 *A function* $X : \Omega \to \mathbb{R}$ *is said to be a random variable defined on a measurable space* $(\Omega, \mathcal{F})$ *if for any* $x \in \mathbb{R}$ *we have* $\{\omega \in \Omega : X(\omega) \le x\} \in \mathcal{F}$. *Such a function is said to be measurable with respect to* $\mathcal{F}$. *In addition, for any fixed* $\omega \in \Omega$, $X(\omega)$ *is called a realization of this random variable.*

As is usually done and also for notational convenience, we suppress the dependence of the random variable on ω if no confusion occurs; that is, we denote $X(\cdot)$ by X. Under this convention, the set $\{\omega \in \Omega : X(\omega) \le x\}$ is simply written as $\{X \le x\}$. In addition, a realization of a random variable is simply denoted by its corresponding lower case letter; for example, a realization of random variable X is denoted by x. We point out that when we consider a sequence of random variables in this monograph, we assume that they are defined on the same probability space.

The σ-algebra generated by the random variable X, often denoted by $\sigma(X)$, is given by

$$\sigma(X) = \{\mathbb{A} \subset \Omega \mid \mathbb{A} = X^{-1}(\mathbb{B}),\ \mathbb{B} \in \mathcal{B}(\mathbb{R})\},$$

where $\mathcal{B}(\mathbb{R})$ is the Borel algebra on $\mathbb{R}$. It is worth noting that this is the smallest σ-algebra with respect to which X is measurable. This means that if X is $\mathcal{F}$-measurable, then $\sigma(X) \subset \mathcal{F}$.

Remark 2.2.1 *The concept of a random variable can be generalized so that its range can be some complicated space rather than $\mathbb{R}$. Let $(\Omega, \mathcal{F}, Prob)$ be a probability space, and $(\mathbb{S}, \mathscr{S})$ be a measurable space. A function $X : \Omega \to \mathbb{S}$ is said to be a random element if for any $\mathbb{B} \in \mathscr{S}$ we have $\{\omega \in \Omega : X(\omega) \in \mathbb{B}\} \in \mathcal{F}$. Specifically, if $\mathbb{S} = \mathbb{R}^m$, then we say that random element X is an m-dimensional random vector.*

2.2.1 Cumulative Distribution Function

For any random variable, there is an associated function called a cumulative distribution function, which is defined as follows.

Definition 2.2.2 *The* cumulative distribution function (CDF) *of random variable X is the function $P : \mathbb{R} \to [0,1]$ defined by*

$$P(x) = Prob\{X \le x\}, \quad x \in \mathbb{R}. \tag{2.10}$$

The cumulative distribution function is sometimes simply called the *distribution function*. It has the following properties (inherited from the probability measure):

(i) P is a right continuous function of x; that is, $\lim_{\Delta x \to 0+} P(x + \Delta x) = P(x)$.

(ii) P is a non-decreasing function of x; that is, if $x_1 \le x_2$, then $P(x_1) \le P(x_2)$.

(iii) $P(-\infty) = 0$, $P(\infty) = 1$.

The last two properties imply that the cumulative distribution function P has bounded variation, where the variation of a function is defined as follows.

Definition 2.2.3 *The m-variation of a real-valued function h on the interval $[\underline{x}, \bar{x}] \subset \mathbb{R}$ is defined as*

$$[h]^{(m)}([\underline{x}, \bar{x}]) = \sup \sum_{j=0}^{l-1} |h(x_{j+1}^l) - h(x_j^l)|^m, \tag{2.11}$$

where the supremum is taken over all partitions $\{x_j^l\}_{j=0}^l$ of $[\underline{x}, \bar{x}]$. For the case $m = 1$ it is simply called variation (it is also called total variation in the literature), and for the case $m = 2$ it is called quadratic variation. If $[h]^{(m)}([\underline{x}, \bar{x}])$ is finite, then we say that h has bounded (or finite) m-variation on the given interval $[\underline{x}, \bar{x}]$.

If h is a function of $x \in \mathbb{R}$, then h is said to have finite m-variation if $[h]^{(m)}([\underline{x}, \bar{x}])$ is finite for any given $\underline{x}$ and $\bar{x}$. In addition, h is said to have bounded m-variation if there exists a constant c_h such that $[h]^{(m)}([\underline{x}, \bar{x}]) < c_h$ for any $\underline{x}$ and $\bar{x}$, where c_h is independent of $\underline{x}$ and $\bar{x}$.

The properties of the cumulative distribution function also imply that P has derivatives almost everywhere with respect to the Lebesgue measure (that is, the set of points at which P is not differentiable is a null set with respect to the Lebesgue measure), but it should be noted that P does not have to be equal to the integral of its derivative. These properties also imply that the cumulative distribution function can only have jump discontinuities and it has at most countably many jumps. Thus, we see that the cumulative distribution function could be a step function (illustrated in the upper left panel of Figure 2.1), a continuous function (illustrated in the upper right panel of Figure 2.1) or a function with a mixture of continuous pieces and jumps (illustrated in the bottom plots of Figure 2.1). The last type of cumulative distribution function could result from a convex combination of the first two types of cumulative distribution functions. In fact, this is how we obtained the cumulative distribution functions demonstrated in the bottom plots of Figure 2.1. Specifically, let P_j be the cumulative distribution function of some random variable, $j = 1, 2, \ldots, m$, and $\varpi_j, j = 1, 2, \ldots, m$, be some non-negative numbers such that $\sum_{j=1}^{m} \varpi_j = 1$. Then we easily see that $\sum_{j=1}^{m} \varpi_j P_j$ is also a cumulative distribution function. This type of distribution is often called a *mixture distribution*.

Remark 2.2.2 *It is worth noting that there are many different ways found in the literature to define the continuity of a random variable. One is based on the range of the random variable. In this case, a <u>discrete random variable</u> is one with its range consisting of a countable subset in $\mathbb{R}$ with either a finite or infinite number of elements. For example, the random variable defined in the experiment of tossing a coin three times is a discrete random variable. A <u>continuous random variable</u> is one that takes values in a continuous interval. For example, the random variable defined in the experiment for recording the reaction time to a certain stimulus is a continuous random variable.*

Another way to define the continuity of random variables is based on the continuity of the cumulative distribution function for a random variable. Specifically, if the cumulative distribution function is continuous (as illustrated in the upper right panel of Figure 2.1), then the associated random variable is said to be continuous. If the cumulative distribution function is a

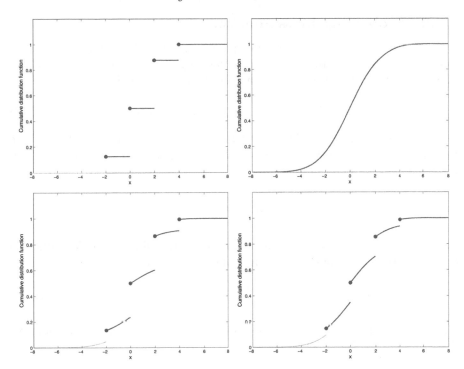

FIGURE 2.1: Cumulative distribution function: (upper left) a step function P_1, (upper right) a continuous function P_2, (lower left) $0.7P_1 + 0.3P_2$, (lower right) $0.4P_1 + 0.6P_2$.

step function (as illustrated in the upper left panel of Figure 2.1), then the associated random variable is said to be discrete. We see that the discrete random variable defined here is equivalent to that in the first case (as the cumulative distribution function has at most countably many jumps). However, the continuous random variable defined here is more restrictive than that in the first case where the cumulative distribution function of a continuous random variable may be either a continuous function or a function with a mixture of continuous pieces and jumps (as illustrated in the bottom plots of Figure 2.1).

The last way to define the continuity of random variables is based on whether or not a random variable has an associated probability density function (discussed blow); specifically, a random variable is said to be continuous if there is a probability density function associated with it. As we shall see below, this definition is even stronger than the second one. In this monograph, we define a random variable to be discrete or continuous based on its range. However, the continuous random variables we will mainly consider in this monograph are those with associated probability density functions.

2.2.2   Probability Mass Function

Any discrete random variable has an associated *probability mass function*. Without loss of generality, we assume the range of discrete random variable X is $\{x_j\}$. Then the probability mass function of X is defined by

$$\Phi(x_j) = \text{Prob}\{X = x_j\}, \quad j = 1, 2, 3, \ldots. \tag{2.12}$$

Hence, we see that the value of the probability mass function at x_j is the probability associated with x_j, $j = 1, 2, 3, \ldots$. In addition, by the definition of probability we know that

$$\sum_j \Phi(x_j) = 1.$$

For example, the probability mass function of the discrete random variable defined in the experiment of tossing a coin three times is

$$\Phi(0) = \frac{1}{8}, \quad \Phi(1) = \frac{3}{8}, \quad \Phi(2) = \frac{3}{8}, \quad \Phi(3) = \frac{1}{8}. \tag{2.13}$$

The relationship between the probability mass function and the cumulative distribution function is given by

$$P(x) = \sum_{\{j:\, x_j \leq x\}} \Phi(x_j), \quad \Phi(x_j) = P(x_j) - \lim_{x \to x_j-} P(x).$$

In such a case, the cumulative distribution function is a step function, and it is said to be discrete. Figure 2.2 illustrates the probability mass function (2.13) of the discrete random variable defined in the experiment of tossing a coin three times as well as the corresponding cumulative distribution function.

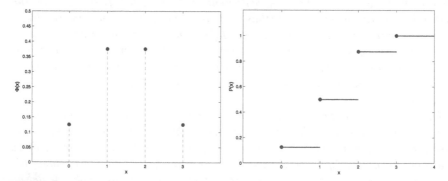

FIGURE 2.2:   (left) The probability mass function of the discrete random variable defined in the experiment of tossing a coin three times; (right) the corresponding cumulative distribution function.

2.2.3 Probability Density Function

If the derivative p of the cumulative distribution function P exists for almost all x, and for all x

$$P(x) = \int_{-\infty}^{x} p(\xi) d\xi,$$

then p is called the *probability density function*. It should be noted that a necessary and sufficient condition for a cumulative distribution function to have an associated probability density function is that P is *absolutely continuous* in the sense that for any positive number ϵ, there exists a positive number c_l such that for any finite collection of disjoint intervals $(x_j, y_j) \subset \mathbb{R}$ satisfying $\sum_j |y_j - x_j| < c_l$ then $\sum_j |P(y_j) - P(x_j)| < \epsilon$. Thus, we see that the requirement for absolute continuity is much stronger than that for continuity. This implies that there are cases in which the cumulative distribution function is continuous but not absolutely continuous. An example is the *Cantor distribution* (also called the Cantor–Lebesgue distribution; e.g., see [10, p. 169] or [17, p. 38]), where the cumulative distribution function P is a constant between the points of the Cantor set (which is an uncountable set in $\mathbb{R}$ that has Lebesgue measure zero). Such a distribution is often called a *singular continuous distribution* (the derivative of its corresponding cumulative distribution function is zero almost everywhere), and the associated random variable is often termed a *singular continuous random variable*. This type of distribution is rarely encountered in practice. Based on Lebesgue's decomposition theorem, any cumulative distribution function can be written as a convex combination of a discrete, an absolutely continuous and a singular continuous cumulative distribution function. We refer the interested reader to [2, Section 31] for more information on this topic.

The name "density function" comes from the fact that the probability of the event $x_1 \leq X \leq x_2$ is given by

$$
\begin{aligned}
\mathrm{Prob}\{x_1 \leq X \leq x_2\} &= \mathrm{Prob}\{X \leq x_2\} - \mathrm{Prob}\{X \leq x_1\} \\
&= P(x_2) - P(x_1) \\
&= \int_{x_1}^{x_2} p(x)\, dx.
\end{aligned}
\tag{2.14}
$$

In addition, the probability density function p satisfies the following properties:

$$p(x) \geq 0, \qquad \int_{-\infty}^{\infty} p(x)\, dx = P(\infty) - P(-\infty) = 1.$$

Figure 2.3 illustrates the probability density function p (shown in the left panel) of a continuous random variable X and the corresponding cumulative distribution function P (shown in the right panel). By (2.14) we know that the shaded area under the probability density function between $x = 0$ and

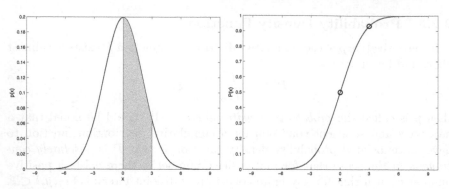

FIGURE 2.3:   (left) Probability density function of a continuous random variable; (right) the corresponding cumulative distribution function.

$x = 3$ is equal to the probability that the value of X is between 0 and 3, and is also equal to $P(3) - P(0)$ (where the points $(0, P(0))$ and $(3, P(3))$ are indicated by the circles in the right panel of Figure 2.3).

For a discrete random variable, the corresponding probability density function does not exist in the ordinary sense. But it can be constructed in the general sense (a generalized function sense, sometimes called a "distributional sense") with the help of Dirac delta "functions" [26]. Specifically, for a discrete random variable X with range $\{x_j\}$, its probability density function can be written as

$$p(x) = \sum_j \Phi(x_j)\delta(x - x_j), \qquad (2.15)$$

where δ is the Dirac delta "function," that is, $\delta(x - x_j) = \begin{cases} 0 & \text{if } x \neq x_j \\ \infty & \text{if } x = x_j \end{cases}$

with the property that $\displaystyle\int_{-\infty}^{\infty} h(x)\delta(x - x_j)dx = h(x_j)$. This construction has many advantages; for example, as we shall see later, it can put the definition of moments in the same framework for a discrete random variable and for a continuous random variable with a probability density function.

2.2.4   Equivalence of Two Random Variables

One can define the equivalence between two random variables in several senses.

Definition 2.2.4 *Two random variables are said to be equal in distribution (or identically distributed) if their associated cumulative distribution functions are equal.*

Definition 2.2.5 *Two random variables X and Y are said to be equal almost surely if $Prob\{X = Y\} = 1$.*

It is worth noting that two random variables that are equal almost surely are equal in distribution. However, the fact that two random variables are equal in distribution does not necessarily imply that they are equal almost surely. For example, in the experiment of tossing a fair coin three times, if we define X as the number of heads obtained in three tosses and Y as the number of tails observed in three tosses, then it can easily be seen that X and Y have the same probability mass function and hence are equal in distribution. However, $X(\omega) \neq Y(\omega)$ for any $\omega \in \Omega$, where Ω is defined in (2.3).

2.2.5 Joint Distribution Function and Marginal Distribution Function

The definition of cumulative distribution and probability density functions can be extended from one random variable to two or more random variables. In this case, the cumulative distribution function is often called the *joint cumulative distribution function* or simply the *joint distribution function*, and the probability density function is often called the *joint probability density function*. For example, the joint distribution function of two random variables X and Y is defined as

$$P(x, y) = \text{Prob}\{X \leq x, Y \leq y\}. \tag{2.16}$$

Similar to the one-dimensional case, the joint distribution function P is non-negative, non-decreasing and right continuous with respect to each variable. In addition,

$$\begin{aligned} P(-\infty, -\infty) &= 0, & P(x, -\infty) &= 0, & P(-\infty, y) &= 0, \\ P(\infty, \infty) &= 1, & P(x, \infty) &= P_X(x), & P(\infty, y) &= P_Y(y). \end{aligned} \tag{2.17}$$

Here P_X and P_Y are, respectively, the cumulative distribution functions of X and Y, and the subscript in P_X (P_Y) is used to emphasize that it is the cumulative distribution function of X (Y). (It should be noted that the subscript in the cumulative distribution function is always suppressed if no confusion occurs. Otherwise, we index a distribution function by the random variable it refers to.) In the context of two or more random variables, each random variable is often called a *marginal variable*, and the cumulative distribution function of each random variable is often called a *marginal distribution function*.

The corresponding joint probability density function of X and Y, if it exists, is defined as

$$p(x, y) = \frac{\partial^2}{\partial x \partial y} P(x, y). \tag{2.18}$$

Similar to the one-dimensional case, the joint probability density function of X and Y is non-negative and

$$\text{Prob}\{x_1 \leq X \leq x_2, y_1 \leq Y \leq y_2\} = \int_{y_1}^{y_2} \int_{x_1}^{x_2} p(x,y)dxdy,$$

$$\int_{\mathbb{R}^2} p(x,y)dxdy = 1. \tag{2.19}$$

By (2.17) and (2.19), we see that the probability density function of X can be derived from the joint probability density function of X and Y, and is given as

$$p_X(x) = \int_{-\infty}^{\infty} p(x,y)dy, \tag{2.20}$$

where the subscript X in p_X is used to emphasize that this is the probability density function of X. (It should be noted that the subscript in the probability density function is always suppressed if no confusion occurs. Otherwise, we index a probability density function by the random variable it refers to.) Again, in the context of two or more random variables, the probability density function of each random variable is called the *marginal probability density function* (or simply the *marginal density function*). Similarly, the probability density function of Y can be derived from the joint probability density function of X and Y,

$$p_Y(y) = \int_{-\infty}^{\infty} p(x,y)dx. \tag{2.21}$$

Again the subscript Y in p_Y is used to emphasize that it is the probability density function of Y. Thus, by (2.20) and (2.21) we see that the probability density function of each random variable can be derived from the joint probability density function of X and Y by integrating across the other variable. However, the converse is usually not true; that is, the joint probability density function usually cannot be derived based on the associated marginal probability density functions.

In general, the joint cumulative distribution function of m random variables $X_1, X_2, \ldots, X_m$ is defined as

$$P(x_1, x_2, \ldots, x_m) = \text{Prob}\{X_1 \leq x_1, X_2 \leq x_2, \ldots, X_m \leq x_m\}, \tag{2.22}$$

and the corresponding joint probability density function of $X_1, X_2, \ldots, X_m$, if it exists, is defined by

$$p(x_1, x_2, \ldots, x_m) = \frac{\partial^m}{\partial x_1 \partial x_2 \ldots \partial x_m} P(x_1, x_2, \ldots, x_m). \tag{2.23}$$

We can view these m random variables $X_1, X_2, \ldots, X_m$ as the components of an m-dimensional random vector (i.e., $\mathbf{X} = (X_1, X_2, \ldots, X_m)^T$). Hence, (2.22) and (2.23) can be respectively viewed as the cumulative distribution function and probability density function of the random vector $\mathbf{X}$. In this case, we write these functions more compactly as $P(\mathbf{x})$ and $p(\mathbf{x})$ with $\mathbf{x} = (x_1, x_2, \ldots, x_m)^T$.

Similar to the two-dimensional case, for any $j \in \{1, 2, \ldots, m\}$, the probability density function of X_j can be derived from the joint probability density function of $X_1, X_2, \ldots, X_m$ by integrating with respect to all the other $m-1$ variables. For example, the probability density function of X_2 is given by

$$p_{X_2}(x_2) = \int_{\mathbb{R}^{m-1}} p(x_1, x_2, \ldots, x_m) dx_1 dx_3 dx_4 \cdots dx_m. \qquad (2.24)$$

Similarly, the joint probability density function of any two or more random variables in the set of $\{X_j\}_{j=1}^m$ can be derived from the joint probability density function of $X_1, X_2, \ldots, X_m$ by integrating with respect to all the rest of the variables. For example, the joint probability density function of X_1 and X_2 is

$$p_{X_1 X_2}(x_1, x_2) = \int_{\mathbb{R}^{m-2}} p(x_1, x_2, \ldots, x_m) dx_3 dx_4 \cdots dx_m.$$

Here the subscript $X_1 X_2$ in $p_{X_1 X_2}$ is used to emphasize that it is the joint probability density function of X_1 and X_2. Again, the (joint) probability density function of any subset of random variables in the set of $\{X_j\}_{j=1}^m$ is often called the *marginal probability density function*.

2.2.6 Conditional Distribution Function

The conditional distribution function of X given $Y = y$ is defined by

$$\mathcal{P}_{X|Y}(x|y) = \text{Prob}\{X \le x | Y = y\}. \qquad (2.25)$$

Then in analogy to (2.5), we have that

$$\mathcal{P}_{X|Y}(x|y) = \frac{\int_{-\infty}^x p_{XY}(\xi, y) d\xi}{p_Y(y)}. \qquad (2.26)$$

Here p_{XY} is the joint probability density function of X and Y, and p_Y is the probability density function of Y. The corresponding *conditional probability density function* (or simply *conditional density function*) of X given $Y = y$, if it exists, is

$$\rho_{X|Y}(x|y) = \frac{d}{dx} \mathcal{P}_{X|Y}(x|y) = \frac{p_{XY}(x, y)}{p_Y(y)}. \qquad (2.27)$$

By (2.27) we see that

$$p_{XY}(x, y) = \rho_{X|Y}(x|y) p_Y(y). \qquad (2.28)$$

Definition 2.2.6 *Two random variables X and Y are said to be independent if the conditional distribution of X given Y is equal to the unconditional distribution of X.*

Definition 2.2.6 implies that if X and Y are independent, then we have $\rho_{X|Y}(x|y) = p_X(x)$. Hence, by (2.28), we obtain

$$p_{XY}(x,y) = p_X(x)p_Y(y), \tag{2.29}$$

which is also used as a definition for the independence of two random variables. Thus, we see that if two random variables are independent, then the joint probability density function can be determined by the marginal probability density functions. It should be noted that the definition for the independence of two general (either discrete or continuous) random variables is based on the independence of two events, and is given as follows.

Definition 2.2.7 *Two random variables X and Y are said to be independent if the σ-algebras they generate, $\sigma(X)$ and $\sigma(Y)$, are independent (that is, for any $\mathbb{A} \in \sigma(X)$ and for any $\mathbb{B} \in \sigma(Y)$, events $\mathbb{A}$ and $\mathbb{B}$ are independent).*

The above definition implies that if X and Y are independent, then their joint distribution function is given by the product of their marginal distribution functions; that is,

$$P_{XY}(x,y) = P_X(x)P_Y(y), \tag{2.30}$$

which is also used as a general definition for the independence of two random variables.

In general, for any positive integer $k \geq 2$, the *conditional distribution function* of X_k given $X_1 = x_1, \ldots, X_{k-1} = x_{k-1}$ is

$$
\begin{aligned}
P_{X_k|X_{k-1},\cdots,X_1}&(x_k|x_{k-1},\cdots,x_1) \\
&= \text{Prob}\{X_k \leq x_k | X_{k-1} = x_{k-1}, \cdots, X_1 = x_1\} \\
&= \frac{\int_{-\infty}^{x_k} p_{X_1,\ldots,X_k}(x_1,\ldots,x_{k-1},\xi)d\xi}{p_{X_1,\ldots,X_{k-1}}(x_1,\ldots,x_{k-1})}.
\end{aligned}
\tag{2.31}
$$

Here $p_{X_1,\ldots,X_{k-1}}$ is the joint probability density function of $X_1,\ldots,X_{k-1}$, and $p_{X_1,\ldots,X_k}$ is the joint probability density function of $X_1,\ldots,X_k$. The corresponding *conditional density function* of X_k given $X_1 = x_1,\ldots,X_{k-1} = x_{k-1}$ is

$$
\begin{aligned}
\rho_{X_k|X_{k-1},\cdots,X_1}&(x_k|x_{k-1},\cdots,x_1) \\
&= \frac{d}{dx_k} P_{X_k|X_{k-1},\cdots,X_1}(x_k|x_{k-1},\cdots,x_1) \\
&= \frac{p_{X_1,\ldots,X_k}(x_1,\ldots,x_k)}{p_{X_1,\ldots,X_{k-1}}(x_1,\ldots,x_{k-1})},
\end{aligned}
\tag{2.32}
$$

which implies that

$$
\begin{aligned}
p_{X_1,\ldots,X_k}&(x_1,\ldots,x_k) \\
&= \rho_{X_k|X_{k-1},\cdots,X_1}(x_k|x_{k-1},\cdots,x_1)p_{X_1,\ldots,X_{k-1}}(x_1,\ldots,x_{k-1}).
\end{aligned}
\tag{2.33}
$$

Hence, by (2.27) and the above equation we see that the joint probability density function $p_{X_1,\dots,X_m}$ of any m random variables $X_1, X_2, \dots, X_m$ can be written as

$$p_{X_1,\dots,X_m}(x_1, x_2, \dots, x_m)$$

$$= p_{X_1}(x_1)\rho_{X_2|X_1}(x_2|x_1)\rho_{X_3|X_2,X_1}(x_3|x_2,x_1) \tag{2.34}$$

$$\cdots \rho_{X_m|X_{m-1},\cdots,X_1}(x_m|x_{m-1},\cdots,x_1).$$

The concept of the independence of two random variables can be extended to the case of a sequence of random variables, and it is given as follows.

Definition 2.2.8 *Random variables $X_1, X_2, \dots, X_m$ are said to be mutually independent if*

$$p_{X_1,\dots,X_m}(x_1, x_2, \dots, x_m) = \prod_{j=1}^{m} p_{X_j}(x_j), \tag{2.35}$$

where $p_{X_1,\dots,X_m}$ is the joint probability density function of m random variables $X_1, X_2, \dots, X_m$, and p_{X_j} is the probability density function of X_j, $j = 1, 2, \dots, m$.

In general, the conditional probability density function of a subset of the coordinates of $(X_1, X_2, \dots, X_m)$ given the values of the remaining coordinates is obtained by dividing the joint probability density function of $(X_1, X_2, \dots, X_m)$ by the marginal probability density function of the remaining coordinates. For example, the conditional probability density function of $(X_{k+1}, \dots, X_m)$ given $X_1 = x_1, \dots, X_k = x_k$ (where k is a positive integer such that $1 < k < m$) is defined as

$$\rho_{X_{k+1},\dots,X_m|X_1,\dots,X_k}(x_{k+1}, \dots, x_m \mid x_1, \dots, x_k) = \frac{p_{X_1,\dots,X_m}(x_1, \dots, x_m)}{p_{X_1,\dots,X_k}(x_1, \dots, x_k)}, \tag{2.36}$$

where $p_{X_1,\dots,X_k}$ is the joint probability density function of $X_1, \dots, X_k$, and $p_{X_1,\dots,X_m}$ is the joint probability density function of $X_1, \dots, X_m$. The concept of mutually independent random variables can also be extended to that of mutually independent random vectors. The definition is given as follows.

Definition 2.2.9 *Random vectors $\mathbf{X}_1, \mathbf{X}_2, \dots, \mathbf{X}_m$ are said to be mutually independent if*

$$p_{\mathbf{X}_1,\dots,\mathbf{X}_m}(\mathbf{x}_1, \mathbf{x}_2, \dots, \mathbf{x}_m) = \prod_{j=1}^{m} p_{\mathbf{X}_j}(\mathbf{x}_j), \tag{2.37}$$

where $p_{\mathbf{X}_1,\dots,\mathbf{X}_m}$ is the joint probability density function of m random vectors $\mathbf{X}_1, \mathbf{X}_2, \dots, \mathbf{X}_m$, and $p_{\mathbf{X}_j}$ is the probability density function of $\mathbf{X}_j$, $j = 1, 2, \dots, m$.

2.2.7   Function of a Random Variable

Let X be a random variable with cumulative distribution function P_X and probability density function p_X. Then by Definition 2.2.1 we know that for any measurable function $\eta : \mathbb{R} \to \mathbb{R}$, the composite function $\eta(X)$ is measurable and indeed is a random variable. Let $Y = \eta(X)$. Then the cumulative distribution function of Y is

$$P_Y(y) = \text{Prob}\{\eta(X) \le y\}. \tag{2.38}$$

If we assume that η is a monotone function, then by (2.38) we have

$$P_Y(y) = \begin{cases} \text{Prob}\{X \le \eta^{-1}(y)\} = P_X(\eta^{-1}(y)), & \eta \text{ is increasing} \\ \text{Prob}\{X \ge \eta^{-1}(y)\} = 1 - P_X(\eta^{-1}(y)), & \eta \text{ is decreasing.} \end{cases} \tag{2.39}$$

If we further assume that η^{-1} is differentiable, then differentiating both sides of (2.39) yields the probability density function of Y

$$p_Y(y) = p_X(\eta^{-1}(y)) \left| \frac{d\eta^{-1}(y)}{dy} \right|. \tag{2.40}$$

In general, we consider an m-dimensional random vector $\mathbf{X}$ with probability density function $p_{\mathbf{X}}$. Let $\mathbf{Y} = \boldsymbol{\eta}(\mathbf{X})$ with $\boldsymbol{\eta} = (\eta_1, \eta_2, \dots, \eta_m)^T$ and $\eta_j : \mathbb{R}^m \to \mathbb{R}$ be a measurable function for all j. Assume that $\boldsymbol{\eta}$ has a unique inverse $\boldsymbol{\eta}^{-1}$. Then the probability density function of the m-dimensional random vector $\mathbf{Y}$ is given by

$$p_{\mathbf{Y}}(\mathbf{y}) = p_{\mathbf{X}}(\boldsymbol{\eta}^{-1}(\mathbf{y}))|\mathcal{J}|, \tag{2.41}$$

where $\mathcal{J}$ is the determinant of the Jacobian matrix $\frac{\partial \mathbf{x}}{\partial \mathbf{y}}$ with its (j, k)th element being $\frac{\partial x_j}{\partial y_k}$.

In the following, whenever we talk about transformation of a random variable or random vector, we always assume that the transformation is measurable so that the resulting function is also a random variable. The following theorem is about the transformation of two independent random variables.

Theorem 2.2.3 *Let X and Z be independent random variables, η_X be a function only of x and η_Z be a function only of z. Then the random variables $U = \eta_X(X)$ and $V = \eta_Z(Z)$ are independent.*

The above theorem is very important in theory, and it can be generalized as follows.

Theorem 2.2.4 *Let $\mathbf{X}_1, \mathbf{X}_2, \dots, \mathbf{X}_m$ be mutually independent random vectors, and η_j be a function only of $\mathbf{x}_j$, $j = 1, 2, \dots, m$. Then random variables $U_j = \eta_j(\mathbf{X}_j)$, $j = 1, 2, \dots, m$, are mutually independent.*

2.3   Statistical Averages of Random Variables

The concepts of moments of a single random variable and the joint moments between any pair of random variables in a multi-dimensional set of random variables are of particular importance in practice. We begin the discussion of these statistical averages by considering first a single random variable X and its cumulative distribution function P. The *expectation* (also called an *expected value* or *mean*) of the random variable X is defined by

$$\mathbb{E}(X) = \int_{-\infty}^{\infty} x \, dP(x). \tag{2.42}$$

Here $\mathbb{E}(\cdot)$ is called the *expectation operator* (or statistical averaging operator), and the integral on the right side is interpreted as a Riemann–Stieltjes integral. We remark that the Riemann–Stieltjes integral is a generalization of the Riemann integral and its definition is given as follows.

Definition 2.3.1 *Let φ and h be real-valued functions defined on $[\underline{x}, \bar{x}] \subset \mathbb{R}$, $\{x_j^l\}_{j=0}^l$ be a partition of $[\underline{x}, \bar{x}]$, $\Delta_l = \max\limits_{0 \le j \le l-1} \{x_{j+1}^l - x_j^l\}$, and $s_j^l \in [x_j^l, x_{j+1}^l]$ denote intermediate points of the partition. Then φ is said to be Riemann–Stieltjes integrable with respect to h on $[\underline{x}, \bar{x}]$ if*

$$\lim_{\substack{l \to \infty \\ \Delta_l \to 0}} \sum_{j=0}^{l-1} \varphi(s_j^l)[h(x_{j+1}^l) - h(x_j^l)] \tag{2.43}$$

exists and the limit is independent of the choice of the partition and their intermediate points. The limit of (2.43) is called the Riemann–Stieltjes integral of φ with respect to h on $[\underline{x}, \bar{x}]$, and is denoted by $\int_{\underline{x}}^{\bar{x}} \varphi(x) \, dh(x)$; that is,

$$\int_{\underline{x}}^{\bar{x}} \varphi(x) \, dh(x) = \lim_{\substack{l \to \infty \\ \Delta_l \to 0}} \sum_{j=0}^{l-1} \varphi(s_j^l)[h(x_{j+1}^l) - h(x_j^l)],$$

where φ and h are called the integrand and the integrator, respectively.

It is worth noting that the Riemann–Stieltjes integral $\int_{\underline{x}}^{\bar{x}} \varphi(x) \, dh(x)$ does not exist for all continuous functions φ on $[\underline{x}, \bar{x}]$ unless h has bounded variation. This is why (2.42) can be interpreted as a Riemann–Stieltjes integral (as P has bounded variation). In general, the Riemann–Stieltjes integral $\int_{\underline{x}}^{\bar{x}} \varphi(x) \, dh(x)$ exists if the following conditions are satisfied:

- The functions φ and h have no discontinuities at the same point $x \in [\underline{x}, \bar{x}]$.

- The function φ has bounded κ_φ-variation on $[\underline{x}, \bar{x}]$ and the function h has bounded κ_h-variation on $[\underline{x}, \bar{x}]$, where κ_φ and κ_h are some positive constants such that $\dfrac{1}{\kappa_\varphi} + \dfrac{1}{\kappa_h} > 1$.

We refer the interested reader to [18, Section 2.1], [21], and the references therein for more information on Riemann–Stieltjes integrals.

If we assume that P is absolutely continuous (that is, the corresponding probability density function p exists), then we can rewrite (2.42) as

$$\mathbb{E}(X) = \int_{-\infty}^{\infty} x p(x) \, dx. \tag{2.44}$$

We note from (2.12) and (2.15) that if X is a discrete random variable with range $\{x_j\}$, then the expectation of X is given by

$$\mathbb{E}(X) = \sum_j x_j \text{Prob}\{X = x_j\}.$$

Thus, we see that with the help of the Dirac delta function one can put the definition of expectation in the same framework for a discrete random variable and for a continuous random variable with a probability density function (as we stated earlier). Since we are mainly interested in discrete random variables and those continuous random variables associated with probability density functions, we will define the statistical average of a random variable in terms of its probability density function in the following presentation.

The expectation of a random variable is also called the *first moment*. In general, the kth moment of a random variable X is defined as

$$\mathbb{E}(X^k) = \int_{-\infty}^{\infty} x^k p(x) \, dx.$$

We can also define the *central moments*, which are the moments of the difference between X and $\mathbb{E}(X)$. For example, the kth central moment of X is defined by

$$\mathbb{E}\left((X - \mathbb{E}(X))^k\right).$$

Of particular importance is the second central moment, called the *variance* of X, which is defined as

$$\sigma^2 = \text{Var}(X) = \mathbb{E}\left((X - \mathbb{E}(X))^2\right) = \int_{-\infty}^{\infty} (x - \mathbb{E}(X))^2 p(x) \, dx. \tag{2.45}$$

The square root σ of the variance of X is called the *standard deviation* of X. Variance is a measure of the "randomness" of the random variable X. It is related to the first and second moments through the relationship

$$
\begin{aligned}
\mathrm{Var}(X) &= \mathbb{E}\left((X - \mathbb{E}(X))^2\right) \\
&= \mathbb{E}\left(X^2 - 2X\mathbb{E}(X) + (\mathbb{E}(X))^2\right) \\
&= \mathbb{E}(X^2) - (\mathbb{E}(X))^2.
\end{aligned}
\tag{2.46}
$$

One of the useful concepts in understanding the variation of a random variable X is the *coefficient of variation* (CV), which is defined as the ratio of the standard deviation to the mean (that is, $\mathrm{CV} = \sqrt{\mathrm{Var}(X)}/\mathbb{E}(X)$). It is the inverse of the so-called *signal-to-noise ratio*. A random variable with $\mathrm{CV} < 1$ is considered to have low variation, while one with $\mathrm{CV} > 1$ is considered to have high variation.

Note that the moments of a function of a random variable, $Y = \eta(X)$, can be defined in the same way as above. For example, the kth moment of Y is

$$
\mathbb{E}(Y^k) = \int_{\Omega_y} y^k p_Y(y) dy,
\tag{2.47}
$$

where Ω_y denotes the range of Y, and p_Y is the probability density function of Y. However, due to the relation (2.40) between p_Y and the probability density function p_X of X, we can also calculate the kth moment of Y by

$$
\mathbb{E}(Y^k) = \mathbb{E}\{\eta^k(X)\} = \int_{-\infty}^{\infty} \eta^k(x) p_X(x) dx.
\tag{2.48}
$$

For any real numbers a and b, by (2.48) we observe that

$$
\mathbb{E}(aX + b) = a\mathbb{E}(X) + b, \quad \mathrm{Var}(aX + b) = a^2 \mathrm{Var}(X).
\tag{2.49}
$$

Remark 2.3.1 *We remark that the infinite sequence of moments is in general not enough to uniquely determine a distribution function. Interested readers can refer to [7, Section 2.3] for a counterexample of two random variables having the same moments but different probability density functions. However, if two random variables have bounded support, then an infinite sequence of moments does uniquely determine the distribution function.*

2.3.1   Joint Moments

The definition of moments and central moments can be extended from one random variable to two or more random variables. In this context, the moments are often called joint moments, and central moments are often called joint central moments. For example, the joint moment of two random variables X and Y is defined as

$$
\mathbb{E}(X^{k_x} Y^{k_y}) = \int_{\mathbb{R}^2} x^{k_x} y^{k_y} p(x, y) dx dy,
\tag{2.50}
$$

where k_x and k_y are positive integers, and p is the joint probability density function of X and Y. The joint central moment of X and Y is given by

$$\mathbb{E}\left((X - \mathbb{E}(X))^{k_x}(Y - \mathbb{E}(Y))^{k_y}\right). \tag{2.51}$$

However, the joint moment that is most useful in practical applications is the *correlation* of two random variables X and Y, defined as

$$\text{Cor}\{X, Y\} = \mathbb{E}(XY) = \int_{\mathbb{R}^2} xyp(x, y)dxdy, \tag{2.52}$$

which implies that the second moment of X is the correlation of X with itself. Also of particular importance is the *covariance* of two random variables X and Y, defined as

$$\text{Cov}\{X, Y\} = \mathbb{E}\left((X - \mathbb{E}(X))(Y - \mathbb{E}(Y))\right) = \mathbb{E}(XY) - \mathbb{E}(X)\mathbb{E}(Y), \tag{2.53}$$

which indicates that the variance of X is the covariance of X with itself. It is worth noting that the correlation of two random variables should not be confused with the *correlation coefficient*, r, which is defined as the covariance of the two random variables divided by the product of their standard deviations.

$$r = \frac{\text{Cov}\{X, Y\}}{\sqrt{\text{Var}(X)\text{Var}(Y)}}.$$

By (2.52) and (2.53) we see that both correlation and covariance are symmetric. That is,

$$\text{Cor}\{X, Y\} = \text{Cor}\{Y, X\}, \quad \text{Cov}\{X, Y\} = \text{Cov}\{Y, X\}.$$

Moreover, they are both linear in each variable. That is, for any real numbers a and b we have

$$\text{Cor}\{aX + bY, Z\} = a\text{Cor}\{X, Z\} + b\text{Cor}\{Y, Z\},$$

$$\text{Cor}\{Z, aX + bY\} = a\text{Cor}\{Z, X\} + b\text{Cor}\{Z, Y\},$$

$$\text{Cov}\{aX + bY, Z\} = a\text{Cov}\{X, Z\} + b\text{Cov}\{Y, Z\},$$

$$\text{Cov}\{Z, aX + bY\} = a\text{Cov}\{Z, X\} + b\text{Cov}\{Z, Y\}.$$

By the above equations, we find that the variance of $aX + bY$ is given by

$$\text{Var}(aX + bY) = a^2\text{Var}(X) + 2ab\text{Cov}\{X, Y\} + b^2\text{Var}(Y).$$

Definition 2.3.2 *Two random variables X and Y are called uncorrelated if $\text{Cov}\{X, Y\} = 0$.*

By the above definition and (2.53) we see that if

$$\mathbb{E}(XY) = \mathbb{E}(X)\mathbb{E}(Y), \tag{2.54}$$

then X and Y are uncorrelated. We observe from (2.29) that if X and Y are independent, then (2.54) also holds. Therefore, the independence of two random variables implies that these two random variables are uncorrelated. However, in general the converse is not true.

2.3.2 Conditional Moments

The kth *conditional moment* of X given $Y = y$ is

$$\mathbb{E}(X^k|Y = y) = \int_{-\infty}^{\infty} x^k \rho_{X|Y}(x|y)dx. \qquad (2.55)$$

This implies that if random variables X and Y are independent (that is, $\rho_{X|Y}(x|y) = p_X(x)$), then

$$\mathbb{E}(X^k|Y = y) = \mathbb{E}(X^k).$$

The first conditional moment of X given $Y = y$ is called the *conditional expectation* of X given $Y = y$.

Observe that $\mathbb{E}(X|Y = y)$ is a function of y. Hence, $\mathbb{E}(X|Y)$ is a random variable, which is called the *conditional expectation* of X given Y. By (2.20), (2.27) and (2.55) we find

$$\begin{aligned}
\mathbb{E}\left(\mathbb{E}(X|Y)\right) &= \int_{-\infty}^{\infty} p_Y(y) \left(\int_{-\infty}^{\infty} x\rho_{X|Y}(x|y)dx \right) dy \\
&= \int_{-\infty}^{\infty} p_Y(y) \left(\int_{-\infty}^{\infty} x \frac{p_{XY}(x,y)}{p_Y(y)} dx \right) dy \\
&= \int_{-\infty}^{\infty} x \left(\int_{-\infty}^{\infty} p_{XY}(x,y)dy \right) dx \\
&= \int_{-\infty}^{\infty} x p_X(x)dx \\
&= \mathbb{E}(X),
\end{aligned}$$

which indicates that the expected value of the conditional expectation of X given Y is the same as the expected value of X. This formula is often called the *law of total expectation*. Similarly, we can show that for any positive integer k we have

$$\mathbb{E}\left(\mathbb{E}(X^k|Y)\right) = \mathbb{E}(X^k). \qquad (2.56)$$

The *conditional variance* of X given Y is defined as

$$\mathrm{Var}(X|Y) = \mathbb{E}\left((X - \mathbb{E}(X|Y))^2|Y\right) = \mathbb{E}(X^2|Y) - (\mathbb{E}(X|Y))^2. \qquad (2.57)$$

By (2.46), (2.56) and (2.57), we find that the unconditional variance is related to the conditional variance by

$$\mathrm{Var}(X) = \mathbb{E}\left(\mathrm{Var}(X|Y)\right) + \mathrm{Var}\left(\mathbb{E}(X|Y)\right), \qquad (2.58)$$

which indicates that the unconditional variance is equal to the sum of the mean of the conditional variance and the variance of the conditional mean. Equation (2.58) is often called the *law of total variance, variance decomposition formula* or *conditional variance formula*.

2.3.3   Statistical Averages of Random Vectors

For an m-dimensional random vector $\mathbf{X} = (X_1, \ldots, X_m)^T$, its *mean vector* and *covariance matrix* are of particular importance. Specifically, the mean vector of $\mathbf{X}$ is given by

$$\boldsymbol{\mu} = \mathbb{E}(\mathbf{X}) = (\mathbb{E}(X_1), \ldots, \mathbb{E}(X_m))^T,$$

and its *covariance matrix* is defined by

$$\boldsymbol{\Sigma} = \text{Var}(\mathbf{X}) = \mathbb{E}\{(\mathbf{X} - \boldsymbol{\mu})(\mathbf{X} - \boldsymbol{\mu})^T\}. \tag{2.59}$$

Hence, we see that $\boldsymbol{\Sigma} \in \mathbb{R}^{m \times m}$ is a non-negative definite matrix with its (k, j)th element being the covariance of random variables X_k and X_j:

$$\text{Cov}\{X_k, X_j\} = \mathbb{E}\left((X_k - \mathbb{E}(X_k))(X_j - \mathbb{E}(X_j))\right)$$

$$= \int_{\mathbb{R}^2} (x_k - \mathbb{E}(X_k))(x_j - \mathbb{E}(X_j)) p_{X_k X_j}(x_k, x_j)\, dx_k dx_j,$$

where $p_{X_k X_j}$ is the joint probability density function of X_k and X_j. For any $\mathcal{A} \in \mathbb{R}^{l \times m}$ and $\mathbf{a} \in \mathbb{R}^l$, it can be easily shown that

$$\mathbb{E}(\mathcal{A}\mathbf{X} + \mathbf{a}) = \mathcal{A}\mathbb{E}(\mathbf{X}) + \mathbf{a}, \quad \text{Var}(\mathcal{A}\mathbf{X} + \mathbf{a}) = \mathcal{A}\text{Var}(\mathbf{X})\mathcal{A}^T. \tag{2.60}$$

Similarly, we can extend the covariance of two random variables to the *cross-covariance matrix* between two random vectors. Let $\mathbf{X} = (X_1, X_2, \ldots, X_{m_x})^T$ and $\mathbf{Y} = (Y_1, Y_2, \ldots, Y_{m_y})^T$, where m_x and m_y are positive integers. Then the *cross-covariance matrix* between $\mathbf{X}$ and $\mathbf{Y}$ is

$$\begin{aligned} \text{Cov}\{\mathbf{X}, \mathbf{Y}\} &= \mathbb{E}\left((\mathbf{X} - \mathbb{E}(\mathbf{X}))(\mathbf{Y} - \mathbb{E}(\mathbf{Y}))^T\right) \\ &= \mathbb{E}(\mathbf{X}\mathbf{Y}^T) - \mathbb{E}(\mathbf{X})(\mathbb{E}(\mathbf{Y}))^T, \end{aligned} \tag{2.61}$$

which implies that $\text{Var}(\mathbf{X}) = \text{Cov}\{\mathbf{X}, \mathbf{X}\}$. By (2.60) it can be easily shown that for any matrix $\mathcal{A} \in \mathbb{R}^{\kappa_x \times m_x}$ and $\mathcal{B} \in \mathbb{R}^{\kappa_y \times m_y}$ (κ_x and κ_y are positive integers) we have

$$\text{Cov}\{\mathcal{A}\mathbf{X}, \mathcal{B}\mathbf{Y}\} = \mathcal{A}\text{Cov}\{\mathbf{X}, \mathbf{Y}\}\mathcal{B}^T. \tag{2.62}$$

The concept of uncorrelatedness of two random variables can also be extended to two random vectors. Specifically, two random vectors $\mathbf{X}$ and $\mathbf{Y}$ are called uncorrelated if $\text{Cov}\{\mathbf{X}, \mathbf{Y}\} = 0$.

2.3.4   Important Inequalities

In this section, we list a few important inequalities that will be used in this monograph.

Theorem 2.3.2 *(Markov's inequality) Let k be any positive integer, and ϵ be a positive real number. Assume that X is a random variable such that $\mathbb{E}(|X|^k)$ exists. Then we have*

$$Prob\left\{|X| \geq \epsilon\right\} \leq \frac{\mathbb{E}(|X|^k)}{\epsilon^k}. \qquad (2.63)$$

Markov's inequality includes Chebyshev's inequality as a special case, where Chebyshev's inequality states that

$$\text{Prob}\left\{|X - \mathbb{E}(X)| \geq \epsilon\right\} \leq \frac{\text{Var}(X)}{\epsilon^2}. \qquad (2.64)$$

Theorem 2.3.3 *(Hölder's inequality) Let k_x and k_y be two positive integers such that $\dfrac{1}{k_x} + \dfrac{1}{k_y} = 1$, and X and Y be two random variables such that $\mathbb{E}(|X|^{k_x})$ and $\mathbb{E}(|Y|^{k_y})$ exist. Then*

$$|\mathbb{E}(XY)| \leq \mathbb{E}\{|XY|\} \leq \left(\mathbb{E}(|X|^{k_x})\right)^{1/k_x} \left(\mathbb{E}(|Y|^{k_y})\right)^{1/k_y}. \qquad (2.65)$$

Hölder's inequality includes the Cauchy–Schwartz inequality as a special case for $k_x = k_y = 2$. The Cauchy–Schwartz inequality states that

$$(\mathbb{E}(XY))^2 \leq \mathbb{E}(X^2)\mathbb{E}(Y^2). \qquad (2.66)$$

Theorem 2.3.4 *(Jensen's inequality) Let ψ be a convex function. Then*

$$\psi(\mathbb{E}(X)) \leq \mathbb{E}(\psi(X)). \qquad (2.67)$$

2.4 Characteristic Functions of a Random Variable

As we remarked in the previous section, the moments are in general not enough to determine a distribution function uniquely. In this section, we introduce another function associated with a random variable, called the *characteristic function*, which uniquely determines the distribution function.

The characteristic function of a random variable X is defined by

$$\Pi(\varsigma) = \mathbb{E}\left(\exp\left(i\varsigma X\right)\right) = \int_{-\infty}^{\infty} \exp\left(i\varsigma x\right) dP(x), \qquad (2.68)$$

where $\varsigma \in \mathbb{R}$, i is the imaginary unit and P is the cumulative distribution function of X. If we assume that the corresponding probability density function exists (that is, P is absolutely continuous), then we can rewrite (2.68) as

$$\Pi(\varsigma) = \mathbb{E}\left(\exp\left(i\varsigma X\right)\right) = \int_{-\infty}^{\infty} \exp\left(i\varsigma x\right) p(x)dx, \qquad (2.69)$$

which implies that the characteristic function Π is essentially the Fourier transform of the probability density function p of X. Hence, p can be uniquely determined from the inverse Fourier transform of its associated characteristic function and is given by

$$p(x) = \frac{1}{2\pi} \int_{-\infty}^{\infty} \exp\left(-i\varsigma x\right) \Pi(\varsigma) d\varsigma. \qquad (2.70)$$

Interested readers are referred to some popular books such as [5, 16, 19] for more information on the Fourier transform and its applications.

The characteristic function has many useful properties. For example, Π is bounded by $\Pi(0) = 1$, which can be directly deduced from (2.69). In addition, the characteristic function of the sum of two independent random variables is equal to the product of the characteristic functions of these two random variables. This is implied by (2.69), Theorem 2.2.3, and the properties of independence of two random variables (2.54). Characteristic functions are also closely related to moments. Specifically, for any positive integer k we have

$$\frac{d^k}{d\varsigma^k} \Pi(0) = i^k \mathbb{E}(X^k). \qquad (2.71)$$

By using the above formula, one can determine any moment of X.

Similarly, the characteristic function of an m-dimensional random vector $\mathbf{X}$ is defined by

$$\Pi(\varsigma) = \mathbb{E}\left(\exp(i\varsigma^T \mathbf{X})\right) = \int_{\mathbb{R}^m} \exp(i\varsigma^T \mathbf{X}) p(\mathbf{x}) dx, \qquad (2.72)$$

where $\varsigma \in \mathbb{R}^m$, and p is the probability density function of $\mathbf{X}$. In the same manner, p can be determined from the inverse Fourier transform of Π and is given by

$$p(\mathbf{x}) = \frac{1}{(2\pi)^m} \int_{\mathbb{R}^m} \exp(-i\varsigma^T \mathbf{X}) \Pi(\varsigma) d\varsigma. \qquad (2.73)$$

2.5   Special Probability Distributions

In this section, we give a brief review of several frequently encountered distributions that are also used in this monograph. Interested readers can refer to Wikipedia as well as some popular textbooks such as [1, 7, 22, 25] for more information on these distributions and for other frequently encountered distributions that are not discussed here.

Before addressing our main task here, we introduce an important definition, *skewness*, which is a measure of the asymmetry of the distribution of a random variable. The skewness of a random variable X is defined as the third standardized moment and is given by

$$\text{Skew}(X) = \mathbb{E}\left(\left(\frac{X-\mu}{\sigma}\right)^3\right) = \frac{\mathbb{E}(X^3) - 3\mu\sigma^2 - \mu^3}{\sigma^3}, \tag{2.74}$$

where μ and σ denote the mean and standard deviation of X, respectively. A positive skew (i.e., $\text{Skew}(X) > 0$) indicates the tail on the right side is longer than that on the left side, and negative skew (i.e., $\text{Skew}(X) < 0$) indicates the tail on the left side is longer than that on the right side. A zero skew (i.e., $\text{Skew}(X) = 0$) indicates that the values are relatively evenly distributed around the mean, and it usually implies the distribution is symmetric.

2.5.1 Poisson Distribution

The Poisson distribution is one of the most widely used discrete distributions in practice. For example, it can be used to model the number of events that happened in a given time interval, such as the number of customers arriving in a bank or the number of traffic accidents occurring at an intersection.

A discrete random variable X is said to be Poisson distributed with rate parameter λ (indicated by $X \sim \text{Poisson}(\lambda)$) if the probability mass function is given by

$$\text{Prob}\{X = k\} = \frac{\lambda^k}{k!}\exp(-\lambda), \tag{2.75}$$

where k is a non-negative integer, and $k!$ is the factorial of k. The mean and variance of X are respectively given by

$$\mathbb{E}(X) = \lambda, \quad \text{Var}(X) = \lambda. \tag{2.76}$$

The Poisson distribution has many nice properties. For example, it can be shown that if $X_j \sim \text{Poisson}(\lambda_j)$, $j = 1, 2$, are independent, then

$$X_1 + X_2 \sim \text{Poisson}(\lambda_1 + \lambda_2).$$

The converse is also true. In other words, if the sum of two independent non-negative random variables is Poisson distributed, then both random variables are Poisson distributed. This result is often referred to as *Raikov's Theorem*, which has implications in reliability theory, telecommunications, nuclear physics and other fields (e.g., see [12] for details).

2.5.2 Uniform Distribution

The probability density function of a uniformly distributed random variable X on $[a, b]$ (denoted by $X \sim \text{Uniform}(a, b)$) is given by

$$p(x) = \begin{cases} 1/(b-a), & \text{if } a \leq x \leq b, \\ 0, & \text{otherwise}, \end{cases} \tag{2.77}$$

and the corresponding cumulative distribution function is

$$P(x) = \begin{cases} 0 & \text{if } x < a, \\ \dfrac{x-a}{b-a} & \text{if } a \le x < b, \\ 1 & \text{if } x \ge b. \end{cases}$$

It should be noted that the values of p at the two boundary points a and b are usually not important as they do not change the value of $\int_a^b p(x)dx$. Hence, in some of the literature $p(a)$ and $p(b)$ are defined to be zero while in other literature they are defined as in (2.77). The plots of the probability density function and the cumulative distribution function of a uniformly distributed random variable are depicted in Figure 2.4.

If $X \sim \text{Uniform}(0,1)$, then X is said to be *standard uniformly distributed*. An interesting property of the standard uniform distribution is

$$\text{if } X \sim \text{Uniform}(0,1), \text{ then } (1-X) \sim \text{Uniform}(0,1), \qquad (2.78)$$

which can be easily derived by (2.40) and (2.77). Let $X \sim \text{Uniform}(a,b)$. Then the mean and the variance of X are respectively given by

$$\mathbb{E}(X) = \frac{a+b}{2}, \quad \text{Var}(X) = \frac{(a-b)^2}{12}.$$

The uniform distribution is widely used in practice, mostly due to its simplicity and an important property demonstrated in the following theorem.

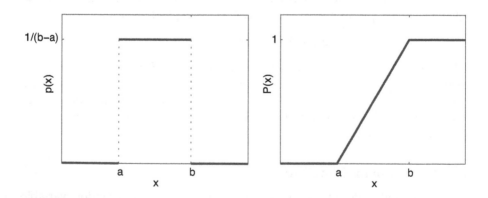

FIGURE 2.4: Plots of the probability density function p and the cumulative distribution function P of a uniform distribution.

Theorem 2.5.1 *Let Y be a continuous random variable with cumulative distribution function P. Then the random variable $X = P(Y)$ is uniformly distributed on $(0,1)$.*

The above theorem is widely used in practice to generate random numbers representing realizations from a particular distribution. For example, if one wants to generate a random number or realization y for a distribution with cumulative distribution function P, one only needs to generate a uniform random number x between 0 and 1 and then solve

$$P(y) = x \tag{2.79}$$

for y. We remark that if P is strictly increasing, then its inverse P^{-1} exists and hence there exists a unique solution y to (2.79) and the solution is given by $y = P^{-1}(x)$. However, if P is not strictly increasing, then (2.79) has many solutions. To avoid this problem, a generalized inverse function defined by

$$P^+(x) = \inf\{y \mid P(y) \geq x\}$$

is used to provide a unique solution $y = P^+(x)$ to (2.79). This technique to generate a random number or realization is often referred to as the *inverse transform method*.

Remark 2.5.2 *There is a discrete version of the inverse transform method that can be used to simulate realizations of a discrete random variable. Let Y be a discrete random variable with range $\{y_j\}$ and the probability associated with y_j being*

$$Prob\{Y = y_j\} = \varpi_j, \quad j = 1, 2, \ldots$$

where $\varpi_j, j = 1, 2, \ldots$ are non-negative numbers such that $\sum_j \varpi_j = 1$. To generate a realization y of Y, we first generate a uniform random number x between 0 and 1, and then set $y = y_m$ if

$$\sum_{j=1}^{m-1} \varpi_j < x \leq \sum_{j=1}^{m} \varpi_j.$$

Here $\sum_{j=1}^{0} \varpi_j$ is defined to be zero. This inverse transform method will be used in Chapter 8 to simulate a continuous time Markov chain.

2.5.3 Normal Distribution

The normal distribution is also called the *Gaussian distribution* (these two names will be used interchangeably in this monograph), and it is one of the most important and widely used distributions. This is partially because of the many useful and important properties it possesses. In addition, under some mild conditions the normal distribution can be used to approximate a large variety of distributions in large samples. This is due to the so-called central limit theorem that will be discussed later.

A continuous random variable X is said to be normally distributed with mean μ and variance σ^2 (commonly denoted by $X \sim \mathcal{N}(\mu, \sigma^2)$) if its probability density function is given by

$$p(x) = \frac{1}{\sqrt{2\pi}\sigma} \exp(-(x - \mu)^2 / (2\sigma^2)). \tag{2.80}$$

Hence, we see that the normal distribution is completely characterized by its mean and variance. If $X \sim \mathcal{N}(0, 1)$, then X is said to have a *standard normal distribution* or *standard Gaussian distribution*. The cumulative distribution function P has the form

$$P(x) = \frac{1}{\sqrt{2\pi}\sigma} \int_{-\infty}^{x} \exp(-(s - \mu)^2 / (2\sigma^2))\, ds = \frac{1}{2}\left[1 + \mathrm{erf}\left(\frac{x - \mu}{\sqrt{2}\sigma}\right)\right],$$

where erf denotes the *error function* and is given by

$$\mathrm{erf}(x) = \frac{2}{\pi} \int_0^x \exp(-s^2)\, ds. \tag{2.81}$$

The probability density function of a standard Gaussian distributed random variable is illustrated in the left panel of Figure 2.5, which indicates p is unimodal and symmetric around its mean value. The corresponding cumulative distribution function is shown in the right panel of Figure 2.5.

The kth central moment of the normally distributed random variable X with mean μ and variance σ^2 is given by the expression

$$\mathbb{E}\{(X - \mu)^k\} = \begin{cases} 1 \cdot 3 \cdots (k - 1)\sigma^k, & k = \text{even}, \\ 0, & k = \text{odd}, \end{cases} \tag{2.82}$$

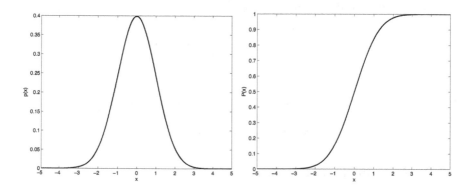

FIGURE 2.5: The probability density function and the cumulative distribution function of a standard Gaussian distributed random variable.

and the kth moments are given in terms of the central moments by

$$\mathbb{E}(X^k) = \sum_{j=0}^{k} \binom{k}{j} \mu^j \mathbb{E}\left((X-\mu)^{k-j}\right) \tag{2.83}$$

where

$$\binom{k}{j} = \frac{k!}{j!(k-j)!}.$$

The normal distribution has a number of useful properties. For example, it is closed under linear transformation. In other words, if $X \sim \mathcal{N}(\mu, \sigma^2)$, then

$$aX + b \sim \mathcal{N}(a\mu + b, a^2\sigma^2)$$

for any real numbers a and b. This implies that if $X \sim \mathcal{N}(\mu, \sigma^2)$, then

$$\frac{X-\mu}{\sigma} \sim \mathcal{N}(0,1).$$

In addition, if $X_1 \sim \mathcal{N}(\mu_1, \sigma_1^2)$ and $X_2 \sim \mathcal{N}(\mu_2, \sigma_2^2)$ are independent, then any linear combination of these two random variables is also normally distributed, that is,

$$aX_1 + bX_2 \sim \mathcal{N}(a\mu_1 + b\mu_2, a^2\sigma_1^2 + b^2\sigma_2^2)$$

for any real numbers a and b.

Remark 2.5.3 *In practice, it is often found useful to restrict the normally distributed random variable in some finite interval. The resulting distribution is called a truncated normal distribution. Let X be a truncated normal random variable on $[\underline{x}, \bar{x}]$, often denoted by $X \sim \mathcal{N}_{[\underline{x},\bar{x}]}(\mu, \sigma^2)$. Then the probability density function of X is given by*

$$p(x) = \frac{\frac{1}{\sigma}\psi\left(\frac{x-\mu}{\sigma}\right)}{\Psi\left(\frac{\bar{x}-\mu}{\sigma}\right) - \Psi\left(\frac{\underline{x}-\mu}{\sigma}\right)}, \quad x \in [\underline{x}, \bar{x}], \tag{2.84}$$

where ψ is the probability density function of the standard normal distribution, and Ψ is the corresponding cumulative distribution function.

2.5.4 Log-Normal Distribution

A widely employed model for biological (and other) phenomena in which the random variable is only allowed to take on positive values is the so-called *log-normal* distribution.

If $\ln(X) \sim \mathcal{N}(\mu, \sigma^2)$, then X has a log-normal distribution with the probability density function given by

$$p(x) = \frac{1}{\sqrt{2\pi}\sigma}\frac{1}{x}\exp\left(-\frac{(\ln x - \mu)^2}{2\sigma^2}\right), \quad 0 < x < \infty, \tag{2.85}$$

with mean and variance

$$\mathbb{E}(X) = \exp\left(\mu + \sigma^2/2\right), \quad \text{Var}(X) = (\exp(\sigma^2) - 1)\exp(2\mu + \sigma^2). \quad (2.86)$$

This relationship between normal and log-normal distributions is heavily used in Chapter 7 to establish the pointwise equivalence between stochastic differential equations and random differential equations.

We observe from (2.86) that $\text{Var}(X)$ is proportional to $(\mathbb{E}(X))^2$ so that the coefficient of variation does not depend on $\mathbb{E}(X)$. In general, the kth moment of a log-normally distributed random variable X is given by

$$\mathbb{E}(X^k) = \exp\left(k\mu + \frac{1}{2}k^2\sigma^2\right). \quad (2.87)$$

By (2.74), (2.86) and (2.87), we find the skewness of this log-normally distributed random variable is given by

$$\text{Skew}(X) = \frac{\exp(3\sigma^2) - 3\exp(\sigma^2) + 2}{[\exp(\sigma^2) - 1]^{3/2}} = [\exp(\sigma^2) + 2]\sqrt{\exp(\sigma^2) - 1}. \quad (2.88)$$

The probability density function for a log-normally distributed random variable $\ln(X) \sim \mathcal{N}(0, \sigma^2)$ with various values for σ is depicted in Figure 2.6 (readily plotted, along with other distributions discussed in this chapter, with MATLAB® commands, in this case using the MATLAB command *lognpdf*). These plots indicate that the log-normal distribution is skewed (asymmetric) with a "long right tail" but becomes more and more symmetric as $\sigma \to 0$. Note that this is consistent with (2.88).

The log-normal distribution has a number of useful properties (inherited from the normal distribution). For example, if $\ln(X) \sim \mathcal{N}(\mu, \sigma^2)$, then for any positive number a we have

$$\ln(aX) \sim \mathcal{N}(\mu + \ln(a), \sigma^2),$$

and for any non-zero real number b we have

$$\ln(X^b) \sim \mathcal{N}(b\mu, b^2\sigma^2).$$

In addition, if X_j, $j = 1, 2, \ldots, m$, are independent and $\ln(X_j) \sim \mathcal{N}(\mu_j, \sigma_j^2)$, $j = 1, 2, \ldots, m$, then

$$\ln\left(\prod_{j=1}^{m} X_j\right) \sim \mathcal{N}\left(\sum_{j=1}^{m} \mu_j, \sum_{j=1}^{m} \sigma_j^2\right).$$

It should be noted that if $\ln(X) \sim \mathcal{N}(\mu, \sigma^2)$, then $X + c$ is said to have a *shifted log-normal distribution*.

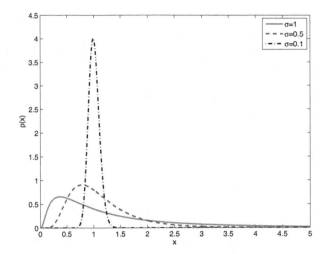

FIGURE 2.6: The plot of the probability density function of a log-normally distributed random variable $\ln(X) \sim \mathcal{N}(0, \sigma^2)$ with various values for σ.

2.5.5 Multivariate Normal Distribution

One of the most often encountered multivariate distributions is the *multivariate normal distribution*. As we shall see in the next chapter, this distribution is incredibly important in statistical *modeling* and *inference*. For example, it arises in discussing asymptotic theory for the construction of confidence intervals for parameter estimators.

Definition 2.5.1 *An m-dimensional random vector $\mathbf{X} = (X_1, \ldots, X_m)^T$ is said to be multivariate normally distributed if every linear combination of its components is normally distributed (that is, for any $\mathbf{a} \in \mathbb{R}^m$, $\mathbf{a}^T \mathbf{X}$ is normally distributed).*

Suppose the m-dimensional random vector $\mathbf{X}$ has a multivariate normal distribution with mean vector $\boldsymbol{\mu}$ and covariance matrix $\boldsymbol{\Sigma}$, denoted by $\mathbf{X} \sim \mathcal{N}(\boldsymbol{\mu}, \boldsymbol{\Sigma})$. If $\boldsymbol{\Sigma}$ is positive definite, then we say that the multivariate normal distribution is *non-degenerate* and the probability density function of $\mathbf{X}$ is given by

$$p(\mathbf{x}) = (2\pi)^{-m/2} |\boldsymbol{\Sigma}|^{-1/2} \exp\left(-(\mathbf{x} - \boldsymbol{\mu})^T \boldsymbol{\Sigma}^{-1}(\mathbf{x} - \boldsymbol{\mu})/2\right),$$

for $\mathbf{x} = (x_1, \ldots, x_m)^T \in \mathbb{R}^m$. Here $|\boldsymbol{\Sigma}|$ is the determinant of covariance matrix $\boldsymbol{\Sigma}$. However, in practice the determinant of the covariance matrix $\boldsymbol{\Sigma}$ may be zero (that is, $|\boldsymbol{\Sigma}| = 0$). In this case, the multivariate normal distribution is said to be *degenerate* (or *singular*), and it does not have a probability density function in the usual sense. A common way to deal with this degenerate case is to use the notion of the so-called generalized inverse and pseudo-determinant to define the probability density function. Let $l = \text{Rank}(\boldsymbol{\Sigma})$ (i.e., $l < m$), and $\boldsymbol{\Sigma} = \mathbf{V}\mathbf{D}\mathbf{V}^T$, where $\mathbf{V} \in \mathbb{R}^{m \times l}$ has rank l and $\mathbf{V}^T\mathbf{V} = \mathbf{I}_l$ with $\mathbf{I}_l \in \mathbb{R}^{l \times l}$ being an identity matrix, and $\mathbf{D} \in \mathbb{R}^{l \times l}$ is a diagonal matrix with its main diagonal entries being the non-zero eigenvalues of $\boldsymbol{\Sigma}$. Then the probability density function for the degenerate case is defined by

$$p(\mathbf{x}) = (2\pi)^{-l/2}|\mathbf{D}|^{-1/2}\exp\left(-(\mathbf{x} - \boldsymbol{\mu})^T\boldsymbol{\Sigma}^+(\mathbf{x} - \boldsymbol{\mu})/2\right), \qquad (2.89)$$

where the generalized inverse $\boldsymbol{\Sigma}^+ = \mathbf{V}\mathbf{D}^{-1}\mathbf{V}^T$. By (2.89) we see that the generalized inverse basically transforms the space from $\mathbb{R}^m$ to $\mathbb{R}^l$ and considers the probability density function in this reduced space. Interested readers can refer to [20] and the references therein for more information on this.

Similar to the normal distribution, the multivariate normal distribution is also closed under linear transformation. In other words, if an m-dimensional random vector $\mathbf{X} \sim \mathcal{N}(\boldsymbol{\mu}, \boldsymbol{\Sigma})$, then by (2.60) we have

$$\mathcal{A}\mathbf{X} + \mathbf{a} \sim \mathcal{N}(\mathcal{A}\boldsymbol{\mu} + \mathbf{a}, \mathcal{A}\boldsymbol{\Sigma}\mathcal{A}^T) \qquad (2.90)$$

for any $\mathcal{A} \in \mathbb{R}^{l \times m}$ and $\mathbf{a} \in \mathbb{R}^l$. This property implies that if the m-dimensional random vector $\mathbf{X} \sim \mathcal{N}(\boldsymbol{\mu}, \boldsymbol{\Sigma})$, then for any $j = 1, 2, \ldots, m$, we have $X_j \sim \mathcal{N}(\mu_j, \Sigma_{jj})$ with μ_j being the jth element of $\boldsymbol{\mu}$ and Σ_{jj} being the (j, j)th entry of $\boldsymbol{\Sigma}$. However, the converse is generally not true, that is, the fact that random variables $X_1, X_2, \ldots X_m$ are normally distributed does not imply that $\mathbf{X}$ is multivariate normally distributed. One exception is when normal random variables $X_1, X_2, \ldots X_m$ are mutually independent. Equation (2.90) also implies that any random vector composed of two or more components of a multivariate normally distributed random vector $\mathbf{X} \sim \mathcal{N}(\boldsymbol{\mu}, \boldsymbol{\Sigma})$ is still multivariate normally distributed. To obtain the mean vector and covariance matrix of this new random vector, we only need to drop the remaining entries from $\boldsymbol{\mu}$ and $\boldsymbol{\Sigma}$. For example, the random vector $(X_j, X_k)^T$ is multivariate normally distributed with mean vector $(\mu_j, \mu_k)^T$ and covariance matrix $\begin{pmatrix} \Sigma_{jj} & \Sigma_{jk} \\ \Sigma_{kj} & \Sigma_{kk} \end{pmatrix}$, where $1 \leq j, k \leq m$.

As we remarked in Section 2.3.1, the fact that two random variables are uncorrelated does not imply that these two random variables are independent. However, for a multivariate normally distributed random vector, if any two or more of its components are uncorrelated, then these components are also independent.

2.5.6   Exponential Distribution

The exponential distribution has many applications in practice, such as in the areas of reliability theory and waiting times. For example, we shall see later in Chapter 7 that the interarrival times (the waiting time between two successive jumps) for a Poisson process are independent and identically distributed with an exponential distribution.

A random variable X is said to exponentially distributed with rate parameter β (indicated by $X \sim \text{Exp}(\beta)$) if its associated probability density function is defined by

$$p(x) = \beta \exp(-\beta x), \quad x \geq 0. \tag{2.91}$$

The corresponding cumulative distribution function can be easily derived from (2.91), and it is given by

$$P(x) = 1 - \exp(-\beta x), \quad x \geq 0. \tag{2.92}$$

Figure 2.7 illustrates the plots for the probability density function and the cumulative distribution function of an exponentially distributed random variable with various values for the rate parameter β.

We observe from the left plot of Figure 2.7 that the plot of the probability density function is skewed with a "long right tail," which is similar to that of the log-normal distribution.

The mean and variance of the exponentially distributed random variable X with rate parameter β are given by, respectively,

$$\mathbb{E}(X) = \frac{1}{\beta}, \quad \text{Var}(X) = \frac{1}{\beta^2}, \tag{2.93}$$

and the kth moment of this random variable is

$$\mathbb{E}(X^k) = \frac{k!}{\beta^k}. \tag{2.94}$$

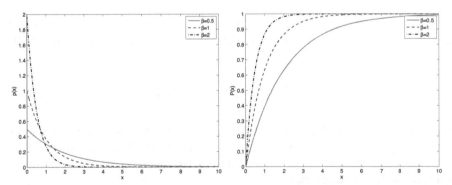

FIGURE 2.7: The plots for the probability density function and the cumulative distribution function of an exponentially distributed random variable with various values for the rate parameter β.

By (2.74), (2.93) and (2.94), we find the skewness of the exponentially distributed random variable is given by

$$\text{Skew}(X) = 2, \tag{2.95}$$

which is consistent with the left plot of Figure 2.7. In addition, (2.95) indicates that the skewness of the exponentially distributed random variable is independent of its rate parameter.

The exponential distribution has a number of important properties. Below we list a few of them, and refer the interested reader to [9, Section 2.1] and [22, Section 5.6] for more information. Let $X \sim \text{Exp}(\beta)$. Then for any positive number a we have

$$aX \sim \text{Exp}(\beta/a), \tag{2.96}$$

which can be easily derived from (2.40) and (2.91). Another important property is about the minimum of a sequence of independent exponentially distributed random variables and the index at which this minimum is achieved. Let $X_j \sim \text{Exp}(\beta_j)$, $j = 1, 2, \ldots, m$, be mutually independent, and

$$X_{\min} = \min_{1 \leq j \leq m} \{X_j\}. \tag{2.97}$$

Then

$$X_{\min} \sim \text{Exp}\left(\sum_{j=1}^{m} \beta_j\right), \tag{2.98}$$

and the probability of X_k being the minimum $X_{\min}$ is given by

$$\text{Prob}\{X_k = X_{\min}\} = \frac{\beta_k}{\sum_{j=1}^{m} \beta_j} \tag{2.99}$$

for $k = 1, 2, \ldots, m$. Let $I_{\min}$ be the index at which the minimum is achieved in (2.97). Then $I_{\min}$ is a random variable with range $\{1, 2, \ldots, m\}$, and (2.99) implies that

$$\text{Prob}\{I_{\min} = k\} = \frac{\beta_k}{\sum_{j=1}^{m} \beta_j}, \quad k = 1, 2, \ldots, m. \tag{2.100}$$

In addition, $I_{\min}$ and $X_{\min}$ are independent. As we shall see later in Chapter 8, the properties of (2.98) and (2.100), as well as the independence of $I_{\min}$ and $X_{\min}$, are important in understanding the algorithms in simulating a continuous time Markov chain.

Perhaps the most important property of the exponential distribution is that it is *memoryless*. A non-negative random variable X is said to be memoryless if

$$\text{Prob}\{X > x + \Delta x \mid X > x\} = \text{Prob}\{X > \Delta x\}, \quad \text{for all } x, \; \Delta x \geq 0. \tag{2.101}$$

If we let X denote the lifetime of an item, then $\text{Prob}\{X > x + \Delta x \mid X > x\}$ indicates the probability that the lifetime of the item exceeds $x + \Delta x$ given that this item is still functioning at x, and $\text{Prob}\{X > \Delta x\}$ represents the probability that the lifetime of an item exceeds Δx. Hence, the memoryless property (2.101) implies that the distribution of additional functional life of an item of age x is the same as that of a new item. In other words, there is no need to remember the age of a functional item since as long as it is still functional it is as good as new. It should be noted that the exponential distribution is the *only* continuous distribution that has this memoryless property. This implies that if a continuous distribution is memoryless, then this distribution must be an exponential distribution.

2.5.7 Gamma Distribution

A random variable X is gamma-distributed with shape parameter υ and rate parameter β, commonly indicated by $X \sim \text{Gamma}(\upsilon, \beta)$, if its probability density function is given by

$$p(x) = \frac{\beta^{\upsilon}}{\Gamma(\upsilon)} x^{\upsilon-1} \exp(-\beta x), \quad x > 0, \tag{2.102}$$

where Γ denotes the *gamma function*, defined as

$$\Gamma(\upsilon) = \int_0^{\infty} x^{\upsilon-1} \exp(-x)\, dx. \tag{2.103}$$

We observe from (2.102) and the above equation that the function Γ is a scaling factor so that $\int_0^{\infty} p(x)dx = 1$. It is because of this that the gamma distribution gets its name. Using integration by parts, it can be shown that

$$\Gamma(\upsilon) = (\upsilon - 1)\Gamma(\upsilon - 1). \tag{2.104}$$

Hence, for any positive integer υ we obtain

$$\Gamma(\upsilon) = (\upsilon - 1)!\,. \tag{2.105}$$

In addition, it can be found that $\Gamma(1/2) = \sqrt{\pi}$ and $\Gamma(3/2) = \sqrt{\pi}/2$.

By (2.102), (2.103) and (2.104), we find that the mean and variance of the gamma-distributed random variable X with shape parameter υ and rate parameter β are respectively given by

$$\mathbb{E}(X) = \frac{\upsilon}{\beta}, \quad \text{Var}(X) = \frac{\upsilon}{\beta^2}, \tag{2.106}$$

and the kth moment of X is given by

$$\mathbb{E}(X^k) = \frac{1}{\beta^k} \left(\prod_{j=0}^{k-1} (\upsilon + j) \right). \tag{2.107}$$

By (2.74), (2.106) and (2.107), we find that the skewness of this gamma-distributed random variable is given by

$$\text{Skew}(X) = \frac{2}{\sqrt{v}}. \tag{2.108}$$

This implies that the skewness of the gamma-distributed random variable depends only on the shape parameter v. Figure 2.8 illustrates the probability density function of a gamma-distributed random variable with rate parameter $\beta = 4$ and various different values for the shape parameter v. We observe from this figure that the probability density function of a gamma-distributed random variable is skewed with a "long right tail," which is consistent with (2.108).

The gamma distribution has a number of useful properties. For example, if $X_j \sim \text{Gamma}(v_j, \beta)$, $j = 1, 2, \ldots, m$, are mutually independent with $v_j, j = 1, 2, \ldots, m$, being positive integers, then we have

$$\sum_{j=1}^{m} X_j \sim \text{Gamma}\left(\sum_{j=1}^{m} v_j, \beta\right).$$

In addition, if $X \sim \text{Gamma}(v, \beta)$, then for any positive number a we have

$$aX \sim \text{Gamma}(v, \beta/a).$$

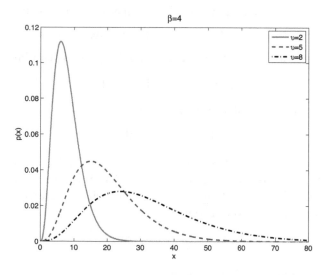

FIGURE 2.8: The plot of the probability density function of a gamma-distributed random variable with rate parameter $\beta = 4$ and various different values for the shape parameter v.

The gamma distribution includes a number of other distributions as special cases. For example, if v is a positive integer, then the gamma distribution becomes the so-called *Erlang distribution*. In addition, if $v = 1$, then the gamma distribution becomes the exponential distribution. Moreover, the exponential distribution has another close relationship with the gamma distribution. Specifically, if $X_j \sim \mathrm{Exp}(\beta)$, $j = 1, 2, \ldots, m$, are mutually independent, then

$$\sum_{j=1}^{m} X_j \sim \mathrm{Gamma}(m, \beta). \tag{2.109}$$

As we shall see in the next section, the gamma distribution also includes the chi-square distribution as a special case.

2.5.8 Chi-Square Distribution

The chi-square distribution is important in statistical analysis of variance (ANOVA) and other statistical procedures [7, 14] based on normally distributed random variables. In particular, the chi-square distribution with k degrees of freedom is the distribution of a sum of the squares of k independent standard normal random variables. That is, if X_j, $j = 1, 2, \ldots, k$, are independent and identically distributed random variables following a standard normal distribution, then the random variable Y defined as $Y = \sum_{j=1}^{k} X_j^2$ has a chi-square distribution with k degrees of freedom (denoted $Y \sim \chi^2(k)$ or $Y \sim \chi_k^2$). The probability density function of Y is then given by

$$p(y) = \frac{1}{2^{k/2}\Gamma(k/2)} y^{k/2-1} \exp(-y/2), \quad y \geq 0, \tag{2.110}$$

where Γ is the gamma function defined by (2.103).

We observe from (2.102) and (2.110) that if we choose $v = \dfrac{k}{2}$ and $\beta = \dfrac{1}{2}$, then the gamma distribution with shape parameter v and rate parameter β becomes the chi-square distribution with k degrees of freedom. By this, (2.106) and (2.107), the mean and variance of a chi-square distributed random variable Y with k degrees of freedom are respectively given by

$$\mathbb{E}(Y) = k, \quad \mathrm{Var}(Y) = 2k,$$

and the kth moment of Y is given as

$$\mathbb{E}(Y^k) = 2^k \left(\prod_{j=0}^{k-1} \left(\frac{k}{2} + j \right) \right).$$

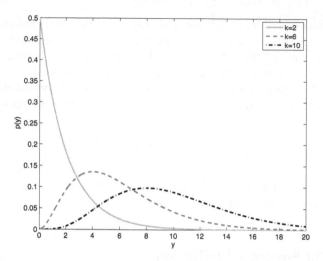

FIGURE 2.9:   The plot of the probability density function of a chi-square distribution for various degrees of freedom k.

In addition, by (2.108) we know that the skewness of Y is given by

$$\text{Skew}(Y) = \sqrt{\frac{8}{k}}. \tag{2.111}$$

The probability density function for Y with several values of k is depicted in Figure 2.9, which indicates that it is skewed with a "long right tail." This is consistent with (2.111). In addition, we observe from Figure 2.9 that the probability density function of Y becomes relatively symmetric when k is sufficiently large.

The chi-square distribution has some useful properties. For example, if $Y_j \sim \chi^2(k_j)$, $j = 1, 2, \ldots, m$, are mutually independent with $k_j, j = 1, 2, \ldots, m$, being some positive integers, then

$$\sum_{j=1}^{m} Y_j \sim \chi^2 \left(\sum_{j=1}^{m} k_j \right). \tag{2.112}$$

2.5.9   Student's t Distribution

If $U \sim \mathcal{N}(0, 1)$ and $V \sim \chi^2(k)$ are independent, then $X = U/\sqrt{V/k}$ has a *Student's t distribution with k degrees of freedom* (denoted by $X \sim t^k$). Its probability density function is

$$p(x) = \frac{\Gamma((k+1)/2)}{\Gamma(k/2)} \frac{1}{\sqrt{k\pi}} \frac{1}{(1+x^2/k)^{(k+1)/2}}, \quad x \in \mathbb{R},$$

where Γ is the gamma function defined by (2.103). The mean and variance are given by

$$\mathbb{E}(X) = 0 \quad \text{if} \quad k > 1 \text{ (otherwise undefined)},$$

and

$$\text{Var}(X) = k/(k-2) \quad \text{if} \quad k > 2 \text{ (otherwise undefined)}.$$

Figure 2.10 compares the probability density function of a Student's t distribution with various degrees of freedom to the probability density function of a standard normal distribution. From this figure we see that the probability density function of a Student's t distribution is symmetric like that of the normal distribution, with "heavier tails," and becomes similar to that of a standard normal distribution when k is sufficiently large.

As we shall see in the next chapter, the Student's t distribution is fundamental to the computation of confidence intervals for parameter estimators using experimental data in inverse problems (where typically $k \gg 2$).

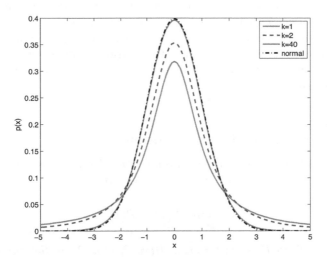

FIGURE 2.10: The plot of the probability density function of a Student's t distributed random variable with various degrees of freedom ($k = 1, 2, 40$), and the plot of the probability density function of a standard normal distribution.

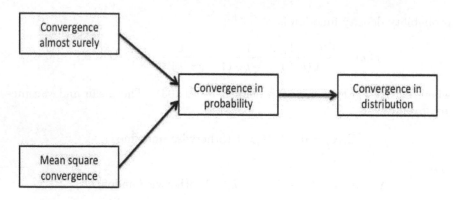

FIGURE 2.11:   Relationship among the four common modes of convergence of random variables.

2.6   Convergence of a Sequence of Random Variables

There are four modes of convergence of random variables that are commonly used in practice. They are *convergence in distribution, convergence in probability, convergence almost surely*, and *mean square convergence*. The relationship among these four modes of convergence is depicted in Figure 2.11, which indicates that both convergence almost surely and mean square convergence imply convergence in probability, and convergence in probability implies convergence in distribution. However, neither convergence almost surely or mean square convergence implies the other. Below we give a brief introduction to these four modes of convergence. Interested readers can refer to [3, 4, 24] for more information.

2.6.1   Convergence in Distribution

Definition 2.6.1 *A sequence of random variables* $\{X_j\}$ *is said to converge in distribution to a random variable X if we have*

$$\lim_{j\to\infty} P_j(x) = P(x) \tag{2.113}$$

for any x at which P is continuous. Here P_j and P are the cumulative distribution functions of X_j and X, respectively.

Convergence in distribution is also called *convergence in law* or *weak convergence*. This type of convergence is often written as

$$X_j \xrightarrow{d} X \quad \text{or} \quad X_j \Rightarrow X.$$

We note from (2.113) that the sequence of random variables $\{X_j\}$ and its limit X are not required to be defined in the same probability space. Even though this is a type of convergence of a sequence of random variables, the definition is in terms of cumulative distribution functions, not in terms of the random variables themselves.

It should be noted that convergence in distribution can be defined in several ways. Specifically, the following statements are equivalent.

- $\{X_j\}$ converge in distribution to X.

- The characteristic functions of X_j, $j = 1, 2, \ldots$ converge to that of X.

- $\lim_{j \to \infty} \mathbb{E}(f(X_j)) = \mathbb{E}(f(X))$ for any bounded continuous function f.

We note that the last statement implies that for any bounded continuous function f we have

$$\lim_{j \to \infty} \int_{-\infty}^{\infty} f(x) dP_j(x) = \int_{-\infty}^{\infty} f(x) dP(x). \tag{2.114}$$

In the language of functional analysis, the above convergence is called *weak star convergence*. However, in the language of probability, it is called *weak convergence* (a misnomer).

The following theorem, known as Slutsky's Theorem, indicates that convergence in distribution is closed under some arithmetic operations in a restrictive sense.

Theorem 2.6.1 *If $X_j \xrightarrow{d} X$ and $Y_j \xrightarrow{d} c$ with c being some constant, then we have*

(i) $X_j + Y_j \xrightarrow{d} X + c$;

(ii) $X_j Y_j \xrightarrow{d} cX$;

(iii) $X_j / Y_j \xrightarrow{d} X/c$ if $Prob\{Y_j = 0\} = 0$ for any j and $c \neq 0$.

One of the most important applications for convergence in distribution is the central limit theorem, which is stated as follows.

Theorem 2.6.2 *(Central limit theorem) Let $\{X_j\}$ be a sequence of independent and identically distributed (i.i.d) random variables with finite mean*

μ and positive standard deviation σ, and define $\bar{X}_m = \dfrac{1}{m}\displaystyle\sum_{j=1}^{m} X_j$. Then, as $m \to \infty$, the limiting distribution of

$$\frac{\bar{X}_m - \mu}{\frac{\sigma}{\sqrt{m}}}$$

is a standard normal distribution.

2.6.2 Convergence in Probability

Definition 2.6.2 *A sequence of random variables $\{X_j\}$ is said to converge in probability to a random variable X if for any $\epsilon > 0$ we have*

$$\lim_{j \to \infty} Prob\{|X_j - X| \ge \epsilon\} = 0. \tag{2.115}$$

This type of convergence is often written as

$$X_j \xrightarrow{\text{p}} X \quad \text{or} \quad \plim_{j \to \infty} X_j = X.$$

We note from (2.115) that to define convergence in probability the sequence of random variables $\{X_j\}$ as well as its limit X must be defined in the same probability space $(\Omega, \mathcal{F}, \text{Prob})$. In addition, condition (2.115) can be equivalently written as

$$\lim_{j \to \infty} Prob\{|X_j - X| < \epsilon\} = 1. \tag{2.116}$$

It is worth noting that convergence in probability implies convergence in distribution. However, the converse is true only when X is a constant.

The limits of convergence in probability are unique in the sense of equivalence almost surely. In other words, if $X_j \xrightarrow{\text{p}} X$ and $X_j \xrightarrow{\text{p}} \tilde{X}$, then X and $\tilde{X}$ are equal almost surely. In addition, similar to the ordinary convergence of real numbers, the following two statements are equivalent:

- $X_j \xrightarrow{\text{p}} X$.

- (Cauchy in probability): $X_j - X_k \xrightarrow{\text{p}} 0$ as $j, k \to \infty$.

Convergence in probability has a number of nice properties. For example, if $X_j \xrightarrow{\text{p}} X$, then for any continuous function f we have $f(X_j) \xrightarrow{\text{p}} f(X)$. In addition, convergence in probability is closed under arithmetic operations. This is indicated in the following theorem.

Theorem 2.6.3 *If $X_j \xrightarrow{p} X$ and $Y_j \xrightarrow{p} Y$, then we have*

(i) $aX_j \xrightarrow{p} aX$ for any a;

(ii) $X_j + Y_j \xrightarrow{p} X + Y$;

(iii) $X_j Y_j \xrightarrow{p} XY$;

(iv) $X_j / Y_j \xrightarrow{p} X/Y$ *if* $Prob\{Y_j = 0\} = 0$ *for any* j *and* $Prob\{Y = 0\} = 0$.

The concept of convergence in probability plays an important role in statistics. For example, it has applications in the consistency of estimators (to be discussed later) and the *weak law of large numbers*, which is stated in the following theorem.

Theorem 2.6.4 *Let* X_1, X_2, ..., *be independent and identically distributed random variables with* $\mathbb{E}(X_j) = \mu$ *and* $Var(X_j) = \sigma^2 < \infty$, *and define* $\bar{X}_m = \frac{1}{m} \sum_{j=1}^{m} X_j$. *Then for any* $\epsilon > 0$ *we have*

$$\lim_{m \to \infty} Prob\{|\bar{X}_m - \mu| \geq \epsilon\} = 0;$$

that is, $\bar{X}_m$ *converges in probability to* μ.

2.6.3 Convergence Almost Surely

Definition 2.6.3 *A sequence of random variables* $\{X_j\}$ *converges almost surely (a.s.) to a random variable* X *if*

$$Prob\left\{ \lim_{j \to \infty} X_j = X \right\} = 1. \qquad (2.117)$$

Convergence almost surely is also called *convergence almost everywhere* (a.e.), *convergence with probability one*, or *strong convergence*. This type of convergence is often written as

$$X_j \xrightarrow{\text{a.s.}} X.$$

We note from (2.117) that to define convergence almost surely the sequence of random variables $\{X_j\}$ as well as its limit X must be defined in the same probability space $(\Omega, \mathcal{F}, \text{Prob})$. Moreover there exists $\mathbb{A} \in \mathcal{F}$ and $\text{Prob}\{\mathbb{A}\} = 1$ such that

$$\lim_{j \to \infty} X_j(\omega) = X(\omega), \quad \omega \in \mathbb{A},$$

and this is a type of pointwise convergence. For simplification, the above equation is often simply written as

$$\lim_{j \to \infty} X_j = X \quad a.s.$$

There are another two alternative and equivalent ways to define convergence almost surely; that is, condition (2.117) can be replaced by either

$$\lim_{m\to\infty} \text{Prob}\left\{ \bigcap_{j\geq m} (|X_j - X| < \epsilon) \right\} = 1, \quad \text{for any positive number } \epsilon,$$

or

$$\lim_{m\to\infty} \text{Prob}\left\{ \bigcup_{j\geq m} (|X_j - X| \geq \epsilon) \right\} = 0, \quad \text{for any positive number } \epsilon.$$

By the above equation, we easily see that convergence a.s. implies convergence in probability, and thus implies convergence in distribution. In general, the converse is not true; that is, convergence in probability does not imply convergence almost surely. However, convergence in probability implies convergence a.s. of a subsequence. In other words, if $\{X_j\}$ converges in probability to X, then there exists a subsequence $\{X_{j_k}\}$ that converges a.s. to the same limit.

The limits in convergence a.s. are unique in the sense of equivalence almost surely. In other words, if $X_j \xrightarrow{\text{a.s.}} X$ and $X_j \xrightarrow{\text{a.s.}} \tilde{X}$, then X and $\tilde{X}$ are equal almost surely. In addition, the following two statements are equivalent:

- $X_j \xrightarrow{\text{a.s.}} X$.

- (Cauchy almost surely): $X_j - X_k \xrightarrow{\text{a.s}} 0$ as $j, k \to \infty$.

Convergence a.s. is also closed under arithmetic operations in the same manner as convergence in probability; that is, Theorem 2.6.3 also holds for convergence almost surely.

The concept of convergence a.s. plays an important role in statistics. For example, it has applications in the consistency of estimators (to be discussed in the next chapter) and the *strong law of large numbers* (SLLN), which is stated as follows.

Theorem 2.6.5 *Let X_1, X_2, ... be independent and identically distributed random variables with $\mathbb{E}(X_j) = \mu$ and $Var(X_j) = \sigma^2 < \infty$, and define $\bar{X}_m = \frac{1}{m}\sum_{j=1}^{m} X_j$. Then*

$$\text{Prob}\left\{ \lim_{m\to\infty} \bar{X}_m = \mu \right\} = 1.$$

That is, $\bar{X}_m$ converges a.s. to μ.

2.6.4 Mean Square Convergence

The concept of mean square convergence plays an important role in studying stochastic and random differential equations (as discussed later in Chapter 7). This concept gives rise to the *mean square calculus* and the *stochastic integral* (to be discussed in Chapter 7). We remark that one particular set of random variables that plays an important role in the mean square calculus is those that have a finite second moment. This type of random variable is called a *second-order random variable*.

Definition 2.6.4 *A sequence of second-order random variables $\{X_j\}$ is said to converge in a mean square (m.s.) sense to a random variable X as $j \to \infty$ if*

$$\lim_{j \to \infty} \mathbb{E}\left((X_j - X)^2\right) = 0. \tag{2.118}$$

Mean square convergence is also referred to as *convergence in quadratic mean* or *second-order convergence*. This type of convergence is often written as

$$X_j \xrightarrow{\text{m.s.}} X \quad \text{or} \quad \underset{j \to \infty}{\text{l.i.m.}} X_j = X.$$

The limits in m.s. convergence are unique in the sense of equivalence almost surely. In other words, if $\underset{j \to \infty}{\text{l.i.m.}} X_j = X$ and $\underset{j \to \infty}{\text{l.i.m.}} X_j = \tilde{X}$, then X and $\tilde{X}$ are equal almost surely. In addition, the following two statements are equivalent:

- $X_j \xrightarrow{\text{m.s.}} X$.

- (Cauchy in mean square): $X_j - X_k \xrightarrow{\text{m.s.}} 0$ as $j, k \to \infty$.

The mean square convergence has some useful properties. For example, l.i.m. and expectation can commute for a sequence of second-order random variables $\{X_j\}$; that is,

$$\text{if} \quad \underset{j \to \infty}{\text{l.i.m.}} X_j = X, \quad \text{then} \quad \lim_{j \to \infty} \mathbb{E}(X_j) = \mathbb{E}(X), \tag{2.119}$$

which is obtained due to the Cauchy–Schwartz inequality (2.66).

It is worth noting that mean square convergence implies convergence in probability (which can be easily derived by Markov's inequality (2.63)), and thus implies convergence in distribution. However, mean square convergence does not imply and also is not implied by convergence almost surely. Below are two examples that can be used to demonstrate this. One shows that mean square convergence does not imply convergence almost surely, while the other one shows that convergence almost surely does not imply mean square convergence.

Example 2.6.1 Consider a sequence of mutually independent random variables $\{X_j\}$, where X_j is a discrete random variable with range $\{0,1\}$ and the probability associated with each value is given by

$$\text{Prob}\{X_j = 0\} = 1 - \frac{1}{j}, \quad \text{Prob}\{X_j = 1\} = \frac{1}{j}. \tag{2.120}$$

It can be shown that $\{X_j\}$ converges in m.s. to 0. However, it does not converge a.s. to 0 (see [24, Section 4.2] for details).

Example 2.6.2 Consider a sequence of random variables $\{X_j\}$, where X_j is a discrete random variable with range $\{0, j\}$ and the probability associated with each value is given by

$$\text{Prob}\{X_j = 0\} = 1 - \frac{1}{j^2}, \quad \text{Prob}\{X_j = j\} = \frac{1}{j^2}. \tag{2.121}$$

It can be shown that $\{X_j\}$ converges a.s. to 0. However, it does not converge in m.s. to 0. We again refer the interested reader to [24, Section 4.2] for details.

References

[1] M.A. Bean, *Probability: The Science of Uncertainty with Applications to Investments, Insurance and Engineering*, American Mathematical Society, Providence, 2009.

[2] P. Billingsley, *Probability and Measure*, 3rd edition, John Wiley & Sons, New York, 1995.

[3] P. Billingsley, *Convergence of Probability Measures*, 2nd edition, John Wiley & Sons, New York, 1999.

[4] B.R. Bhat, *Modern Probability Theory*, 3rd edition, New Age Internal Limited, New Delhi, 2004.

[5] R.N. Bracewell, *The Fourier Transform and Its Applications*, 3rd edition, McGraw-Hill, Boston, 2000.

[6] R.J. Carroll and D. Ruppert, *Transformation and Weighting in Regression*, Chapman & Hall, New York, 1988.

[7] G. Casella and R. L. Berger, *Statistical Inference*, Duxbury, Pacific Grove, California, 2002.

[8] M. Davidian and D. Giltinan, *Nonlinear Models for Repeated Measurement Data*, Chapman & Hall, London, 1998.

[9] R. Durrett, *Essentials of Stochastic Processes*, http://www.math.duke.edu/%7Ertd/EOSP/EOSP2E.pdf.

[10] G.A. Edgar, *Integral, Probability, and Fractal Measures*, Springer-Verlag, New York, 1998.

[11] A. Friedman, *Foundations of Modern Analysis*, Dover Publications, New York, 1982.

[12] J. Galambos, Raikov's theorem, in *Encyclopedia of Statistical Sciences*, Wiley, 2006; http://dx.doi.org/10.1002/0471667196.ess2160.pub2.

[13] A.R. Gallant, *Nonlinear Statistical Models*, Wiley, New York, 1987.

[14] F. Graybill, *Theory and Application of the Linear Model*, Duxbury, North Scituate, MA, 1976.

[15] M. Grigoriu, *Stochastic Calculus: Applications in Science and Engineering*, Birkhauser, Boston, 2002.

[16] D.W. Kammler, *A First Course in Fourier Analysis*, Cambridge University Press, New York, 2007.

[17] F. Klebaner, *Introduction to Stochastic Calculus with Applications*, 2nd edition, Imperial College Press, London, 2006.

[18] T. Mikosch, *Elementary Stochastic Calculus*, World Scientific, Singapore, 1998.

[19] A. Papoulis, *The Fourier Integral and Its Applications*, McGraw-Hill, New York, 1962.

[20] C.R. Rao and S.K. Mitra, *Generalized Inverse of a Matrix and Its Applications*, John Wiley & Sons, New York, 1971.

[21] P. Protter, *Stochastic Integration and Differential Equations*, Springer-Verlag, New York, 1990.

[22] S.M. Ross, *Introduction to Probability and Statistics for Engineers and Scientists*, 4th edition, Elsevier Academic Press, Burlington, MA, 2009.

[23] H.L. Royden, *Real Analysis*, 3rd edition, Prentice Hall, Englewood Cliffs, NJ, 1988.

[24] T.T. Soong, *Random Differential Equations in Science and Engineering*, Academic Press, New York and London, 1973.

[25] R.E. Walpole and R.H. Myers, *Probability and Statistics for Engineers and Scientists*, 3rd edition, Macmillan Publishing Company, New York, 1985.

[26] J. Wloka, *Partial Differential Equations*, Cambridge University Press, Cambridge, 1987.

Chapter 3

Mathematical and Statistical
Aspects of Inverse Problems

In inverse or parameter estimation problems, as discussed in the Introduction, an important but practical question is how successful the mathematical model is in describing the physical or biological phenomena represented by the experimental data. In general, it is very unlikely that the residual sum of squares (RSS) in the least-squares formulation is zero. Indeed, due to measurement noise as well as modeling error, there may not be a "true" set of parameters so that the mathematical model will provide an exact fit to the experimental data.

Even if one begins with a deterministic model and has no initial interest in uncertainty or stochasticity, as soon as one employs experimental data in the investigation, one is led to uncertainty that should not be ignored. In fact, all measurement procedures contain error or uncertainty in the data collection process and hence statistical questions arise regarding that sampling error. To correctly formulate, implement and analyze the corresponding inverse problems, one requires a framework entailing a *statistical model* as well as a *mathematical model*.

In this chapter we discuss mathematical, statistical and computational aspects of inverse or parameter estimation problems for deterministic dynamical systems based on the Ordinary Least Squares (OLS), Weighted Least Squares (WLS) and Generalized Least Squares (GLS) methods with appropriate corresponding data noise assumptions of constant variance and non-constant variance (e.g., relative error). Among the topics included are the interplay between the mathematical model, the statistical model, and observation or data assumptions, and some techniques (residual plots) for analyzing the uncertainties associated with inverse problems employing experimental data. We also outline a standard theory underlying the construction of confidence intervals for parameter estimators. This asymptotic theory for confidence intervals can be found in Seber and Wild [35]. Finally, we also compare this asymptotic error approach to the popular "bootstrapping" approach.

3.1   Least Squares Inverse Problem Formulations

3.1.1   The Mathematical Model

We consider inverse or parameter estimation problems in the context of a parameterized (with vector parameter $q \in \mathbb{R}^{\kappa_q}$) n-dimensional vector dynamical system or **mathematical model**

$$\frac{d\boldsymbol{x}}{dt}(t) = \boldsymbol{g}(t, \boldsymbol{x}(t), \boldsymbol{q}), \tag{3.1}$$

$$\boldsymbol{x}(t_0) = \boldsymbol{x}_0, \tag{3.2}$$

with **observation process**

$$\boldsymbol{f}(t; \boldsymbol{\theta}) = \mathcal{C}\boldsymbol{x}(t; \boldsymbol{\theta}), \tag{3.3}$$

where $\boldsymbol{\theta} = (\boldsymbol{q}^T, \tilde{\boldsymbol{x}}_0^T)^T \in \mathbb{R}^{\kappa_q + \tilde{n}} = \mathbb{R}^{\kappa_\theta}, \tilde{n} \leq n$, and the observation operator $\mathcal{C}$ maps $\mathbb{R}^n$ to $\mathbb{R}^m$. In most of the discussions below we assume without loss of generality that some subset $\tilde{\boldsymbol{x}}_0$ of the initial values $\boldsymbol{x}_0$ are also unknown.

The mathematical model is a deterministic system—in this chapter we primarily treat ordinary differential equations (ODE), but our discussions are relevant to problems involving parameter-dependent partial differential equations (PDE), delay differential equations, etc., as long as the system is assumed to be well-posed (i.e., to possess unique solutions that depend smoothly on the parameters and initial data). For example, in the CFSE example of this chapter we shall also consider partial differential equation systems such as

$$\frac{\partial \boldsymbol{u}}{\partial t}(t, x) = \mathcal{G}\left(t, x, \boldsymbol{u}, \frac{\partial \boldsymbol{u}}{\partial x}, \frac{\partial^2 \boldsymbol{u}}{\partial x^2}, \boldsymbol{q}\right), \quad t \in [t_0, t_f], x \in [x_0, x_f] \tag{3.4}$$

$$\boldsymbol{u}(t_0, x) = \boldsymbol{u}_0(x)$$

with appropriate boundary conditions.

Following usual conventions (which correspond to the form of data usually available from experiments), we assume a discrete form of the observations in which one has N longitudinal observations $\mathbf{y}_j$ corresponding to

$$\boldsymbol{f}(t_j; \boldsymbol{\theta}) = \mathcal{C}\boldsymbol{x}(t_j; \boldsymbol{\theta}), \quad j = 1, \ldots, N. \tag{3.5}$$

In general, the corresponding observations or data $\{\mathbf{y}_j\}$ will not be exactly $\boldsymbol{f}(t_j; \boldsymbol{\theta})$. Due to the nature of the phenomena leading to this discrepancy, we treat this uncertainty pertaining to the observations with a statistical model for the observation process.

3.1.2   The Statistical Model

In our discussions here we consider a **statistical model** of the form

$$\boldsymbol{Y}_j = \boldsymbol{f}(t_j; \boldsymbol{\theta}_0) + \boldsymbol{h}_j \circ \tilde{\boldsymbol{\mathcal{E}}}_j, \quad j = 1, \ldots, N, \tag{3.6}$$

where $\boldsymbol{f}(t_j; \boldsymbol{\theta}) = \mathcal{C}\boldsymbol{x}(t_j; \boldsymbol{\theta})$, $j = 1, \ldots, N$, and $\mathcal{C}$ is an $m \times n$ matrix. This corresponds to the observed part of the solution of the mathematical model (3.1)–(3.2) at the jth covariate or observation time for a particular vector of parameters $\boldsymbol{\theta} \in \mathbb{R}^{\kappa_q + \tilde{n}} = \mathbb{R}^{\kappa_\theta}$. Here the m-vector function $\boldsymbol{h}_j$ is defined by

$$
\boldsymbol{h}_j = \begin{cases} (1, \ldots, 1)^T & \text{for the vector OLS case} \\ (w_{1,j}, \ldots, w_{m,j})^T & \text{for the vector WLS case} \\ (f_1^\gamma(t_j; \boldsymbol{\theta}_0), \ldots, f_m^\gamma(t_j; \boldsymbol{\theta}_0))^T & \text{for the vector GLS case,} \end{cases} \tag{3.7}
$$

for $j = 1, \ldots, N$, and $\boldsymbol{h}_j \circ \widetilde{\boldsymbol{\mathcal{E}}}_j$ denotes the component-wise multiplication of the vectors $\boldsymbol{h}_j$ and $\widetilde{\boldsymbol{\mathcal{E}}}_j$. The vector $\boldsymbol{\theta}_0$ represents the "truth" parameter that generates the observations $\{\mathbf{Y}_j\}_{j=1}^N$. (The existence of a truth parameter $\boldsymbol{\theta}_0$ is a standard assumption in statistical formulations and this along with the assumption that the means $\mathbb{E}(\widetilde{\boldsymbol{\mathcal{E}}}_j)$ are zero yields implicitly that (3.1)–(3.2) is a correct description of the process being modeled.) The terms $\boldsymbol{h}_j \circ \widetilde{\boldsymbol{\mathcal{E}}}_j$ are random variables which can represent observation or measurement error, "system fluctuations" or other phenomena that cause observations to not fall exactly on the points $\boldsymbol{f}(t_j; \boldsymbol{\theta}_0)$ from the smooth path $\boldsymbol{f}(t, \boldsymbol{\theta}_0)$. Since these fluctuations are unknown to the modeler, we will assume that realizations $\widetilde{\boldsymbol{\epsilon}}_j$ of $\widetilde{\boldsymbol{\mathcal{E}}}_j$ are generated from a probability distribution which reflects the assumptions regarding these phenomena. Thus specific data (*realizations*) corresponding to (3.6) will be represented by

$$
\boldsymbol{y}_j = \boldsymbol{f}(t_j; \boldsymbol{\theta}_0) + \boldsymbol{h}_j \circ \widetilde{\boldsymbol{\epsilon}}_j, \quad j = 1, \ldots, N. \tag{3.8}
$$

We make standard assumptions about the $\widetilde{\boldsymbol{\mathcal{E}}}_j$ in that they are independent and identically distributed with mean zero and constant covariance matrix. This model (3.8) allows for a fairly wide range of error models, including the usual *absolute* (or *constant variance*) error model, when $\gamma = 0$ (the OLS case), as well as the *relative (or constant coefficient of variation) error* model when $\gamma = 1$. For instance, in a statistical model for pharmacokinetics of drugs in human blood samples, a natural choice for the statistical model might be $\gamma = 0$ and a multivariate normal distribution for $\widetilde{\boldsymbol{\mathcal{E}}}_j$. In observing (counting) populations, the error may depend on the size of the population itself (i.e., $\gamma = 1$) while studies of flow cytometry data [9] have revealed that a choice of $\gamma = \dfrac{1}{2}$ may be most appropriate. Each of these cases will be further discussed below.

The purpose of our presentation in this chapter is to discuss methodology related to estimates $\hat{\boldsymbol{\theta}}$ for the true value of the parameter $\boldsymbol{\theta}_0$ from a set Ω_θ of admissible parameters, and the dependence of this methodology on what is assumed about the choice of γ and the covariance matrices of the errors $\widetilde{\boldsymbol{\mathcal{E}}}_j$. We discuss a class of inverse problem methodologies that can be used to calculate estimates $\hat{\boldsymbol{\theta}}$ for $\boldsymbol{\theta}_0$: the ordinary, the weighted and the generalized least-squares formulations.

We are interested in situations (as is the case in most applications) where the error distribution is <u>unknown</u> to the modeler beyond the assumptions on $\mathbb{E}(\boldsymbol{Y}_j)$ embodied in the model and the assumptions made on $\mathrm{Var}(\widetilde{\boldsymbol{\mathcal{E}}}_j)$. We seek to explore how one should proceed in estimating $\boldsymbol{\theta}_0$ and the covariance matrix in these circumstances.

3.2   Methodology: Ordinary, Weighted and Generalized Least Squares

3.2.1   Scalar Ordinary Least Squares

To simplify notation, we first consider the absolute error statistical model ($\gamma = 0$) in the scalar case. This then takes the form

$$Y_j = f(t_j; \boldsymbol{\theta}_0) + \widetilde{\mathcal{E}}_j, \ j = 1, \ldots, N, \tag{3.9}$$

where the variance $\mathrm{Var}(\widetilde{\mathcal{E}}_j) = \sigma_0^2$ is assumed to be unknown to the modeler. (Note also that the distribution of the error need not be specified.) It is assumed that the observation errors are independent across j (i.e., time), which may be a reasonable one when the observations are taken with sufficient intermittency or when the primary source of error is measurement error. If we define

$$\boldsymbol{\theta}_{\mathrm{OLS}} = \boldsymbol{\theta}_{\mathrm{OLS}}^N(\boldsymbol{Y}) = \arg\min_{\boldsymbol{\theta}\in\Omega_\theta} \sum_{j=1}^{N}[Y_j - f(t_j; \boldsymbol{\theta})]^2, \tag{3.10}$$

where $\boldsymbol{Y} = (Y_1, Y_2, \ldots, Y_N)^T$, then $\boldsymbol{\theta}_{\mathrm{OLS}}$ can be viewed as minimizing the distance between the data and model where all observations are treated as being of equal importance. We note that minimizing the functional in (3.10) corresponds to solving for $\boldsymbol{\theta}$ in

$$\sum_{j=1}^{N}[Y_j - f(t_j; \boldsymbol{\theta})]\nabla f(t_j; \boldsymbol{\theta}) = 0, \tag{3.11}$$

the so-called *normal equations* or *estimating equations*. We point out that $\boldsymbol{\theta}_{\mathrm{OLS}}$ is a *random vector* (because $\widetilde{\mathcal{E}}_j = Y_j - f(t_j; \boldsymbol{\theta})$ is a random variable); hence if $\{y_j\}_{j=1}^N$ are realizations of the *random variables* $\{Y_j\}_{j=1}^N$ then solving

$$\hat{\boldsymbol{\theta}}_{\mathrm{OLS}} = \hat{\boldsymbol{\theta}}_{\mathrm{OLS}}^N = \arg\min_{\boldsymbol{\theta}\in\Omega_\theta} \sum_{j=1}^{N}[y_j - f(t_j; \boldsymbol{\theta})]^2 \tag{3.12}$$

provides a realization for $\boldsymbol{\theta}_{\mathrm{OLS}}$.

We remark on our notation, which is adopted from [16]. For a random vector or estimator $\boldsymbol{\theta}_{\text{OLS}}$, we will always denote a corresponding realization or estimate with an <u>over hat</u>; e.g., $\hat{\boldsymbol{\theta}}_{\text{OLS}}$ is an estimate for $\boldsymbol{\theta}_0$. Thus we have here abandoned the usual convention of capital letters for random variables and a lower case letter for a corresponding realization. We sometimes suppress the dependence on N unless it is specifically needed. Finally, we drop the subscript OLS for the estimates when it is clearly understood in context.

Returning to (3.10) and (3.12) and noting that

$$\sigma_0^2 = \frac{1}{N} \mathbb{E} \left(\sum_{j=1}^{N} [Y_j - f(t_j; \boldsymbol{\theta}_0)]^2 \right), \tag{3.13}$$

we see that once we have solved for $\hat{\boldsymbol{\theta}}_{\text{OLS}}$ in (3.12), we may readily obtain an estimate $\hat{\sigma}_{\text{OLS}}^2$ for σ_0^2. (Recall that $\mathbb{E}$ denotes the expectation operator.)

Even though the distribution of the error random variables is not specified, we can use asymptotic theory to approximate the mean and covariance of the random vector $\boldsymbol{\theta}_{\text{OLS}}$ [35]. As will be explained in more detail below, as $N \to \infty$, we have that

$$\boldsymbol{\theta}_{\text{OLS}} = \boldsymbol{\theta}_{\text{OLS}}^N \sim \mathcal{N}(\boldsymbol{\theta}_0, \Sigma_0^N) \approx \mathcal{N}(\boldsymbol{\theta}_0, \sigma_0^2 [F_{\boldsymbol{\theta}}^N(\boldsymbol{\theta}_0)^T F_{\boldsymbol{\theta}}^N(\boldsymbol{\theta}_0)]^{-1}), \tag{3.14}$$

where the sensitivity matrix $F_{\boldsymbol{\theta}}(\boldsymbol{\theta}) = F_{\boldsymbol{\theta}}^N(\boldsymbol{\theta}) = \left((F_{\boldsymbol{\theta}}^N)_{jk}(\boldsymbol{\theta}) \right)$ is defined by

$$(F_{\boldsymbol{\theta}}^N)_{jk}(\boldsymbol{\theta}) = \frac{\partial f(t_j; \boldsymbol{\theta})}{\partial \theta_k}, \quad j = 1, \ldots, N, \quad k = 1, \ldots, \kappa_\theta, \tag{3.15}$$

and

$$\Sigma_0^N \equiv \sigma_0^2 [N\Omega_0]^{-1}, \tag{3.16}$$

with

$$\Omega_0 \equiv \lim_{N \to \infty} \frac{1}{N} F_{\boldsymbol{\theta}}^N(\boldsymbol{\theta}_0)^T F_{\boldsymbol{\theta}}^N(\boldsymbol{\theta}_0), \tag{3.17}$$

where the limit is assumed to exist (see [7, 10, 35]). Thus we have that $\boldsymbol{\theta}_{\text{OLS}}$ is approximately distributed as a multivariate normal random variable with mean $\boldsymbol{\theta}_0$ and covariance matrix Σ_0^N. However, $\boldsymbol{\theta}_0$ and σ_0^2 are generally unknown, and therefore one usually uses instead the *realization* $\mathbf{y} = (y_1, \ldots, y_N)^T$ of the random vector $\mathbf{Y}$ to obtain the estimate $\hat{\boldsymbol{\theta}}_{\text{OLS}}$ given by (3.12) and the *bias adjusted* approximation for σ_0^2:

$$\hat{\sigma}_{\text{OLS}}^2 = \frac{1}{N - \kappa_\theta} \sum_{j=1}^{N} [y_j - f(t_j; \hat{\boldsymbol{\theta}}_{\text{OLS}})]^2. \tag{3.18}$$

Both $\hat{\boldsymbol{\theta}}_{\text{OLS}}$ and $\hat{\sigma}_{\text{OLS}}^2$ are then used as approximations in (3.14). That is, both $\hat{\boldsymbol{\theta}} = \hat{\boldsymbol{\theta}}_{\text{OLS}}$ and $\hat{\sigma}^2 = \hat{\sigma}_{\text{OLS}}^2$ will then be used to approximate the covariance matrix

$$\Sigma_0^N \approx \hat{\Sigma}^N \equiv \hat{\sigma}^2 [F_{\boldsymbol{\theta}}^N(\hat{\boldsymbol{\theta}})^T F_{\boldsymbol{\theta}}^N(\hat{\boldsymbol{\theta}})]^{-1}. \tag{3.19}$$

We can obtain the standard errors $\mathrm{SE}_k(\hat{\boldsymbol{\theta}}_{\mathrm{OLS}})$ (discussed in more detail below) for the kth element of $\hat{\boldsymbol{\theta}}_{\mathrm{OLS}}$ by calculating $\mathrm{SE}_k(\hat{\boldsymbol{\theta}}_{\mathrm{OLS}}) \approx \sqrt{\hat{\Sigma}^N_{kk}}$.

We note that (3.18) represents the estimate for σ_0^2 of (3.13) with the factor $\dfrac{1}{N}$ replaced by the factor $\dfrac{1}{N - \kappa_\theta}$. In the linear case the estimate with $\dfrac{1}{N}$ can be shown to be biased downward (i.e., biased too low) and the same behavior can be observed in the general nonlinear case – see Chapter 12 of [35] and p. 28 of [22]. The subtraction of κ_θ degrees of freedom reflects the fact that $\hat{\boldsymbol{\theta}}$ has been computed to satisfy the κ_θ normal equations (3.11). We also remark that (3.13) is true even in the general nonlinear case—it does not rely on any asymptotic theories, although it does depend on the assumption of constant variance being correct.

3.2.2   Vector Ordinary Least Squares

We next consider the more general case in which we have a **vector of observations** for the jth covariate t_j. If we still assume the variance is constant in longitudinal data, then the statistical model is reformulated as

$$\boldsymbol{Y}_j = \boldsymbol{f}(t_j; \boldsymbol{\theta}_0) + \widetilde{\boldsymbol{\mathcal{E}}}_j, \tag{3.20}$$

where $\boldsymbol{f}(t_j; \boldsymbol{\theta}_0) \in \mathbb{R}^m$ and $\widetilde{\boldsymbol{\mathcal{E}}}_j$, $j = 1, \ldots, N$ are independent and identically distributed with zero mean and covariance matrix given by

$$V_0 = \mathrm{Var}(\widetilde{\boldsymbol{\mathcal{E}}}_j) = \mathrm{diag}(\sigma_{0,1}^2, \ldots, \sigma_{0,m}^2), \tag{3.21}$$

for $j = 1, \ldots, N$. Here we have allowed for the possibility that the observation coordinates $\boldsymbol{Y}_j$ may have different *constant* variances $\sigma_{0,i}^2$, i.e., $\sigma_{0,i}^2$ does not necessarily have to equal $\sigma_{0,k}^2$. We note that this formulation also can be used to treat the case where V_0 is used to simply scale the observations (i.e., $V_0 = \mathrm{diag}(v_1, \ldots, v_m)$ is known). In this case the formulation is simply a *vector OLS* (sometimes also called a *weighted least squares* (WLS)). The problem will consist of finding the minimizer

$$\boldsymbol{\theta}_{\mathrm{OLS}} = \arg \min_{\boldsymbol{\theta} \in \Omega_\theta} \sum_{j=1}^{N} [\boldsymbol{Y}_j - \boldsymbol{f}(t_j, \boldsymbol{\theta})]^T V_0^{-1} [\boldsymbol{Y}_j - \boldsymbol{f}(t_j, \boldsymbol{\theta})], \tag{3.22}$$

where the procedure weights elements of the vector $\boldsymbol{Y}_j - \boldsymbol{f}(t_j, \boldsymbol{\theta})$ according to their variability. (Some authors refer to (3.22) as a generalized least squares (GLS) procedure, but we will make use of this terminology in a different formulation in subsequent discussions). Just as in the scalar OLS case, $\boldsymbol{\theta}_{\mathrm{OLS}}$ is a *random vector* (again, because $\widetilde{\boldsymbol{\mathcal{E}}}_j = \boldsymbol{Y}_j - \boldsymbol{f}(t_j, \boldsymbol{\theta})$ is a random vector); hence if $\{\boldsymbol{y}_j\}_{j=1}^{N}$ is a collection of realizations of the *random vectors* $\{\boldsymbol{Y}_j\}_{j=1}^{N}$, then solving

$$\hat{\boldsymbol{\theta}}_{\mathrm{OLS}} = \arg \min_{\boldsymbol{\theta} \in \Omega_\theta} \sum_{j=1}^{N} [\boldsymbol{y}_j - \boldsymbol{f}(t_j, \boldsymbol{\theta})]^T V_0^{-1} [\boldsymbol{y}_j - \boldsymbol{f}(t_j, \boldsymbol{\theta})] \tag{3.23}$$

provides a realization $\hat{\boldsymbol{\theta}} = \hat{\boldsymbol{\theta}}_{\text{OLS}}$ for $\boldsymbol{\theta}_{\text{OLS}}$. By the definition of the covariance matrix we have

$$V_0 = \text{diag} \, \mathbb{E} \left(\frac{1}{N} \left(\sum_{j=1}^{N} [\boldsymbol{Y}_j - \boldsymbol{f}(t_j, \boldsymbol{\theta}_0)][\boldsymbol{Y}_j - \boldsymbol{f}(t_j, \boldsymbol{\theta}_0)]^T \right)_{ii} \right).$$

Thus an unbiased approximation for V_0 is given by

$$\hat{V} = \text{diag} \left(\frac{1}{N - \kappa_\theta} \left(\sum_{j=1}^{N} [\boldsymbol{y}_j - \boldsymbol{f}(t_j, \hat{\boldsymbol{\theta}})][\boldsymbol{y}_j - \boldsymbol{f}(t_j, \hat{\boldsymbol{\theta}})]^T \right)_{ii} \right). \quad (3.24)$$

However, the estimate $\hat{\boldsymbol{\theta}}$ of (3.23) requires the (generally unknown) matrix V_0, and V_0 requires the unknown vector $\boldsymbol{\theta}_0$, so we will instead use the following expressions to calculate $\hat{\boldsymbol{\theta}}$ and $\hat{V}$:

$$\boldsymbol{\theta}_0 \approx \hat{\boldsymbol{\theta}} = \arg \min_{\boldsymbol{\theta} \in \Omega_\theta} \sum_{j=1}^{N} [\boldsymbol{y}_j - \boldsymbol{f}(t_j, \boldsymbol{\theta})]^T \hat{V}^{-1} [\boldsymbol{y}_j - \boldsymbol{f}(t_j, \boldsymbol{\theta})] \quad (3.25)$$

$$V_0 \approx \hat{V} = \text{diag} \left(\frac{1}{N - \kappa_\theta} (\sum_{j=1}^{N} [\boldsymbol{y}_j - \boldsymbol{f}(t_j; \hat{\boldsymbol{\theta}})][\boldsymbol{y}_j - \boldsymbol{f}(t_j; \hat{\boldsymbol{\theta}})]^T)_{ii} \right). \quad (3.26)$$

Note that the expressions for $\hat{\boldsymbol{\theta}}$ and $\hat{V}$ constitute a coupled system of equations that will require greater effort in implementing a numerical scheme, as discussed in the next section.

Just as in the scalar case, we can determine the asymptotic properties of the OLS estimator (3.22). As $N \to \infty$, $\boldsymbol{\theta}_{\text{OLS}}$ has the following asymptotic properties [22, 35]:

$$\boldsymbol{\theta}_{\text{OLS}} \sim \mathcal{N}(\boldsymbol{\theta}_0, \Sigma_0^N), \quad (3.27)$$

where

$$\Sigma_0^N \approx \left(\sum_{j=1}^{N} D_j^T(\boldsymbol{\theta}_0) V_0^{-1} D_j(\boldsymbol{\theta}_0) \right)^{-1}, \quad (3.28)$$

and the $m \times \kappa_\theta$ matrix $D_j(\boldsymbol{\theta}_0) = D_j^N(\boldsymbol{\theta}_0)$ is given by

$$\begin{pmatrix} \dfrac{\partial f_1(t_j; \boldsymbol{\theta}_0)}{\partial \theta_1} & \dfrac{\partial f_1(t_j; \boldsymbol{\theta}_0)}{\partial \theta_2} & \cdots & \dfrac{\partial f_1(t_j; \boldsymbol{\theta}_0)}{\partial \theta_{\kappa_\theta}} \\ \vdots & \vdots & & \vdots \\ \dfrac{\partial f_m(t_j; \boldsymbol{\theta}_0)}{\partial \theta_1} & \dfrac{\partial f_m(t_j; \boldsymbol{\theta}_0)}{\partial \theta_2} & \cdots & \dfrac{\partial f_m(t_j; \boldsymbol{\theta}_0)}{\partial \theta_{\kappa_\theta}} \end{pmatrix}.$$

Since the true values of the parameters $\boldsymbol{\theta}_0$ and V_0 are unknown, their estimates $\hat{\boldsymbol{\theta}}$ and $\hat{V}$ are used to approximate the asymptotic properties of the least-squares estimator $\boldsymbol{\theta}_{\mathrm{OLS}}$:

$$\boldsymbol{\theta}_{\mathrm{OLS}} \sim \mathcal{N}(\boldsymbol{\theta}_0, \Sigma_0^N) \approx \mathcal{N}(\hat{\boldsymbol{\theta}}, \hat{\Sigma}^N), \tag{3.29}$$

where

$$\Sigma_0^N \approx \hat{\Sigma}^N = \left(\sum_{j=1}^{N} D_j^T(\hat{\boldsymbol{\theta}}) \hat{V}^{-1} D_j(\hat{\boldsymbol{\theta}}) \right)^{-1}. \tag{3.30}$$

The standard errors $\mathrm{SE}_k(\hat{\boldsymbol{\theta}}_{\mathrm{OLS}})$ can then be calculated for the kth element of $\hat{\boldsymbol{\theta}}_{\mathrm{OLS}}$ by $\mathrm{SE}_k(\hat{\boldsymbol{\theta}}_{\mathrm{OLS}}) \approx \sqrt{\hat{\Sigma}_{kk}^N}$.

3.2.3   Numerical Implementation of the Vector OLS Procedure

In the scalar statistical model (3.9), the estimates $\hat{\boldsymbol{\theta}}$ and $\hat{\sigma}$ can be solved separately (this is also true of the vector statistical model in the case $V_0 = \sigma_0^2 \mathbf{I}_m$, where $\mathbf{I}_m$ is the $m \times m$ identity matrix) and thus the numerical implementation is straightforward. First determine $\hat{\boldsymbol{\theta}}_{\mathrm{OLS}}$ according to (3.12) and then calculate $\hat{\sigma}_{\mathrm{OLS}}^2$ according to (3.18). However, as already noted, the estimates $\hat{\boldsymbol{\theta}}$ and $\hat{V}$ in the case of the general vector statistical model (3.20) require more effort since the equations (3.25)–(3.26) are coupled. To solve this coupled system the following iterative process can be used:

1. Set $\hat{V} = \hat{V}^{(0)} = \mathbf{I}_m$ and solve for the initial estimate $\hat{\boldsymbol{\theta}}^{(0)}$ using (3.25). Set $l = 0$.

2. Use $\hat{\boldsymbol{\theta}}^{(l)}$ to calculate $\hat{V}^{(l+1)}$ using (3.26).

3. Re-estimate $\hat{\boldsymbol{\theta}}$ by solving (3.25) with $\hat{V} = \hat{V}^{(l+1)}$ to obtain $\hat{\boldsymbol{\theta}}^{(l+1)}$.

4. Set $l = l + 1$ and return to step 2. Terminate the process and set $\hat{\boldsymbol{\theta}}_{\mathrm{OLS}} = \hat{\boldsymbol{\theta}}^{(l+1)}$ when two successive estimates for $\hat{\boldsymbol{\theta}}$ are sufficiently close to one another.

3.2.4   Weighted Least Squares (WLS)

Although in the above discussion the measurement error's distribution remained unspecified, we did require that the measurement error remain constant in variance in longitudinal data. That assumption may not be appropriate for data sets whose measurement error is not constant in a longitudinal sense. A common *weighted error* model, in which the error is weighted according to some *known* weights, an assumption which might be reasonable when one has data that varies widely in the scale of observations that experimentalists must use for the scalar observation case is

$$Y_j = f(t_j; \boldsymbol{\theta}_0) + w_j \widetilde{\mathcal{E}}_j. \tag{3.31}$$

Here $\mathbb{E}(Y_j) = f(t_j; \boldsymbol{\theta}_0)$ and $\mathrm{Var}(Y_j) = \sigma_0^2 w_j^2$, which derives from the assumptions that $\mathbb{E}(\widetilde{\mathcal{E}}_j) = 0$ and $\mathrm{Var}(\widetilde{\mathcal{E}}_j) = \sigma_0^2$. We see that the variance generated in this fashion generally is longitudinally non-constant. In many situations where the observation process is well understood, the weights $\{w_j\}$ may be known.

The WLS estimator is defined here by

$$\boldsymbol{\theta}_{\mathrm{WLS}} = \arg \min_{\boldsymbol{\theta} \in \Omega_\theta} \sum_{j=1}^{N} w_j^{-2} [Y_j - f(t_j; \boldsymbol{\theta})]^2, \tag{3.32}$$

with corresponding estimate

$$\hat{\boldsymbol{\theta}}_{\mathrm{WLS}} = \arg \min_{\boldsymbol{\theta} \in \Omega_\theta} \sum_{j=1}^{N} w_j^{-2} [y_j - f(t_j; \boldsymbol{\theta})]^2. \tag{3.33}$$

This special form of the WLS estimate can be thought of as minimizing the distance between the data and model while taking into account the known but unequal quality of the observations [22].

The WLS estimator $\boldsymbol{\theta}_{\mathrm{WLS}} = \boldsymbol{\theta}_{\mathrm{WLS}}^N$ has the following asymptotic properties [21, 22]:

$$\boldsymbol{\theta}_{\mathrm{WLS}} \sim \mathcal{N}(\boldsymbol{\theta}_0, \Sigma_0^N), \tag{3.34}$$

where

$$\Sigma_0^N \approx \sigma_0^2 \left(F_{\boldsymbol{\theta}}^T(\boldsymbol{\theta}_0) W F_{\boldsymbol{\theta}}(\boldsymbol{\theta}_0) \right)^{-1}, \tag{3.35}$$

the sensitivity matrix is given by

$$F_{\boldsymbol{\theta}}(\boldsymbol{\theta}) = F_{\boldsymbol{\theta}}^N(\boldsymbol{\theta}) = \begin{pmatrix} \dfrac{\partial f(t_1; \boldsymbol{\theta})}{\partial \theta_1} & \dfrac{\partial f(t_1; \boldsymbol{\theta})}{\partial \theta_2} & \cdots & \dfrac{\partial f(t_1; \boldsymbol{\theta})}{\partial \theta_{\kappa_\theta}} \\ \vdots & & & \vdots \\ \dfrac{\partial f(t_N; \boldsymbol{\theta})}{\partial \theta_1} & \dfrac{\partial f(t_N; \boldsymbol{\theta})}{\partial \theta_2} & \cdots & \dfrac{\partial f(t_N; \boldsymbol{\theta})}{\partial \theta_{\kappa_\theta}} \end{pmatrix} \tag{3.36}$$

and the matrix W is defined by $W^{-1} = \mathrm{diag}\left(w_1^2, \dots, w_N^2 \right)$. Note that because $\boldsymbol{\theta}_0$ and σ_0^2 are unknown, the estimates $\hat{\boldsymbol{\theta}} = \hat{\boldsymbol{\theta}}_{\mathrm{WLS}}$ and $\hat{\sigma}^2 = \hat{\sigma}_{\mathrm{WLS}}^2$ will be used in (3.35) to calculate

$$\Sigma_0^N \approx \hat{\Sigma}^N = \hat{\sigma}^2 \left(F_{\boldsymbol{\theta}}^T(\hat{\boldsymbol{\theta}}) W F_{\boldsymbol{\theta}}(\hat{\boldsymbol{\theta}}) \right)^{-1},$$

where we take the approximation

$$\sigma_0^2 \approx \hat{\sigma}_{\mathrm{WLS}}^2 = \frac{1}{N - \kappa_\theta} \sum_{j=1}^N \frac{1}{w_j^2} [y_j - f(t_j; \hat{\theta})]^2.$$

We can then approximate the standard errors of $\boldsymbol{\theta}_{\mathrm{WLS}}$ by taking the square roots of the diagonal elements of $\hat{\Sigma}^N$.

3.2.5   Generalized Least Squares Definition and Motivation

A method motivated by the WLS (as we have presented it above) involves the so-called Generalized Least Squares (GLS) estimator. To define the *random vector* $\boldsymbol{\theta}_{\mathrm{GLS}}$ [21, Chapter 3] and [35, p. 69], the following *normal equations* are solved for the estimator $\boldsymbol{\theta}_{\mathrm{GLS}}$:

$$\sum_{j=1}^N f^{-2\gamma}(t_j; \boldsymbol{\theta}_{\mathrm{GLS}})[Y_j - f(t_j; \boldsymbol{\theta}_{\mathrm{GLS}})]\nabla f(t_j; \boldsymbol{\theta}_{\mathrm{GLS}}) = \mathbf{0}_{\kappa_\theta}, \qquad (3.37)$$

where Y_j satisfies

$$Y_j = f(t_j; \boldsymbol{\theta}_0) + f^\gamma(t_j; \boldsymbol{\theta}_0)\widetilde{\mathcal{E}}_j,$$

and

$$\nabla f(t_j; \boldsymbol{\theta}) = \left(\frac{\partial f(t_j; \boldsymbol{\theta})}{\partial \theta_1}, \dots, \frac{\partial f(t_j; \boldsymbol{\theta})}{\partial \theta_{\kappa_\theta}} \right)^T.$$

The quantity $\boldsymbol{\theta}_{\mathrm{GLS}}$ is a random vector; hence if $\{y_j\}_{j=1}^N$ is a *realization* of $\{Y_j\}_{j=1}^N$, then solving

$$\sum_{j=1}^N f^{-2\gamma}(t_j; \boldsymbol{\theta})[y_j - f(t_j; \boldsymbol{\theta})]\nabla f(t_j; \boldsymbol{\theta}) = \mathbf{0}_{\kappa_\theta} \qquad (3.38)$$

for $\boldsymbol{\theta}$ will provide an estimate for $\boldsymbol{\theta}_{\mathrm{GLS}}$.

The GLS equation (3.38) can be motivated by examining the special weighted least-squares estimate

$$\hat{\boldsymbol{\theta}}_{\mathrm{WLS}} = \arg \min_{\boldsymbol{\theta} \in \Omega_\theta} \sum_{j=1}^N w_j [y_j - f(t_j; \boldsymbol{\theta})]^2 \qquad (3.39)$$

for a *given* $\{w_j\}$. If we differentiate the sum of squares in (3.39) with respect to $\boldsymbol{\theta}$ and *then* choose $w_j = f^{-2\gamma}(t_j; \boldsymbol{\theta})$, an estimate $\hat{\boldsymbol{\theta}}_{\mathrm{GLS}}$ is obtained by solving

$$\sum_{j=1}^N w_j [y_j - f(t_j; \boldsymbol{\theta})]\nabla f(t_j; \boldsymbol{\theta}) = \mathbf{0}_{\kappa_\theta}$$

for $\boldsymbol{\theta}$, i.e., solving (3.38). However, we note the GLS relationship (3.38) does *not* follow from minimizing the weighted least squares with weights chosen as $w_j = f^{-2\gamma}(t_j; \boldsymbol{\theta})$ (see p. 89 of [35]).

Another motivation for the GLS estimating equations (3.37) and (3.38) can be found in [18]. In that text, Carroll and Ruppert claim that if the data are distributed according to the gamma distribution, then the *maximum-likelihood estimate* for $\boldsymbol{\theta}$ (a standard approach when one assumes that the distribution for the measurement error is completely known—to be discussed later) is the solution to

$$\sum_{j=1}^{N} f^{-2}(t_j; \boldsymbol{\theta})[y_j - f(t_j; \boldsymbol{\theta})]\nabla f(t_j; \boldsymbol{\theta}) = \mathbf{0}_{\kappa_\theta},$$

which is equivalent to the corresponding GLS estimating equations (3.38) with $\gamma = 1$. (We refer the reader to Chapter 4 on this as well as the maximum likelihood estimation method.) The connection between the maximum likelihood estimation method and our GLS method is reassuring, but it also poses another interesting question: what if the variance of the data is assumed to be independent of the model output $f(t_j; \boldsymbol{\theta})$ but dependent on some other function $h(t_j; \boldsymbol{\theta})$ (i.e., $\text{Var}(Y_j) = \sigma_0^2 h^2(t_j; \boldsymbol{\theta})$)? Is there a corresponding maximum likelihood estimator of $\boldsymbol{\theta}$ whose form is equivalent to the appropriate GLS estimating equation

$$\sum_{j=1}^{N} h^{-2}(t_j; \boldsymbol{\theta})[Y_j - f(t_j; \boldsymbol{\theta})]\nabla f(t_j; \boldsymbol{\theta}) = \mathbf{0}_{\kappa_\theta} \ ? \tag{3.40}$$

In their text, Carroll and Ruppert [18] briefly describe how distributions belonging to the exponential family of distributions (to be discussed later) generate maximum-likelihood estimating equations equivalent to (3.40).

The GLS estimator $\boldsymbol{\theta}_{\text{GLS}} = \boldsymbol{\theta}_{\text{GLS}}^N$ has the following asymptotic properties [22, 35]:

$$\boldsymbol{\theta}_{\text{GLS}} \sim \mathcal{N}(\boldsymbol{\theta}_0, \Sigma_0^N), \tag{3.41}$$

where

$$\Sigma_0^N \approx \sigma_0^2 \left(F_{\boldsymbol{\theta}}^T(\boldsymbol{\theta}_0) W(\boldsymbol{\theta}_0) F_{\boldsymbol{\theta}}(\boldsymbol{\theta}_0) \right)^{-1}, \tag{3.42}$$

the sensitivity matrix is given by (3.36) and the matrix $W(\boldsymbol{\theta})$ is defined by $W^{-1}(\boldsymbol{\theta}) = \text{diag}\left(f^{2\gamma}(t_1; \boldsymbol{\theta}), \dots, f^{2\gamma}(t_N; \boldsymbol{\theta}) \right)$. Note that because $\boldsymbol{\theta}_0$ and σ_0^2 are unknown, the estimates $\hat{\boldsymbol{\theta}} = \hat{\boldsymbol{\theta}}_{\text{GLS}}$ and $\hat{\sigma}^2 = \hat{\sigma}_{\text{GLS}}^2$ will again be used in (3.42) to calculate

$$\Sigma_0^N \approx \hat{\Sigma}^N = \hat{\sigma}^2 \left(F_{\boldsymbol{\theta}}^T(\hat{\boldsymbol{\theta}}) W(\hat{\boldsymbol{\theta}}) F_{\boldsymbol{\theta}}(\hat{\boldsymbol{\theta}}) \right)^{-1},$$

where we take the approximation

$$\sigma_0^2 \approx \hat{\sigma}_{\text{GLS}}^2 = \frac{1}{N - \kappa_\theta} \sum_{j=1}^{N} \frac{1}{f^{2\gamma}(t_j; \hat{\boldsymbol{\theta}})} [y_j - f(t_j; \hat{\boldsymbol{\theta}})]^2.$$

We can then approximate the standard errors of $\boldsymbol{\theta}_{\mathrm{GLS}}$ by taking the square roots of the diagonal elements of $\hat{\Sigma}^N$.

3.2.6   Numerical Implementation of the GLS Procedure

We note that an estimate $\hat{\boldsymbol{\theta}}_{\mathrm{GLS}}$ can be solved either directly according to (3.38) or iteratively using an algorithm similar in spirit to that in Section 3.2.3. This iterative procedure as described in [22] (often referred to as the "GLS algorithm") is summarized below:

1. Solve for the initial estimate $\hat{\boldsymbol{\theta}}^{(0)}$ obtained using the OLS minimization (3.12). Set $l = 0$.

2. Form the weights $\hat{w}_j = f^{-2\gamma}(t_j; \hat{\boldsymbol{\theta}}^{(l)})$.

3. Re-estimate $\hat{\boldsymbol{\theta}}$ by solving

$$\sum_{j=1}^{N} \hat{w}_j \left(y_j - f\left(t_j, \boldsymbol{\theta}\right) \right) \nabla f(t_j; \boldsymbol{\theta}) = \mathbf{0}_{\kappa_\theta} \qquad (3.43)$$

to obtain $\hat{\boldsymbol{\theta}}^{(l+1)}$.

4. Set $l = l + 1$ and return to step 2. Terminate the process and set $\hat{\boldsymbol{\theta}}_{\mathrm{GLS}} = \hat{\boldsymbol{\theta}}^{(l+1)}$ when two of the successive estimates are sufficiently close.

We note that the above iterative procedure was formulated by the equivalent of minimizing for a given $\tilde{\boldsymbol{\theta}}$:

$$\sum_{j=1}^{N} f^{-2\gamma}(t_j; \tilde{\boldsymbol{\theta}})[y_j - f(t_j; \boldsymbol{\theta})]^2$$

(over $\boldsymbol{\theta} \in \Omega_\theta$) and then updating the weights $w_j = f^{-2\gamma}(t_j; \tilde{\boldsymbol{\theta}})$ after each iteration. One would hope that after a sufficient number of iterations $\hat{w}_j$ would converge to $f^{-2\gamma}(t_j; \hat{\boldsymbol{\theta}}_{\mathrm{GLS}})$. Fortunately, under reasonable conditions [22], if the process enumerated above is continued a sufficient number of times, then $\hat{w}_j \to f^{-2\gamma}(t_j; \hat{\boldsymbol{\theta}}_{\mathrm{GLS}})$.

3.3   Asymptotic Theory: Theoretical Foundations

Finally we turn to the general least squares problems discussed in previous sections where the statistical models have one of the forms given in (3.6).

For ease in notation we discuss only versions of these models for scalar observation cases in the OLS and WLS formulations; the reader can easily use the vector extensions discussed earlier in this chapter to treat vector observations. We first explain the *asymptotic theory for OLS formulations*, i.e., for an absolute error model ($\gamma = 0$ in the general formulation) in the scalar case. We then explain the extensions to more general error models, again in the scalar observation case. Thus we consider the statistical model (3.6) with $m = 1$ and $\gamma = 0$ given by

$$Y_j = f(t_j; \boldsymbol{\theta}_0) + \widetilde{\mathcal{E}}_j \,, j = 1, \ldots, N. \tag{3.44}$$

In the discussions below, for notational simplicity the dependence of the estimators and cost functions on N may be suppressed, e.g., using $J_{\text{OLS}}(\boldsymbol{\theta})$ or $J_{\text{OLS}}(\boldsymbol{\theta}; \boldsymbol{Y})$ instead of

$$J_{\text{OLS}}^N(\boldsymbol{\theta}) = J_{\text{OLS}}^N(\boldsymbol{\theta}; \boldsymbol{Y}) = \sum_{j=1}^{N}(Y_j - f(t_j; \boldsymbol{\theta}))^2.$$

This is the case particularly during the discussions of the cell proliferation example in Section 3.5.3 below.

We present existing results on the asymptotic properties of non-linear ordinary least squares estimators. The notation and organization of this section closely follows that of Banks and Fitzpatrick [7] and the extension in [10]. The work of Banks and Fitzpatrick was partially inspired by the work of Gallant [25], and many of their results and assumptions are similar. In the discussion that follows, we comment on any differences between the two approaches that we feel are noteworthy. While Gallant's work is remarkably general, allowing for a misspecified model and a general estimation procedure (both least mean distance estimators and method of moments estimators are included), we do not consider such generalities here. The comments below are limited to ordinary and weighted least squares estimation with a correctly specified model (recall the discussion above).

We first follow [7] in focusing exclusively on asymptotic properties for *i.i.d.* (absolute) error models. The theoretical results of Gallant similarly focus on *i.i.d.* errors, though some mathematical tools are discussed [25, Chapter 2, p. 156–157] which help to address more general error models. In fact, these tools are used in a rigorous fashion in the next section to extend the results of [7].

It is assumed that the error random variables $\{\widetilde{\mathcal{E}}_j\}$ are defined on some probability space $(\Omega, \mathcal{F}, \text{Prob})$ and take their values in Euclidean space $\mathbb{R}$. By construction, it follows that the data $\boldsymbol{Y}$ as well as the estimators $\boldsymbol{\theta}_{\text{WLS}}^N$ and $\boldsymbol{\theta}_{\text{OLS}}^N$ are random vectors defined on this probability space, and hence are functions of $\omega \in \Omega$ so that we may write $Y_j(\omega)$, $\boldsymbol{\theta}_{\text{OLS}}^N(\omega)$, etc., as necessary.

For a given *sampling set* $\{t_j\}_{j=1}^N$, one can define the empirical distribution function

$$\mu_N(t) = \frac{1}{N} \sum_{j=1}^{N} \Delta_{t_j}(t), \qquad (3.45)$$

where Δ_{t_j} is the Dirac measure with atom at t_j, that is,

$$\Delta_{t_j}(t) = \begin{cases} 0, & t < t_j \\ 1, & \text{otherwise.} \end{cases}$$

Clearly, $\mu_N \in \mathbb{P}([t_0, t_f])$, the space of probability measures (or, equivalently, the cumulative distribution functions) on $[t_0, t_f]$. Following popular conventions we will not always distinguish between probability measures and their associated cumulative distribution functions.

Again, the results presented below are paraphrased from [7, 10], and comments have been included to indicate the alternative approach of [25]. No proofs are given here, though the interested reader can find a complete set of proofs in [7, 10] and [25]. First, we consider the following set of assumptions:

(A1) The random variables $\{\widetilde{\mathcal{E}}_j\}$ are independent and identically distributed random variables with distribution function P. Moreover, $\mathbb{E}(\widetilde{\mathcal{E}}_j) = 0$ and $\text{Var}(\widetilde{\mathcal{E}}_j) = \sigma_0^2$.

(A2) Ω_θ is a compact, separable, finite-dimensional Euclidean space (i.e., $\Omega_\theta \subset \mathbb{R}^{\kappa_\theta}$) with $\boldsymbol{\theta}_0 \in int(\Omega_\theta)$.

(A3) $[t_0, t_f]$ is a compact subset of $\mathbb{R}$ (i.e., t_0, t_f are both finite).

(A4) The function $f(\cdot; \cdot) \in C([t_0, t_f], C^2(\Omega_\theta))$.

(A5) There exists a finite measure μ on $[t_0, t_f]$ such that the sampling sets $\{t_j\}_{j=1}^N$ satisfy as $N \to \infty$

$$\frac{1}{N} \sum_{j=1}^{N} h(t_j) = \int_{t_0}^{t_f} h(t) d\mu_N(t) \to \int_{t_0}^{t_f} h(t) d\mu(t)$$

for *all continuous functions* h, where μ_N is the finite distribution function as defined in (3.45). That is, μ_N converges to μ in the weak* topology (where $\mathbb{P}([t_0, t_f])$ is viewed as a subset of $C^*([t_0, t_f])$, the topological dual of the space of continuous functions $C([t_0, t_f])$). (Note that this means that the data must be taken in a way such that in the limit it "fills up" the interval $[t_0, t_f]$.)

(A6) The functional

$$J^0(\boldsymbol{\theta}) = \sigma_0^2 + \int_{t_0}^{t_f} (f(t; \boldsymbol{\theta}_0) - f(t; \boldsymbol{\theta}))^2 \, d\mu(t)$$

has a unique minimizer at $\boldsymbol{\theta}_0 \in \Omega_\theta$.

(A7) The matrix

$$
\begin{aligned}
\mathcal{J} &= \frac{\partial^2 J^0(\boldsymbol{\theta}_0)}{\partial \boldsymbol{\theta}^2} \\
&= 2 \int_{t_0}^{t_f} \left(\frac{\partial f(t; \boldsymbol{\theta}_0)}{\partial \boldsymbol{\theta}} \left(\frac{\partial f(t; \boldsymbol{\theta}_0)}{\partial \boldsymbol{\theta}} \right)^T \right) d\mu(t)
\end{aligned}
$$

is positive definite.

Remark: The most notable difference between the assumptions above (which are those of [7]) and those of [25] is assumption (A5). In its place, Gallant states the following.

(A5′) Define the probability measure

$$
\nu(S) = \int_{t_0}^{t_f} \int_E I_S(\mathcal{E}, t) dP(\mathcal{E}) d\mu(t)
$$

for an indicator function I_S and set $S \subset E \times [t_0, t_f]$ and μ defined as above. Then almost every realized pair (e_j, t_j) is a *Cesaro sum generator* with respect to ν and a dominating function $b(\mathcal{E}, t)$ satisfying $\int_{[t_0, t_f] \times E} b d\nu < \infty$. That is,

$$
\lim_{N \to \infty} \frac{1}{N} \sum_{j=1}^{N} h(\epsilon_j, t_j) = \int_{t_0}^{t_f} \int_E h(\mathcal{E}, t) d\nu(\mathcal{E}, t)
$$

almost always, for all continuous functions h such that $|h(\mathcal{E}, t)| < b(\mathcal{E}, t)$. Moreover, it is assumed that for each $t \in [t_0, t_f]$, there exists a neighborhood $\mathbb{O}_t$ such that

$$
\int_E \sup_{\mathbb{O}_t} b(\mathcal{E}, t) dP(\mathcal{E}) < \infty.
$$

The assumption (A5′) is stronger than the assumption (A5) as it supposes not only the existence of a dominating function, but also involves the behavior of the probability distribution P, *which is generally unknown in practice.* The practical importance of the dominating function b arises in the proof of consistency for the least squares estimator (see Theorem 3.3.1 below).

As has been noted elsewhere (see, e.g., [1], [35, Chapter 12]), the strong consistency of the estimator is proved by arguing that $J^0(\boldsymbol{\theta})$ is the almost sure limit of $J_{\text{OLS}}^N(\boldsymbol{\theta}; \boldsymbol{Y})$. Thus, if $J_{\text{OLS}}^N(\boldsymbol{\theta}; \boldsymbol{Y})$ is "close" to $J^0(\boldsymbol{\theta})$ and $J^0(\boldsymbol{\theta})$ is uniquely minimized by $\boldsymbol{\theta}_0$, it makes intuitive sense that $\boldsymbol{\theta}_{\text{OLS}}^N$, which minimizes $J_{\text{OLS}}^N(\boldsymbol{\theta}; \boldsymbol{Y})$, should be close to $\boldsymbol{\theta}_0$. This task is made difficult by the fact that the null set (from the "almost sure" statement) may depend on the parameter

$\boldsymbol{\theta}$. In [7], the almost sure convergence of $J_{\text{OLS}}^N(\boldsymbol{\theta}; \boldsymbol{Y})$ to $J^0(\boldsymbol{\theta})$ is demonstrated constructively, that is, by building a set $\mathbb{A} \in \mathcal{F}$ (which does not depend on $\boldsymbol{\theta}$) with $\text{Prob}\{\mathbb{A}\} = 1$ such that $J_{\text{OLS}}^N(\boldsymbol{\theta}; \boldsymbol{Y}) \to J^0(\boldsymbol{\theta})$ for each $\omega \in \mathbb{A}$ and for each $\boldsymbol{\theta} \in \Omega_\theta$. This construction relies upon the separability of the parameter space Ω_θ (assumption (A2)) as well as the compactness of the space $[t_0, t_f]$ (assumption (A3)). The alternative approach of Gallant uses a consequence of the Glivenko–Cantelli theorem [25, p. 158] to demonstrate a uniform (with respect to $\boldsymbol{\theta}$) strong law of large numbers. The proof relies upon the dominated convergence theorem [32, p. 246], and hence the dominating function b. As a result, Gallant does not need the space Ω_θ to be separable or the interval $[t_0, t_f]$ to be compact. It should be noted, however, that in most practical applications of interest Ω_θ and $[t_0, t_f]$ are compact subsets of Euclidean space so that relaxing these assumptions provides little advantage.

While the list of assumptions above is extensive, we remark that the set is not overly restrictive. Assumptions (A2) and (A3) are naturally satisfied for most problem formulations (although the requirement $\boldsymbol{\theta}_0 \in int(\Omega_\theta)$ may be occasionally problematic [7, Remark 4.4]). Assumption (A4) is easily checked. Though assumption (A1) may be difficult to verify, it is much less restrictive than, say, a complete likelihood specification. Moreover, residual plots (see [16, Chapter 3] and the discussions below) can aid in assessing the reliability of the assumption.

Assumption (A5) is more difficult to check in practice as one does not know the limiting distribution μ. Of course, this is simply an assumption regarding *the manner in which data is sampled* (in the independent variable space $[t_0, t_f]$). Namely, it *must be taken in a way that "fills up the space" in an appropriate sense.* Similarly, assumptions (A6) and (A7) cannot be verified directly as one knows neither μ nor $\boldsymbol{\theta}_0$. In many practical applications of interest, μ is Lebesgue measure, and one can assess the assumptions at $\hat{\boldsymbol{\theta}}_{\text{OLS}}^N$ (which is hopefully close to $\boldsymbol{\theta}_0$). Of course, if assumption (A7) holds, then assumption (A6) must hold at least for a small region around $\boldsymbol{\theta}_0$ – though possibly not on all of Ω_θ.

Assumption (A7) is not strictly necessary if one uses assumption (A5$'$) in the place of (A5). Given assumptions (A2)–(A4), it follows that the function b (and its relevant derivatives) is bounded (and hence dominated by a ν-measurable function) provided the *space* E in which the random variables $\widetilde{\mathcal{E}}_j$ take their values is *bounded*. On one hand, this has the desirable effect of weakening the assumptions placed on the Hessian matrix $\mathcal{J}$. Yet the assumption that E is bounded *precludes* certain error models, in particular *normally distributed errors*.

We now give several results which summarize the asymptotic properties of the ordinary least squares estimator.

Theorem 3.3.1 *Given the probability space* $(\Omega, \mathcal{F}, \text{Prob})$ *as defined above describing the space of sampling points and assumptions (A1)–(A6),* $\boldsymbol{\theta}_{OLS}^N \to \boldsymbol{\theta}_0$

with probability one or almost surely. That is,

$$Prob\left(\left\{\omega \in \Omega \,\middle|\, \lim_{N\to\infty} \boldsymbol{\theta}^N_{OLS}(\omega) = \boldsymbol{\theta}_0\right\}\right) = 1.$$

This theorem states that the ordinary least squares estimator is consistent. We remark that the finite dimensionality of the parameter space Ω_θ (see assumption (A2)) is not necessary in the proof of this theorem, and it is sufficient for the function f to be continuous from Ω_θ into $C([t_0, t_f])$ rather than to be twice continuously differentiable.

Given Theorem 3.3.1, the following theorem may also be proven.

Theorem 3.3.2 *Given assumptions (A1)–(A7), as $N \to \infty$*

$$\sqrt{N}\left(\boldsymbol{\theta}^N_{OLS} - \boldsymbol{\theta}_0\right) \xrightarrow{d} Z \sim \mathcal{N}\left(\mathbf{0}_{\kappa_\theta}, 2\sigma_0^2 \mathcal{J}^{-1}\right), \tag{3.46}$$

that is, the convergence is in the sense of convergence in distribution.

To reconcile these asymptotic results with the approximations used in practice given in (3.14)–(3.17) and (3.19), we argue as follows. Define

$$\mathcal{J}^N = 2\int_{t_0}^{t_f} \left(\frac{\partial f(t; \boldsymbol{\theta}_0)}{\partial \boldsymbol{\theta}} \left(\frac{\partial f(t; \boldsymbol{\theta}_0)}{\partial \boldsymbol{\theta}}\right)^T\right) d\mu_N(t), \tag{3.47}$$

which from the definition in (3.45) is the same as

$$2\frac{1}{N}\sum_{j=1}^{N}\left(\frac{\partial f(t_j; \boldsymbol{\theta}_0)}{\partial \boldsymbol{\theta}} \left(\frac{\partial f(t_j; \boldsymbol{\theta}_0)}{\partial \boldsymbol{\theta}}\right)^T\right) = 2\frac{1}{N}(F^N_{\boldsymbol{\theta}}(\boldsymbol{\theta}_0))^T F^N_{\boldsymbol{\theta}}(\boldsymbol{\theta}_0). \tag{3.48}$$

Recalling (3.17) and using (A5), we have

$$\frac{1}{2}\mathcal{J}^N = \frac{1}{N}(F^N_{\boldsymbol{\theta}}(\boldsymbol{\theta}_0))^T F^N_{\boldsymbol{\theta}}(\boldsymbol{\theta}_0) \to \Omega_0 = \frac{1}{2}\mathcal{J}. \tag{3.49}$$

Thus the convergence statement in (3.46) is

$$\sqrt{N}\left(\boldsymbol{\theta}^N_{\text{OLS}} - \boldsymbol{\theta}_0\right) \xrightarrow{d} Z \sim \mathcal{N}\left(\mathbf{0}_{\kappa_\theta}, \sigma_0^2 \Omega_0^{-1}\right). \tag{3.50}$$

Given (3.19), in practice we make an approximation in (3.50) given by

$$\mathcal{N}\left(\mathbf{0}_{\kappa_\theta}, \sigma_0^2 \Omega_0^{-1}\right) \approx \mathcal{N}\left(\mathbf{0}_{\kappa_\theta}, \sigma_0^2 N[(F^N_{\boldsymbol{\theta}}(\boldsymbol{\theta}_0))^T F^N_{\boldsymbol{\theta}}(\boldsymbol{\theta}_0)]^{-1}\right),$$

which, in light of (3.50), leads to the approximation

$$\left(\boldsymbol{\theta}^N_{\text{OLS}} - \boldsymbol{\theta}_0\right) \xrightarrow{d} Z_1 \sim \mathcal{N}\left(\mathbf{0}_{\kappa_\theta}, \sigma_0^2[(F^N_{\boldsymbol{\theta}}(\boldsymbol{\theta}_0))^T F^N_{\boldsymbol{\theta}}(\boldsymbol{\theta}_0)]^{-1}\right), \tag{3.51}$$

which is indeed (3.14).

3.3.1   Extension to Weighted Least Squares

The results presented above are quite useful, but they only apply to the class of problems in which the measurement error random variables are independent and identically distributed with constant variance σ_0^2. While the assumption of independence is common, there are many practical cases of interest in which these random variables are not identically distributed. In many cases one encounters a weighted least squares problem in which $\mathbb{E}(\tilde{\mathcal{E}}_j) = 0$ for all j but $\mathrm{Var}(Y_j) = \sigma_0^2 w_j^2$. As discussed previously, in such cases, the results above (in particular, assumption (A1)) fail to apply directly. Here we return to the weighted least squares model (3.31) for the scalar observation case. Thus we recall our problem of interest involves the *weighted least squares cost*

$$J_{\mathrm{WLS}}^N(\boldsymbol{\theta}) = \sum_{j=1}^N \left(\frac{Y_j - f(t_j; \boldsymbol{\theta})}{w(t_j)} \right)^2, \tag{3.52}$$

where $w_j = w(t_j)$. The *weighted least squares estimator* is defined as the random vector $\boldsymbol{\theta}_{\mathrm{WLS}}^N$ which minimizes the weighted least squares cost function for a given set of random variables $\{Y_j\}$. Hence

$$\boldsymbol{\theta}_{\mathrm{WLS}}^N = \arg \min_{\boldsymbol{\theta} \in \Omega_\theta} J_{\mathrm{WLS}}^N(\boldsymbol{\theta}). \tag{3.53}$$

In order to extend the results presented above (Theorems 3.3.1 and 3.3.2) to independent, heteroscedastic error models (and hence, to general weighted least squares problems), we turn to a technique suggested by Gallant [25, p. 124] in which one defines a change of variables in an attempt to normalize the heteroscedasticity of the random variables. As has been noted previously, Gallant used this technique under a different set of assumptions in order to obtain results similar to those presented above. This change of variables technique will allow us to extend the results above, originally from [7], in a rigorous fashion, as given in [10].

Consider the following assumptions.

(A1′a) The error random variables $\{\tilde{\mathcal{E}}_j\}$ are independent, have central moments which satisfy $\mathbb{E}(\tilde{\mathcal{E}}_j) = 0$ and $\mathrm{Var}(\tilde{\mathcal{E}}_j) = \sigma_0^2$ and yield observations Y_j satisfying

$$\mathrm{Cov}(\vec{Y}) = \sigma_0^2 W = \sigma_0^2 \mathrm{diag}(w_1^2, \ldots, w_N^2),$$

where $\vec{Y} = (Y_1, Y_2, \ldots, Y_N)^T$.

(A1′b) The function w satisfies $w \in C([t_0, t_f], \mathbb{R}^+)$ and $w(t) \neq 0$ for $t \in [t_0, t_f]$.

(A7′) The matrix

$$\tilde{\mathcal{J}} = 2 \int_{t_0}^{t_f} \frac{1}{w^2(t)} \left(\frac{\partial f(t; \boldsymbol{\theta}_0)}{\partial \boldsymbol{\theta}} \left(\frac{\partial f(t; \boldsymbol{\theta}_0)}{\partial \boldsymbol{\theta}} \right)^T \right) d\mu(t)$$

is positive definite.

Theorem 3.3.3 *Under assumptions (A1′a), (A1′b), (A2)–(A6) and (A7′),*

1. $\boldsymbol{\theta}_{WLS}^N \to \boldsymbol{\theta}_0$ *with probability one, and*

2. $\sqrt{N}\left(\boldsymbol{\theta}_{WLS}^N - \boldsymbol{\theta}_0\right) \xrightarrow{d} \tilde{Z} \sim \mathcal{N}\left(\mathbf{0}_{\kappa_\theta}, 2\sigma_0^2 \tilde{\mathcal{J}}^{-1}\right).$

PROOF   We first recall the statistical model (3.31)

$$Y_j = f(t_j; \boldsymbol{\theta}_0) + w_j \tilde{\mathcal{E}}_j,$$

which can be written in vector form as

$$\vec{Y} = \vec{f}(\boldsymbol{\theta}_0) + \vec{\mathcal{E}}, \tag{3.54}$$

where

$$\vec{f}(\boldsymbol{\theta}_0) = (f(t_1; \boldsymbol{\theta}_0), f(t_2; \boldsymbol{\theta}_0), \ldots, f(t_N; \boldsymbol{\theta}_0))^T$$

and

$$\vec{\mathcal{E}} = (w_1 \tilde{\mathcal{E}}_1, w_2 \tilde{\mathcal{E}}_2 \ldots, w_N \tilde{\mathcal{E}}_N)^T.$$

Thus we have $\mathbb{E}(\vec{\mathcal{E}}) = \mathbf{0}_N$ and $\mathrm{Cov}(\vec{\mathcal{E}}) = \sigma_0^2 W$. Let $L = \mathrm{diag}(w_1, \ldots, w_N)$. It follows that $LL^T = W$ (L is the Cholesky decomposition of W). Also, L^{-1} exists and can be applied to both sides of (3.54) to obtain

$$L^{-1}\vec{Y} = L^{-1}\vec{f}(\boldsymbol{\theta}_0) + L^{-1}\vec{\mathcal{E}}$$

or

$$\vec{Z} = \vec{h}(\boldsymbol{\theta}_0) + \vec{\eta}, \tag{3.55}$$

where $\vec{Z}$, $\vec{h}$ and $\vec{\eta}$ have the obvious definitions. The OLS cost functional for the transformed model and data (3.55) is

$$\begin{aligned}
J_{\mathrm{OLS}}^N(\boldsymbol{\theta}) &= \left(\vec{Z} - \vec{h}(\boldsymbol{\theta})\right)^T \left(\vec{Z} - \vec{h}(\boldsymbol{\theta})\right) \\
&= \left(L^{-1}\vec{Y} - L^{-1}\vec{f}(\boldsymbol{\theta})\right)^T \left(L^{-1}\vec{Y} - L^{-1}\vec{f}(\boldsymbol{\theta})\right) \\
&= \left(\vec{Y} - \vec{f}(\boldsymbol{\theta})\right)^T L^{-1}L^{-1} \left(\vec{Y} - \vec{f}(\boldsymbol{\theta})\right) \\
&= \left(\vec{Y} - \vec{f}(\boldsymbol{\theta})\right)^T W^{-1} \left(\vec{Y} - \vec{f}(\boldsymbol{\theta})\right).
\end{aligned}$$

In other words, the ordinary least squares cost function with respect to the transformed model and data (3.55) is exactly the weighted least squares cost function (3.52) for the original model and data. Thus, in order to prove Theorem 3.3.3, we will show that the rescaled model (3.55) satisfies the assumptions of Theorems 3.3.1–3.3.2.

Clearly, $\mathbb{E}(\vec{\eta}) = \mathbf{0}_N$ and $\mathrm{Var}(\vec{\eta}) = \sigma_0^2 L^{-1}WL^{-1} = \sigma_0^2 \mathbf{I}_N$. Thus the random variables η_k (the components of $\vec{\eta}$) are independent and identically distributed (by assumption (A1$'$a)) with constant variance, and thus assumption (A1) is satisfied. Assumptions (A2), (A3) and (A5) are unchanged. For assumption (A4), we must show that $h \in C([t_0, t_f], C^2(\Omega_\theta))$ where h is given by $h(t; \theta) = f(t; \theta)/w(t)$. This follows from assumption (A1$'$b). For the analogue of assumption (A6), we must show

$$\tilde{J}^*(\theta) = \sigma_0^2 + \int_{t_0}^{t_f} (h(t; \theta_0) - h(t; \theta))^2 \, d\mu(t)$$

$$= \sigma_0^2 + \int_{t_0}^{t_f} \left(\frac{f(t; \theta_0) - f(t; \theta)}{w(t)} \right)^2 \, d\mu(t)$$

has a unique minimizer at $\theta = \theta_0$. Clearly, $\tilde{J}^*(\theta_0) = \sigma_0^2$. Since the function J^* (see assumption (A6)) has a unique minimizer at $\theta = \theta_0$ and $w(t) > 0$, it follows immediately that $\tilde{J}^*(\theta) > \sigma_0^2$ if $\theta \neq \theta_0$ so that $\tilde{J}^*(\theta)$ has a unique minimizer at $\theta = \theta_0$. Assumption (A7) is satisfied for the formulation (3.55) directly by assumption (A7$'$). $\qquad\Box$

In fact, the proof of Theorem 3.3.3 applies to any set of observations in which a change of variables can be used to produce a set of error random variables which are independent and identically distributed. The weighted least squares problem addressed in the above discussion arises from an observation process in which the measurement errors are assumed to be independent but are not necessarily identically distributed. By rescaling the observations in accordance with their variances (which are assumed to be known) one obtains error random variables which are identically distributed as well as independent. Even more generally, it is not strictly necessary that the observations be independent. For instance, one might have observations generated by an autoregressive process of order r [35]. Then, by definition, some linear combination of r observational errors will give rise to errors which are independent and identically distributed. This linear combination is exactly the change of variables necessary to obtain a model which is suitable for ordinary least squares. Thus, even in the most general situation, when one has a general covariance matrix R, one may still use the Cholesky decomposition in the manner discussed above, provided one has sufficient assumptions regarding the underlying error process. See [25, Chapter 2] for details.

Analogous with (A7) above, the assumption (A7$'$) is the most problematic to verify in practice. In the proof above for the weighted least squares problem, the assumption (A1) has been replaced with assumptions (A1$'$a)–(A1$'$b), a change which merely accounts for the heteroscedastic statistical model. Then the assumptions (A2)–(A6) for the rescaled model (3.55) can be verified directly from the original assumptions (A2)–(A6) for the ordinary least squares formulation, as shown above. The only exception is the assumption (A7), which cannot be established directly from the ordinary least squares assumptions; hence the need for assumption (A7$'$). On one hand, the existence of a unique minimizer (assumption (A6)) is sufficient to prove that the matrix $\mathcal{J}$ or $\tilde{\mathcal{J}}$ must be positive *semi*-definite, so that the assumption (A7) or (A7$'$) may not be overly restrictive. Alternatively, as has been noted before, one can relax the assumptions (A7) or (A7$'$) by assuming the existence of a dominating function b. Moreover, provided the weights $w(t)$ satisfy the requirement $w(t) \geq \bar{w} > 0$ for all $t \in [t_0, t_f]$, then one can use the dominating function b (from the ordinary least squares problem) to obtain a new dominating function $\tilde{b}(\mathcal{E}, t) = \bar{w}^{-1} b(\mathcal{E}, t)$ which is also ν-integrable. Even in this case, though, one must still make the additional assumption that $\tilde{\mathcal{J}}$ is invertible.

We note that asymptotic results similar to those of Theorems 3.3.1 and 3.3.2 above can be given for the GLS estimators defined in the estimating equation (3.37) and their analogues for the general vector observation case – see p. 88–89 of [35]. All of the results above also readily extend to systems governed by partial differential equations; the essential elements of the theory are the form of the discrete observation operator for the dynamical system and the statistical model as given in (3.6).

3.4 Computation of $\hat{\Sigma}^N$, Standard Errors and Confidence Intervals

We return to the case of N scalar longitudinal observations and consider the OLS case of Section 3.2 (the extension of these ideas to vectors is completely straightforward). Recall that in the ordinary least squares approach, we seek to use a realization $\{y_j\}$ of the observation process $\{Y_j\}$ along with the model to determine a vector $\hat{\boldsymbol{\theta}}_{\mathrm{OLS}}^N$ where

$$\hat{\boldsymbol{\theta}}_{\mathrm{OLS}}^N = \arg\min_{\boldsymbol{\theta} \in \Omega_\theta} J_{\mathrm{OLS}}^N(\boldsymbol{\theta}; \mathbf{y}) = \arg\min_{\boldsymbol{\theta} \in \Omega_\theta} \sum_{j=1}^N [y_j - f(t_j; \boldsymbol{\theta})]^2. \qquad (3.56)$$

Since Y_j is a random variable, the corresponding estimator $\boldsymbol{\theta}_{\mathrm{OLS}}^N$ (here we wish to emphasize the dependence on the sample size N) is also a random vector with a distribution called the *sampling distribution*. Knowledge of this sampling distribution provides uncertainty information (e.g., standard errors) for the numerical values of $\hat{\boldsymbol{\theta}}^N$ obtained using a specific data set $\{y_j\}$. In particular, loosely speaking, the sampling distribution characterizes the distribution of possible values the estimator could take on across all possible realizations with data of size N that could be collected. The standard errors thus approximate the extent of variability in possible parameter values across all possible realizations, and hence provide a measure of the extent of uncertainty involved in estimating $\boldsymbol{\theta}$ using a specific estimator and sample size N in actual data collection.

74   *Modeling and Inverse Problems in the Presence of Uncertainty*

The uncertainty quantifications (as embodied in standard errors) we discuss here are given in terms of standard nonlinear regression approximation theory ([7, 10, 22, 25, 28], and Chapter 12 of [35]) for **asymptotic** (as $N \to \infty$) **distributions**. As we have already mentioned for N large, and a corresponding data set $\mathbf{Y} = (Y_1, \ldots, Y_N)^T$, the sampling distribution satisfies the approximation (see (3.14))

$$\theta_{\text{OLS}}^N(\mathbf{Y}) \sim \mathcal{N}(\theta_0, \Sigma_0^N) \approx \mathcal{N}(\theta_0, \sigma_0^2 [F_{\theta}^N(\theta_0)^T F_{\theta}^N(\theta_0)]^{-1}). \tag{3.57}$$

We thus see that the quantity F_{θ} is the fundamental entity in computational aspects of this theroy. There are typically several ways to compute the matrix F_{θ} (which actually is composed of the well known **sensitivity functions** widely used in applied mathematics and engineering—e.g., see the discussions in [3, 4, 6] and the references therein). First, the elements of the matrix $F_{\theta} = (F_{\theta jk})$ can always be estimated using the forward difference

$$F_{\theta jk}(\theta) = \frac{\partial f(t_j; \theta)}{\partial \theta_k} \approx \frac{f(t_j; \theta + h_k) - f(t_j; \theta)}{|h_k|},$$

where h_k is a κ_θ-vector with a non-zero entry in only the kth component which is chosen "small" and $|\cdot|$ is the Euclidean norm in $\mathbb{R}^{\kappa_\theta}$. But the choice of h_k can be problematic in practice, i.e., what does "small" mean, especially when the parameters may vary by orders of magnitude? Of course, in some cases the function $f(t_j; \theta)$ may be sufficiently simple so as to allow one to derive analytical expressions for the components of F_{θ}.

Alternatively, if the $f(t_j; \theta)$ correspond to longitudinal observations $f(t_j; \theta) = \mathcal{C}x(t_j; \theta)$ of solutions to a parameterized n-vector differential equation system $\dot{x} = g(t, x(t), q)$ as in (3.1)–(3.2), then one can use the $n \times \kappa_\theta$ matrix **sensitivity equations** (see [4, 6] and the references therein)

$$\frac{d}{dt}\left(\frac{\partial x}{\partial \theta}\right) = \frac{\partial g}{\partial x}\frac{\partial x}{\partial \theta} + \frac{\partial g}{\partial \theta} \tag{3.58}$$

to obtain

$$\frac{\partial f(t_j; \theta)}{\partial \theta_k} = \mathcal{C}\frac{\partial x(t_j, \theta)}{\partial \theta_k}.$$

To be a little more specific, we may examine the variations in the output of a model f resulting from variations in the parameters q and the initial conditions $\tilde{x}_0$. In this section, for notational convenience, we temporarily assume $\tilde{x}_0 = x_0$, i.e., we assume we estimate all of the initial conditions in our discussions of the sensitivity equations.

In order to quantify the variation in the state variable $x(t)$ with respect to changes in the parameters q and the initial conditions x_0, we are naturally led to consider the individual *(traditional) sensitivity functions (TSF)* defined by the derivatives

$$s_{q_k}(t) = \frac{\partial x}{\partial q_k}(t) = \frac{\partial x}{\partial q_k}(t, \theta), \qquad k = 1, \ldots, \kappa_q, \tag{3.59}$$

and

$$r_{x_{0l}}(t) = \frac{\partial x}{\partial x_{0l}}(t) = \frac{\partial x}{\partial x_{0l}}(t, \theta), \qquad l = 1, \ldots, n, \qquad (3.60)$$

where x_{0l} is the lth component of the initial condition x_0. If the function g is sufficiently regular, the solution x is differentiable with respect to q_k and x_{0l}, and therefore the sensitivity functions s_{q_k} and $r_{x_{0l}}$ are well defined.

Often in practice, the model under investigation is simple enough to allow us to combine the sensitivity functions (3.59) and (3.60), as is the case with the logistic growth population example discussed below. However, when one deals with a more complex model, it is often preferable to consider these sensitivity functions separately for clarity purposes.

Because they are defined by partial derivatives which have a *local* character, the sensitivity functions are also local in nature. Thus sensitivity and insensitivity ($s_{q_k} = \partial x/\partial q_k$ not close to zero and very close to zero, respectively) depend on the time interval, the state values x and the values of θ for which they are considered. Thus, for example, in a certain time subinterval we might find s_{q_k} small so that the state variable x is *insensitive* to the parameter q_k on that particular interval. The same function s_{q_k} can take large values on a different subinterval, indicating to us that the state variable x is *very sensitive* to the parameter q_k on the latter interval. From the sensitivity analysis theory for dynamical systems, one finds that $s = (s_{q_1}, \ldots, s_{q_{\kappa_q}})$ is an $n \times \kappa_q$ vector function that satisfies the ODE system

$$\dot{s}(t) = \frac{\partial g}{\partial x}(t, x(t; \theta), q)s(t) + \frac{\partial g}{\partial q}(t, x(t; \theta), q), \qquad (3.61)$$

$$s(t_0) = \mathbf{0}_{n \times \kappa_q},$$

which is obtained by differentiating (3.1)–(3.2) with respect to q. Here the dependence of s on $(t, x(t; \theta))$ as well as q is readily apparent.

In a similar manner, the sensitivity functions with respect to the components of the initial condition x_0 define an $n \times n$ vector function $r = (r_{x_{01}}, \ldots, r_{x_{0n}})$, which satisfies

$$\dot{r}(t) = \frac{\partial g}{\partial x}(t, x(t; \theta), q)r(t), \qquad (3.62)$$

$$r(t_0) = \mathbf{I}_n.$$

This is obtained by differentiating (3.1)–(3.2) with respect to the initial conditions x_0. Equations (3.61) and (3.62) are used in conjunction with (i.e., usually solved simultaneously with) Equations (3.1)–(3.2) to numerically compute the sensitivities s and r for general cases when the function g is sufficiently complicated to prohibit a closed form solution by direct integration. These can be succinctly written as a system for $\frac{\partial x}{\partial \theta} = \left(\frac{\partial x}{\partial q}, \frac{\partial x}{\partial x_0} \right)$ given by (3.58).

Because the parameters may have different units and the state variables may have varying orders of magnitude, sometimes in practice it is more convenient to work with the normalized version of the TSF, referred to as *relative sensitivity functions* (RSF). However, since in this monograph we are using the standard error approach to analyze the performance of the least squares algorithm in estimating the true parameter values, we will focus solely on the non-scaled sensitivities, i.e., the TSF.

Finally, a third possibility for computation of the sensitivity functions is **automatic differentiation**. This involves several numerical tools to help calculate the solutions of the sensitivity equations when it is unreasonable to analytically or computationally determine the coefficients (e.g., for large systems of differential equations). One such tool for MATLAB, the Automatic Differentiation (AD) package [24], determines the analytic derivatives $\dfrac{\partial g}{\partial x}$ and $\dfrac{\partial g}{\partial q}$ of a vector system of ODEs that use only scalar operations by constructing functions based on the derivatives of each operator in g. These derivatives of g are then used in the sensitivity equations (3.58) in the same manner as if one had manually computed the derivatives. This technique is particularly beneficial when the number of state-parameter combinations (and thus the number of sensitivity equations being computed) becomes large (e.g., in a system with 38 states and more than 120 parameters [12]) and manually computing derivatives of g would be unwieldy or error-prone based on the form of the model.

As we have already noted, since θ_0 and σ_0 are unknown, we will use their estimates to make the approximation

$$\Sigma_0^N \approx \sigma_0^2 [F_{\boldsymbol{\theta}}^N(\boldsymbol{\theta}_0)^T F_{\boldsymbol{\theta}}^N(\boldsymbol{\theta}_0)]^{-1} \approx \hat{\Sigma}^N(\hat{\boldsymbol{\theta}}_{\text{OLS}}^N) = \hat{\sigma}^2 [F_{\boldsymbol{\theta}}^N(\hat{\boldsymbol{\theta}}_{\text{OLS}}^N)^T F_{\boldsymbol{\theta}}^N(\hat{\boldsymbol{\theta}}_{\text{OLS}}^N)]^{-1},$$

$$(3.63)$$

where the approximation $\hat{\sigma}^2$ of σ_0^2, as discussed earlier, is given by

$$\sigma_0^2 \approx \hat{\sigma}^2 = \frac{1}{N - \kappa_\theta} \sum_{j=1}^N [y_j - f(t_j; \hat{\boldsymbol{\theta}}_{\text{OLS}}^N)]^2. \qquad (3.64)$$

Standard errors to be used in the confidence interval calculations are given by $\text{SE}_k(\hat{\boldsymbol{\theta}}^N) = \sqrt{\hat{\Sigma}_{kk}^N(\hat{\boldsymbol{\theta}}^N)}$, $k = 1, 2, \ldots, \kappa_\theta$ (see [20]).

To compute the confidence intervals (at the $100(1 - \alpha)\%$ level) for the estimated parameters in our example, we define the confidence intervals associated with the estimated parameters so that

$$\text{Prob}\{\theta_k^N - t_{1-\alpha/2}\text{SE}_k(\hat{\boldsymbol{\theta}}^N) < \theta_{0k} < \theta_k^N + t_{1-\alpha/2}\text{SE}_k(\hat{\boldsymbol{\theta}}^N)\} = 1 - \alpha, \; (3.65)$$

where $\alpha \in [0, 1]$ and $t_{1-\alpha/2} \in \mathbb{R}^+$. For a realization $\boldsymbol{y}$ and estimates $\hat{\boldsymbol{\theta}}^N$, the corresponding confidence intervals are given by

$$[\hat{\theta}_k^N - t_{1-\alpha/2}\text{SE}_k(\hat{\boldsymbol{\theta}}^N), \hat{\theta}_k^N + t_{1-\alpha/2}\text{SE}_k(\hat{\boldsymbol{\theta}}^N)]. \qquad (3.66)$$

Given a small α value (e.g., $\alpha = 0.05$ for 95% confidence intervals), the critical value $t_{1-\alpha/2}$ is computed from the *Student's* t distribution $t^{N-\kappa_\theta}$ with $N - \kappa_\theta$ degrees of freedom. The value of $t_{1-\alpha/2}$ is determined by $\text{Prob}\{T \geq t_{1-\alpha/2}\} = \alpha/2$ where $T \sim t^{N-\kappa_\theta}$. In general, a confidence interval is constructed so that, if the confidence interval could be constructed for each possible realization of data of size N that could have been collected, $100(1-\alpha)\%$ of the intervals so constructed would contain the true value θ_{0k}. Thus, a confidence interval provides further information on the extent of uncertainty involved in estimating θ_0 using the given estimator and sample size N.

Remark 3.4.1 *We turn to a further comment on the use of the Student's t distribution in computing confidence intervals. We have already argued $\theta^N_{OLS} \sim \mathcal{N}(\theta_0, \Sigma_0^N)$. We can further establish*

$$\frac{(N-\kappa_\theta)S^2}{\sigma_0^2} \sim \chi^2_{N-\kappa_\theta}$$

where

$$S^2 = \frac{J^N_{OLS}(\theta^N_{OLS}; \mathbf{Y})}{N - \kappa_\theta}$$

and χ^2_ν is a chi-square distribution with ν degrees of freedom. Moreover, θ^N_{OLS} and S^2 are independent.

Some comments about these three statements are in order. The first distributional statement has already been established. The third statement regarding independence is exact for linear regression problems. A proof of the statement can be found in [34]. In the non-linear case, the statement is true to within the order of the linear approximation. The second distributional statement requires some explanation. We can argue

$$\frac{(N-\kappa_\theta)S^2}{\sigma_0^2} = \frac{1}{\sigma_0^2}\sum_{j=1}^N \left(Y_j - f(t_j; \theta^N_{OLS})\right)^2$$

$$\approx \frac{1}{\sigma_0^2}\sum_{j=1}^N \widetilde{\mathcal{E}}_j^2$$

$$= \sum_{j=1}^N \left(\frac{\widetilde{\mathcal{E}}_j}{\sigma_0}\right)^2. \tag{3.67}$$

If the errors $\widetilde{\mathcal{E}}_j$ are assumed to be independent and normally distributed with zero mean and constant variance σ_0^2, then the final expression above is (by definition) distributed as $\chi^2_{N-\kappa_\theta}$. The missing κ_θ degrees of freedom follow because the estimator θ^N_{OLS} must satisfy the κ_θ normal equations (3.11). If the errors are not assumed to be normally distributed, one has to modify the *arguments. However, we are only interested in behavior as $N \to \infty$. If we assume only that the $\tilde{\mathcal{E}}_j$ are independent and identically distributed with constant variance (as is commonly the case), the Central Limit Theorem (i.e., Theorem 2.6.2) applies and the distribution (whether chi-squared or not) will tend toward the normal. Even in the more general case of non-constant variance, a simple change of variables can be used to reduce to the constant variance case. Hence we really only need the variance terms to be bounded above.*

Next, consider only a single element, θ_i^N, of the estimator. Then we have as $N \to \infty$

$$\frac{\theta_i^N - \theta_{0,l}}{\sqrt{\Sigma_{0,ll}^N}} \sim \mathcal{N}(0,1).$$

Now define

$$T = \left(\frac{\theta_i^N - \theta_{0,l}}{\sqrt{\Sigma_{0,ll}^N}} \right) \left(\frac{(N - \kappa_\theta)S^2}{(N - \kappa_\theta)\sigma_0^2} \right)^{-1/2}$$

The first factor above is distributed as a standard normal variable. The second factor above is a chi-squared random variable rescaled by its degrees of freedom, and then raised to the $-1/2$ power. Thus T is distributed as $t^{N-\kappa_\theta}$. Moreover, by simple algebra (and using (3.16)),

$$T = \frac{\theta_i^N - \theta_{0,l}}{S\sqrt{(N\Omega_0)_{ll}^{-1}}}.$$

One then makes standard arguments (i.e., approximation of Ω_0—see (3.17)) to arrive at the usual confidence interval calculations.

When one is taking longitudinal samples corresponding to solutions of a dynamical system, the $N \times \kappa_\theta$ sensitivity matrix depends explicitly on where in time the scalar observations are taken when $f(t_j; \boldsymbol{\theta}) = C\boldsymbol{x}(t_j; \boldsymbol{\theta})$, as mentioned above. That is, the sensitivity matrix (3.15) depends on the number N and the nature (for example, how they are taken) of the sampling times $\{t_j\}$. Moreover, it is the matrix $[F_{\boldsymbol{\theta}}^T F_{\boldsymbol{\theta}}]^{-1}$ in (3.63) and the parameter $\hat{\sigma}^2$ in (3.64) that ultimately determine the standard errors and confidence intervals. At first investigation of (3.64), it appears that an increased number N of samples might drive $\hat{\sigma}^2 \to \sigma_0^2$ and hence drive the standard error (SE) to zero as long as this is done in a way to maintain a bound on the residual sum of squares in (3.64). However, we observe that the *condition number* of the Fisher information matrix $F_{\boldsymbol{\theta}}^T F_{\boldsymbol{\theta}}$ is also very important in these considerations and increasing the sampling could potentially adversely affect the numerical inversion of $F_{\boldsymbol{\theta}}^T F_{\boldsymbol{\theta}}$. In this regard, we note that among the important hypotheses in the asymptotic statistical theory (see p. 571 of [35]) is the existence of a matrix function $\Omega(\boldsymbol{\theta})$ such that

$$\frac{1}{N} F_{\boldsymbol{\theta}}^N(\boldsymbol{\theta})^T F_{\boldsymbol{\theta}}^N(\boldsymbol{\theta}) \to \Omega(\boldsymbol{\theta}) \quad \text{uniformly in } \boldsymbol{\theta} \text{ as } N \to \infty,$$

with $\Omega_0 = \Omega(\boldsymbol{\theta}_0)$ being a **non-singular** matrix. It is this condition that is rather easily violated in practice when one is dealing with data from differential equation systems, especially near an equilibrium or steady state (see the examples of [6] where this rather common phenomenon is illustrated).

Since the computations for standard errors and confidence intervals (and also *model comparison tests* to be discussed in a subsequent chapter) depend on *an asymptotic limit distribution theory*, one should interpret the findings as sometimes crude indicators of uncertainty inherent in the inverse problem findings. Nonetheless, it is useful to consider the formal mathematical requirements underpinning these techniques. We offer the following summary of possibilities:

(1) Among the more readily checked hypotheses are those of the statistical model requiring that the errors $\widetilde{\mathcal{E}}_j$, $j = 1, 2, \ldots, N$, are independent and identically distributed (*i.i.d.*) random variables with mean $\mathbb{E}(\widetilde{\mathcal{E}}_j) = 0$ and constant variance $\mathrm{Var}(\widetilde{\mathcal{E}}_j) = \sigma_0^2$. After carrying out the estimation procedures, one can readily plot the *residuals* $r_j = y_j - f(t_j; \hat{\boldsymbol{\theta}}_{OLS}^N)$ *vs. time* t_j and the *residuals vs. the resulting estimated model/observation* $f(t_j; \hat{\boldsymbol{\theta}}_{OLS}^N)$ *values.* A random pattern for the first is strong support for the validity of the independence assumption; a random pattern for the latter suggests the assumption of constant variance may be reasonable. This will be further explained in the next section.

(2) The underlying assumption that the sampling size N must be large (recall the theory is asymptotic in that it holds as $N \to \infty$) is not so readily "verified" and is often ignored (albeit at the user's peril in regard to the quality of the uncertainty findings). Often asymptotic results provide remarkably good approximations to the true sampling distributions for finite N. However, in practice there is no way to ascertain whether all assumptions for the theory hold and N is sufficiently large for a specific example.

All of the above theory readily generalizes to vector systems with partial, non-scalar observations. For example, suppose now we have the vector system (3.1) with partial vector observations given by Equation (3.5). That is, suppose we have m coordinate observations where $m \leq n$. In this case, we have

$$\frac{d\boldsymbol{x}}{dt}(t) = \boldsymbol{g}(t, \boldsymbol{x}(t), \boldsymbol{q}) \tag{3.68}$$

and

$$\boldsymbol{Y}_j = \boldsymbol{f}(t_j; \boldsymbol{\theta}_0) + \widetilde{\boldsymbol{\mathcal{E}}}_j = \mathcal{C}\boldsymbol{x}(t_j, \boldsymbol{\theta}_0) + \widetilde{\boldsymbol{\mathcal{E}}}_j, \tag{3.69}$$

where $\mathcal{C}$ is an $m \times n$ matrix and $\boldsymbol{f}(t_j; \boldsymbol{\theta}_0) \in \mathbb{R}^m, \boldsymbol{x} \in \mathbb{R}^n$. As already explained in Section 3.2, if we assume that different observation coordinates Y_j may have different variances $\sigma_{0,j}^2$ associated with different coordinates of the errors $\widetilde{\boldsymbol{\mathcal{E}}}_j$, then we have that $\widetilde{\boldsymbol{\mathcal{E}}}_j$ is an m-dimensional random vector with

$$\mathbb{E}(\widetilde{\boldsymbol{\mathcal{E}}}_j) = \mathbf{0}_m, \quad \text{Var}(\widetilde{\boldsymbol{\mathcal{E}}}_j) = V_0,$$

where $V_0 = \text{diag}(\sigma_{0,1}^2, ..., \sigma_{0,m}^2)$, and we may follow a similar asymptotic theory to calculate approximate covariances, standard errors and confidence intervals for parameter estimates.

3.5   Investigation of Statistical Assumptions

The form of error in the data (which of course is rarely known) dictates which method from those discussed above one should choose. The OLS method is most appropriate for constant variance observations of the form $Y_j = f(t_j; \boldsymbol{\theta}_0) + \widetilde{\mathcal{E}}_j$ whereas the GLS should be used for problems in which we have non-constant variance observations $Y_j = f(t_j; \boldsymbol{\theta}_0) + f^\gamma(t_j; \boldsymbol{\theta}_0)\widetilde{\mathcal{E}}_j$.

We emphasize that to obtain *the correct standard errors* in an inverse problem calculation, the OLS method (and *corresponding asymptotic formulas*) must be used with constant variance generated data, while the GLS method (and *corresponding asymptotic formulas*) should be applied to non-constant variance generated data.

Not doing so can lead to *incorrect conclusions*. In either case, the standard error calculations are not valid unless the correct formulas (which depend on the error structure) are employed. Unfortunately, it is very difficult to ascertain the structure of the error, and hence the correct method to use, without a priori information. Although the error structure cannot definitively be determined, the two residual tests can be performed *after* the estimation procedure has been completed to assist in concluding whether or not the correct asymptotic statistics were used.

3.5.1   Residual Plots

One can carry out simulation studies with a proposed mathematical model to assist in understanding the behavior of the model in inverse problems with different types of data with respect to misspecification of the statistical model. For example, we consider a statistical model with constant variance (CV) noise ($\gamma = 0$)

$$Y_j = f(t_j; \boldsymbol{\theta}_0) + \widetilde{\mathcal{E}}_j, \qquad \text{Var}(Y_j) = \sigma_0^2,$$

and another with non-constant variance (NCV) noise ($\gamma = 1$)

$$Y_j = f(t_j; \boldsymbol{\theta}_0)(1 + \widetilde{\mathcal{E}}_j), \qquad \text{Var}(Y_j) = \sigma_0^2 f^2(t_j; \boldsymbol{\theta}_0).$$

We obtain a data set by considering a *realization* $\{y_j\}_{j=1}^N$ of the random variables $\{Y_j\}_{j=1}^N$ through a realization of $\{\tilde{\mathcal{E}}_j\}_{j=1}^N$, and then calculate an estimate $\hat{\boldsymbol{\theta}}$ of $\boldsymbol{\theta}_0$ using the OLS or GLS procedure. Other values for γ could also readily be analyzed and in fact will be compared in an example below (in Section 3.5.3) on cell proliferation models. Here we focus on two of the more prominent statistical models for absolute error vs. relative error.

We will then use the *residuals* $r_j = y_j - f(t_j; \hat{\boldsymbol{\theta}})$ to test whether the data set is *i.i.d.* and possesses the assumed variance structure. If a data set has constant variance, then

$$Y_j = f(t_j; \boldsymbol{\theta}_0) + \tilde{\mathcal{E}}_j \quad \text{or} \quad \tilde{\mathcal{E}}_j = Y_j - f(t_j; \boldsymbol{\theta}_0),$$

and hence the residuals r_j are approximations to realizations of the errors $\tilde{\mathcal{E}}_j$ (when it is tacitly assumed that $\hat{\boldsymbol{\theta}} \approx \boldsymbol{\theta}_0$). As we have discussed above and want to summarize again, since it is assumed that the errors $\tilde{\mathcal{E}}_j$ are *i.i.d.*, a plot of the residuals $r_j = y_j - f(t_j; \hat{\boldsymbol{\theta}})$ vs. t_j should be random (and neither increasing nor decreasing with time). Also, the error in the constant variance case does not depend on $f(t_j; \boldsymbol{\theta}_0)$, and so a plot of the residuals $r_j = y_j - f(t_j; \hat{\boldsymbol{\theta}})$ vs. $f(t_j; \hat{\boldsymbol{\theta}})$ should also be random (and neither increasing nor decreasing). Therefore, *if* the error has constant variance, then a plot of the residuals $r_j = y_j - f(t_j; \hat{\boldsymbol{\theta}})$ against t_j and against $f(t_j; \hat{\boldsymbol{\theta}})$ should both be random. If not, then the constant variance assumption is suspect.

We next turn to questions of what to expect if this residual test is applied to a data set that has non-constant variance (NCV) generated error. That is, we wish to investigate what happens if the data are incorrectly assumed to have CV error when in fact they have NCV error. Since in the NCV example, $R_j = Y_j - f(t_j; \boldsymbol{\theta}_0) = f(t_j; \boldsymbol{\theta}_0)\tilde{\mathcal{E}}_j$ depends upon the deterministic model $f(t_j; \boldsymbol{\theta}_0)$, we should expect that a plot of the residuals $r_j = y_j - f(t_j; \hat{\boldsymbol{\theta}})$ vs. t_j should exhibit some type of pattern. Also, the residuals actually depend on $f(t_j; \hat{\boldsymbol{\theta}})$ in the NCV case, and so as $f(t_j; \hat{\boldsymbol{\theta}})$ increases the variation of the residuals $r_j = y_j - f(t_j; \hat{\boldsymbol{\theta}})$ should increase as well. Thus $r_j = y_j - f(t_j; \hat{\boldsymbol{\theta}})$ vs. $f(t_j; \hat{\boldsymbol{\theta}})$ should have a fan shape in the NCV case.

If a data set has non-constant variance generated data, then

$$Y_j = f(t_j; \boldsymbol{\theta}_0) + f(t_j; \boldsymbol{\theta}_0)\tilde{\mathcal{E}}_j \quad \text{or} \quad \tilde{\mathcal{E}}_j = \frac{Y_j - f(t_j; \boldsymbol{\theta}_0)}{f(t_j; \boldsymbol{\theta}_0)}.$$

If the distributions of $\tilde{\mathcal{E}}_j$ are *i.i.d.*, then a plot of the *modified residuals* $r_j^m = (y_j - f(t_j; \hat{\boldsymbol{\theta}}))/f(t_j; \hat{\boldsymbol{\theta}})$ vs. t_j should be random for non-constant variance generated data. A plot of $r_j^m = (y_j - f(t_j; \hat{\boldsymbol{\theta}}))/f(t_j; \hat{\boldsymbol{\theta}})$ vs. $f(t_j; \hat{\boldsymbol{\theta}})$ should also be random.

Another question of interest concerns the case in which the data are incorrectly assumed to have non-constant variance error when in fact they have constant variance error. Since $Y_j - f(t_j; \boldsymbol{\theta}_0) = \tilde{\mathcal{E}}_j$ in the constant variance case, we should expect that a plot of $r_j^m = (y_j - f(t_j; \hat{\theta}))/f(t_j; \hat{\theta})$ vs. t_j as well as that for $r_j^m = (y_j - f(t_j; \hat{\theta}))/f(t_j; \hat{\theta})$ vs. $f(t_j; \hat{\theta})$ will possess some distinct pattern (such as a fan shape).

There are two further issues regarding residual plots. As we shall see by examples, some data sets might have values that are repeated or nearly repeated a large number of times (for example, when sampling near an equilibrium of a mathematical model or when sampling a periodic system over many periods). If a certain value is repeated numerous times (e.g., f_{repeat}), then any plot with $f(t_j; \hat{\theta})$ along the horizontal axis should have a cluster of values along the vertical line $x = f_{\text{repeat}}$. This feature can easily be removed by excluding the data points corresponding to these high frequency values (or simply excluding the corresponding points in the residual plots). Another common technique when plotting against model predictions is to plot against $\ln(f(t_j; \hat{\theta}))$ instead of $f(t_j; \hat{\theta})$ itself, which has the effect of "stretching out" plots at the ends. Also, note that the model value $f(t_j; \hat{\theta})$ could possibly be zero or very near zero, in which case the modified residuals $r_j^m = (y_j - f(t_j; \hat{\theta}))/f(t_j; \hat{\theta})$ would be undefined or extremely large. To remedy this situation one might exclude values very close to zero (in either the plots or in the data themselves). We chose here to reduce the data sets (although this sometimes could lead to a deterioration in the estimation results obtained). In our examples below, estimates obtained using a truncated data set will be denoted by, for example, $\hat{\theta}_{\text{OLS}}^{\text{TCV}}$ for constant variance data using OLS procedures, $\hat{\theta}_{\text{OLS}}^{\text{TNCV}}$ for non-constant variance data using OLS procedures, $\hat{\theta}_{\text{GLS}}^{\text{TCV}}$ for constant variance data using GLS procedures and $\hat{\theta}_{\text{GLS}}^{\text{TNCV}}$ for non-constant variance data using GLS procedures.

3.5.2   An Example Using Residual Plots: Logistic Growth

We illustrate residual plot techniques by exploring a widely used model – the logistic population growth model of Verhulst/Pearl [29]

$$\dot{x} = rx\left(1 - \frac{x}{K}\right), \quad x(0) = x_0. \tag{3.70}$$

Here K is the population's carrying capacity, r is the intrinsic growth rate and x_0 is the initial population size. This well-known logistic model describes how populations grow when constrained by resources or competition. (We discussed this model, its derivation and properties in more detail in [16, Chapter 9]). The closed form solution of this simple model is given by

$$x(t) = \frac{K x_0 e^{rt}}{K + x_0 \left(e^{rt} - 1\right)}. \tag{3.71}$$

The left plot in Figure 3.1 depicts the solution of the logistic model with $K = 17.5$, $r = 0.7$ and $x_0 = 0.1$ for $0 \le t \le 25$. If high frequency repeated or

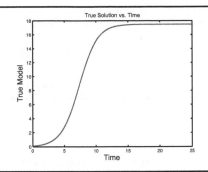
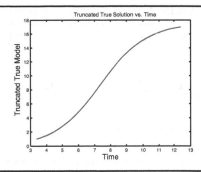

FIGURE 3.1: Original and truncated logistic curve with $K = 17.5$, $r = 0.7$ and $x_0 = 0.1$.

nearly repeated values (i.e., near the initial value x_0 or near the asymptote $x = K$) are removed from the original plot, the resulting truncated plot is given in the right panel of Figure 3.1 (there are no near zero values for this function).

For this example we generated both CV and NCV noisy data (we sampled from $\mathcal{N}(0, 25 \times 10^{-4})$ distributed random variables to obtain realizations of $\tilde{\mathcal{E}}_j$) and obtained estimates $\hat{\boldsymbol{\theta}}$ of $\boldsymbol{\theta}_0$ by applying either the OLS or GLS method to a realization $\{y_j\}_{j=1}^N$ of the random variables $\{Y_j\}_{j=1}^N$. The initial guesses $\boldsymbol{\theta}_{init} = \hat{\boldsymbol{\theta}}^{(0)}$ along with estimates for each method and error structure are given in Tables 3.1–3.4. As expected, both methods do a good job of estimating $\boldsymbol{\theta}_0$; however, the error structure was not always correctly specified since incorrect asymptotic formulas were used in some cases.

When the OLS method was applied to non-constant variance data and the GLS method was applied to constant variance data, the residual plots given below do reveal that the error structure was misspecified. For instance, the plot of the residuals for $\hat{\boldsymbol{\theta}}_{\text{OLS}}^{\text{NCV}}$ given in Figures 3.4 and 3.5 reveal a fan shaped pattern, which indicates the constant variance assumption is suspect. In addition, the plot of the residuals for $\hat{\boldsymbol{\theta}}_{\text{GLS}}^{\text{CV}}$ given in Figures 3.6 and 3.7 reveals an inverted fan shaped pattern, which indicates the non-constant variance assumption is suspect. As expected, when the correct error structure is

TABLE 3.1: Estimation using the OLS procedure with CV data.

$\boldsymbol{\theta}_{\text{init}}$	$\boldsymbol{\theta}_0$	$\hat{\boldsymbol{\theta}}_{\text{OLS}}^{\text{CV}}$	$\text{SE}(\hat{\boldsymbol{\theta}}_{\text{OLS}}^{\text{CV}})$	$\hat{\boldsymbol{\theta}}_{\text{OLS}}^{\text{TCV}}$	$\text{SE}(\hat{\boldsymbol{\theta}}_{\text{OLS}}^{\text{TCV}})$
17	17.5	1.7500e+001	1.5800e-003	1.7494e+001	6.4215e-003
.8	.7	7.0018e-001	4.2841e-004	7.0062e-001	6.5796e-004
1.2	.1	9.9958e-002	3.1483e-004	9.9702e-002	4.3898e-004

TABLE 3.2:   Estimation using the GLS procedure with CV data.

θ_{init}	θ_0	$\hat{\theta}_{\text{GLS}}^{\text{CV}}$	$\text{SE}(\hat{\theta}_{\text{GLS}}^{\text{CV}})$	$\hat{\theta}_{\text{GLS}}^{\text{TCV}}$	$\text{SE}(\hat{\theta}_{\text{GLS}}^{\text{TCV}})$
17	17.5	1.7500e+001	1.3824e-004	1.7494e+001	9.1213e-005
.8	.7	7.0021e-001	7.8139e-005	7.0060e-001	1.6009e-005
1.2	.1	9.9938e-002	6.6068e-005	9.9718e-002	1.2130e-005

TABLE 3.3:   Estimation using the OLS procedure with NCV data.

θ_{init}	θ_0	$\hat{\theta}_{\text{OLS}}^{\text{NCV}}$	$\text{SE}(\hat{\theta}_{\text{OLS}}^{\text{NCV}})$	$\hat{\theta}_{\text{OLS}}^{\text{TNCV}}$	$\text{SE}(\hat{\theta}_{\text{OLS}}^{\text{TNCV}})$
17	17.5	1.7499e+001	2.2678e-002	1.7411e+001	7.1584e-002
.8	.7	7.0192e-001	6.1770e-003	7.0955e-001	7.6039e-003
1.2	.1	9.9496e-002	4.5115e-003	9.4967e-002	4.8295e-003

TABLE 3.4:   Estimation using the GLS procedure with NCV data.

θ_{init}	θ_0	$\hat{\theta}_{\text{GLS}}^{\text{NCV}}$	$\text{SE}(\hat{\theta}_{\text{GLS}}^{\text{NCV}})$	$\hat{\theta}_{\text{GLS}}^{\text{TNCV}}$	$\text{SE}(\hat{\theta}_{\text{GLS}}^{\text{TNCV}})$
17	17.5	1.7498e+001	9.4366e-005	1.7411e+001	3.1271e-004
.8	.7	7.0217e-001	5.3616e-005	7.0959e-001	5.7181e-005
1.2	.1	9.9314e-002	4.4976e-005	9.4944e-002	4.1205e-005

specified, the *i.i.d.* test and the model dependence test each displays a random pattern (Figures 3.2, 3.3 and Figures 3.8, 3.9).

Also, included in the right panel of Figures 3.2 – 3.9 are the residual plots with the truncated data sets. In those plots only model values between 1 and 17 were considered (i.e., $1 \leq y_j \leq 17$). Doing so removed the dense vertical lines in the plots with $f(t_j; \hat{\theta})$ along the x-axis. Nonetheless, the conclusions regarding the error structure remain the same.

In addition to the residual plots, we can also compare the standard errors obtained for each simulation. With a quick glance at Tables 3.1 – 3.4, we see that the standard error of the parameter K in the truncated data set is larger than the standard error of K in the original data set. This behavior is expected. If we remove the "flat" region in the logistic curve, we actually discard measurements with high information content about the carrying capacity K—see [6]. Doing so reduces the quality of the estimator for K. Another interesting observation is that the standard errors of the GLS estimate are more optimistic than that of the OLS estimate, even when the non-constant variance assumption is wrong. This example further solidifies the conclusion that before one reports an estimate and corresponding standard errors, there needs to be some assurance that the proper error structure has been specified.

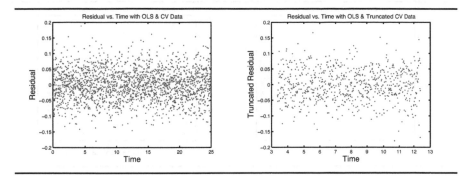

FIGURE 3.2: Residual vs. time plots in tests for independence: Original and truncated logistic curve for $\hat{\theta}_{\mathrm{OLS}}^{\mathrm{CV}}$.

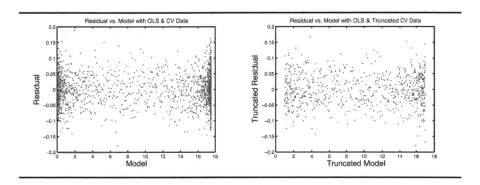

FIGURE 3.3: Residual vs. model plots in tests of form of variance: Original and truncated logistic curve for $\hat{\theta}_{\mathrm{OLS}}^{\mathrm{CV}}$.

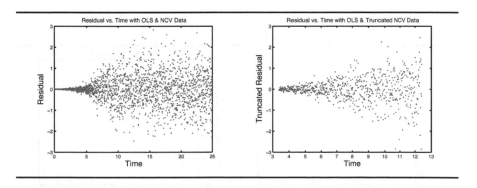

FIGURE 3.4: Residual vs. time plots in tests for independence: Original and truncated logistic curve for $\hat{\theta}_{\mathrm{OLS}}^{\mathrm{NCV}}$.

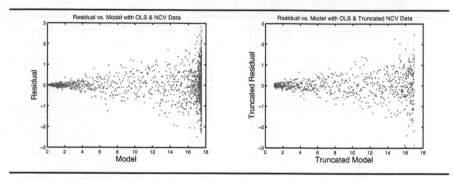

FIGURE 3.5:   Residual vs. model plots in tests of form of variance: Original and truncated logistic curve for $\hat{\boldsymbol{\theta}}^{\mathrm{NCV}}_{\mathrm{OLS}}$.

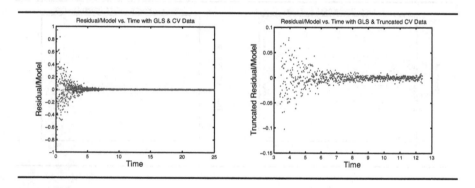

FIGURE 3.6:   Residual vs. time plots in tests for independence: Original and truncated logistic curve for $\hat{\boldsymbol{\theta}}^{\mathrm{CV}}_{\mathrm{GLS}}$.

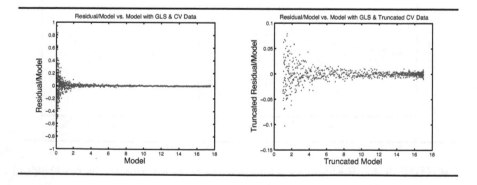

FIGURE 3.7:   Modified residual vs. model plots in tests of form of variance: Original and truncated logistic curve for $\hat{\boldsymbol{\theta}}^{\mathrm{CV}}_{\mathrm{GLS}}$.

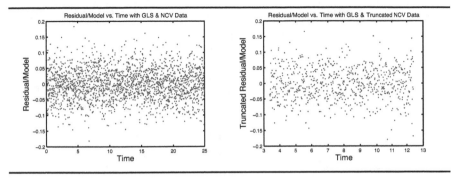

FIGURE 3.8: Modified residual vs. time plots in tests for independence: Original and truncated logistic curve for $\hat{\theta}_{\text{GLS}}^{\text{NCV}}$.

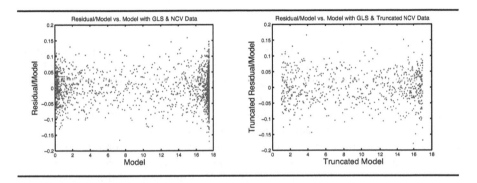

FIGURE 3.9: Modified residual vs. model plots in tests of form of variance: Original and truncated logistic curve for $\hat{\theta}_{\text{GLS}}^{\text{NCV}}$.

3.5.3 A Second Example Using Residual Plot Analysis: Cell Proliferation

We now consider a practical application using residual plots to determine the appropriate error model for a weighted least squares problem. We consider the problem of modeling flow cytometry data for a dividing population of lymphocytes labeled with the intracellular dye CFSE. As cells divide, the highly fluorescent intracellular CFSE is partitioned evenly between two daughter cells. A flow cytometer measures the CFSE fluorescence intensity (FI) of labeled cells as a surrogate for the mass of CFSE within a cell, thus providing an indication of the number of times a cell has divided. Most commonly, the data is presented in histograms showing the distribution of CFSE FI in the measured population of cells at each measurement time. A sample data set is shown in Figure 3.10. See [15, 37] and the references therein for a detailed overview of the experimental procedure.

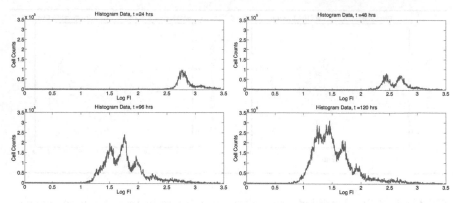

FIGURE 3.10:   Typical histogram data for a CFSE-based proliferation assay showing cell count vs. log fluorescence intensity; originally from [30].

The mathematical modeling problem is to develop a system of equations which accounts for the manner in which the population distribution of CFSE evolves as cells divide, die and slowly lose fluorescence intensity as a result of protein turnover. In doing so, one must account for not only the division and death dynamics of the population of cells, but also how these processes relate to the observed distribution of fluorescence intensities. Let $\tilde{n}_i(t,\tilde{x})$ represent the label-structured density (cells per unity of intensity, cells/UI) of cells at time t having completed i divisions and having fluorescence intensity $\tilde{x}$ from *intracellular CFSE* (that is, not including the contribution of cellular autofluorescence). Then the corresponding total cell density in the population is

$$\tilde{n}(t,\tilde{x}) = \Sigma_{i=0}^{\infty}\tilde{n}_i(t,\tilde{x}).$$

It can be shown [26, 33] that these densities can be factored (modeled) in terms of two independently functioning components, one factor $N_i(t)$ accounting for the division dynamics of the population of cells, and one $\bar{n}_i(t,\tilde{x})$ accounting for the intracellular processing of CFSE. It has been shown that the slow loss of fluorescence intensity as a result of intracellular protein turnover can be well modeled by a Gompertz growth/decay process [2, 13, 37]. Moreover, it has also been established that the highly flexible cyton model [27] can very accurately describe the evolving generation structure (cells per number of divisions undergone) [15]. Thus we consider the model

$$\tilde{n}_i(t,\tilde{x}) = N_i(t)\bar{n}_i(t,\tilde{x}). \tag{3.72}$$

The functions $\bar{n}_i(t,\tilde{x})$ each satisfies the partial differential equation

$$\frac{\partial \bar{n}_i(t,\tilde{x})}{\partial t} - ce^{-kt}\frac{\partial[\tilde{x}\bar{n}_i(t,\tilde{x})]}{\partial \tilde{x}} = 0, \tag{3.73}$$

with initial condition

$$\bar{n}_i(0, \tilde{x}) = \frac{2^i \Phi(2^i \tilde{x})}{N_{00}},$$

where $\dfrac{d\tilde{x}}{dt} = v(t, \tilde{x}) = c\tilde{x}e^{-kt}$ is the Gompertzian velocity of label decay, Φ is the label-structured density of the cells at $t = 0$ and $N_0(0) = N_{00}$ is the number of cells in the population at $t = 0$, all of which are assumed to be undivided. The functions $N_i(t)$ are described by the cyton model,

$$N_0(t) = N_{00} - \int_0^t \left(n_0^{div}(s) - n_0^{die}(s) \right) ds$$

$$N_i(t) = \int_0^t \left(2n_{i-1}^{div}(s) - n_i^{div}(s) - n_i^{die}(s) \right) ds, \qquad (3.74)$$

where $n_i^{div}(t)$ and $n_i^{die}(t)$ indicate the numbers per unit time of cells having undergone i divisions that divide and die, respectively, at time t. In this regard, the cyton model (and hence the mathematical model as a whole) can be considered as a large class of models differentiated by the mechanisms one uses to describe the terms $n_i^{div}(t)$ and $n_i^{die}(t)$, which we consider in more detail below. We remark that the mathematical model (3.72) can be equivalently characterized by the system of partial differential equations

$$\frac{\partial \tilde{n}_0}{\partial t}(t, \tilde{x}) - ce^{-kt}\frac{\partial [\tilde{x}\tilde{n}_0]}{\partial \tilde{x}}(t, \tilde{x}) = \left(n_0^{div}(t) - n_0^{die}(t) \right) \bar{n}_0(t, \tilde{x})$$

$$\frac{\partial \tilde{n}_1}{\partial t}(t, \tilde{x}) - ce^{-kt}\frac{\partial [\tilde{x}\tilde{n}_1]}{\partial \tilde{x}}(t, \tilde{x}) = \left(2n_0^{div}(t) - n_1^{div}(t) - n_1^{die}(t) \right) \bar{n}_1(t, \tilde{x}) \quad (3.75)$$

$$\vdots$$

Given the solution $\tilde{n}(t, \tilde{x}) = \sum_i \tilde{n}_i(t, \tilde{x})$ in terms of the fluorescence resulting from intracellular CFSE, one must add in the contribution of cellular auto-fluorescence to obtain a population density structured by total fluorescence intensity, which is the quantity measured by a flow cytometer. This density can be computed as

$$n(t, x) = \int_0^\infty \tilde{n}(t, \tilde{x})p_{x_a}(x - \tilde{x})d\tilde{x}, \qquad (3.76)$$

where p_{x_a} is a probability density function describing the distribution of cellular autofluorescence [26].

Let ϕ_0 and ψ_0 be probability density functions (in units 1/hr) for the time to first division and time to death, respectively, for an undivided cell. Let F_0, the initial precursor fraction, be the fraction of undivided cells which would hypothetically divide in the absence of any cell death. It follows that

$$n_0^{div}(t) = F_0 N_{00} \left(1 - \int_0^t \psi_0(s)ds\right) \phi_0(t)$$

$$n_0^{die}(t) = N_{00} \left(1 - F_0 \int_0^t \phi_0(s)ds\right) \psi_0(t). \tag{3.77}$$

Similarly, one can define probability density functions ϕ_i and ψ_i for times to division and death, respectively, for cells having undergone i divisions (the cellular "clock" begins at the completion of the previous division), as well as the progressor fractions F_i of cells which would complete the ith division in the absence of cell death. Then

$$n_i^{div}(t) = 2F_i \int_0^t n_{i-1}^{div}(s) \left(1 - \int_0^{t-s} \psi_i(\xi)d\xi\right) \phi_i(t-s)ds$$

$$n_i^{die}(t) = 2 \int_0^t n_{i-1}^{div}(s) \left(1 - F_i \int_0^{t-s} \phi_i(\xi)d\xi\right) \psi_i(t-s)ds. \tag{3.78}$$

It is assumed that the probability density functions for times to cell division are log-normal, and that undivided cells and divided cells have two separate density functions. Similarly, it is assumed that divided cells die with a log-normal probability density. Thus

$$\phi_0(t) = \frac{1}{t\sigma_0^{div}\sqrt{2\pi}} \exp\left(-\frac{(\ln t - \mu_0^{div})^2}{2(\sigma_0^{div})^2}\right)$$

$$\phi_i(t) = \frac{1}{t\sigma_i^{div}\sqrt{2\pi}} \exp\left(-\frac{(\ln t - \mu_i^{div})^2}{2(\sigma_i^{div})^2}\right) \quad (i \geq 1)$$

$$\psi_i(t) = \frac{1}{t\sigma_i^{die}\sqrt{2\pi}} \exp\left(-\frac{(\ln t - \mu_i^{die})^2}{2(\sigma_i^{die})^2}\right) \quad (i \geq 1).$$

We treat cell death for undivided cells as a special case. It is assumed that the density function describing cell death for undivided cells is

$$\psi_0(t) = \frac{F_0}{t\sigma_0^{die}\sqrt{2\pi}} \exp\left(-\frac{(\ln t - \mu_0^{die})^2}{2(\sigma_0^{die})^2}\right) + (1 - p_{idle})(1 - F_0)\beta e^{-\beta t}, \tag{3.79}$$

where p_{idle} is the fraction of cells which will neither die nor divide over the course of the experiment. The form of the function above arises from the assumption that progressing cells die with a log-normal probability density. Non-progressing cells die at an exponential rate, except for the fraction of idle cells. A more detailed discussion of the mathematical model can be found in [14].

With the mathematical model established, we now turn our attention to a statistical model of the data. All of our discussions above have dealt with ODE models, whereas here we need to treat a PDE example where the data collection is in terms of times t_j and label intensities x_k or, after a transformation, z_k. All of the above stated results for inverse problems hold for PDE problems after one re-indexes the data collection points $\{(t_j, z_k)\}_{j,k=1}^{N_t, N_z}$ by $\{t_i\}_{i=1}^{N_t \times N_z}$. We tacitly assume this has been done when applying the results from previous sections. Let $\boldsymbol{\theta}$ be the vector of parameters of the model (3.76), so that we may rewrite $n(t, x) = n(t, x; \boldsymbol{\theta})$. We first transform the FI axis (x) to a logarithmic coordinate $z = \log_{10} x$. Let N_k^j be a random variable representing the number of cells measured at time t_j with log-fluorescence intensity in the region $[z_k, z_{k+1})$. Define the operator

$$I[n](t_j, z_k; \boldsymbol{\theta}) = \int_{z_k}^{z_{k+1}} 10^z \ln(10) n(t, 10^z; \boldsymbol{\theta}) dz,$$

which represents the computation of cell counts from the structured density $n(t, x)$ transformed to the logarithmic coordinate $z = \log_{10} x$. Then a statistical model linking the random variables N_k^j to the mathematical model is

$$N_k^j = \lambda_j I[n](t_j, z_k^j; \boldsymbol{\theta}_0) + \lambda_j \left(I[n](t_j, z_k^j; \boldsymbol{\theta}_0) \right)^\gamma \widetilde{\mathcal{E}}_{kj} \qquad (3.80)$$

where the λ_j are scaling factors [37, Chapter 4] and the random variables $\widetilde{\mathcal{E}}_{kj}$ satisfy assumption (A1). In principle, it is possible to use a multi-stage estimation procedure to determine the values of the statistical model parameters λ_j and γ [22], but we do not consider that here. It is assumed that $\lambda_j = 1$ for all j.

In the case that $\gamma = 0$, one has precisely absolute error and a statistical model suitable for the ordinary least squares cost function and estimator (3.10) to estimate the unknown parameter $\boldsymbol{\theta}_0$. The results of using the OLS estimator can be found in Figure 3.11.

It follows from the form of Equation (3.80) that when $\gamma = 0$ the residuals $r_k^j = I[n](t_j, z_k^j; \hat{\boldsymbol{\theta}}_{\mathrm{OLS}}) - n_k^j$ (where the n_k^j are realizations of the data random variables N_k^j) correspond to realizations of the error terms $\widetilde{\mathcal{E}}_{kj}$. As such, these residuals should appear random when plotted as a function of the magnitude of the model solution, as discussed in Section 3.5.1. However, we see in Figure 3.12 (top) that this is not the case. There is a noticeable increase in the variance of the residuals as the magnitude of the model solution increases. Thus we must conclude that the constant variance model $\gamma = 0$ is not correct. A constant coefficient of variation (CCV) model ($\gamma = 1$) leading to a generalized least squares problem formulation was also considered in [37]. Again residual analysis—see Figure 3.12 (bottom)—suggested that this was not the correct formulation.

In fact, the most appropriate theoretical value of γ appears to be $\gamma = 1/2$ [37], so the generalized least squares estimating equations defined via (3.38) with $\gamma = 1/2$ should be used. Because the "weights" in the GLS algorithm must be bounded away from zero, we define the weights to be used in (3.38)

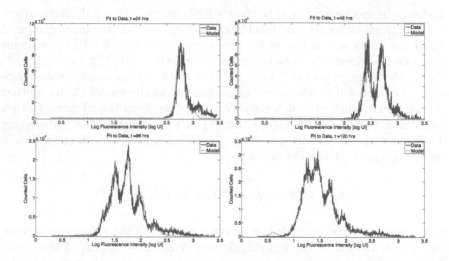

FIGURE 3.11: Results of fitting the model (3.76) using ordinary least squares. The fits to data are superb, and it is difficult to ascertain the correctness or not of the underlying statistical model.

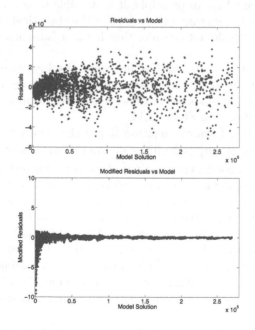

FIGURE 3.12: Residual plots for models $N_k^j = I[n](t_j, z_k^j; \boldsymbol{\theta}_0) + (I[n](t_j, z_k^j; \boldsymbol{\theta}_0))^\gamma \widetilde{\mathcal{E}}_{kj}$ with $\gamma = 0$ (top) vs. $\gamma = 1$ (bottom).

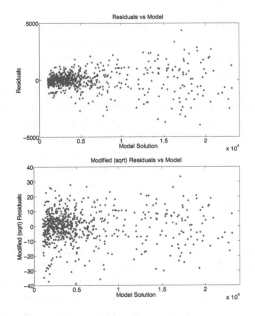

FIGURE 3.13: Truncated residual plots for models $N_k^j = I[n](t_j, z_k^j; \boldsymbol{\theta}_0) + (I[n](t_j, z_k^j; \boldsymbol{\theta}_0))^\gamma \widetilde{\mathcal{E}}_{kj}$ with $\gamma = 0$ (top) vs. $\gamma = 1/2$ (bottom) where $\gamma = 1/2$ works well!!

by

$$\hat{w}_{j,k} = \begin{cases} \left(I[n](t_j, z_k^j; \hat{\boldsymbol{\theta}}^{(l)}) \right)^{1/2}, & I[n](t_j, z_k^j; \hat{\boldsymbol{\theta}}^{(l)}) > I_0 \\ (I_0)^{1/2}, & I[n](t_j, z_k^j; \hat{\boldsymbol{\theta}}^{(l)}) \leq I_0. \end{cases} \quad (3.81)$$

The cutoff value $I_0 > 0$ is determined by the experimenter so that the resulting residuals appear random. In the work that follows, $I_0 = 1 \times 10^4$.

Truncated residual plots obtained using the theoretical value of $\gamma = 1/2$ are shown in Figure 3.13 (bottom). (The corresponding least squares fit to data is not shown, as the results are sufficiently similar to the ordinary least squares fits of Figure 3.11.) While the variance of the standard residuals is observed to increase with the magnitude of the model solution, the variance of the modified residuals is approximately constant, providing some evidence that the statistical model (3.80) with $\gamma = 1/2$ is correct.

3.6 Bootstrapping vs. Asymptotic Error Analysis

In the above discussions we used asymptotic theory to compute uncertainty features for parameter estimates. One popular alternative to the asymptotic

94   *Modeling and Inverse Problems in the Presence of Uncertainty*

theory is *bootstrapping* wherein one uses the residuals from an initial estimation to construct a family of samples or simulated data sets. One then uses these samples to construct an empirical distribution for the parameters from which the means, standard errors and hence the associated confidence intervals can be readily obtained for the underlying true parameters $\boldsymbol{\theta}_0$. In this section we investigate computationally and compare the bootstrapping approach and the asymptotic theory approach to computing parameter standard errors corresponding to data with various noise levels (denoted by σ_0). We consider both absolute error (with constant variance) measurement data and relative error (and hence non-constant variance) measurement data sets in obtaining parameter estimates. We compare and contrast parameter estimates, standard errors, confidence intervals and computational times for both bootstrapping and asymptotic theory approaches. Our goal here is to investigate and illustrate some possible advantages and/or disadvantages of each approach in treating problems for non-linear dynamical systems, focusing on computational methods. We discuss these in the context of a simple example (the *logistic growth population example* of (3.70) in Section 3.5.2) for which an analytical solution is available to provide readily obvious distinct qualitative behaviors. We chose this example because its solutions have many features found in those of much more complicated models: regions of rapid change as well as a region near an equilibrium with very little change and different regions of relative insensitivity of solutions to different parameters.

We first give detailed algorithms for computing the bootstrapping estimates for both constant variance data and non-constant variance data. We then illustrate these ideas in the context of the logistic population growth model. This is followed by a report of the numerical results using correct statistical models, and then numerical results for data sets where the incorrect variance models are assumed. Finally, we present remarks on the corrective nature of the bootstrapping approach, followed by a short summary of our findings.

3.6.1 Bootstrapping Algorithm: Constant Variance Data

Assume we are given experimental data $(t_1, y_1), \ldots, (t_N, y_N)$ for a dynamical system (e.g., the logistic growth model) from an underlying observation process

$$Y_j = f(t_j; \boldsymbol{\theta}_0) + \widetilde{\mathcal{E}}_j, \quad j = 1, \ldots, N, \tag{3.82}$$

where the $\widetilde{\mathcal{E}}_j$ are independent and identically distributed *(i.i.d.)* with mean zero ($\mathbb{E}(\mathcal{E}_j) = 0$) and constant variance σ_0^2, and $\boldsymbol{\theta}_0$ is the "true value" hypothesized to exist in statistical treatments of data. Associated corresponding realizations $\{y_j\}$ of the random variables $\{Y_j\}$ are given by

$$y_j = f(t_j; \boldsymbol{\theta}_0) + \widetilde{\epsilon}_j.$$

The following algorithm [18, 19, 21, p. 285–287] can be used to compute the *bootstrapping estimate* $\hat{\boldsymbol{\theta}}_{\text{BOOT}}$ of $\boldsymbol{\theta}_0$ and its empirical distribution.

1. First estimate $\hat{\boldsymbol{\theta}}^0$ from the entire sample $\{y_j\}_{j=1}^N$ using OLS.

2. Using this estimate, define the standardized residuals

$$\bar{r}_j = \sqrt{\frac{N}{N-\kappa_\theta}}\left(y_j - f(t_j;\hat{\boldsymbol{\theta}}^0)\right)$$

for $j = 1,\ldots,N$. Set $m = 0$.

3. Create a bootstrapping sample of size N using random sampling with replacement from the data (realizations) $\{\bar{r}_1,\ldots,\bar{r}_N\}$ to form a bootstrapping sample $\{r_1^m,\ldots,r_N^m\}$.

4. Create bootstrap sample points

$$y_j^m = f(t_j;\hat{\boldsymbol{\theta}}^0) + r_j^m,$$

where $j = 1,\ldots,N$.

5. Obtain a new estimate $\hat{\boldsymbol{\theta}}^{m+1}$ from the bootstrapping sample $\{y_j^m\}$ using OLS.

6. Set $m = m+1$ and repeat steps 3–5 until $m \geq M$ (e.g., typically $M = 1000$, as in our calculations below).

We then calculate the mean, standard error and confidence intervals using the formulae

$$\hat{\boldsymbol{\theta}}_{\text{BOOT}} = \frac{1}{M}\sum_{m=1}^M \hat{\boldsymbol{\theta}}^m,$$

$$\text{Var}(\boldsymbol{\theta}_{\text{BOOT}}) = \frac{1}{M-1}\sum_{m=1}^M (\hat{\boldsymbol{\theta}}^m - \hat{\boldsymbol{\theta}}_{\text{BOOT}})^T(\hat{\boldsymbol{\theta}}^m - \hat{\boldsymbol{\theta}}_{\text{BOOT}}), \qquad (3.83)$$

$$\text{SE}_k(\hat{\boldsymbol{\theta}}_{\text{BOOT}}) = \sqrt{\text{Var}(\boldsymbol{\theta}_{\text{BOOT}})_{kk}},$$

where $\boldsymbol{\theta}_{\text{BOOT}}$ denotes the bootstrapping estimator.

Remark 3.6.1 *In the above procedures, the $\{\bar{r}_1,\ldots,\bar{r}_N\}$ are realizations of i.i.d. random variables $\bar{R}_j$ with the empirical distribution function F_N. It can be shown that*

$$\mathbb{E}(\bar{R}_j|F_N) = N^{-1}\sum_{j=1}^N \bar{r}_j = 0, \quad Var(\bar{R}_j|F_N) = N^{-1}\sum_{j=1}^N \bar{r}_j^2 = \hat{\sigma}^2.$$

3.6.2   Bootstrapping Algorithm: Non-Constant Variance Data

We suppose now that we are given experimental data $(t_1, y_1), \ldots, (t_N, y_N)$ from the underlying observation process

$$Y_j = f(t_j; \boldsymbol{\theta}_0)(1 + \widetilde{\mathcal{E}}_j), \qquad (3.84)$$

where $j = 1, \ldots, N$ and the $\widetilde{\mathcal{E}}_j$ are *i.i.d.* with mean zero and constant variance σ_0^2. Then we see that $\mathbb{E}(Y_j) = f(t_j; \boldsymbol{\theta}_0)$ and $\mathrm{Var}(Y_j) = \sigma_0^2 f^2(t_j, \boldsymbol{\theta}_0)$, with associated corresponding realizations of Y_j given by

$$y_j = f(t_j; \boldsymbol{\theta}_0)(1 + \widetilde{\epsilon}_j).$$

Once again, a standard algorithm can be used to compute the corresponding *bootstrapping estimate* $\hat{\boldsymbol{\theta}}_{\mathrm{BOOT}}$ of $\boldsymbol{\theta}_0$ and its empirical distribution. We treat the general case for non-linear dependence of the model output on the parameters $\boldsymbol{\theta}$. (For the special case of linear dependence, one can consult [18, 19].)

The algorithm is given as follows.

1. First obtain the estimate $\hat{\boldsymbol{\theta}}^0$ from the entire sample $\{y_j\}$ using the GLS given in (3.38) with $\gamma = 1$. An estimate $\hat{\boldsymbol{\theta}}_{\mathrm{GLS}}$ can be solved for iteratively as described in Section 3.2.6.

2. Define the non-constant variance standardized residuals

$$\bar{s}_j = \frac{y_j - f(t_j; \hat{\boldsymbol{\theta}}^0)}{f(t_j; \hat{\boldsymbol{\theta}}^0)}, \quad j = 1, 2, \ldots, N.$$

   Set $m = 0$.

3. Create a bootstrapping sample of size N using random sampling with replacement from the data (realizations) $\{\bar{s}_1, \ldots, \bar{s}_N\}$ to form a bootstrapping sample $\{s_1^m, \ldots, s_N^m\}$.

4. Create bootstrapping sample points

$$y_j^m = f(t_j; \hat{\boldsymbol{\theta}}^0) + f(t_j; \hat{\boldsymbol{\theta}}^0)s_j^m,$$

   where $j = 1, \ldots, N$.

5. Obtain a new estimate $\hat{\boldsymbol{\theta}}^{m+1}$ from the bootstrapping sample $\{y_j^m\}$ using GLS.

6. Set $m = m + 1$ and repeat steps 3–5 until $m \geq M$ where M is large (e.g., $M = 1000$).

We then calculate the mean, standard error and confidence intervals using the same formulae (3.83) as before.

Remark 3.6.2 *The $\{\bar{s}_1 \ldots \bar{s}_N\}$ defined above are realizations of i.i.d. random variables $\bar{S}_j$ with empirical distribution function F_N. In this non-linear weighted case the desired mean and variance conditions only hold approximately.*

$$\mathbb{E}(\bar{S}_j|F_N) = N^{-1}\sum_{j=1}^{N}\bar{s}_j \approx 0, \quad Var(\bar{S}_j|F_N) = N^{-1}\sum_{j=1}^{N}\bar{s}_j^2 \approx \hat{\sigma}^2.$$

If bootstrapping samples $\{y_1^m, ..., y_N^m\}$ resemble the data $\{y_1, ..., y_N\}$ in terms of the empirical distribution function, F_N, then the parameter estimators are expected to be consistent. The modification of the standardized residuals allows each of the bootstrapping samples to have an empirical distribution with the same mean and variance as the original F_N. In a previous study [8], we used both definitions (normalized and non-normalized) for the standardized residuals (for linear and non-linear models) when carrying out our analysis. Although the definition of the standardized residual for the non-linear model gives an approximation to the conditions needed to hold, it performs comparably to the standardized residual as defined for the linear model. As a result, we use the non-linear standardized residuals for our simulation comparison studies summarized below.

3.6.3   Results of Numerical Simulations

We report on calculations reported in [8] for the logistic model. For that report, we created noisy data sets using model simulations and a time vector of length $N = 50$. The underlying logistic model with the true parameter values $\boldsymbol{\theta}_0 = (17.5, 0.7, 0.1)^T$ was solved for $f(t_j; \boldsymbol{\theta}_0) = x(t_j; \boldsymbol{\theta}_0)$ using the MATLAB function *ode45* where $\boldsymbol{\theta} = (K, r, x_0)^T$. A noise vector of length N with noise level σ_0 was taken from a random number generator for $\mathcal{N}(0, \sigma_0^2)$. The constant variance data sets were obtained from the equation

$$y_j = f(t_j; \boldsymbol{\theta}_0) + \tilde{\epsilon}_j.$$

Similarly, for non-constant variance data sets we used

$$y_j = f(t_j; \boldsymbol{\theta}_0)(1 + \tilde{\epsilon}_j).$$

Constant variance and non-constant variance data sets were created for 1%, 5% and 10% noise, i.e., $\sigma_0 = 0.01, 0.05$ and 0.1.

3.6.3.1   Constant Variance Data with OLS

We used the constant variance (CV) data with OLS to carry out the parameter estimation calculations. The bootstrapping estimates were computed with $M = 1000$. We use $M = 1000$ because we are computing confidence

TABLE 3.5:   Asymptotic and bootstrap OLS
estimates for CV data, $\sigma_0 = 0.01$.

θ	$\hat{\theta}$	SE($\hat{\theta}$)	95% CI
$\hat{K}_{asy}$	17.498576	0.002021	(17.494615, 17.502537)
$\hat{r}_{asy}$	0.700186	0.000553	(0.699103, 0.701270)
$(\hat{x}_0)_{asy}$	0.100044	0.000407	(0.099247, 0.100841)
$\hat{K}_{boot}$	17.498464	0.001973	(17.494597, 17.502331)
$\hat{r}_{boot}$	0.700193	0.000548	(0.699118, 0.701268)
$(\hat{x}_0)_{boot}$	0.100034	0.000399	(0.099252, 0.100815)

TABLE 3.6:   Asymptotic and bootstrap OLS
estimates for CV data, $\sigma_0 = 0.05$.

θ	$\hat{\theta}$	SE($\hat{\theta}$)	95% CI
$\hat{K}_{asy}$	17.486571	0.010269	(17.466444, 17.506699)
$\hat{r}_{asy}$	0.702352	0.002825	(0.696815, 0.707889)
$(\hat{x}_0)_{asy}$	0.098757	0.002050	(0.0947386, 0.102775)
$\hat{K}_{boot}$	17.489658	0.010247	(17.469574, 17.509742)
$\hat{r}_{boot}$	0.702098	0.002938	(0.696339, 0.707857)
$(\hat{x}_0)_{boot}$	0.0990520	0.002152	(0.094834, 0.103270)

TABLE 3.7:   Asymptotic and bootstrap OLS
estimates for CV data, $\sigma_0 = 0.1$.

θ	$\hat{\theta}$	SE($\hat{\theta}$)	95% CI
$\hat{K}_{asy}$	17.528701	0.019091	(17.491283, 17.566120)
$\hat{r}_{asy}$	0.699335	0.005201	(0.689140, 0.709529)
$(\hat{x}_0)_{asy}$	0.100650	0.003851	(0.093103, 0.108198)
$\hat{K}_{boot}$	17.523374	0.019155	(17.485829, 17.560918)
$\hat{r}_{boot}$	0.699803	0.005092	(0.689824, 0.709783)
$(\hat{x}_0)_{boot}$	0.100317	0.003800	(0.092869 0.107764)

intervals and not only estimates and standard errors, and Efron and Tibshirani [23] recommend that $M = 1000$ when confidence intervals are to be computed. The standard errors $\text{SE}_k(\hat{\theta})$ and corresponding confidence intervals $[\hat{\theta}_k - 1.96\text{SE}_k(\hat{\theta}), \hat{\theta}_k + 1.96\text{SE}_k(\hat{\theta})]$ are listed in Tables 3.5, 3.6 and 3.7. In Figure 3.14 we plot the empirical distributions for the case $\sigma_0 = 0.05$ corresponding to the results in Table 3.6; plots in the other two cases are quite similar.

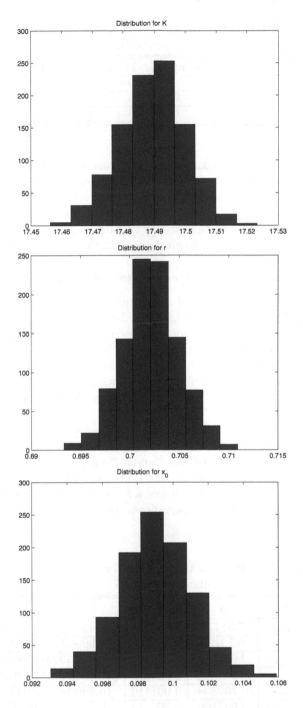

FIGURE 3.14: Bootstrap parameter distributions corresponding to 5% noise with CV.

TABLE 3.8: Computational times (sec) for asymptotic theory vs. bootstrapping.

Noise Level	Asymptotic Theory	Bootstrapping
1%	0.017320	4.285640
5%	0.009386	4.625428
10%	0.008806	4.914146

The parameter estimates and standard errors are comparable between the asymptotic theory and the bootstrapping theory for this case of constant variance. However, the computational times (given in Table 3.8) are two to three orders of magnitude greater for the bootstrapping method as compared to those for the asymptotic theory. For this reason, the asymptotic approach would appear to be the more advantageous method for this simple example.

3.6.3.2   Non-Constant Variance Data with GLS

We carried out a similar set of computations for the case of non-constant variance (NCV) data using a GLS formulation (based on previous computational tests, in these calculations we used only one GLS iteration). The bootstrap estimates were again computed with $M = 1000$. Standard errors and corresponding confidence intervals are listed in Tables 3.9, 3.10 and 3.11. In Figure 3.15 we plot the empirical distributions for the case $\sigma_0 = 0.05$ corresponding to the results in Table 3.10. Again, plots in the other two cases are quite similar.

We observe that the standard errors computed from the bootstrapping method are very similar to the standard errors computed using asymptotic theory. In each of these cases the standard errors for the parameter K are one to two orders of magnitude greater than the standard errors for r and x_0. The computational time is also slower for the bootstrapping method (as we can see from Table 3.12); thus asymptotic theory may again be the method of choice.

TABLE 3.9: Asymptotic and bootstrap GLS estimates for NCV data, $\sigma_0 = 0.01$.

θ	$\hat{\theta}$	SE($\hat{\theta}$)	95% CI
$\hat{K}_{asy}$	17.514706	0.028334	(17.459171, 17.570240)
$\hat{r}_{asy}$	0.70220	0.001156	(0.699934, 0.704465)
$(\hat{x}_0)_{asy}$	0.099145	0.000435	(0.098292, 0.099999)
$\hat{K}_{boot}$	17.515773	0.027923	(17.461045, 17.570502)
$\hat{r}_{boot}$	0.702136	0.001110	(0.699960, 0.704311)
$(\hat{x}_0)_{boot}$	0.099160	0.000416	(0.098344, 0.099976)

TABLE 3.10: Asymptotic and bootstrap GLS estimates for NCV data, $\sigma_0 = 0.05$.

θ	$\hat{\theta}$	SE($\hat{\theta}$)	95% CI
$\hat{K}_{asy}$	17.322554	0.148891	(17.030728,17.614380)
$\hat{r}_{asy}$	0.699744	0.006126	(0.687736, 0.711752)
$(\hat{x}_0)_{asy}$	0.099256	0.002313	(0.094723, 0.103790)
$\hat{K}_{boot}$	17.329282	0.146030	(17.043064,17.615500)
$\hat{r}_{boot}$	0.700060	0.006003	(0.688294,0.711825)
$(\hat{x}_0)_{boot}$	0.099210	0.002329	(0.094645,0.103775)

TABLE 3.11: Asymptotic and bootstrap GLS estimates for NCV data, $\sigma_0 = 0.1$.

θ	$\hat{\theta}$	SE($\hat{\theta}$)	95% CI
$\hat{K}_{asy}$	17.233751	0.294422	(16.656683,17.810818)
$\hat{r}_{asy}$	0.676748	0.011875	(0.653473, 0.700024)
$(\hat{x}_0)_{asy}$	0.109710	0.005015	(0.099880, 0.119540)
$\hat{K}_{boot}$	17.241977	0.275328	(16.702335, 17.781619)
$\hat{r}_{boot}$	0.676694	0.011845	(0.653479, 0.699909)
$(\hat{x}_0)_{boot}$	0.109960	0.005031	(0.100098, 0.119821)

3.6.4 Using Incorrect Assumptions on Errors

As we pointed out earlier, in practice one rarely knows the form of the statistical error with any degree of certainty, so that the assumed models (3.82) and (3.84) may well be incorrect for a given data set. To obtain some information on the effect, if any, of incorrect error model assumptions on comparisons between bootstrapping and use of asymptotic theory in computing standard errors, we carried out further computations.

To illustrate our concerns, we repeat the comparisons for asymptotic theory and bootstrapping generated standard errors, but with incorrect assumptions about the error (constant or non-constant variance). Using the same data sets as created previously, we computed parameter estimates and standard errors for the constant variance data for asymptotic theory and bootstrapping using a GLS formulation, which usually is employed if non-constant variance data is suspected. Similarly, we estimated parameters and standard errors for both approaches in an OLS formulation with the non-constant variance data.

In calculations presented below we see that if incorrect assumptions are made about the statistical model for error contained within the data, one cannot ascertain whether the correct assumption about the error has been made simply from examining the estimated standard errors. The residual plots must be examined to determine if the error model for constant or non-constant variance is reasonable. This is discussed more fully in [3] and [16, Chapter 3] as well as in the previous section on the logistic example.

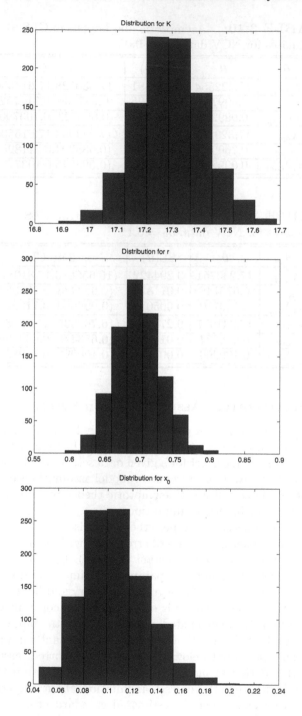

FIGURE 3.15:   Bootstrap parameter distributions for 5% noise with NCV.

TABLE 3.12:   Computational times (sec) for asymptotic theory vs. bootstrapping.

Noise Level	Asymptotic Theory	Bootstrapping
1%	0.030065	16.869108
5%	0.032161	21.549255
10%	0.037183	24.530157

3.6.4.1   Constant Variance Data Using GLS

The bootstrap estimates were computed with $M = 1000$ and one GLS iteration, with the findings reported in the tables below in a format similar to those in the previous sections. No empirical distribution plots are given here because we found that they added little in the way of notable new information.

As can be summarized from the results obtained, when the data has constant variance, but the incorrect assumption is made about the error, the asymptotic theory error estimates are again comparable to the bootstrapping error estimates. Also, K has large standard error as σ_0 increases (specifically it goes from approximately 0.043 to 0.5 for noise levels 1% to 10%, respectively). The r and x_0 standard errors also increase as the noise levels increase. These incorrectly obtained standard error estimates are larger in comparison to those obtained using an OLS formulation with the correct assumption about the error, comparing Tables 3.13–3.15 to Tables 3.5–3.7, respectively. Again, computational times for bootstrapping are considerably longer (see Table 3.16) than those for the asymptotic theory.

3.6.4.2   Non-Constant Variance Data Using OLS

The bootstrap estimates were again computed with $M = 1000$. When the error estimates for data sets with non-constant variance are estimated using

TABLE 3.13:   Asymptotic and bootstrap GLS estimates for CV data, $\sigma_0 = 0.01$.

θ	$\hat{\theta}$	SE($\hat{\theta}$)	95% CI
$\hat{K}_{asy}$	17.473446	0.043866	(17.387468, 17.559424)
$\hat{r}_{asy}$	0.706720	0.001802	(0.703188, 0.710252)
$(\hat{x}_0)_{asy}$	0.096926	0.000662	(0.095628, 0.098224)
$\hat{K}_{boot}$	17.471946	0.041954	(17.389716, 17.554176)
$\hat{r}_{boot}$	0.706738	0.001766	(0.703277, 0.710199)
$(\hat{x}_0)_{boot}$	0.096932	0.000655	(0.095649, 0.098215)

TABLE 3.14:   Asymptotic and bootstrap GLS estimates for CV data, $\sigma_0 = 0.05$.

θ	$\hat{\theta}$	SE($\hat{\theta}$)	95% CI
$\hat{K}_{asy}$	17.486405	0.169916	(17.153369, 17.819441)
$\hat{r}_{asy}$	0.696663	0.006939	(0.683063, 0.710263)
$(\hat{x}_0)_{asy}$	0.103291	0.002722	(0.097956, 0.108625)
$\hat{K}_{boot}$	17.477246	0.165693	(17.152487, 17.802004)
$\hat{r}_{boot}$	0.696828	0.006770	(0.683558, 0.710098)
$(\hat{x}_0)_{boot}$	0.103210	0.002548	(0.098215, 0.108205)

TABLE 3.15:   Asymptotic and bootstrap GLS estimates for CV data, $\sigma_0 = 0.1$.

θ	$\hat{\theta}$	SE($\hat{\theta}$)	95% CI
$\hat{K}_{asy}$	17.648739	0.504870	(16.659193, 18.638285)
$\hat{r}_{asy}$	0.680706	0.019755	(0.641985, 0.719427)
$(\hat{x}_0)_{asy}$	0.106576	0.008143	(0.090616, 0.122537)
$\hat{K}_{boot}$	17.668510	0.490496	(16.707137, 18.629882)
$\hat{r}_{boot}$	0.681164	0.018922	(0.644077, 0.718251)
$(\hat{x}_0)_{boot}$	0.106784	0.007835	(0.091427, 0.122140)

OLS, the estimates are comparable for the asymptotic theory and bootstrapping methods. In comparing these error estimates to the error estimates obtained under accurate assumptions (comparing Tables 3.17–3.19 to Tables 3.9–3.11, respectively), we observe that under the incorrect error model assumption, the standard error for K is always smaller (though comparable) regardless of the noise level, while the corresponding r and x_0 standard errors are always larger. For x_0, the standard error under OLS is an order of magnitude larger as compared to the GLS case. Corresponding computational times are shown in Table 3.20.

TABLE 3.16:   Computational times (sec) for asymptotic theory vs. bootstrapping.

Noise Level	Asymptotic Theory	Bootstrapping
1%	0.021631	17.416247
5%	0.019438	21.073697
10%	0.044052	27.152972

TABLE 3.17: Asymptotic and bootstrap OLS estimates for NCV data, $\sigma_0 = 0.01$.

θ	$\hat{\theta}$	SE($\hat{\theta}$)	95% CI
$\hat{K}_{asy}$	17.509101	0.024315	(17.461445, 17.556757)
$\hat{r}_{asy}$	0.706845	0.006750	(0.693615, 0.720076)
$(\hat{x}_0)_{asy}$	0.095928	0.004757	(0.086605, 0.105252)
$\hat{K}_{boot}$	17.511800	0.023465	(17.465808, 17.557791)
$\hat{r}_{boot}$	0.706690	0.006891	(0.693184, 0.720195)
$(\hat{x}_0)_{boot}$	0.096219	0.004871	(0.086671, 0.105767)

TABLE 3.18: Asymptotic and bootstrap OLS estimates for NCV data, $\sigma_0 = 0.05$.

θ	$\hat{\theta}$	SE($\hat{\theta}$)	95% CI
$\hat{K}_{asy}$	17.301393	0.122316	(17.061653, 17.541133)
$\hat{r}_{asy}$	0.694052	0.033396	(0.628596, 0.759508)
$(\hat{x}_0)_{asy}$	0.105893	0.025879	(0.055171, 0.156615)
$\hat{K}_{boot}$	17.291459	0.118615	(17.058973, 17.523945)
$\hat{r}_{boot}$	0.697367	0.034670	(0.629413, 0.765320)
$(\hat{x}_0)_{boot}$	0.106239	0.026765	(0.053780, 0.158697)

TABLE 3.19: Asymptotic and bootstrap OLS estimates for NCV data, $\sigma_0 = 0.1$.

θ	$\hat{\theta}$	SE($\hat{\theta}$)	95% CI
$\hat{K}_{asy}$	17.081926	0.262907	(16.566629, 17.597223)
$\hat{r}_{asy}$	0.727602	0.078513	(0.573717, 0.881487)
$(\hat{x}_0)_{asy}$	0.082935	0.047591	(-0.010343, 0.176213)
$\hat{K}_{boot}$	17.095648	0.250940	(16.603807, 17.587490)
$\hat{r}_{boot}$	0.733657	0.081852	(0.573228, 0.894087)
$(\hat{x}_0)_{boot}$	0.094020	0.054849	(-0.013484, 0.201524)

TABLE 3.20: Computational times (sec) for asymptotic theory vs. bootstrapping.

Noise Level	Asymptotic Theory	Bootstrapping
1%	0.008697	5.075417
5%	0.009440	6.002897
10%	0.013305	6.903042

3.7 The "Corrective" Nature of Bootstrapping Covariance Estimates and Their Effects on Confidence Intervals

Although in our example results above there are minimal differences between the two estimates obtained via bootstrapping and asymptotic theory (when used correctly) in the OLS case, Carroll, Wu and Ruppert [19] suggest bootstrapping should be preferred when estimating weights that have a large effect on the parameter estimates (i.e., particularly in the GLS case). They attribute this at least in part to the effect of a corrective term in the bootstrap estimate of the covariance matrix as given by

$$\text{Var}(\boldsymbol{\theta}_{\text{BOOT}}) = \hat{\Sigma}^N(\hat{\boldsymbol{\theta}}_{\text{BOOT}}) + N^{-1}\sigma_0^2\Lambda(F) + O_p(N^{-3/2}), \qquad (3.85)$$

where $\Lambda(F)$ is an unknown positive-definite matrix depending on the distribution F for the *i.i.d.* errors $\mathcal{E}_j$ and O_p is a (bounded) order term. As a result, the bootstrapping covariance matrix is generally thought to be more accurate than the asymptotic theory covariance estimate due to the corrective term $N^{-1}\sigma_0^2\Lambda(F)$ even though when estimating $\boldsymbol{\theta}_0$ using GLS, there may be more error in the estimated covariance matrix, $\hat{\Sigma}^N(\hat{\boldsymbol{\theta}})$, due to the estimation of weights. The corrective term, originally given in [31, p. 815], is discussed for linear models in [19, p. 1050], and for nonlinear models in [18, p. 28]. This corrective term does not arise with the OLS method where no weights are being estimated. Therefore, as might be expected, there seems to be no advantage in implementing the bootstrapping method over asymptotic theory for estimation of parameters and standard errors from constant variance data [21, p. 287].

To examine the effects of this second-order correction, we re-ran our analysis for $N = 10$ data points for non-constant variance (10% noise) using GLS. The results were similar comparing asymptotic theory to bootstrapping; however, the standard errors from bootstrapping were slightly smaller (less conservative) than the standard errors from asymptotic theory for all parameters. This is unexpected due to the correction term described above. We note that in repeated simulations different parameter estimates resulted in smaller (less conservative) SE some of the time, and for other times resulted in larger (more conservative) SE for bootstrapping than for the corresponding asymptotically computed SE. However, the bootstrapping standard errors were never all larger (more conservative) than the asymptotic theory standard errors. (We note that in general one desires small standard errors in assessing uncertainty in estimates but *only if they are accurate*; in some cases—essentially when they are more accurate—*more conservative* (larger) SE are desirable.) For fewer than $N = 10$ data points (as compared to $N = 50$ used in earlier calculations), the inverse problem using bootstrapping did not converge in a reasonable time period, though it did for asymptotic theory. This may be due to the resampling with replacement of the residuals causing the bootstrap samples to sometimes produce a large estimate of K, which in turn makes the estimated GLS weights very small.

In simulations reported in previous efforts [8], we found that our results did not necessarily meet expectations consistent with the theory presented by Carroll, Wu and Ruppert [19]. After further examination of the theory, we found that the theory is presented and discussed in detail for a linear model. To further explore this, we linearized our original model $f(t; \boldsymbol{\theta})$ about $\boldsymbol{\theta} = \boldsymbol{\theta}_0$, and re-ran our simulations. As a result of the linearization we have the following new model.

$$x_{lin}(t) = \chi_0(t) + \chi_1(t)(K - K_0) + \chi_2(t)(r - r_0) + \chi_3(t)(x_0 - x_{00}), \quad (3.86)$$

where $\chi_0 = f(t, \boldsymbol{\theta}_0)$, $\chi_1(t) = \dfrac{\partial f(t, \boldsymbol{\theta}_0)}{\partial K}$, $\chi_2(t) = \dfrac{\partial f(t, \boldsymbol{\theta}_0)}{\partial r}$, and $\chi_3(t) = \dfrac{\partial f(t, \boldsymbol{\theta}_0)}{\partial x_0}$. Note that $\boldsymbol{\theta}_0 = (K_0, r_0, x_{00}) = (17.5, 0.7, 0.1)$. We considered this new model and performed the same computational analysis as described earlier. At each simulation, while the values for the bootstrapping standard errors are similar, their specific values would vary in comparison to the asymptotic theory standard errors, which are the same for repeated simulations using the same set of simulated data each time. For example, during the first run all bootstrapping computed standard errors for (K, r, x_0) would be larger in comparison to the asymptotic theory estimates, while on the next run only K, r would have corresponding larger estimates. We performed a Monte Carlo analysis, for 1000 trials, to determine if the corrective nature of the bootstrap was present on average. For each of the Monte Carlo trials we computed the estimates and standard errors using asymptotic theory and bootstrapping with $M = 250$. We performed these Monte Carlo simulations on the same time interval $[0, 20]$, at 10% noise with relative error, generating new simulated data for each Monte Carlo simulation. Each Monte Carlo analysis used a fixed N for the 1000 trials, and the analysis was repeated for $N = 20, 25, 30, 35, 40, 45$ and 50 time points. The results for these Monte Carlo simulations are given in Tables 3.21–3.27.

When comparing this average bootstrapping standard error with the average asymptotic standard error for (K, r, x_0) at $N = 20, 25, 30, 35, 40, 45$ and 50, we observe that the bootstrapping estimates are larger (as expected with the correction term) for parameters K, r, but slightly smaller (not expected) for the parameter x_0 (see Tables 3.21–3.27). The relative error model has very little noise around $t = 0$, due to the model having the smallest function value in that region. Therefore, we do not expect the weights from GLS to have a large impact on the standard errors for x_0. This may explain why we do not observe a correction from the bootstrapping method for x_0.

We would expect that the theory presented in [19] would be observed at lower values of N due to the correction term $N^{-1}\sigma_0^2 \Lambda(F)$; however, the inverse problem was very unstable at this sample size because it appeared that at least

TABLE 3.21: Average of 1000 Monte Carlo trials of asymptotic theory and bootstrap ($M = 250$) estimates, and standard errors, using GLS estimates for NCV data, $\sigma_0 = 0.1$ and $N = 20$.

θ	$\hat{\theta}_{avg}$	$SE_{avg}(\hat{\theta})$
$\hat{K}_{asy}$	17.212625	0.533831
$\hat{r}_{asy}$	0.750631	0.021148
$(\hat{x}_0)_{asy}$	0.119780	0.008379
$\hat{K}_{boot}$	17.210893	0.556522
$\hat{r}_{boot}$	0.750690	0.022585
$(\hat{x}_0)_{boot}$	0.119753	0.008229

TABLE 3.22: Average of 1000 Monte Carlo trials of asymptotic theory and bootstrap ($M = 250$) estimates, and standard errors, using GLS estimates for NCV data, $\sigma_0 = 0.1$ and $N = 25$.

θ	$\hat{\theta}_{avg}$	$SE_{avg}(\hat{\theta})$
$\hat{K}_{asy}$	17.233834	0.479995
$\hat{r}_{asy}$	0.750298	0.019315
$(\hat{x}_0)_{asy}$	0.120060	0.007727
$\hat{K}_{boot}$	17.234113	0.501584
$\hat{r}_{boot}$	0.750304	0.020572
$(\hat{x}_0)_{boot}$	0.120055	0.007571

TABLE 3.23: Average of 1000 Monte Carlo trials of asymptotic theory and bootstrap ($M = 250$) estimates, and standard errors, using GLS estimates for NCV data, $\sigma_0 = 0.1$ and $N = 30$.

θ	$\hat{\theta}_{avg}$	$SE_{avg}(\hat{\theta})$
$\hat{K}_{asy}$	17.168470	0.436761
$\hat{r}_{asy}$	0.750720	0.017809
$(\hat{x}_0)_{asy}$	0.119995	0.007166
$\hat{K}_{boot}$	17.167208	0.456785
$\hat{r}_{boot}$	0.750792	0.018936
$(\hat{x}_0)_{boot}$	0.119964	0.007025

9 data points are needed for each parameter that is being estimated in this problem. We remark that in general the correlation between the number of data points required per parameter estimated is not easy to compute and is very much problem dependent. This is only one of a number of difficulties arising in such problems—see [17, p. 87] for discussions. Indeed, this problem

TABLE 3.24: Average of 1000 Monte Carlo trials of asymptotic theory and bootstrap ($M = 250$) estimates, and standard errors, using GLS estimates for NCV data, $\sigma_0 = 0.1$ and $N = 35$.

θ	$\hat{\theta}_{avg}$	$SE_{avg}(\hat{\theta})$
$\hat{K}_{asy}$	17.205855	0.403379
$\hat{r}_{asy}$	0.750168	0.016522
$(\hat{x}_0)_{asy}$	0.119871	0.006678
$\hat{K}_{boot}$	17.206793	0.423147
$\hat{r}_{boot}$	0.750167	0.017589
$(\hat{x}_0)_{boot}$	0.119886	0.006548

TABLE 3.25: Average of 1000 Monte Carlo trials of asymptotic theory and bootstrap ($M = 250$) estimates, and standard errors, using GLS estimates for NCV data, $\sigma_0 = 0.1$ and $N = 40$.

θ	$\hat{\theta}_{avg}$	$SE_{avg}(\hat{\theta})$
$\hat{K}_{asy}$	17.216539	0.379682
$\hat{r}_{asy}$	0.750290	0.015638
$(\hat{x}_0)_{asy}$	0.1198878	0.006340
$\hat{K}_{boot}$	17.215171	0.398576
$\hat{r}_{boot}$	0.750355	0.016641
$(\hat{x}_0)_{boot}$	0.119875	0.006212

TABLE 3.26: Average of 1000 Monte Carlo trials of asymptotic theory and bootstrap ($M = 250$) estimates, and standard errors, using GLS estimates for NCV data, $\sigma_0 = 0.1$ and $N = 45$.

θ	$\hat{\theta}_{avg}$	$SE_{avg}(\hat{\theta})$
$\hat{K}_{asy}$	17.220679	0.359348
$\hat{r}_{asy}$	0.750266	0.014858
$(\hat{x}_0)_{asy}$	0.119829	0.006038
$\hat{K}_{boot}$	17.219544	0.377269
$\hat{r}_{boot}$	0.750351	0.015801
$(\hat{x}_0)_{boot}$	0.119809	0.005919

is one of active research interest in the areas of sensitivity functions [4, 5, 6, 36] and design of experiments. We expected to observe theory-consistent SE at the higher values of N ($N = 20$), because we have three parameter estimates. For the parameters that did show a correction from bootstrapping, $\{K, r\}$, Table 3.28 shows the average correction term for each N, given by $SE_{avg}(\hat{\theta}_{BOOT}) - SE_{avg}(\hat{\theta}_{asy})$.

TABLE 3.27: Average of 1000 Monte Carlo trials of asymptotic theory and bootstrap ($M = 250$) estimates, and standard errors, using GLS estimates for NCV data, $\sigma_0 = 0.1$ and $N = 50$.

θ	$\hat{\theta}_{avg}$	$\text{SE}_{avg}(\hat{\theta})$
$\hat{K}_{asy}$	17.187670	0.337956
$\hat{r}_{asy}$	0.749043	0.014056
$(\hat{x}_0)_{asy}$	0.120329	0.005734
$\hat{K}_{boot}$	17.187487	0.3539129
$\hat{r}_{boot}$	0.749089	0.014936
$(\hat{x}_0)_{boot}$	0.120308	0.005619

TABLE 3.28: Average correction term: $\text{SE}_{avg}(\hat{\theta}_{boot}) - \text{SE}_{avg}(\hat{\theta}_{asy})$, from the average standard errors of 1000 Monte Carlo trials of asymptotic theory and bootstrap method ($M = 250$) using GLS estimates for NCV data, $\sigma_0 = 0.1$, for $\theta = \{K, r\}$, and N time points.

n	$\text{SE}_{avg}(\hat{K}_{boot}) - \text{SE}_{avg}(\hat{K}_{asy})$	$\text{SE}_{avg}(\hat{r}_{boot}) - \text{SE}_{avg}(\hat{r}_{asy})$
20	0.022691	0.001437
25	0.021588	0.001257
30	0.020024	0.001127
35	0.019769	0.001067
40	0.018893	0.001003
45	0.017921	0.000943
50	0.015957	0.000880

Table 3.28 confirms that there is a larger correction for smaller N. This is expected given the correction term $N^{-1}\sigma_0^2\Lambda(F)$ in (3.85). We also observe that the correction for K is consistently larger than the correction for r. This implies that the estimation of GLS weights has a greater effect on the standard errors of K than those for r.

To better understand the fluctuations in bootstrapping standard errors among the three parameters, we ran the Monte Carlo simulations for the estimation of just one parameter. When only the parameter r was estimated, the bootstrapping standard errors were larger than those of asymptotic theory for every case, using $N = 10, 15, 20$ and 25 time points. These simulations supported the theory of the corrective nature of bootstrapping. We also examined the standard errors when only estimating x_0, again for $N = 10, 15, 20$ and 25. For x_0, the standard errors were consistently smaller by an order of magnitude for the bootstrapping method, as compared to asymptotic theory, thus not displaying the corrective nature of the bootstrap. This example provides some evidence that the corrective nature of the bootstrap may only be realized in certain situations.

Carroll and Ruppert [18, p. 15–16] describe the following result for the behavior of GLS with a particular non-constant variance model. If the data points have higher variance in a neighborhood of importance (i.e., high sensitivity of observations to the specific parameter) for estimating a specific parameter (for our example the region of rapidly increasing slope is important for estimating r), then the weights in GLS heavily influence the estimate and the standard error. Based on our computational findings, we are inclined to think that this local variance also influences whether or not the bootstrapping estimate will be "corrective." If the GLS weights are not as important, then the bootstrap error estimate will not exhibit the corrective term properly.

We ran some simulations for both the linearized and non-linear models to test this hypothesis. For our example, in the solution to the logistic equation there are three regions, each of which corresponds to importance for estimating one of the parameters (i.e, exhibits great sensitivity to one of the parameters). Region I is important for estimation of x_0 and is the region where $x(t)$ is near x_0, located near the initial time points before the slope begins to increase significantly. Region II data heavily influences the estimation of r and is located in the area of the increasing slope. Region III is located where the solution approaches the steady state $x(\infty) = K$. Due to the manner in which our simulated non-constant variance data was modeled, Region I had little variation in data in comparison to Regions II and III. This low variation in Region I led us to believe that increasing the variance in this region would demonstrate the corrective nature of the bootstrapping estimate. We subsequently created data sets (linear and non-linear) with larger variance in the data in Region I while the variance remained the same in Regions II and III. For these new data sets all of the bootstrapping standard errors were larger than the asymptotic standard errors; however, the estimates for the intrinsic growth rate, r, became unreasonable. Although it appears the theoretically predicted corrective nature of the bootstrap is exhibited, there is a breakdown in the inverse problem due to (or perhaps it leads to) the unreasonable estimate for r. This breakdown occurs as a result of the added data variance in Region I, thus changing the assumptions on the variance model for the inverse problem, which are now incorrect. While there are issues in observing the effect of the addition of data variance into regions with small local sensitivity, the regions that naturally have a larger amount of local sensitivity exhibit the corrective nature of the bootstrap. This leads us to believe that *local data variability* strongly influences the presence or absence of the corrective nature of the bootstrapping estimate.

3.8   Some Summary Remarks on Asymptotic Theory vs. Bootstrapping

A brief survey of the limited theoretical foundations for the bootstrapping approach is given in an appendix to [8]. Based on our limited computational experience with the above known example, we make some summary remarks/ conclusions on asymptotic theory vs. bootstrapping, which are distinct methods for quantifying uncertainty in parameter estimates. Asymptotic theory is generally faster computationally than bootstrapping (certainly in simpler examples for which little difficulty arises in computing the sensitivity matrices needed for the covariance matrix calculations); however, this is the only definitive comparison that can be made. Depending on the type of data set, one method may be preferred over the other. For constant variance data using OLS there is no clear advantage in using bootstrapping over asymptotic theory. However, for a complex system it may be too complicated to compute the sensitivities needed for asymptotic theory; then bootstrapping may be more reasonable for estimating the standard error. If computation time is already a concern, but sensitivities can be computed, then asymptotic theory may be advantageous.

Given that the statistical model correctly assumes non-constant variance and the parameters are estimated using GLS, the asymptotic theory error estimates and the bootstrapping error estimates are comparable. If local variation in the data points is large in a region of importance for estimation of a parameter, then the bootstrapping covariance estimate may contain a corrective term of significance. In that situation, the standard errors for bootstrapping will be *larger* than those of asymptotic theory, and hence *more conservative* and also more *accurate*. If the local variation in the data is low in a region of importance for the estimation of a parameter, then there will be insignificant correction in the bootstrap's estimate of standard error over that of the asymptotic theory. Thus for non-constant variance data, the choice between bootstrapping and asymptotic theory depends on *local variation* in the data in regions of importance (high model sensitivity) for the estimation of the parameters.

References

[1] Takeshi Amemiya, Nonlinear regression models, Chapter 6 in *Handbook of Econometrics, Volume I*, Z. Griliches and M.D. Intriligator, Eds. North Holland, Amsterdam (1983), 333–389.

[2] H.T. Banks, Amanda Choi, Tori Huffman, John Nardini, Laura Poag and W. Clayton Thompson, Modeling CFSE label decay in flow cytometry data, CRSC-TR12-20, November, 2012; *Applied Mathematical Letters*, **26** (2013), 571–577.

[3] H. T. Banks, M. Davidian, J.R. Samuels, Jr. and K.L. Sutton, An in-

verse problem statistical methodology summary, CRSC-TR08-01, January, 2008; Chapter 11 in *Statistical Estimation Approaches in Epidemiology* (edited by Gerardo Chowell, Mac Hyman, Nick Hengartner, Luis M.A Bettencourt and Carlos Castillo-Chavez), Springer, Berlin, Heidelberg, New York, 2009, 249–302.

[4] H.T. Banks, S. Dediu and S.E. Ernstberger, Sensitivity functions and their uses in inverse problems, *J. Inverse and Ill-posed Problems*, **15** (2007), 683–708.

[5] H. T. Banks, S. Dediu, S.L. Ernstberger and F. Kappel, A new approach to optimal design problems, CRSC-TR08-12, September, 2008.

[6] H. T. Banks, S.L. Ernstberger and S.L. Grove, Standard errors and confidence intervals in inverse problems: sensitivity and associated pitfalls, *J. Inverse and Ill-posed Problems*, **15** (2007), 1–18.

[7] H.T. Banks and B.G. Fitzpatrick, Statistical methods for model comparison in parameter estimation problems for distributed systems, CAMS Tech. Rep. 89-4, September, 1989, University of Southern California; *J. Math. Biol.*, **28** (1990), 501–527.

[8] H.T. Banks, K. Holm and D. Robbins, Standard error computations for uncertainty quantification in inverse problems: Asymptotic theory vs. bootstrapping, CRSC-TR09-13, June, 2009; Revised August, 2009; Revised, May, 2010; *Mathematical and Computer Modeling*, **52** (2010), 1610–1625.

[9] H.T. Banks, D.F. Kapraun, W. Clayton Thompson, Cristina Peligero, Jordi Argilaguet and Andreas Meyerhans, A novel statistical analysis and interpretation of flow cytometry data, CRSC-TR12-23, December, 2012; *J. Biological Dynamics*, **7** (2013), 96–132.

[10] H.T. Banks, Z.R. Kenz and W.C. Thompson, An extension of RSS-based model comparison tests for weighted least squares, CRSC-TR12-18, August, 2012; *Intl. J. Pure and Appl. Math.*, **79** (2012), 155–183.

[11] H.T. Banks and K. Kunisch, *Estimation Techniques for Distributed Parameter Systems*, Birkhäuser, Boston, 1989.

[12] H.T. Banks, L. Potter and K.L. Rehm, Modeling plant growth using a system of enzyme kinetic equations, to appear.

[13] H.T. Banks, Karyn L. Sutton, W. Clayton Thompson, G. Bocharov, Marie Doumic, Tim Schenkel, Jordi Argilaguet, Sandra Giest, Cristina Peligero and Andreas Meyerhans, A new model for the estimation of cell proliferation dynamics using CFSE data, CRSC-TR11-05, Revised July 2011; *J. Immunological Methods*, **373** (2011), 143–160.

[14] H.T. Banks and W. Clayton Thompson, A division-dependent compartmental model with cyton and intracellular label dynamics, CRSC-TR12-

12, Revised August 2012; *Intl. J. Pure and Appl. Math*, **77** (2012), 119–147.

[15] H.T. Banks and W. Clayton Thompson, Mathematical models of dividing cell populations: Application to CFSE data, CRSC-TR12-10, April 2012; *J. Math. Modeling of Natural Phenomena*, **7** (2012), 24–52.

[16] H.T. Banks and H.T. Tran, *Mathematical and Experimental Modeling of Physical and Biological Processes*, CRC Press, Boca Raton, FL, 2009.

[17] D. M. Bates, *Nonlinear Regression Analysis and Its Applications*, John Wiley & Sons, Somerset, NJ, 1988.

[18] R.J. Carroll and D. Ruppert, *Transformation and Weighting in Regression*, Chapman & Hall, New York, 1988.

[19] R.J. Carroll, C.F.J. Wu and D. Ruppert, The effect of estimating weights in Weighted Least Squares, *J. Amer. Statistical Assoc.*, **83** (1988), 1045–1054.

[20] G. Casella and R.L. Berger, *Statistical Inference*, Duxbury, Pacific Grove, California, 2002.

[21] M. Davidian, *Nonlinear Models for Univariate and Multivariate Response*, ST 762 Lecture Notes, Chapters 2, 3, 9 and 11, 2007; http://www4.stat.ncsu.edu/ davidian/courses.html.

[22] M. Davidian and D. Giltinan, *Nonlinear Models for Repeated Measurement Data*, Chapman & Hall, London, 1998.

[23] B. Efron and R. J. Tibshirani, *An Introduction to the Bootstrap*, Chapman & Hall/CRC Press, Boca Raton, FL, 1998.

[24] M. Fink. myAD, Retrieved August 2011, from http://www.mathworks.com/matlabcentral/fileexchange/15235-automatic-differentiation-for-matlab.

[25] A.R. Gallant, *Nonlinear Statistical Models*, Wiley, New York, 1987.

[26] J. Hasenauer, D. Schittler and F. Allgöwer, A computational model for proliferation dynamics of division- and label-structured populations, **arXive.org**, arXiv:1202.4923v1, 22 February 2012.

[27] E.D. Hawkins, M.L. Turner, M.R. Dowling, C. van Gend and P.D. Hodgkin, A model of immune regulation as a consequence of randomized lymphocyte division and death times, *Proc. Natl. Acad. Sci. U.S.A.*, **104** (2007), 5032–5037.

[28] R.I. Jennrich, Asymptotic properties of non-linear least squares estimators, *Ann. Math. Statist.*, **40** (1969), 633–643.

[29] M. Kot, *Elements of Mathematical Ecology*, Cambridge University Press, Cambridge, 2001.

[30] T. Luzyanina, D. Roose, T. Schenkel, M. Sester, S. Ehl, A. Meyerhans and G. Bocharov, Numerical modelling of label-structured cell population growth using CFSE distribution data, *Theoretical Biology and Medical Modelling*, **4** (2007), published online.

[31] T. J. Rothenberg, Approximate normality of generalized least squares estimates, *Econometrica*, **52**(4) (1984), 811–825.

[32] W. Rudin, *Principles of Mathematical Analysis*, 2nd edition, McGraw-Hill, New York, 1964.

[33] D. Schittler, J. Hasenauer and F. Allgöwer, A generalized model for cell proliferation: Integrating division numbers and label dynamics, *Proc. Eighth International Workshop on Computational Systems Biology (WCSB 2011)*, June 2011, Zurich, Switzerland, p. 165–168.

[34] G.A.F. Seber and A.J. Lee, *Linear Regression Analysis*, J. Wiley & Sons, Hoboken, NJ, 2003.

[35] G.A.F. Seber and C.J. Wild, *Nonlinear Regression*, J. Wiley & Sons, Hoboken, NJ, 2003.

[36] K. Thomaseth and C. Cobelli, Generalized sensitivity functions in physiological system identification, *Annals of Biomedical Engineering*, **27** (1999), 607–616.

[37] W. Clayton Thompson, *Partial Differential Equation Modeling of Flow Cytometry Data from CFSE-based Proliferation Assays*, Ph.D. Dissertation, Dept. of Mathematics, North Carolina State University, Raleigh, December, 2011.

Chapter 4

Model Selection Criteria

Mathematical models play an important and indispensable role in studying and understanding the mechanisms, pathways, etc., in a complex dynamical system. These models are often described by ordinary differential equations, partial differential equations or delay differential equations. In practice, one may have a set of candidate mathematical models used to describe the dynamical system. The question is, which model is the best one to describe the investigated phenomena of interest based on the available experimental data? This question has received considerable interest from researchers, especially those in the statistics community. In this chapter, we give an introductory review of some widely used model selection criteria in the literature in the context of mathematical models.

4.1 Introduction

Before addressing our main task, we give some background material on model selection criteria. Basically, what does "model" mean in the context of model selection criteria, and on what principles are these selection criteria based?

4.1.1 Statistical and Probability Distribution Models

In the context of model selection criteria, "model" refers to a statistical model or a probability distribution model. Below we give a short introduction to these two types of models.

The statistical model refers to a model used to describe the observation process of a given dynamical system described by a mathematical model. It is used to make inferences on parameters of interest in the model. For example, if we have a scalar observation Y_j at each measurement time point t_j, then the statistical model is given by

$$Y_j = f(t_j; \mathbf{q}) + \mathcal{E}_j, \quad j = 1, 2, \ldots, N. \tag{4.1}$$

Here $f(t_j; \mathbf{q})$ corresponds to the observed part of the solution of a mathematical model with unknown model parameters $\mathbf{q} \in \mathbb{R}^{1 \times \kappa_q}$ at the measurement

117

time point t_j (without loss of generality, here we assume that the initial condition for the mathematical model is known and focus on the mathematical model parameters $\mathbf{q}$), $\mathcal{E}_j$ is the measurement error (random variable) at measurement time point t_j and N is the total number of observations. Equation (4.1) along with the specification of measurement errors is called the *statistical model* for the given system.

As discussed in the previous chapter, the measurement error $\mathcal{E}_j$ is a random variable. Hence, Y_j is also a random variable (and it is often referred to as the *response variable* in the statistics literature). In practice, it is often assumed that the mean of the measurement error is zero; that is, $\mathbb{E}(\mathcal{E}_j) = 0$, $j = 1, 2, \ldots, N$. This implies that the mean of the response variables is the observed part of the solution of the adopted mathematical model; that is,

$$\mathbb{E}(Y_j) = f(t_j; \mathbf{q}), \quad j = 1, 2, \ldots, N.$$

Hence, for this assumption the adopted mathematical model is implicitly assumed to be correct in describing the modeled system. In addition, it is often assumed that measurement errors $\mathcal{E}_j$, $j = 1, 2, \ldots, N$, are independent, especially in the case where one has little knowledge about the measurement errors. It should be noted that the measurement errors could be either independent of the mean of the response variables or dependent on the mean of the response variable (e.g., the simplest case is that the measurement errors are directly proportional to the mean of the response variables, i.e., $\mathcal{E}_j = \tilde{\mathcal{E}}_j f(t_j; \mathbf{q})$ with $\tilde{\mathcal{E}}_j, j = 1, 2, \ldots, N$, being random variables with constant variance).

If we know the probability distribution form of the measurement errors, then we can write down the probability density function for the observations. In this chapter, this probability density function is referred to as a *probability distribution model*. For example, if we assume that the measurement errors $\mathcal{E}_j$, $j = 1, 2, \ldots, N$, in (4.1) are i.i.d. $\mathcal{N}(0, \sigma^2)$ with σ being some unknown positive constant, then the probability density function p of observations $\mathbf{Y} = (Y_1, Y_2, \ldots, Y_N)^T$, given σ and mathematical model parameters $\mathbf{q}$, is

$$p(\mathbf{y}|\mathbf{q}, \sigma) = \prod_{j=1}^{N} \left[\frac{1}{\sqrt{2\pi}\sigma} \exp\left(-\frac{(y_j - f(t_j; \mathbf{q}))^2}{2\sigma^2} \right) \right], \qquad (4.2)$$

where $\mathbf{y} = (y_1, y_2, \ldots, y_N)^T$ is a realization of $\mathbf{Y}$. Hence, we see that the probability distribution model is just another way to describe the observation process (if we know the probability distribution form of the measurement errors). In this chapter, the parameters used to describe the measurement errors are termed *statistical model parameters* (e.g., for the above example (4.2), σ is the only statistical model parameter considered), and $\boldsymbol{\theta} \in \mathbb{R}^{\kappa_\theta}$ is used to denote a column vector of mathematical model parameters $\mathbf{q}$ and statistical model parameters that need to be estimated (e.g., for the above example (4.2), $\boldsymbol{\theta} = (\mathbf{q}, \sigma)^T$), where $\kappa_\theta > \kappa_q$ with $\kappa_\theta - \kappa_q$ being the dimension of the unknown statistical model parameters (note that $\boldsymbol{\theta}$ is defined differently from the one defined in the previous chapter, where it does not include the statistical model parameters as components but may include unknown initial conditions as well as mathematical model parameter **q**). We remark that these statistical model parameters are often called *nuisance parameters* (parameters that are needed to be estimated but are not the focus of one's study) in the statistics literature.

It is worth noting here that the probability density function $p(\mathbf{y}|\boldsymbol{\theta})$ can also be interpreted as a function of $\boldsymbol{\theta}$ given the sample outcomes $\mathbf{y}$. In this interpretation it is referred to as a *likelihood function* and is often denoted by $\mathcal{L}(\boldsymbol{\theta}|\mathbf{y})$. In other words, $\mathcal{L}(\boldsymbol{\theta}|\mathbf{y}) = p(\mathbf{y}|\boldsymbol{\theta})$. In practice, it is often found to be more convenient to work with the natural logarithm of the likelihood function (i.e., the *log-likelihood function*) than with the likelihood function. We remark that likelihood functions play an important role in the so-called maximum likelihood estimation method, which is another widely used parameter estimation method in the literature (with the obtained estimates referred to as *maximum likelihood estimates*). A short introduction to this estimation method will be given below in Section 4.2.2.

4.1.2 Risks Involved in the Process of Model Selection

There are roughly three types of errors often encountered during the process of choosing the best approximating model from a prior set of candidate models with the given experimental data. One is *numerical error* arising from numerically solving the mathematical model. We assume here that the numerical scheme employed is sufficiently accurate so that this type of error is negligible. Another error is *"modeling" error*. Two sources could be attributed to this type of error: one is mathematical model error, and the other is due to misspecification of measurement errors (i.e., specifying an inappropriate form for the probability distribution of measurement errors). The last type of error is *estimation error* caused by estimation of parameters in the model.

To have a basic and intuitive illustration of modeling error (often referred to as *bias* in the statistics literature) and estimation error (often referred to as *variance* in the statistics literature), we assume that the true model (i.e., the model that actually generates the data) is characterized by some parameter ϑ, and the approximating model is a special case of the true model. Then the total error between the true model and the approximating model can be characterized by (as demonstrated in Figure 4.1)

$$\|\vartheta - \hat{\boldsymbol{\theta}}_{\mathrm{MLE}}\|^2 = \underbrace{\|\vartheta - \boldsymbol{\theta}\|^2}_{\text{bias}} + \underbrace{\|\boldsymbol{\theta} - \hat{\boldsymbol{\theta}}_{\mathrm{MLE}}\|^2}_{\text{variance}}.$$

Here $\boldsymbol{\theta}$ is the projection of ϑ onto the parameter space $\Omega_\theta \subset \mathbb{R}^{\kappa_\theta}$ of the approximating model (this implies that bias decreases as the number of parame-

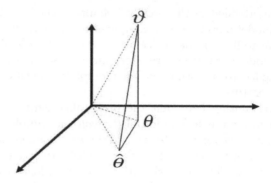

FIGURE 4.1:   A simple illustration of modeling error and estimation error, where $\hat{\theta} = \hat{\theta}_{\mathrm{MLE}}$.

ters κ_θ in the approximating model increases), and $\hat{\theta}_{\mathrm{MLE}}$ represents the maximum likelihood estimate of θ, with $\theta_{\mathrm{MLE}}(\mathbf{Y})$ denoting the maximum likelihood estimator of θ (i.e., $\hat{\theta}_{\mathrm{MLE}}$ is a realization of $\theta_{\mathrm{MLE}}(\mathbf{Y})$). For sufficiently large sample size N, $N\|\theta - \theta_{\mathrm{MLE}}(\mathbf{Y})\|^2$ is approximately chi-squared distributed with κ_θ degrees of freedom [15]. Note that the mean value of a chi-squared distributed random variable with κ_θ degrees of freedom is equal to κ_θ. Hence, for a fixed sample size, the variance (i.e., estimation error) increases as the number of parameters increases. This implies that we have less confidence in the accuracy of parameter estimates with more parameters involved in the model.

4.1.3   Model Selection Principle

There are numerous model selection criteria in the literature that can be used to select a best approximating model from the prior set of candidate models. These criteria are based on either hypothesis testing (e.g., log-likelihood ratio test, and residual sum of squares based model selection criterion, which will be discussed in Section 4.5) or mean squared error (e.g., Mallows' C_p [31]) or Bayes factors (such as Bayesian information criterion [34]) or information theory (e.g., Akaike information criterion as well as its variations [14, 15, 16, 27], which will be discussed in Section 4.2). But they all are based to some extent on the *principle of parsimony* (e.g., see [17]), which states there should be a trade off between lack-of-fit and complexity. The complexity reflects the difficulty in parameter estimation, and it can be viewed through the number of parameters in the model (the more parameters the

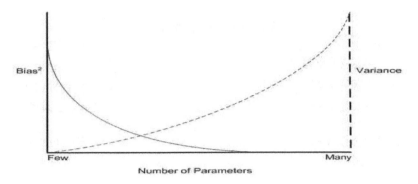

FIGURE 4.2: Principle of parsimony, where the complexity of the model is viewed through the number of parameters in the model.

more complex the model is) or through the complexity of the functional form or both. Figure 4.2 demonstrates the principle of parsimony, where the complexity is viewed through the number of the parameters in the model. From this figure, we see that model accuracy increases (i.e., bias decreases) as the number of parameters increases, which reflects the common understanding that the more complex the model is the more accurate the model is (with the tacit assumption that the approximation models are special cases of the unknown true model). But we also see from this figure that as the number of parameters increases we have less confidence in the estimate of the parameters. Thus, the goal in model selection is to minimize both modeling error (bias) and estimation error (variance).

4.2 Likelihood Based Model Selection Criteria – Akaike Information Criterion and Its Variations

Most of model selection criteria in the literature are based on the likelihood function. These types of model selection criteria are referred to as likelihood based model selection criteria, and include one of the most widely used model selection criteria, the Akaike information criterion (AIC). In 1973, Akaike found a relationship between Kullback–Leibler information and the maximum value of the log-likelihood function of a given approximating model, and this relationship is referred to as the *Akaike information criterion*. Before we introduce the AIC, we give a brief review of Kullback–Leibler information and maximum likelihood estimation.

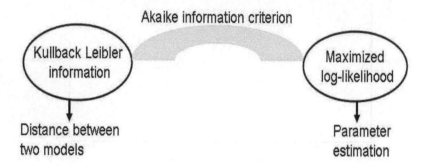

FIGURE 4.3: Akaike information criterion: relationship between Kullback–Leibler information and the maximum value of the log-likelihood function of a given approximating model.

4.2.1 Kullback–Leibler Information

Kullback–Leibler (K–L) information [29] is a well-known measure of "distance" between two probability distribution models. Hence, it can be used to measure information lost when an approximating probability distribution model is used to approximate the true probability distribution model. Let p_0 denote the probability distribution model that actually generates the data (that is, p_0 is the true probability density function of observations **Y**), and p be a probability distribution model that is presumed to generate the data. In addition, p is assumed to be dependent on parameter vector $\boldsymbol{\theta} \in \mathbb{R}^{\kappa_\theta}$ (that is, $p(\cdot|\boldsymbol{\theta})$ is the specified probability density function of observations **Y**, and is used to approximate the true model p_0). Then the K–L information between these two models is given by

$$
\begin{aligned}
\mathcal{I}(p_0, p(\cdot|\boldsymbol{\theta})) &= \int_{\Omega_y} p_0(\mathbf{y}) \ln\left(\frac{p_0(\mathbf{y})}{p(\mathbf{y}|\boldsymbol{\theta})}\right) d\mathbf{y} \\
&= \int_{\Omega_y} p_0(\mathbf{y}) \ln(p_0(\mathbf{y})) d\mathbf{y} - \int_{\Omega_y} p_0(\mathbf{y}) \ln(p(\mathbf{y}|\boldsymbol{\theta})) d\mathbf{y},
\end{aligned}
\tag{4.3}
$$

where Ω_y denotes the set of all possible values for **y**, and the second term on the right side, $\int_{\Omega_y} p_0(\mathbf{y}) \ln(p(\mathbf{y}|\boldsymbol{\theta})) d\mathbf{y}$, is referred to as the *relative K–L information*.

K–L information has a number useful properties. For example, for any probability density function p with given $\boldsymbol{\theta}$ we have

$$
\mathcal{I}(p_0, p(\cdot|\boldsymbol{\theta})) \geq 0,
$$
$$
\mathcal{I}(p_0, p(\cdot|\boldsymbol{\theta})) = 0 \text{ if and only if } p_0 = p(\cdot|\boldsymbol{\theta}) \text{ a.e.}
\tag{4.4}
$$

This property implies that a good approximating model is one with low K–L information. The property of $\mathcal{I}(p_0, p(\cdot|\boldsymbol{\theta})) \geq 0$ can be easily derived based on Jensen's inequality (2.67). The proof is given as follows. Let $\psi(z) = -\ln(z)$, and let the random variable $Z = \dfrac{p(\mathbf{Y}|\boldsymbol{\theta})}{p_0(\mathbf{Y})}$. Then it is easy to verify that ψ is a convex function. Note that

$$\psi(\mathbb{E}(Z)) = -\ln\left(\int_{\Omega_y} p_0(\mathbf{y})\frac{p(\mathbf{y}|\boldsymbol{\theta})}{p_0(\mathbf{y})}\,d\mathbf{y}\right) = -\ln(1) = 0,$$

and

$$\mathbb{E}(\psi(Z)) = -\int_{\Omega_y} p_0(\mathbf{y})\ln\left(\frac{p(\mathbf{y}|\boldsymbol{\theta})}{p_0(\mathbf{y})}\right)\,d\mathbf{y} = \mathcal{I}(p_0, p(\cdot|\boldsymbol{\theta})),$$

where the expectation $\mathbb{E}$ in the above equations is taken with respect to the true probability density function p_0 of observations $\mathbf{Y}$. Hence, by Jensen's inequality we have $\mathcal{I}(p_0, p(\cdot|\boldsymbol{\theta})) \geq 0$.

It should be noted that K–L information is not symmetric (that is, $\mathcal{I}(p_0, p(\cdot|\boldsymbol{\theta})) \neq \mathcal{I}(p(\cdot|\boldsymbol{\theta}), p_0)$), and it does not satisfy the usual triangle inequality (e.g., see [38]). Hence, K–L information does not provide a metric on the space of probability density functions and thus does not give a real "distance" between two probability distribution models.

4.2.2 Maximum Likelihood Estimation

In reality, K–L information cannot be directly calculated, as the true model p_0 is unknown and the parameter $\boldsymbol{\theta}$ in p must be estimated from the experimental data. As we remarked earlier, maximum likelihood estimation is another widely used parameter estimation method in the literature. Basically this method obtains the parameter estimate by maximizing the likelihood function or log-likelihood function with respect to the parameter $\boldsymbol{\theta}$ (note that maximizing the likelihood function is equivalent to maximizing the log-likelihood function due to the monotonicity of the logarithm function). Hence, we see that for this method one must know the probability density function of observations $\mathbf{Y}$. This is unfortunately often unavailable to investigators. However, under certain regularity conditions the maximum likelihood estimator $\boldsymbol{\theta}_{\mathrm{MLE}}(\mathbf{Y})$ of $\boldsymbol{\theta}$ (based on the specified model p and observations $\mathbf{Y}$) has a number of important and useful limiting properties, such as asymptotic normality and consistency even when the probability distribution model of observations is misspecified (i.e., when there is no $\boldsymbol{\theta}$ such that $p(\cdot|\boldsymbol{\theta}) \equiv p_0$). It is worth noting here that under the misspecified case the likelihood function is sometimes referred to as a *quasi-likelihood function*.

For the misspecified case the *consistency* property means that the maximum likelihood estimator converges in probability to the so-called *pseudo-true value*, which is defined by

$$\boldsymbol{\theta}_{\mathrm{PT}} = \arg\min_{\boldsymbol{\theta}\in\Omega_\theta} \mathcal{I}(p_0, p(\cdot|\boldsymbol{\theta})), \qquad (4.5)$$

that is, the pseudo-true value is the one that minimizes K–L information. (It is worth noting that this was observed by Akaike when he derived the AIC.) Here $\Omega_\theta = \Omega_q \times \Omega_{sp}$, with Ω_q being a compact set in $\mathbb{R}^{\kappa_q}$ and Ω_{sp} being a compact set in $\mathbb{R}^{\kappa_\theta - \kappa_q}$, and $\boldsymbol{\theta}_{\mathrm{PT}}$ is assumed to be an interior point of Ω_θ. It should be noted that $\boldsymbol{\theta}_{\mathrm{PT}}$ depends on the sample size, as p_0 is the true probability density function of a finite number of observations and $p(\cdot|\boldsymbol{\theta})$ is the corresponding specified one. But for simplicity, we suppress this dependency.

The *asymptotic normality* property in the misspecified case states that $\boldsymbol{\theta}_{\mathrm{MLE}}(\mathbf{Y})$ is asymptotically normally distributed with mean and variance respectively given by

$$\mathbb{E}(\boldsymbol{\theta}_{\mathrm{MLE}}(\mathbf{Y})) = \boldsymbol{\theta}_{\mathrm{PT}}, \quad \mathrm{Var}(\boldsymbol{\theta}_{\mathrm{MLE}}(\mathbf{Y})) = (H(\boldsymbol{\theta}_{\mathrm{PT}}))^{-1} F(\boldsymbol{\theta}_{\mathrm{PT}})(H(\boldsymbol{\theta}_{\mathrm{PT}}))^{-1}.$$
$$(4.6)$$

Here $F(\boldsymbol{\theta}_{\mathrm{PT}}) \in \mathbb{R}^{\kappa_\theta \times \kappa_\theta}$ and $H(\boldsymbol{\theta}_{\mathrm{PT}}) \in \mathbb{R}^{\kappa_\theta \times \kappa_\theta}$ are given by

$$F(\boldsymbol{\theta}_{\mathrm{PT}}) = \mathbb{E}_{\mathbf{Y}}\left(\left(\frac{\partial}{\partial\boldsymbol{\theta}} \ln(p(\mathbf{Y}|\boldsymbol{\theta}_{\mathrm{PT}})) \right) \left(\frac{\partial}{\partial\boldsymbol{\theta}} \ln(p(\mathbf{Y}|\boldsymbol{\theta}_{\mathrm{PT}})) \right)^T \right), \qquad (4.7)$$

and

$$H(\boldsymbol{\theta}_{\mathrm{PT}}) = \mathbb{E}_{\mathbf{Y}}\left(-\left(\frac{\partial^2 \ln(p(\mathbf{Y}|\boldsymbol{\theta}_{\mathrm{PT}}))}{\partial\theta_i\partial\theta_j} \right)_{\kappa_\theta \times \kappa_\theta} \right), \qquad (4.8)$$

where $\frac{\partial}{\partial\boldsymbol{\theta}} \ln(p(\mathbf{X}|\boldsymbol{\theta}_{\mathrm{PT}}))$ denotes the gradient of $\ln(p(\mathbf{X}|\boldsymbol{\theta}_{\mathrm{PT}}))$ with respect to $\boldsymbol{\theta}$; that is, it is a κ_θ-dimensional column vector with its ith element defined by $\frac{\partial}{\partial\theta_i} \ln(p(\mathbf{X}|\boldsymbol{\theta}_{\mathrm{PT}}))$, $i = 1, 2, 3 \ldots, \kappa_\theta$. The expectation $\mathbb{E}_{\mathbf{Y}}$ in (4.7) and (4.8) is used to emphasize that the expectation is taken with respect to the true probability density function p_0 of observations $\mathbf{Y}$.

It is worth noting that if the specified probability distribution model $p(\cdot|\boldsymbol{\theta})$ for the observations $\mathbf{Y}$ is correctly specified (i.e., there exists a unique $\boldsymbol{\theta}_0$ such that $p(\cdot|\boldsymbol{\theta}_0) \equiv p_0$), then under certain regularity conditions it can be shown that the pseudo-true value $\boldsymbol{\theta}_{\mathrm{PT}}$ is equal to $\boldsymbol{\theta}_0$ and $H(\boldsymbol{\theta}_{\mathrm{PT}}) = F(\boldsymbol{\theta}_{\mathrm{PT}})$. By (4.6) we see that for this case we have $\boldsymbol{\theta}_{\mathrm{MLE}}(\mathbf{Y})$ is asymptotically normally distributed with mean and variance respectively given by

$$\mathbb{E}(\boldsymbol{\theta}_{\mathrm{MLE}}(\mathbf{Y})) = \boldsymbol{\theta}_{\mathrm{PT}}, \quad \mathrm{Var}(\boldsymbol{\theta}_{\mathrm{MLE}}(\mathbf{Y})) = (H(\boldsymbol{\theta}_{\mathrm{PT}}))^{-1} = (F(\boldsymbol{\theta}_{\mathrm{PT}}))^{-1}. \quad (4.9)$$

The interested reader can refer to [20, 26] for more information on maximum likelihood estimation under correct model specification, and to [22, 24, 37, 38], [23, Chapter 3] and [17, Chapter 7] for more information on parameter estimation under model misspecification.

4.2.3 A Large Sample AIC

As we remarked earlier, the AIC is based on K–L information theory, which measures the "distance" between two probability distribution models. Hence, in establishing the AIC, the maximum likelihood estimation method is used for parameter estimation. Note that the K–L information $\mathcal{I}(p_0, p(\cdot|\boldsymbol{\theta}_{\mathrm{MLE}}(\mathbf{Y})))$ is a random variable (inherited from the fact that $\boldsymbol{\theta}_{\mathrm{MLE}}(\mathbf{Y})$ is a random vector). Hence, we need to use expected K–L information $\mathbb{E}_{\mathbf{Y}}\left(\mathcal{I}(p_0, p(\cdot|\boldsymbol{\theta}_{\mathrm{MLE}}(\mathbf{Y})))\right)$ to measure the "distance" between p and p_0, where $\mathbb{E}_{\mathbf{Y}}$ denotes the expectation with respect to the true probability density function p_0 of observations $\mathbf{Y}$. Based on the K–L information property (4.4), the selected best approximating model is the one that solves

$$\min_{p \in \mathbb{P}} \mathbb{E}_{\mathbf{Y}}\left(\mathcal{I}(p_0, p(\cdot|\boldsymbol{\theta}_{\mathrm{MLE}}(\mathbf{Y})))\right) \qquad (4.10)$$

where $\mathbb{P}$ is a given prior set of candidate models (defined by the presumed probability density function of observations $\mathbf{Y}$). Note that we can rewrite expected K–L information as

$$\mathbb{E}_{\mathbf{Y}}\left(\mathcal{I}(p_0, p(\cdot|\boldsymbol{\theta}_{\mathrm{MLE}}(\mathbf{Y})))\right) = \int_{\Omega_y} p_0(\mathbf{x}) \ln(p_0(\mathbf{x})) d\mathbf{x}$$
$$- \int_{\Omega_y} p_0(\mathbf{y}) \left[\int_{\Omega_y} p_0(\mathbf{x}) \ln(p(\mathbf{x}|\boldsymbol{\theta}_{\mathrm{MLE}}(\mathbf{y}))) d\mathbf{x}\right] d\mathbf{y},$$

where $\boldsymbol{\theta}_{\mathrm{MLE}}(\mathbf{Y})$ is the MLE estimator of $\boldsymbol{\theta}$ based on model p and observations $\mathbf{Y}$, and $\boldsymbol{\theta}_{\mathrm{MLE}}(\mathbf{y})$ is the corresponding MLE estimate of $\boldsymbol{\theta}$ (that is, $\boldsymbol{\theta}_{\mathrm{MLE}}(\mathbf{y})$ is a realization of $\boldsymbol{\theta}_{\mathrm{MLE}}(\mathbf{Y})$). We further observe that the first term on the right side of the above equation is an unknown constant (i.e., it does not depend on p). Hence, our model selection target can be equivalently written as

$$\max_{p \in \mathbb{P}} \mathbb{E}_{\mathbf{Y}}\mathbb{E}_{\mathbf{X}}\left(\ln(p(\mathbf{X}|\boldsymbol{\theta}_{\mathrm{MLE}}(\mathbf{Y})))\right),$$

where

$$\mathbb{E}_{\mathbf{Y}}\mathbb{E}_{\mathbf{X}}\left(\ln(p(\mathbf{X}|\boldsymbol{\theta}_{\mathrm{MLE}}(\mathbf{Y})))\right) = \int_{\Omega_y} p_0(\mathbf{y}) \left[\int_{\Omega_y} p_0(\mathbf{x}) \ln(p(\mathbf{x}|\boldsymbol{\theta}_{\mathrm{MLE}}(\mathbf{y}))) d\mathbf{x}\right] d\mathbf{y}.$$

It was shown (e.g., see [17] for details) that for a *large sample* and *"good"* *model* (a model that is close to p_0 in the sense of having a small K–L value) we have

$$\mathbb{E}_{\mathbf{Y}}\mathbb{E}_{\mathbf{X}}\left(\ln(p(\mathbf{X}|\boldsymbol{\theta}_{\mathrm{MLE}}(\mathbf{Y})))\right) \approx \ln(\mathcal{L}(\hat{\boldsymbol{\theta}}_{\mathrm{MLE}}|\mathbf{y})) - \kappa_\theta. \qquad (4.11)$$

Here $\hat{\boldsymbol{\theta}}_{\mathrm{MLE}}$ is the maximum likelihood estimate of $\boldsymbol{\theta}$ given sample outcomes $\mathbf{y}$ (that is, $\hat{\boldsymbol{\theta}}_{\mathrm{MLE}} = \boldsymbol{\theta}_{\mathrm{MLE}}(\mathbf{y})$), $\mathcal{L}(\hat{\boldsymbol{\theta}}_{\mathrm{MLE}}|\mathbf{y})$ represents the likelihood of $\hat{\boldsymbol{\theta}}_{\mathrm{MLE}}$ given sample outcomes $\mathbf{y}$ (that is, $\mathcal{L}(\hat{\boldsymbol{\theta}}_{\mathrm{MLE}}|\mathbf{y}) = p(\mathbf{y}|\hat{\boldsymbol{\theta}}_{\mathrm{MLE}})$) and κ_θ is the total

number of estimated parameters (including mathematical model parameters
q and statistical model parameters). It is worth pointing out here that having
a *large sample* and *"good" model* are used to ensure that the estimate $\hat{\boldsymbol{\theta}}_{\text{MLE}}$
provides a good approximation to the true value ϑ (involving the consistency
property of the maximum likelihood estimator).

For historical reasons, Akaike multiplied (4.11) by -2 to obtain his criterion,
which is given by

$$\text{AIC} = -2\ln\mathcal{L}(\hat{\boldsymbol{\theta}}_{\text{MLE}}|\mathbf{y}) + 2\kappa_\theta. \tag{4.12}$$

We note that the first term in the AIC is a measure of the goodness-of-fit of the
approximating model, and the second term gives a measure of the complexity
of the approximating model (i.e., the reliability of the parameter estimation
of the model). Thus, we see that for the AIC the complexity of a model is
viewed as the number of parameters in the model.

Based on the above discussion, we see that to use the AIC to select a best
approximating model from a given prior set of candidate models, we need to
calculate the AIC value for each model in the set, and the "best" model is the
one with the minimum AIC value. Note that the value of the AIC depends on
data, which implies that we may select a different best approximating model
if a different data set arising from the same experiment is used. Hence, the
AIC values must be calculated for all the models being compared by using
the *same data set*.

4.2.4   A Small Sample AIC

The discussion in Section 4.2.3 reveals that one of the assumptions made
in the derivation of the AIC is that the sample size must be large. Hence, the
AIC may perform poorly if the sample size N is small relative to the total
number of estimated parameters. It is suggested in [17] that the AIC can
be used only if the sample size N is at least 40 times the total number of
estimated parameters (that is, $N/\kappa_\theta \geq 40$). In this section, we introduce a
small sample AIC (denoted by AIC_c) that can be used in the case where N
is small relative to κ_θ.

4.2.4.1   Univariate Observations

The AIC_c was originally proposed in [35] for a scalar linear regression model,
and then was extended in [27] for a scalar non-linear regression model based
on asymptotic theory. In deriving the AIC_c, it was assumed in [27] that the
measurement errors $\mathcal{E}_j$, $j = 1, 2, \ldots, N$, are independent and normally distributed with mean zero and variance σ^2. In addition, the true model p_0 was
assumed to be known, with measurement errors being independent and normally distributed with zero mean and variance σ_0^2. With these assumptions,
the small sample AIC is given by

$$\text{AIC}_c = \text{AIC} + \frac{2\kappa_\theta(\kappa_\theta + 1)}{N - \kappa_\theta - 1}, \tag{4.13}$$

where the last term on the right-hand side of the above equation is often referred to as the *bias-correction* term. We observe that as the sample size $N \to \infty$ this bias-correction term approaches zero, and the resultant criterion is just the usual AIC.

It should be noted that the bias-correction term in (4.13) changes if a different probability distribution (e.g., exponential, Poisson) is assumed for the measurement errors. However, it was suggested in [17] that in practice AIC_c given by (4.13) is generally suitable unless the underlying probability distribution is extremely non-normal, especially in terms of being strongly skewed.

4.2.4.2 Multivariate Observations

The AIC_c in the multivariate observation case was derived in [14]. All the specified statistical models are assumed to have the following form:

$$\mathbf{Y} = \mathbf{A}\boldsymbol{\beta} + \boldsymbol{\mathcal{E}},$$

where $\mathbf{Y} \in \mathbb{R}^{N \times \nu}$, with ν being the total number of observed components and N being the number of observations for each observed component (that is, all the components have the same number of observations). Here $\mathbf{A} \in \mathbb{R}^{N \times \tilde{k}}$ is a known matrix of covariate values, and $\boldsymbol{\beta} \in \mathbb{R}^{\tilde{k} \times \nu}$ is a matrix of unknown model parameters (i.e., there are $\tilde{k}\nu$ unknown mathematical model parameters). Moreover, $\boldsymbol{\mathcal{E}} \in \mathbb{R}^{N \times \nu}$ is the error matrix with the rows assumed to be independent and multivariate normally distributed with mean zero and covariance matrix $\boldsymbol{\Sigma}$. That is, $\boldsymbol{\mathcal{E}}_i \sim \mathcal{N}(\mathbf{0}_\nu, \boldsymbol{\Sigma})$, $i = 1, 2, \ldots, N$, are independent, where $\boldsymbol{\mathcal{E}}_i$ is the ith row of $\boldsymbol{\mathcal{E}}$, and $\mathbf{0}_\nu$ is a ν-dimensional column zero vector. In addition, it is assumed that the true statistical model has the form

$$\mathbf{Y} = \mathbf{A}_0\boldsymbol{\beta}_0 + \boldsymbol{\mathcal{E}}_0,$$

where $\mathbf{A}_0 \in \mathbb{R}^{N \times \tilde{k}_0}$ is a known matrix of covariate values, $\boldsymbol{\beta}_0 \in \mathbb{R}^{\tilde{k}_0 \times \nu}$ is a matrix of true model parameters, and $\boldsymbol{\mathcal{E}}_0 \in \mathbb{R}^{N \times \nu}$ is the error matrix. Its rows are assumed to be independent and multivariate normally distributed with mean zero and covariance matrix $\boldsymbol{\Sigma}_0$. That is, $\boldsymbol{\mathcal{E}}_{0i} \sim \mathcal{N}(\mathbf{0}_\nu, \boldsymbol{\Sigma}_0)$, $i = 1, 2, \ldots, N$, are independent, where $\boldsymbol{\mathcal{E}}_{0i}$ is the ith row of $\boldsymbol{\mathcal{E}}_0$, and $\mathbf{0}_\nu$ is a ν-dimensional column zero vector. With these assumptions on the approximating models and the true model, as derived in [14] the AIC_c is given by

$$\text{AIC}_c = \text{AIC} + 2\frac{\kappa_\theta(\tilde{k} + 1 + \nu)}{N - \tilde{k} - 1 - \nu}, \tag{4.14}$$

where κ_θ is the total number of unknown parameters. This is given by $\kappa_\theta = \tilde{k}\nu + \nu(\nu + 1)/2$, where $\tilde{k}\nu$ is the total number of mathematical model parameters (i.e., the number of elements in $\boldsymbol{\beta}$), and $\nu(\nu + 1)/2$ is the total number of statistical model parameters (as $\boldsymbol{\Sigma} \in \mathbb{R}^{\nu \times \nu}$ is symmetric). We see that if $\nu = 1$ (i.e., the case with univariate observations), then $\kappa_\theta = \tilde{k} + 1$ and (4.14) becomes

$$\text{AIC}_c = \text{AIC} + 2\frac{\kappa_\theta(\tilde{k} + 1 + 1)}{N - \tilde{k} - 1 - 1},$$

which is just (4.13) since $\kappa_\theta = \tilde{k} + 1$. Thus, (4.14) includes (4.13) as a special case with $\nu = 1$.

Although the AIC_c in (4.14) is derived based on the assumption that all the approximating models are multivariate linear regression models, Bedrick and Tsai [14] claimed that this result can be extended to the multivariate nonlinear regression model.

4.2.5   Takeuchi's Information Criterion

In Akaike's original derivation of the AIC, he made the assumption that the candidate model set $\mathbb{P}$ includes the true data-generation model p_0. This strong assumption has been the subject of criticism. In 1976, Takeuchi gave a general derivation of the AIC. In his derivation, he did not make this strong assumption (that is, p_0 is not required to be in the set $\mathbb{P}$), and the resulting formula is now known as *Takeuchi's information criterion* (TIC). This criterion was thought to be useful in cases where the approximation models are not particularly close to the true model p_0.

The discussion in Section 4.2.3 revealed that the selected model is the one that results from finding

$$\max_{p \in \mathbb{P}} \mathbb{E}_\mathbf{Y} \mathbb{E}_\mathbf{X} \left(\ln(p(\mathbf{X}|\boldsymbol{\theta}_{\text{MLE}}(\mathbf{Y})))\right).$$

It was found (e.g., see [17]) by Takeuchi that an approximately unbiased estimate of $\mathbb{E}_\mathbf{Y} \mathbb{E}_\mathbf{X} \left(\ln(p(\mathbf{X}|\boldsymbol{\theta}_{\text{MLE}}(\mathbf{Y})))\right)$ for a *large sample* is given by

$$\ln(\mathcal{L}(\hat{\boldsymbol{\theta}}_{\text{MLE}}|\mathbf{y})) - \text{Tr}(F(\boldsymbol{\theta}_{\text{PT}})H(\boldsymbol{\theta}_{\text{PT}})^{-1}).$$

Here $\boldsymbol{\theta}_{\text{PT}}$ denotes the pseudo-true value of $\boldsymbol{\theta}$, matrix functions F and H are given by (4.7) and (4.8), respectively, and $\text{Tr}(\cdot)$ denotes the trace of a matrix. We note that $H(\boldsymbol{\theta}_{\text{PT}})$ and $F(\boldsymbol{\theta}_{\text{PT}})$ cannot be calculated directly, as both p_0 and $\boldsymbol{\theta}_{\text{PT}}$ are unknown. However, if the sample size is large enough, then $F(\boldsymbol{\theta}_{\text{PT}})$ and $H(\boldsymbol{\theta}_{\text{PT}})$ can be approximated by

$$\widehat{F}(\hat{\boldsymbol{\theta}}_{\text{MLE}}) = \left[\frac{\partial}{\partial\boldsymbol{\theta}}\ln(p(\mathbf{y}|\hat{\boldsymbol{\theta}}_{\text{MLE}}))\right]\left[\frac{\partial}{\partial\boldsymbol{\theta}}\ln(p(\mathbf{y}|\hat{\boldsymbol{\theta}}_{\text{MLE}}))\right]^T$$

and

$$\widehat{H}(\hat{\boldsymbol{\theta}}_{\text{MLE}}) = -\left(\frac{\partial^2 \ln(p(\mathbf{y}|\hat{\boldsymbol{\theta}}_{\text{MLE}}))}{\partial\theta_i\theta_j}\right)_{\kappa_\theta \times \kappa_\theta},$$

respectively. With these approximations, Takeuchi obtained his criterion:

$$\text{TIC} = -2\ln(\mathcal{L}(\hat{\boldsymbol{\theta}}_{\text{MLE}}|\mathbf{y})) + 2\text{Tr}(\widehat{F}(\hat{\boldsymbol{\theta}}_{\text{MLE}})[\widehat{H}(\hat{\boldsymbol{\theta}}_{\text{MLE}})]^{-1}). \qquad (4.15)$$

For a detailed derivation of the TIC, the interested reader can refer to [17, Section 7.2].

As we remarked in Section 4.2.2, if $\boldsymbol{\theta}_{\text{PT}}$ satisfies $p(\cdot|\boldsymbol{\theta}_{\text{PT}}) \equiv p_0$, then we have $H(\boldsymbol{\theta}_{\text{PT}}) = F(\boldsymbol{\theta}_{\text{PT}})$, which implies that $\text{Tr}(F(\boldsymbol{\theta}_{\text{PT}})H(\boldsymbol{\theta}_{\text{PT}})^{-1}) = \kappa_\theta$. Hence, if $p(\cdot|\boldsymbol{\theta}_{\text{PT}})$ provides a good approximation to the true model p_0 (in terms of K–L information), then $\text{Tr}(F(\boldsymbol{\theta}_{\text{PT}})H(\boldsymbol{\theta}_{\text{PT}})^{-1}) \approx \kappa_\theta$. Thus, for this case the AIC provides a good approximation to the TIC. This is why the assumption of p_0 being in the candidate model set $\mathbb{P}$ to obtain the AIC is relaxed to the assumption that the approximating model is close to the true model p_0 (the assumption made in Section 4.2.3).

It is worth noting here that even though the TIC is very attractive in theory (one does not need to worry whether or not the approximation models are close to the true model), it is rarely used in practice because one needs a very large sample size to obtain good estimates for both $F(\boldsymbol{\theta}_{\text{PT}})$ and $H(\boldsymbol{\theta}_{\text{PT}})$. It was found in [17] that the AIC often provides a good approximation to the TIC when the model structure and assumed error distribution are not drastically different from the truth. In addition, if the candidate model set $\mathbb{P}$ contains one or more good approximating models besides some poor approximating models, then one is still able to use the AIC for model selection; it was remarked in [17] that when the approximating model is poor the value of $-2\ln(\mathcal{L}(\hat{\boldsymbol{\theta}}_{\text{MLE}}|\mathbf{y}))$ tends to be very large and dominate the AIC value, and hence it will not be chosen as the best approximating model.

4.2.6 Remarks on Akaike Information Criterion and Its Variations

The Akaike information criterion and its variations (AIC_c and TIC) attempt to select a best approximating model among a prior set of candidate models (recall all the models in the set are written in terms of probability density functions) based on Kullback–Leibler information. There are several advantages in using the Akaike information criterion or its variations: (1) it is valid for both *nested models* (e.g., two models are said to be nested if one model is a special case of the other one) and non-nested models; (2) it can be used to compare mathematical models with different measurement error distributions; (3) it can avoid multiple testing issues, a problem involved in model selection criteria based on hypothesis testing (as we will demonstrate in Section 4.5), where only two models can be compared at a time. In the remainder of this section, we will use the "AIC" as a generic term to denote the Akaike information criterion and its variations.

4.2.6.1 Candidate Models

All the candidate models should have correct physical or biological meaning. Under this restrictive assumption, the prior set of candidate models can be constructed in one of the following ways.

First, they could be constructed based on different mathematical models with the same measurement error distribution. In this case, one is trying to select a best approximating mathematical model to describe the given system with the tacit assumption that the probability distribution of the measurement errors is known. This is the scenario often explored in the applied mathematics community, where researchers tend to be more interested in the appropriateness of mathematical models in describing a given complex dynamical system. However, this assumption is usually invalid in practice, as it is often the case that one may have either no information or partial information on the measurement errors.

The prior set of candidate models could also be constructed based on the same mathematical models with different measurement error distributions. For this case, one is trying to select a best approximating distribution model for the measurement errors under the implicit assumption that the correct mathematical model is used to describe the system. This assumption is also not true, as stated by the famous quote (George Box) that "all mathematical models are wrong, but some are more useful than others."

Based on the above discussion, one needs to consider the utility of probability distribution forms of measurement errors as well as mathematical models when one tries to understand the underlying observation process of the given complex system. Therefore, when it is possible, one needs to construct a prior set of candidate models that contains different mathematical models (based on known hypotheses and knowledge) with different possible probability distribution forms of measurement errors.

4.2.6.2 The Selected Best Model

The selected model is the one with the minimum AIC value. It should be noted that the selected model is specific to the set of candidate models. It is also specific to the given data set. In other words, if one has a different set of experimental data arising from the same experiment, one may select a different model. Hence, in practice, the absolute size of the AIC value may have limited use in supporting the chosen best approximating model. In addition, the AIC value is an estimate of the expected relative K–L information (hence, the actual value of the AIC is meaningless). Thus, one may often employ other related values, such as the Akaike difference and Akaike weights.

The Akaike difference is defined by

$$\Delta_i = \text{AIC}_i - \text{AIC}_{\min}, \quad i = 1, 2, \dots l, \tag{4.16}$$

where AIC_i is the AIC value of the ith model in the set, $\text{AIC}_{\min}$ denotes the AIC value for the best model in the set and l is the total number of models in the set. We see that the selected model is the one with zero Akaike difference. The larger Δ_i, the less plausible it is that the ith model is the best approximating model given the data set.

Akaike weights are defined by

$$w_i = \frac{\exp(-\frac{1}{2}\Delta_i)}{\sum_{r=1}^{l} \exp(-\frac{1}{2}\Delta_r)}, \quad i = 1, 2, \ldots l. \tag{4.17}$$

The Akaike weight w_i is similar to the relative frequency for the ith model selected as the best approximating model by using the bootstrapping method. It can also be interpreted (in a Bayesian framework) as the actual posterior probability that the ith model is the best approximating model given the data. We refer the interested reader to [17, Section 2.13] for details. Akaike weights are also used as a heuristic way to construct the 95% confidence set for the selected model by summing the Akaike weights from largest to smallest until the sum is ≥ 0.95. The corresponding subset of models is the confidence set for the selected model. Interested readers can refer to [17] for other heuristic approaches for construction of the confidence set.

4.2.6.3   Pitfalls When Using the AIC

Below we list a few pitfalls when using the AIC (the interested reader can refer to [17, Sections 2.11, 6.7, 6.8] for details).

- The AIC cannot be used to compare models for different data sets.

  For example, if a model is fit to a data set with $N = 140$ observations, one cannot validly compare it with another model when 7 outliers have been deleted, leaving only $N = 133$.

- The same response variables should be used for all the candidate models.

  For example, if there was interest in the normal and log-normal model forms, the models would have to be expressed, respectively, as

$$p_1(y|\mu, \sigma) = \frac{1}{\sqrt{2\pi}\sigma} \exp\left(-\frac{(y-\mu)^2}{2\sigma^2}\right),$$

$$p_2(y|\mu, \sigma) = \frac{1}{y\sqrt{2\pi}\sigma} \exp\left(-\frac{(\ln(y)-\mu)^2}{2\sigma^2}\right), \tag{4.18}$$

instead of

$$p_1(y|\mu, \sigma) = \frac{1}{\sqrt{2\pi}\sigma} \exp\left(-\frac{(y-\mu)^2}{2\sigma^2}\right),$$

$$p_2(\ln(y)|\mu, \sigma) = \frac{1}{\sqrt{2\pi}\sigma} \exp\left(-\frac{(\ln(y)-\mu)^2}{2\sigma^2}\right).$$

- Null hypothesis testing should not be mixed with the information criterion in reporting the results.

  - The information criterion is not a "test," so one should avoid use of "significant" and "not significant," or "rejected" and "not rejected" in reporting results.
  - One should not use the AIC to rank models in the set and then test whether the best model is "significantly better" than the second-best model.

- All the components in each log-likelihood function should be retained when comparing models with different probability distribution forms.

  For example, if we want to use the AIC to compare the normal model and the log-normal model in (4.18), we need to keep all the components in their corresponding log-likelihood functions, including all the constants (as the constants are not the same for these two models).

4.3   The AIC under the Framework of Least Squares Estimation

In Chapter 3, we introduced the least squares estimation method for parameter estimation. Hence, a natural question to ask is whether or not one can still use the AIC if the least squares method is used for parameter estimation. It turns out that there are several cases where one can establish the equivalence between maximum likelihood estimates and least squares estimates. Hence, for these cases one can still use the AIC for model selection.

4.3.1   Independent and Identically Normally Distributed Observations

Under the assumption of $\mathcal{E}_j$, $j = 1, 2, \ldots, N$, being i.i.d. $\mathcal{N}(0, \sigma^2)$ for the statistical model (4.1), we see that $\{Y_j\}_{j=1}^{N}$ are independent and normally distributed random variables with mean and variance given by

$$\mathbb{E}(Y_j) = f(t_j; \mathbf{q}), \quad \text{Var}(Y_j) = \sigma^2, \quad j = 1, 2, \ldots, N.$$

Hence, under this case $\boldsymbol{\theta} = (\mathbf{q}, \sigma)^T$ and the likelihood function of $\boldsymbol{\theta}$ given the sample outcomes $\mathbf{y}$ is

$$\mathcal{L}(\boldsymbol{\theta}|\mathbf{y}) = \frac{1}{\left(\sqrt{2\pi}\sigma\right)^N} \exp\left(-\frac{\sum_{j=1}^{N}(y_j - f(t_j; \mathbf{q}))^2}{2\sigma^2}\right),$$

which implies that

$$\ln(\mathcal{L}(\boldsymbol{\theta}|\mathbf{y})) = -\frac{N}{2}\ln(2\pi) - N\ln(\sigma) - \frac{\sum_{j=1}^{N}(y_j - f(t_j;\mathbf{q}))^2}{2\sigma^2}. \qquad (4.19)$$

We note that maximizing the above log-likelihood function over Ω_q to obtain the maximum likelihood estimate $\hat{\mathbf{q}}_{\text{MLE}}$ of $\mathbf{q}$ is the same as minimizing the cost function

$$J(\mathbf{q}) = \sum_{j=1}^{N}(y_j - f(t_j;\mathbf{q}))^2 \qquad (4.20)$$

over Ω_q to obtain the ordinary least squares estimate $\hat{\mathbf{q}}_{\text{OLS}}$ of $\mathbf{q}$; that is,

$$\hat{\mathbf{q}}_{\text{OLS}} = \hat{\mathbf{q}}_{\text{MLE}}.$$

After we obtain the maximum likelihood estimate $\hat{\mathbf{q}}_{\text{MLE}}$ of $\mathbf{q}$, we can solve the following equation for the maximum likelihood estimate $\hat{\sigma}_{\text{MLE}}$ of σ:

$$\frac{\partial\ln(\mathcal{L}(\boldsymbol{\theta}|\mathbf{y}))}{\partial\sigma}\bigg|_{\boldsymbol{\theta}=\hat{\boldsymbol{\theta}}_{\text{MLE}}} = 0. \qquad (4.21)$$

Note that

$$\frac{\partial\ln(\mathcal{L}(\boldsymbol{\theta}|\mathbf{y}))}{\partial\sigma} = -\frac{N}{\sigma} + \frac{\sum_{j=1}^{N}(y_j - f(t_j;\mathbf{q}))^2}{\sigma^3}.$$

Hence, by the above equation and (4.21) the maximum likelihood estimate $\hat{\sigma}_{\text{MLE}}$ of σ is given by

$$\hat{\sigma}_{\text{MLE}}^2 = \frac{\sum_{j=1}^{N}(y_j - f(t_j;\hat{\mathbf{q}}_{\text{MLE}}))^2}{N}.$$

By (3.18) and the above equation, we see that the maximum likelihood estimate for σ is different from that obtained using the least squares method. Substituting $\hat{\mathbf{q}}_{\text{MLE}}$ for $\mathbf{q}$ and $\hat{\sigma}_{\text{MLE}}$ for σ into (4.19) yields

$$\ln(\mathcal{L}(\hat{\boldsymbol{\theta}}_{\text{MLE}}|\mathbf{y})) = -\frac{N}{2}\ln(2\pi) - \frac{N}{2}\ln\left(\frac{\sum_{j=1}^{N}(y_j - f(t_j;\hat{\mathbf{q}}_{\text{MLE}}))^2}{N}\right) - \frac{N}{2}.$$

Then by the above equation and (4.12) we find that the AIC value is given by

$$\text{AIC} = N\left[1 + \ln(2\pi)\right] + N\ln\left(\frac{\sum_{j=1}^{N}(y_j - f(t_j;\hat{\mathbf{q}}_{\text{MLE}}))^2}{N}\right) + 2(\kappa_q + 1).$$

Note that the first term on the right-hand side of the above equation is just some constant related to sample size N. Hence, if all the probability distribution models considered are constructed based on the assumption of $\mathcal{E}_j$, $j = 1, 2, \ldots, N$, being i.i.d. $\mathcal{N}(0, \sigma^2)$ and the parameters are estimated by using the ordinary least squares (OLS) method with the cost function defined by (4.20) for all the models, then we can omit this constant (recall that the AIC value is an estimate of the expected relative K–L information, hence constants do not matter if they are the same across all the models). The AIC value is then given by

$$\text{AIC} = N \ln \left(\frac{\sum_{j=1}^{N} (y_j - f(t_j; \hat{\mathbf{q}}_{\text{OLS}}))^2}{N} \right) + 2(\kappa_q + 1). \qquad (4.22)$$

Thus, under this case what we select is the best approximating mathematical model from a given set of mathematical models (as the probability distribution forms of measurement errors are assumed to be the same across all these models).

4.3.2 Independent Multivariate Normally Distributed Observations

In this section, we consider the multivariate case with the statistical model given by

$$\mathbf{Y}_j = \mathbf{f}(t_j; \mathbf{q}) + \mathcal{E}_j, \quad j = 1, 2, \dots, N, \qquad (4.23)$$

where $\mathbf{Y}_j = (Y_{1j}, Y_{2j}, \dots, Y_{\nu j})^T$ is a random vector with its realization represented by $\mathbf{y}_j = (y_{1j}, y_{2j}, \dots, y_{\nu j})^T$, and $\mathbf{f} = (f_1, f_2, \dots, f_\nu)^T$. We assume that the measurement errors $\mathcal{E}_j = (\mathcal{E}_{1j}, \mathcal{E}_{2j}, \dots, \mathcal{E}_{\nu j})^T$, $j = 1, 2, \dots, N$, are independent and $\mathcal{E}_j \sim \mathcal{N}(\mathbf{0}_\nu, \sigma^2 \mathbf{I}_\nu), j = 1, 2, \dots, N$, where $\mathbf{0}_\nu$ is a ν-dimensional column zero vector and $\mathbf{I}_\nu$ is a $\nu \times \nu$ identity matrix. With this assumption, we see that $\{\mathbf{Y}_j\}_{j=1}^{N}$ are independent and multivariate normally distributed random vectors with mean vector and covariance matrix respectively given by

$$\mathbb{E}(\mathbf{Y}_j) = \mathbf{f}(t_j; \mathbf{q}), \quad \text{Var}(\mathbf{Y}_j) = \sigma^2 \mathbf{I}_\nu, \quad j = 1, 2, \dots, N.$$

Hence, for this setting the components of $\mathbf{Y}_j$ are assumed to be independent and have the same constant variance σ^2.

We observe that for this case $\boldsymbol{\theta} = (\mathbf{q}, \sigma)^T$, and the likelihood function of $\boldsymbol{\theta}$ given the sample outcomes

$$\mathbf{y} = (y_{11}, y_{12}, \dots, y_{1N}, y_{21}, y_{22}, \dots, y_{2N}, \dots, y_{\nu 1}, y_{\nu 2}, \dots, y_{\nu N})$$

is

$$\mathcal{L}(\boldsymbol{\theta} | \mathbf{y}) = \frac{1}{(\sqrt{2\pi}\sigma)^{N\nu}} \exp \left(-\frac{\sum_{j=1}^{N} (\mathbf{y}_j - \mathbf{f}(t_j; \mathbf{q}))^T (\mathbf{y}_j - \mathbf{f}(t_j; \mathbf{q}))}{2\sigma^2} \right),$$

which implies that

$$\ln(\mathcal{L}(\boldsymbol{\theta} | \mathbf{y})) = -\frac{N\nu}{2} \ln (2\pi) - N\nu \ln (\sigma)$$

$$- \frac{\sum_{j=1}^{N} (\mathbf{y}_j - \mathbf{f}(t_j; \mathbf{q}))^T (\mathbf{y}_j - \mathbf{f}(t_j; \mathbf{q}))}{2\sigma^2}. \qquad (4.24)$$

We note that maximizing the above log-likelihood function over the set Ω_q to obtain the maximum likelihood estimate $\hat{\mathbf{q}}_{\text{MLE}}$ of $\mathbf{q}$ is the same as minimizing the cost function

$$J(\mathbf{q}) = \sum_{j=1}^{N} (\mathbf{y}_j - \mathbf{f}(t_j; \mathbf{q}))^T (\mathbf{y}_j - \mathbf{f}(t_j; \mathbf{q})) \qquad (4.25)$$

over the set Ω_q to obtain the ordinary least squares estimate $\hat{\mathbf{q}}_{\text{OLS}}$. Hence,

$$\hat{\mathbf{q}}_{\text{OLS}} = \hat{\mathbf{q}}_{\text{MLE}}.$$

After we obtain the maximum likelihood estimate $\hat{\mathbf{q}}_{\text{MLE}}$ of $\mathbf{q}$, we can solve the following equation for the maximum likelihood estimate $\hat{\sigma}_{\text{MLE}}$ of σ:

$$\left. \frac{\partial \ln(\mathcal{L}(\boldsymbol{\theta}|\mathbf{y}))}{\partial \sigma} \right|_{\boldsymbol{\theta}=\hat{\boldsymbol{\theta}}_{\text{MLE}}} = 0. \qquad (4.26)$$

Note that

$$\frac{\partial \ln(\mathcal{L}(\boldsymbol{\theta}|\mathbf{y}))}{\partial \sigma} = -\frac{N\nu}{\sigma} + \frac{\sum_{j=1}^{N} (\mathbf{y}_j - \mathbf{f}(t_j; \mathbf{q}))^T (\mathbf{y}_j - \mathbf{f}(t_j; \mathbf{q}))}{\sigma^3}.$$

Hence, by the above equation and (4.26) the maximum likelihood estimate $\hat{\sigma}_{\text{MLE}}$ of σ is given by

$$\hat{\sigma}^2_{\text{MLE}} = \frac{\sum_{j=1}^{N} (\mathbf{y}_j - \mathbf{f}(t_j; \hat{\mathbf{q}}_{\text{MLE}}))^T (\mathbf{y}_j - \mathbf{f}(t_j; \hat{\mathbf{q}}_{\text{MLE}}))}{N\nu}.$$

Substituting $\hat{\mathbf{q}}_{\text{MLE}}$ for $\mathbf{q}$ and $\hat{\sigma}_{\text{MLE}}$ for σ in (4.24) yields

$$\ln(\mathcal{L}(\hat{\boldsymbol{\theta}}_{\text{MLE}}|\mathbf{y})) = -\frac{N\nu}{2} \ln(2\pi) - \frac{N\nu}{2}$$

$$- \frac{N\nu}{2} \ln \left(\frac{\sum_{j=1}^{N} (\mathbf{y}_j - \mathbf{f}(t_j; \hat{\mathbf{q}}_{\text{MLE}}))^T (\mathbf{y}_j - \mathbf{f}(t_j; \hat{\mathbf{q}}_{\text{MLE}}))}{N\nu} \right).$$

Then by the above equation and (4.12) we find that the AIC value is given by

$$\text{AIC} = N\nu \ln \left(\frac{\sum_{j=1}^{N} (\mathbf{y}_j - \mathbf{f}(t_j; \hat{\mathbf{q}}_{\text{MLE}}))^T (\mathbf{y}_j - \mathbf{f}(t_j; \hat{\mathbf{q}}_{\text{MLE}}))}{N\nu} \right)$$

$$+ N\nu \left[1 + \ln(2\pi) \right] + 2(\kappa_q + 1).$$

As before, note that the second term on the right-hand side of the above equation is just some constant related to sample size N and the dimension of $\mathbf{f}$. Hence, if all the probability distribution models considered are constructed based on the assumption that $\mathcal{E}_j$, $j = 1, 2, \ldots, N$, are independent and multivariate normally distributed $\mathcal{N}(\mathbf{0}_\nu, \sigma^2 \mathbf{I}_\nu)$ and the parameters are estimated by using the OLS method with the cost function defined by (4.25) for all the models, then we can omit this constant. The AIC value is then given by

$$\text{AIC} = N\nu \ln\left(\frac{\sum_{j=1}^{N}(\mathbf{y}_j - \mathbf{f}(t_j; \hat{\mathbf{q}}_{\text{OLS}}))^T(\mathbf{y}_j - \mathbf{f}(t_j; \hat{\mathbf{q}}_{\text{OLS}}))}{N\nu}\right) + 2(\kappa_q + 1).$$

$$(4.27)$$

It is worth noting that if $\nu = 1$, then the above equation is the same as (4.22).

4.3.2.1   Unequal Number of Observations for Different Observed Components

In practice, it is often the case that we may have different numbers of measurements for each observed component. Under this case, the statistical model has the following form:

$$Y_{ij} = f_i(t_j; \mathbf{q}) + \mathcal{E}_{ij}, \quad j = 1, 2, \ldots, N_i, \; i = 1, 2, \ldots, \nu, \qquad (4.28)$$

where the measurement errors $\mathcal{E}_{ij}$, $j = 1, 2, \ldots, N_i$, $i = 1, 2, \ldots, \nu$, are assumed to be independent and $\mathcal{E}_{ij} \sim \mathcal{N}(0, \sigma^2)$. This implies that Y_{ij}, $j = 1, 2, \ldots, N_i$, $i = 1, 2, \ldots, \nu$, are independent and normally distributed with mean and variance given by

$$\mathbb{E}(Y_{ij}) = f_i(t_j; \mathbf{q}), \quad \text{Var}(Y_{ij}) = \sigma^2, \quad j = 1, 2, \ldots, N_i, \; i = 1, 2, \ldots, \nu.$$

Hence, under this case $\boldsymbol{\theta} = (\mathbf{q}, \sigma)^T$, and the likelihood function of $\boldsymbol{\theta}$ given the sample outcomes

$$\mathbf{y} = (y_{11}, y_{12}, \ldots, y_{1,N_1}, y_{21}, y_{22}, \ldots, y_{2,N_2}, \ldots, y_{\nu 1}, y_{\nu 2}, \ldots, y_{\nu,N_\nu})$$

is

$$\mathcal{L}(\boldsymbol{\theta}|\mathbf{y}) = \frac{1}{(\sqrt{2\pi}\sigma)^{\sum_{i=1}^{\nu} N_i}} \exp\left(-\sum_{i=1}^{\nu}\sum_{j=1}^{N_i} \frac{(y_{ij} - f_i(t_j; \mathbf{q}))^2}{2\sigma^2}\right),$$

which implies that

$$\ln(\mathcal{L}(\boldsymbol{\theta}|\mathbf{y})) = -\frac{\sum_{i=1}^{\nu} N_i}{2} \ln(2\pi) - \left(\sum_{i=1}^{\nu} N_i\right) \ln(\sigma)$$

$$- \sum_{i=1}^{\nu}\sum_{j=1}^{N_i} \frac{(y_{ij} - f_i(t_j; \mathbf{q}))^2}{2\sigma^2}.$$

$$(4.29)$$

We note that maximizing the above log-likelihood function over the set Ω_q to obtain the maximum likelihood estimate $\hat{\mathbf{q}}_{\text{MLE}}$ of $\mathbf{q}$ is the same as minimizing the cost function

$$J(\mathbf{q}) = \sum_{i=1}^{\nu}\sum_{j=1}^{N_i}(y_{ij} - f_i(t_j; \mathbf{q}))^2 \qquad (4.30)$$

over the set Ω_q to obtain the ordinary least squares estimate $\hat{\mathbf{q}}_{\mathrm{OLS}}$. Hence, we have

$$\hat{\mathbf{q}}_{\mathrm{OLS}} = \hat{\mathbf{q}}_{\mathrm{MLE}}.$$

Note that

$$\frac{\partial \ln(\mathcal{L}(\boldsymbol{\theta}|\mathbf{y}))}{\partial \sigma} = -\frac{\sum_{i=1}^{\nu} N_i}{\sigma} + \frac{\sum_{i=1}^{\nu} \sum_{j=1}^{N_i} (y_{ij} - f_i(t_j; \mathbf{q}))^2}{\sigma^3}.$$

Hence, by the above equation we see that the maximum likelihood estimate $\hat{\sigma}_{\mathrm{MLE}}$ of σ is given by

$$\hat{\sigma}_{\mathrm{MLE}}^2 = \frac{\sum_{i=1}^{\nu} \sum_{j=1}^{N_i} (y_{ij} - f_i(t_j; \hat{\mathbf{q}}_{\mathrm{MLE}}))^2}{\sum_{i=1}^{\nu} N_i}.$$

Substituting $\hat{\mathbf{q}}_{\mathrm{MLE}}$ for $\mathbf{q}$ and $\hat{\sigma}_{\mathrm{MLE}}$ for σ in (4.29) yields

$$\ln(\mathcal{L}(\hat{\boldsymbol{\theta}}_{\mathrm{MLE}}|\mathbf{y})) = -\frac{\sum_{i=1}^{\nu} N_i}{2} \ln(2\pi) - \frac{\sum_{i=1}^{\nu} N_i}{2}$$
$$- \left(\sum_{i=1}^{\nu} \frac{N_i}{2} \right) \ln \left(\frac{\sum_{i=1}^{\nu} \sum_{j=1}^{N_i} (y_{ij} - f_i(t_j; \hat{\mathbf{q}}_{\mathrm{MLE}}))^2}{\sum_{i=1}^{\nu} N_i} \right).$$

Then by the above equation and (4.12) we find that the AIC value is given by

$$\mathrm{AIC} = \left(\sum_{i=1}^{\nu} N_i \right) \ln \left(\frac{\sum_{i=1}^{\nu} \sum_{j=1}^{N_i} (y_{ij} - f_i(t_j; \hat{\mathbf{q}}_{\mathrm{MLE}}))^2}{\sum_{i=1}^{\nu} N_i} \right)$$
$$+ \sum_{i=1}^{\nu} N_i \left[1 + \ln(2\pi) \right] + 2(\kappa_q + 1).$$

Note that the second term on the right-hand side of the above equation is just some constant related to sample sizes $N_1, N_2, \ldots, N_\nu$. Hence, if all the probability distribution models considered are constructed based on the assumption of $\mathcal{E}_{ij}$, $j = 1, 2, \ldots, N_i$, $i = 1, 2, \ldots, \nu$, being independent and normally distributed $\mathcal{N}(0, \sigma^2)$ and the parameters are estimated using the OLS method with the cost function defined by (4.30) for all the models, then we can omit this constant; that is, the AIC value is given by

$$\mathrm{AIC} = \left(\sum_{i=1}^{\nu} N_i \right) \ln \left(\frac{\sum_{i=1}^{\nu} \sum_{j=1}^{N_i} (y_{ij} - f_i(t_j; \hat{\mathbf{q}}_{\mathrm{OLS}}))^2}{\sum_{i=1}^{\nu} N_i} \right) + 2(\kappa_q + 1). \quad (4.31)$$

Thus, in this case what we select is the best approximating mathematical model from a given set of mathematical models (as the probability distribution form of measurement errors is assumed to be the same across all these models). It is worth noting that if $N_i = N$, $i = 1, 2, \ldots, \nu$, then (4.31) is just (4.27).

4.3.3   Independent Gamma Distributed Observations

Assume $\{Y_j\}_{j=1}^N$ are independent gamma distributed random variables with mean and variance given by

$$\mathbb{E}(Y_j) = f(t_j; \mathbf{q}), \quad \mathrm{Var}(Y_j) = \sigma^2 f^2(t_j; \mathbf{q}), \quad j = 1, 2, \ldots, N, \qquad (4.32)$$

where σ is some unknown positive constant. In other words,

$$Y_j \sim \mathrm{Gamma}\left(\frac{1}{\sigma^2}, \frac{1}{\sigma^2 f(t_j; \mathbf{q})}\right), \quad j = 1, 2, \ldots, N.$$

This assumption implies that the sample outcomes $y_j, j = 1, 2, \ldots, N$, should be positive, and f is a positive function. In addition, we see that with this assumption the statistical model takes the form

$$Y_j = f(t_j; \mathbf{q})(1 + \tilde{\mathcal{E}}_j), \quad j = 1, 2, \ldots, N, \qquad (4.33)$$

where $\tilde{\mathcal{E}}_j$, $j = 1, 2, \ldots, N$, are independent and identically distributed random variables with zero mean and constant variance σ^2. In other words, the measurement errors take the form

$$\mathcal{E}_j = \tilde{\mathcal{E}}_j f(t_j; \mathbf{q}), \quad j = 1, 2, \ldots, N.$$

It is worth noting here that even though the observations are gamma distributed, the measurement errors and $\tilde{\mathcal{E}}_j$ are not.

We see that for this case $\boldsymbol{\theta} = (\mathbf{q}, \sigma)^T$ and the likelihood function of $\boldsymbol{\theta}$ given the sample outcomes $\mathbf{y} = (y_1, y_2, y_3, \ldots, y_N)$ is given by

$$\mathcal{L}(\boldsymbol{\theta}|\mathbf{y}) = \prod_{j=1}^N \frac{\sigma^{-2/\sigma^2}}{\Gamma(1/\sigma^2)} y_j^{\frac{1}{\sigma^2}-1} \exp\left(-\frac{1}{\sigma^2}\frac{y_j}{f(t_j; \mathbf{q})} - \frac{1}{\sigma^2}\ln(f(t_j; \mathbf{q}))\right),$$

where Γ is the gamma function defined by (2.103). By the above equation, we find that the corresponding log-likelihood function is

$$\ln(\mathcal{L}(\boldsymbol{\theta}|\mathbf{y})) = \sum_{j=1}^N \ln\left(\frac{\sigma^{-2/\sigma^2}}{\Gamma(1/\sigma^2)} y_j^{\frac{1}{\sigma^2}-1}\right) \\ -\frac{1}{\sigma^2}\sum_{j=1}^N\left[\frac{y_j}{f(t_j; \mathbf{q})} + \ln(f(t_j; \mathbf{q}))\right]. \qquad (4.34)$$

The maximum likelihood estimate $\hat{\boldsymbol{\theta}}_{\mathrm{MLE}} = (\hat{\mathbf{q}}_{\mathrm{MLE}}, \hat{\sigma}_{\mathrm{MLE}})^T$ of $\boldsymbol{\theta}$ can be obtained by solving the following equations:

$$\frac{\partial \ln(\mathcal{L}(\boldsymbol{\theta}|\mathbf{y}))}{\partial q_i}\bigg|_{\boldsymbol{\theta}=\hat{\boldsymbol{\theta}}_{\mathrm{MLE}}} = 0, \quad i = 1, 2, \ldots \kappa_q, \qquad (4.35)$$

and

$$\frac{\partial \ln(\mathcal{L}(\boldsymbol{\theta}|\mathbf{y}))}{\partial \sigma}\bigg|_{\boldsymbol{\theta}=\hat{\boldsymbol{\theta}}_{\text{MLE}}} = 0. \tag{4.36}$$

It is worth noting here that the gradient of the log-likelihood function with respect to $\boldsymbol{\theta}$ is called the *score* for $\boldsymbol{\theta}$ in the statistics literature. Hence, the above equations are often referred to as *score equations*.

Note that for $i = 1, 2, \ldots, \kappa_q$ we have

$$\frac{\partial \ln(\mathcal{L}(\boldsymbol{\theta}|\mathbf{y}))}{\partial q_i} = -\frac{1}{\sigma^2} \sum_{j=1}^{N} \left[-\frac{y_j}{f^2(t_j; \mathbf{q})} \frac{\partial f(t_j; \mathbf{q})}{\partial q_i} + \frac{1}{f(t_j; \mathbf{q})} \frac{\partial f(t_j; \mathbf{q})}{\partial q_i} \right]$$

$$= \frac{1}{\sigma^2} \sum_{j=1}^{N} \frac{1}{f^2(t_j; \mathbf{q})} [y_j - f(t_j; \mathbf{q})] \frac{\partial f(t_j; \mathbf{q})}{\partial q_i}.$$

Hence, by (4.35) and the above equality we know that the maximum likelihood estimate $\hat{\mathbf{q}}_{\text{MLE}}$ satisfies the following equations:

$$\sum_{j=1}^{N} \frac{1}{f^2(t_j; \mathbf{q})} [y_j - f(t_j; \mathbf{q})] \nabla f(t_j; \mathbf{q}) = 0, \tag{4.37}$$

where $\nabla f(t_j; \mathbf{q}) = \left(\frac{\partial f(t_j; \mathbf{q})}{\partial q_1}, \frac{\partial f(t_j; \mathbf{q})}{\partial q_2}, \ldots, \frac{\partial f(t_j; \mathbf{q})}{\partial q_{\kappa_q}} \right)^T$. Recall from Chapter 3 that (4.37) is a *generalized least squares equation*. Thus, we see that in this case the maximum likelihood estimate $\hat{\mathbf{q}}_{\text{MLE}}$ of $\mathbf{q}$ is equivalent to the generalized least squares estimate $\hat{\mathbf{q}}_{\text{GLS}}$ of $\mathbf{q}$.

After we obtain the maximum likelihood estimate of $\mathbf{q}$, we can obtain the maximum likelihood estimate $\hat{\sigma}_{\text{MLE}}$ of σ by solving (4.36). Note that

$$\frac{\partial \ln(\mathcal{L}(\boldsymbol{\theta}|\mathbf{y}))}{\partial \sigma} = N \left(\frac{4}{\sigma^3} \ln(\sigma) - \frac{2}{\sigma^3} \right) - \frac{2}{\sigma^3} \sum_{j=1}^{N} \ln(y_j)$$

$$+ \frac{2}{\sigma^3} \sum_{j=1}^{N} \left[\frac{y_j}{f(t_j; \mathbf{q})} + \ln(f(t_j; \mathbf{q})) \right] - N \frac{d}{d\sigma} \ln \left(\Gamma(1/\sigma^2) \right),$$

which implies that we are unable to analytically determine the maximum likelihood estimate $\hat{\sigma}_{\text{MLE}}$. Hence, one needs to numerically solve the above equation for $\hat{\sigma}_{\text{MLE}}$. With this $\hat{\sigma}_{\text{MLE}}$ and $\hat{\mathbf{q}}_{\text{MLE}}$, the AIC value is given by

$$\text{AIC} = -2 \sum_{j=1}^{N} \ln \left(\frac{\hat{\sigma}_{\text{MLE}}^{-2/\hat{\sigma}_{\text{MLE}}^2}}{\Gamma(1/\hat{\sigma}_{\text{MLE}}^2)} y_j^{\frac{1}{\hat{\sigma}_{\text{MLE}}^2}-1} \right) + 2(\kappa_q + 1)$$

$$+ \frac{2}{\hat{\sigma}_{\text{MLE}}^2} \sum_{j=1}^{N} \left[\frac{y_j}{f(t_j; \hat{\mathbf{q}}_{\text{MLE}})} + \ln(f(t_j; \hat{\mathbf{q}}_{\text{MLE}})) \right]. \tag{4.38}$$

For the special case where one knows the value of σ (i.e., one does not need to estimate σ), then the AIC value can be easily written down in terms of the general least squares estimate. This is given by

$$\text{AIC} = -2\sum_{j=1}^{N}\ln\left(\frac{\sigma^{-2/\sigma^2}}{\Gamma(1/\sigma^2)}y_j^{\frac{1}{\sigma^2}-1}\right)$$

$$+\frac{2}{\sigma^2}\sum_{j=1}^{N}\left[\frac{y_j}{f(t_j;\hat{\mathbf{q}}_{\text{MLE}})}+\ln(f(t_j;\hat{\mathbf{q}}_{\text{MLE}}))\right]+2\kappa_q. \tag{4.39}$$

Note that the first term on the right-hand side of the above equation is just a constant (relating to the sample size N, σ, and the experimental data). Hence, if all the probability distribution models considered are constructed based on the assumption of independent gamma distributed observations with the same σ and the parameters are estimated by using the generalized least squares method for all the models, then we can omit this constant; that is, the AIC value is given by

$$\text{AIC} = \frac{2}{\sigma^2}\sum_{j=1}^{N}\left[\frac{y_j}{f(t_j;\hat{\mathbf{q}}_{\text{GLS}})}+\ln(f(t_j;\hat{\mathbf{q}}_{\text{GLS}}))\right]+2\kappa_q. \tag{4.40}$$

Remark 4.3.1 *If we just assume that $\{Y_j\}_{j=1}^{N}$ are independent random variables with mean and variance given by (4.32), then Y_j, $j=1,2,\ldots,N$, could also be log-normally distributed; that is,*

$$\ln(Y_j) \sim \mathcal{N}\left(\ln(f(t_j;\mathbf{q})) - \frac{1}{2}\ln(\sigma^2+1), \ln(\sigma^2+1)\right).$$

In practice, for this scenario one often takes the natural logarithm of the variable to simplify the problem. In other words, we consider the statistical model having the following form:

$$\tilde{Y}_j = \ln(f(t_j;\mathbf{q})) + \eta_j, \quad j=1,2,\ldots,N, \tag{4.41}$$

where $\tilde{Y}_j = \ln(Y_j) + \frac{1}{2}\ln(\sigma^2+1) = \ln\left(Y_j\sqrt{\sigma^2+1}\right)$, $j=1,2,\ldots,N$, and η_j, $j=1,2,\ldots,N$, are independent and identically normally distributed with zero mean and variance $\ln(\sigma^2+1)$. With this statistical model, one can use the ordinary least squares method to estimate the mathematical model parameters $\mathbf{q}$, and hence the computations involved are much easier.

4.3.4 General Remarks

In the above sections we have discussed several special cases (normally distributed observations, multivariate normally distributed observations and gamma distributed observations) where we can establish the equivalence between maximum likelihood estimates of mathematical model parameters $\mathbf{q}$ and least squares estimates of $\mathbf{q}$ and hence for these scenarios one can use the AIC under the least squares framework. It is worth noting here that for all cases considered here the value of the statistical parameter σ does not affect the estimate of $\mathbf{q}$ and hence estimation of the statistical parameter can be treated separately from estimation of the mathematical model parameters (this is indeed the case under the least squares framework).

In general, if the probability density function of $\mathbf{Y} = (Y_1, Y_2, \ldots, Y_N)^T$ is a member of the regular *exponential family* of distributions,

$$
p(\mathbf{y}|\boldsymbol{\theta}) = \left[\prod_{j=1}^{N} \varphi(y_j) \right] \exp \left(\sum_{j=1}^{N} y_j \phi(f(t_j; \mathbf{q})) - \sum_{j=1}^{N} \psi(f(t_j; \mathbf{q})) \right),
$$

with φ, ϕ and ψ some known functions, then the maximum likelihood estimates of $\mathbf{q}$ can be found using the method of weighted least squares. The relationship between the mean and variance of Y_j and functions ϕ and ψ is given by (e.g., see [18])

$$
\mathbb{E}(Y_j) = \frac{\psi'(f(t_j; \mathbf{q}))}{\phi'(f(t_j; \mathbf{q}))} = f(t_j; \mathbf{q}), \quad \mathrm{Var}(Y_j) = \frac{1}{\phi'(f(t_j; \mathbf{q}))}, \quad j = 1, 2, \ldots, N,
$$

where $\phi'(z)$ and $\psi'(z)$, respectively, denote the derivatives of ϕ and ψ with respect to z. The exponential family of distributions includes a number of well-known distributions, such as the normal, gamma, and Poisson distributions. For more information on the equivalence of least squares and maximum likelihood estimates in the exponential family, interested readers are referred to [18] and the references therein. In fact, for some cases where the likelihood function even does not have the exponential form, one is still able to use weighted least squares to find the maximum likelihood estimates (e.g., see [32] and the references therein).

Theoretically, for the scenarios where one can establish the equivalence between maximum likelihood estimates and least squares estimates, one is still able to use the AIC if parameters are estimated by using the corresponding least squares method. However, in practice one may often find it is difficult to use the AIC under the least squares framework for those scenarios where the statistical model parameters are unknown. This is because there are many cases where one is unable to find the analytical maximum likelihood estimates for the statistical model parameters (e.g., σ, as demonstrated in Section 4.3.3). Hence, for these cases if one still wants to use the AIC under a least squares framework, one needs to resort to the maximum likelihood estimation method to obtain the numerical estimates for the statistical model parameters (recall that estimation of statistical model parameters by the least squares method is different from that obtained by the maximum likelihood estimation method).

4.4   Example: CFSE Label Decay

We return to the CFSE cell proliferation example of Section 3.5.3 and in particular Equation (3.73) characterizing CFSE label decay as part of the cell proliferation system (3.75). Current mathematical models, such as those described in Section 3.5.3, allow one to estimate proliferation and death rates in terms of a CFSE florescence intensity (FI) structure variable as a surrogate for division number, so the manner in which CFSE naturally decays directly affects the cell turnover parameter estimates. Thus, the loss rate function, $\dot{x} = v(t, x)$, is of vital importance to the partial differential equation model formulations such as (3.75) and subsequently developed models.

It was hypothesized in [10, 30] that an exponential rate of loss

$$\dot{x} = cx, \tag{4.42}$$

is sufficient to model the label loss observed in the data, but was later shown [11] that a Gompertz decay rate (see (3.73)),

$$\dot{x} = cx \exp(-kt), \tag{4.43}$$

gave a significantly better fit to the flow cytometry data sets. We remark that (4.43) is a generalization of (4.42), the latter being the limiting value (as $k \to 0$) of the former. It appears from the data fits in Figure 4.4 that the *rate* of exponential decay of a label may itself decrease as a function of time (and this can be readily modeled by the Gompertz decay process).

One would like to determine the most appropriate functional forms which might be used in order to quantify the label loss observed in the data and that would also correspond to plausible biological assumptions. Through examination of the chemical properties and biological processes associated with carboxyfluorescein diacetate succinimidyl ester (CFDA-SE) and its derivatives, namely, CFSE, and the intermolecular conjugates CF-R1 and CF-R2, the authors of [2, 19] attempted to understand and model natural label decay in cells treated with CFDA-SE. This allows one to test hypotheses regarding underlying mechanisms and formulate new assumptions about CFSE decay, future data and the larger biological process of cell division. This also answered to a large extent the question raised in the recent overview [25, p. 2695] regarding the *need for basic understanding of label decay* and *appropriate forms for its mathematical representations* in general proliferation assay models.

In [2, 19] the authors summarized biological understanding of mechanisms related to CFSE labeling and subsequent loss in cells, as depicted schematically in Figure 4.5. Several biological models (increasing in simplicity) were formulated in detail in [19] and findings with the corresponding mathematical

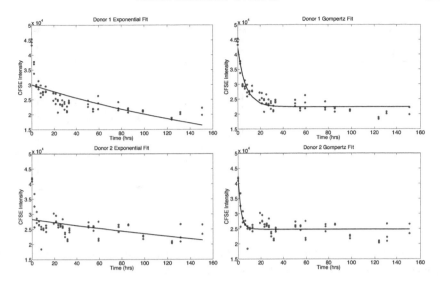

FIGURE 4.4: Results of fitting the exponential model (4.42) and the Gompertz model (4.43) to mean CFSE data. For two different blood donors, Donor 1 (top) and Donor 2 (bottom), we see that the Gompertz model provides better fits to the observed data.

models were discussed. These models were based on compartmental models ranging from 5 to 2 compartments and employed combinations of exponential (Malthusian: $\dot{\xi} = c\xi$) and/or Michaelis–Menten kinetic rates (see [1, 13, 21, 28, 33]).

$$\dot{\xi} = \frac{V_{max}\xi}{K_M + \xi}.$$

In other types of very simple biological models tested, the authors considered decay represented by a single velocity term for total fluorescence loss and modeled these with exponential, logistic

$$\dot{x} = \alpha x \left(1 - \frac{x}{K}\right),$$

and Gompertz

$$\dot{x} = \alpha x \ln\left(\frac{K}{x}\right) \tag{4.44}$$

decay mechanism fits-to-data, respectively. Recall that (4.43) is also a Gompertz model. It has dramatically different form from that in (4.44), which is the one that is often employed in the literature. However, one can easily show that (4.43) with initial condition $x(0) = x_0$ has the same solution as (4.44) with the same initial condition if we set $k = \alpha$ and $c = \alpha(\ln(K) - \ln(x_0))$.

A total of 12 different dynamical loss models was considered. For ease in reference, in the tables below these models are numbered according to the

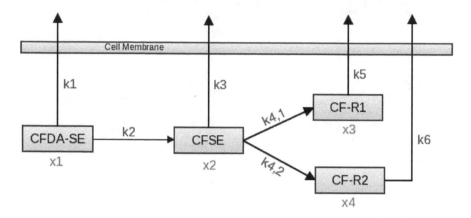

FIGURE 4.5:   A schematic of the *Basic Biological Model*, with independent reaction rates from CFSE to CF-R1 and CF-R2.

numbering scheme in [2, 19]. After careful examination of related chemical processes and biological processes, these 12 different mathematical models were formulated and analyzed for the rate of label decay in a cell treated with CDSA-SE. The first model, the Basic Biological Model of Figure 4.5, represents our best understanding of label staining, internal biochemistry and loss. An esterase reaction was hypothesized to specifically involve the enzyme acetylesterase, which binds to acetic ester and water, yielding alcohol and acetate as products. This knowledge led to incorporation of Michaelis–Menten kinetics to support the hypothesis that inflow of CFDA-SE and the rate of its conversion to CFSE can be ignored in model fitting to the data.

In order to better understand the process of label decay in cells treated with CFSE, a series of ordinary least squares inverse problems for the different mathematical models of interest were carried out with the assumption that measurement errors are independent and identically normally distributed with mean zero and variance σ^2. The residual plots [19] suggest that the measurement errors appear to be independent and identically distributed. This implies that our assumption on the measurement errors is reasonable.

With our assumption on the measurement errors and the discussion in Section 4.3.1, we can use the AIC to evaluate the models. It is worth noting that under this case what we select is the best approximating mathematical model from the given set of mathematical models (as all the models are constructed based on the same assumption on the measurement errors). Since the CFSE decay data has only 24 time points, by the discussion in Section 4.2.4 we will use the AIC_c to evaluate the models. By (4.22) and (4.13) we find

$$\text{AIC}_c = N \ln \left(\frac{J(\hat{\mathbf{q}}_{\text{OLS}})}{N} \right) + 2\kappa_\theta + \frac{2\kappa_\theta(\kappa_\theta + 1)}{N - \kappa_\theta - 1},$$

TABLE 4.1: Statistical results for a first data set from Donor 2, with best w_i in bold. Models are numbered to be consistent with their designations in [2, 19].

Model	κ_θ	$J(\hat{\mathbf{q}}_{\text{OLS}})$	AIC_c	w_i
1.1	8	1.223×10^8	396.25	0.0000
2.1	8	1.108×10^8	393.89	0.0001
2.2	9	1.075×10^8	398.43	0.0000
2.3	11	3.270×10^8	438.26	0.0000
3.1	6	1.165×10^8	386.44	0.0048
3.2	8	1.096×10^8	393.62	0.0001
4.1	4	1.110×10^8	**378.43**	**0.2638**
4.2	5	1.097×10^8	381.37	0.0605
5.1	3	1.110×10^8	**378.43**	**0.2634**
6.1	3	3.671×10^8	404.23	0.000
6.2	3	1.251×10^8	**378.41**	**0.2664**
6.3	3	1.320×10^8	**379.69**	**0.1403**

where J is the cost function defined by (4.20), $\ddot{\mathbf{q}}_{\text{OLS}}$ is the ordinary least squares estimate of the mathematical model parameter $\mathbf{q}$ and $\kappa_\theta = \kappa_q + 1$ (as for this example there is only one statistical model parameter, σ, which needs to be estimated).

The calculated AIC_c and the Akaike weights w_i for several of the data sets are given in Tables 4.1 and 4.2. Although no specific model was the best fit for every data set, these results suggest that any models which have *multiple loss rate mechanisms* (specifically, models labeled below as 4.1, 5.1, 6.2 and

TABLE 4.2: Statistical results for a second data set from Donor 2, with best w_i in bold. Again, models are numbered to be consistent with their designations in [2, 19].

Model	κ_θ	$J(\hat{\mathbf{q}}_{\text{OLS}})$	AIC_c	w_i
1.1	8	1.143×10^8	394.64	0.0002
2.1	8	1.221×10^8	396.21	0.0001
2.2	9	1.212×10^8	401.29	0.0000
2.3	11	4.5191×10^8	446.02	0.0000
3.1	6	1.230×10^8	387.74	0.0050
3.2	8	1.101×10^8	393.73	0.0002
4.1	4	1.167×10^8	**379.64**	**0.2862**
4.2	5	1.215×10^8	383.76	0.0365
5.1	3	1.167×10^8	**376.43**	**0.2863**
6.1	3	3.765×10^8	404.84	0.0000
6.2	3	1.331×10^8	**379.88**	**0.2538**
6.3	3	1.406×10^8	**381.19**	**0.1317**

6.3) are the models that most closely match the data. *All of these models involve either multiple rates of label leaking/decay or a time-dependent rate for the decay of a single label quantity.* These findings provide a reasonable explanation in support of the need for multiple or time variable decay rates involving CFSE labeling based on current best biological understanding of the labeling and loss processes.

4.5   Residual Sum of Squares Based Model Selection Criterion

As we remarked earlier in Section 4.3.4, it is generally difficult to use the Akaike information criterion and its variations under the least squares framework. Since the least squares method is a widely used parameter estimation method, a natural question to ask is whether or not there exists a model selection criterion designed specifically for the least squares case. This question was considered at length by Gallant [23], where the results are based on *hypothesis testing* as well as *asymptotic theory* (discussed in Section 3.3) for non-linear least squares. Subsequently, these ideas were reconsidered by Banks and Fitzpatrick [3] under a slightly different set of assumptions to develop a hypothesis test which could be computed from the residual sum of squares (RSS) after fitting the models to a data set in a least squares framework. Here the observations were assumed to be independent and identically distributed. This RSS based model selection criterion was later extended in [8] to the case where the measurement errors are not identically distributed.

The RSS based model selection criterion can be used as a tool for model comparison for certain classes of models, in which potentially extraneous mechanisms can be eliminated from the model by a simple restriction on the underlying parameter space while the form of the mathematical model remains unchanged. In other words, it can be used to compare two nested mathematical models where the parameter set Ω_q^H (this notation will be defined explicitly in Section 4.5.1) for the restricted model can be identified as a linearly constrained subset of the admissible parameter set Ω_q of the unrestricted model. Thus, the RSS based model selection criterion is a useful tool to determine whether or not certain terms in the mathematical models are statistically important in describing the given experimental data.

4.5.1   Ordinary Least Squares

We now turn to the statistical model (4.1), where the measurement errors are assumed to be independent and identically distributed with zero mean and constant variance σ^2. In addition, we assume that there exists $\mathbf{q}_0$ such that the statistical model

$$Y_j = f(t_j; \mathbf{q}_0) + \mathcal{E}_j, \quad j = 1, 2, \ldots, N \tag{4.45}$$

correctly describes the observation process. In other words, (4.45) is the true model, and $\mathbf{q}_0$ is the true value of the mathematical model parameter $\mathbf{q}$.

With our assumption on measurement errors, the mathematical model parameter $\mathbf{q}$ can be estimated by using the ordinary least squares method; that is, the ordinary least squares estimator of $\mathbf{q}$ is obtained by solving

$$\mathbf{q}_{\text{OLS}}^N = \arg \min_{\mathbf{q} \in \Omega_q} J_{\text{OLS}}^N(\mathbf{q}; \mathbf{Y}).$$

Here $\mathbf{Y} = (Y_1, Y_2, \ldots, Y_N)^T$, and the cost function J_{OLS}^N is defined as

$$J_{\text{OLS}}^N(\mathbf{q}; \mathbf{Y}) = \frac{1}{N} \sum_{k=1}^{N} (Y_k - f(t_k; \mathbf{q}))^2 .$$

The corresponding realization $\hat{\mathbf{q}}_{\text{OLS}}^N$ of $\mathbf{q}_{\text{OLS}}^N$ is obtained by solving

$$\hat{\mathbf{q}}_{\text{OLS}}^N = \arg \min_{\mathbf{q} \in \Omega_q} J_{\text{OLS}}^N(\mathbf{q}; \mathbf{y}),$$

where $\mathbf{y}$ is a realization of $\mathbf{Y}$ (that is, $\mathbf{y} = (y_1, y_2, \ldots, y_N)^T$).

As alluded to in the introduction, we might also consider a restricted version of the mathematical model in which the unknown true parameter is assumed to lie in a subset $\Omega_q^H \subset \Omega_q$ of the admissible parameter space. We assume this restriction can be written as a linear constraint, $\mathcal{H}\mathbf{q}_0 = \mathbf{h}$, where $\mathcal{H} \in \mathbb{R}^{\kappa_r \times \kappa_q}$ is a matrix having rank κ_r (that is, κ_r is the number of constraints imposed), and $\mathbf{h}$ is a known vector. Thus the restricted parameter space is

$$\Omega_q^H = \{\mathbf{q} \in \Omega_q : \mathcal{H}\mathbf{q} = \mathbf{h}\} .$$

Then the null and alternative hypotheses are

$$H_0 : \quad \mathbf{q}_0 \in \Omega_q^H$$
$$H_A : \quad \mathbf{q}_0 \notin \Omega_q^H .$$

We may define the restricted parameter estimator as

$$\mathbf{q}_{\text{OLS}}^{N,H} = \arg \min_{\mathbf{q} \in \Omega_q^H} J_{\text{OLS}}^N(\mathbf{q}; \mathbf{Y}),$$

and the corresponding realization is denoted by $\hat{\mathbf{q}}_{\text{OLS}}^{N,H}$. Since $\Omega_q^H \subset \Omega_q$, it is clear that

$$J_{\text{OLS}}^N(\hat{\mathbf{q}}_{\text{OLS}}^N; \mathbf{y}) \leq J_{\text{OLS}}^N(\hat{\mathbf{q}}_{\text{OLS}}^{N,H}; \mathbf{y}).$$

This fact forms the basis for a model selection criterion based upon the residual sum of squares.

We remark that while we only consider the linear restriction $\mathcal{H}\mathbf{q}_0 = \mathbf{h}$, it is possible to extend the results above to include non-linear restrictions of the form $\tilde{H}(\mathbf{q}_0) = \tilde{\mathbf{h}}$ [8]. In such a case, one is interested in the restricted parameter space

$$\tilde{\Omega}_q^H = \left\{ \mathbf{q} \in \Omega_q : \tilde{H}(\mathbf{q}) = \tilde{\mathbf{h}} \right\}.$$

Assuming the null hypothesis is true, one can construct the linearization

$$\Psi(\mathbf{q}) = \tilde{H}(\mathbf{q}_0) + \frac{\partial \tilde{H}(\mathbf{q}_0)}{\partial \mathbf{q}}(\mathbf{q} - \mathbf{q}_0).$$

Let $\tilde{\mathbf{h}} = \tilde{H}(\mathbf{q}_0)$. Then

$$\tilde{\Omega}_q^H \approx \left\{ \mathbf{q} \in \Omega_q : \Psi(\mathbf{q}) = \tilde{\mathbf{h}} \right\}.$$

But the condition $\Psi(\mathbf{q}) = \tilde{\mathbf{h}}$ is equivalent to the condition that

$$\frac{\partial \tilde{H}(\mathbf{q}_0)}{\partial \mathbf{q}}\mathbf{q} = \frac{\partial \tilde{H}(\mathbf{q}_0)}{\partial \mathbf{q}}\mathbf{q}_0.$$

Thus, under the null hypothesis, the non-linear restriction $\tilde{H}(\mathbf{q}_0) = \tilde{\mathbf{h}}$ is asymptotically locally equivalent to the linear restriction with

$$\mathcal{H} = \frac{\partial \tilde{H}(\mathbf{q}_0)}{\partial \mathbf{q}}, \quad \mathbf{h} = \frac{\partial \tilde{H}(\mathbf{q}_0)}{\partial \mathbf{q}}\mathbf{q}_0.$$

As such, we only consider the problem of testing a linear hypothesis. For non-linear hypotheses, the results presented here are accurate to the order of the linear approximations above.

Using the assumptions (A1)–(A7) given in Section 3.3, one can establish asymptotic convergence for the *test statistic* (which is a function of observations and is used to determine whether or not the null hypothesis is rejected)

$$U_{\text{OLS}}^N = \frac{N\left(J_{\text{OLS}}^N(\mathbf{q}_{\text{OLS}}^{N,H}; \mathbf{Y}) - J_{\text{OLS}}^N(\mathbf{q}_{\text{OLS}}^N; \mathbf{Y})\right)}{J_{\text{OLS}}^N(\mathbf{q}_{\text{OLS}}^N; \mathbf{Y})},$$

where the corresponding realization $\hat{U}_{\text{OLS}}^N$ is defined as

$$\hat{U}_{\text{OLS}}^N = \frac{N\left(J_{\text{OLS}}^N(\hat{\mathbf{q}}_{\text{OLS}}^{N,H}; \mathbf{y}) - J_{\text{OLS}}^N(\hat{\mathbf{q}}_{\text{OLS}}^N; \mathbf{y})\right)}{J_{\text{OLS}}^N(\hat{\mathbf{q}}_{\text{OLS}}^N; \mathbf{y})}. \tag{4.46}$$

This asymptotic convergence result is summarized in the following theorem.

Theorem 4.5.1 *Under assumptions (A1)–(A7) given in Section 3.3 and assuming the null hypothesis H_0 is true, then U_{OLS}^N converges in distribution (as $N \to \infty$) to a random variable U having a chi-square distribution with κ_r degrees of freedom.*

TABLE 4.3: $\chi^2(1)$ values.

α	τ	confidence level
0.25	1.32	75%
0.1	2.71	90%
0.05	3.84	95%
0.01	6.63	99%
0.001	10.83	99.9%

The above theorem suggests that if the sample size N is sufficiently large, then U_{OLS}^N is approximately chi-square distributed with κ_r degrees of freedom. We use this fact to determine whether or not the null hypothesis H_0 is rejected. To do that, we choose a *significance level* α (usually chosen to be 0.05 for a 95% confidence interval) and use χ^2 tables to obtain the corresponding *threshold* value τ so that $\mathrm{Prob}(U > \tau) = \alpha$. We next compute $\hat{U}_{\mathrm{OLS}}^N$ and compare it to τ. If $\hat{U}_{\mathrm{OLS}}^N > \tau$, then we *reject* the null hypothesis H_0 with confidence level $(1 - \alpha)100\%$; otherwise, we do not reject it. We emphasize that care should be taken in stating conclusions: we either reject or do not reject H_0 at the specified level of confidence. Table 4.3 illustrates the threshold value for $\chi^2(1)$ with the given significance level. This can be found in any elementary statistics text or online or calculated by some software package such as MATLAB, and is given here for illustrative purposes and also for use in the examples demonstrated below.

In the framework of hypothesis testing, we may often hear another terminology, *p-value*, which is defined as the minimum significance level at which H_0 can be rejected. For our application here, the *p*-value is calculated as

$$p\text{-value} = 1 - P_U(\hat{U}_{\mathrm{OLS}}^N), \qquad (4.47)$$

where P_U denotes the cumulative distribution function of a chi-square distributed random variable U with κ_r degrees of freedom. If the *p*-value is less than the given significance level, then we reject the null hypothesis. Thus, the smaller the *p*-value, the stronger the evidence in the data in support of rejecting the null hypothesis. We implement this as follows: we first compute $\hat{U}_{\mathrm{OLS}}^N$, and then obtain the *p*-value using (4.47). The fact that the *p*-value is the minimum significance level at which H_0 can be rejected implies that the null hypothesis can be rejected at any confidence level less than $(1-p\text{-value})100\%$. For example, if for a computed $\hat{U}_{\mathrm{OLS}}^N$ we find a *p*-value $= 0.0182$, then the null hypothesis H_0 can be rejected at any confidence level less than 98.18%. For more information, the reader can consult analysis of variance discussions in any good statistics book.

4.5.2 Application: Cat Brain Diffusion/Convection Problem

We now illustrate the RSS based model selection criterion in the case of ordinary least squares by considering a mathematical model for a diffusion-convection process. This model was proposed for use with experiments designed to study substance (labeled sucrose) transport in cat brains, which are heterogeneous, containing grey and white matter [5]. In general, the transport of a substance in cats' brains can be described by a partial differential equation having the form

$$\frac{\partial u}{\partial t} + \mathcal{V}\frac{\partial u}{\partial x} = \mathcal{D}\frac{\partial^2 u}{\partial x^2},$$

where $u(t,x)$ represents the concentration of a substance being transported in the brain at location x at time t, $\mathcal{D}$ is the diffusion coefficient and $\mathcal{V}$ is the bulk velocity of a transporting fluid.

A primary question concerning transport mechanisms in brain tissue is whether diffusion alone or diffusion and convection are responsible for transport in gray and white matter. Thus, given data involving the transport of labeled sucrose, one would like to compare the accuracy of two mathematical models. In the first, both convection and diffusion play a role and one must estimate the parameters $\mathcal{V}$ and $\mathcal{D}$ in an inverse problem. In the second, it is assumed that convection is not present ($\mathcal{V} = 0$), and only the diffusion coefficient $\mathcal{D}$ is estimated. Thus, in effect one has two mathematical models which must be compared in terms of how accurately and parsimoniously they describe an available data set. Equivalently, this problem can be cast in the form of a hypothesis test: one would like to test the null hypothesis $\mathcal{V} = 0$ against the alternative hypothesis that $\mathcal{V} \neq 0$. Under the null hypothesis, the constrained mathematical model (that is, the model with $\mathcal{V} = 0$) is a special case of the mathematical model subject to the unconstrained parameter space (in the sense that estimating both $\mathcal{V}$ and $\mathcal{D}$ simultaneously includes the possibility that $\mathcal{V} = 0$). For this reason, the unconstrained model must necessarily fit the data (in an appropriate sense) at least as accurately as the constrained model. The role of the hypothesis test is to compare these two models on a quantitative basis to investigate whether the model with $\mathcal{V} \neq 0$ provides a significantly (in some sense) better fit to the data.

There were three sets of experimental data examined. For Data Set 1, we have eight data points (that is, $N = 8$) and found that

$$J_{\text{OLS}}^N(\hat{\mathbf{q}}_{\text{OLS}}^N;\mathbf{y}) = 106.15 \quad \text{and} \quad J_{\text{OLS}}^N(\hat{\mathbf{q}}_{\text{OLS}}^{N,H};\mathbf{y}) = 180.17.$$

By (4.46), we find that for this case $\hat{U}_{\text{OLS}}^N \approx 5.579$. Since there is only one constraint ($\mathcal{V} = 0$), $\kappa_r = 1$. Then by (4.47) we find that the p-value $= 0.0182$. Thus, the null hypothesis in this case can be rejected at *any* confidence level less than 98.18%, which suggests convection is important in describing this data set.

For Data Set 2, we also have eight data points (that is, $N = 8$) and found that

$$J_{\text{OLS}}^N(\hat{\mathbf{q}}_{\text{OLS}}^N;\mathbf{y}) = 14.68 \quad \text{and} \quad J_{\text{OLS}}^N(\hat{\mathbf{q}}_{\text{OLS}}^{N,H};\mathbf{y}) = 15.35,$$

and thus, in this case, we have $\hat{U}_{\text{OLS}}^N \approx 0.365$, which implies we *do not reject* the null hypothesis with *high degrees of confidence* (p-value very high). This suggests $\mathcal{V} = 0$, which is completely opposite to the findings for Data Set 1.

For the final set (Data Set 3) we again have eight data points and found

$$J_{\text{OLS}}^N(\hat{\mathbf{q}}_{\text{OLS}}^N; \mathbf{y}) = 7.8 \quad \text{and} \quad J_{\text{OLS}}^N(\hat{\mathbf{q}}_{\text{OLS}}^{N,H}; \mathbf{y}) = 22.7,$$

which yields in this case $\hat{U}_{\text{OLS}}^N = 15.28$. This, as in the case of the first data set, suggests (with $p\text{-}value < 0.001$) that $\mathcal{V} \neq 0$ is important in describing the given data.

The difference in conclusions between the first and last sets and that of the second set is interesting and perhaps at first puzzling. However, when discussed with the doctors who provided the data, it was discovered that the first and last set were taken from the *white matter* of the brain, while the other was taken from the *grey matter*. This latter finding was consistent with observed microscopic tests on the various matter (micro channels in white matter that promote convective "flow"). Thus, it can be suggested with a reasonably high degree of confidence that white matter exhibits convective transport, while grey matter does not.

Remark 4.5.2 *We note that our use of the RSS based model selection criterion here is in keeping with practices frequently encountered in statistical analysis of experimental data, heuristic in several senses. First, we do not (and usually cannot) verify that the assumptions (some of them rather strong!) of Theorem 4.5.1 hold. Furthermore, we compute the statistic U_{OLS}^N for a fixed value of N, but we use the limiting chi-square distribution for setting the rejection threshold in the test. In practice N may be rather small (like the example demonstrated here), but, we hope, sufficiently large so that the distribution of U_{OLS}^N is reasonably approximated by a chi-square distribution. Nonetheless, with these caveats in mind, the example demonstrated here shows that the RSS based model selection criterion can be quite useful as an aid in evaluating various questions arising in inverse problems.*

4.5.3   Weighted Least Squares

The results presented in Section 4.5.1 can be extended to statistical model (4.1) in which measurement errors are independent with $\mathbb{E}(\mathcal{E}_k) = 0$ and $\text{Var}(\mathcal{E}_k) = \sigma^2 w^2(t_k)$, $k = 1, 2, \ldots, N$, where w is some known real-valued function with $w(t) \neq 0$ for any t. This is achieved through rescaling the observations in accordance with their variance (as discussed in Section 3.3.1) so that the resulting (transformed) observations are identically distributed as well as independent.

With this assumption on the measurement errors, the mathematical model parameters $\mathbf{q}$ can be estimated by the weighted least squares method with cost function

$$J_{\text{WLS}}^N(\mathbf{q}; \mathbf{Y}) = \frac{1}{N} \sum_{k=1}^N \left(\frac{Y_k - f(t_k; \mathbf{q})}{w(t_k)} \right)^2,$$

where $\mathbf{Y} = (Y_1, Y_2, \ldots, Y_N)^T$. In other words, the *weighted least squares estimator* $\mathbf{q}_{\text{WLS}}^N$ is obtained by solving

$$\mathbf{q}_{\text{WLS}}^N = \arg \min_{\mathbf{q} \in \Omega_q} J_{\text{WLS}}^N(\mathbf{q}; \mathbf{Y}),$$

and its corresponding realization is obtained by solving

$$\hat{\mathbf{q}}_{\text{WLS}}^N = \arg \min_{\mathbf{q} \in \Omega_q} J_{\text{WLS}}^N(\mathbf{q}; \mathbf{y}),$$

where $\mathbf{y}$ is a realization of $\mathbf{Y}$. Similarly, the restricted weighted least squares estimator $\mathbf{q}_{\text{WLS}}^{N,H}$ (or estimate $\hat{\mathbf{q}}_{\text{WLS}}^{N,H}$) can be obtained by minimizing $J_{\text{WLS}}^N(\mathbf{q}; \mathbf{Y})$ (or $J_{\text{WLS}}^N(\mathbf{q}; \mathbf{y})$) over Ω_q^H, as we did for the ordinary least squares case.

Using assumptions (A1′a), (A1′b), (A2)–(A6) and (A7′) given in Section 3.3, one can establish asymptotic convergence results for the test statistics

$$U_{\text{WLS}}^N = \frac{N \left(J_{\text{WLS}}^N(\mathbf{q}_{\text{WLS}}^{N,H}; \mathbf{Y}) - J_{\text{WLS}}^N(\mathbf{q}_{\text{WLS}}^N; \mathbf{Y}) \right)}{J_{\text{WLS}}^N(\mathbf{q}_{\text{WLS}}^N; \mathbf{Y})},$$

where the corresponding realization is given by

$$\hat{U}_{\text{WLS}}^N = \frac{N \left(J_{\text{WLS}}^N(\hat{\mathbf{q}}_{\text{WLS}}^{N,H}; \mathbf{y}) - J_{\text{WLS}}^N(\hat{\mathbf{q}}_{\text{WLS}}^N; \mathbf{y}) \right)}{J_{\text{WLS}}^N(\hat{\mathbf{q}}_{\text{WLS}}^N; \mathbf{y})}. \tag{4.48}$$

This asymptotic convergence result is summarized in the following theorem (see [8] for details of the proof).

Theorem 4.5.3 *Under assumptions (A1′a), (A1′b), (A2)–(A6) and (A7′) given in Section 3.3, and assuming that the null hypothesis H_0 is true, then U_{WLS}^N converges in distribution (as $N \to \infty$) to a random variable U having a chi-square distribution with κ_r degrees of freedom.*

4.5.4   Summary Remarks

The RSS based model selection criterion is based on hypothesis testing. Hence, it is valid only for nested models, and it can only compare two models at a time. In addition, the RSS based model selection criterion is derived based on the strong assumption that the unconstrained model is the true model. However, this criterion is designed specifically for least squares type estimation (where only the first two moments of the measurement errors need to be specified), and the test statistic depends only on the RSS for the constrained and unconstrained models (which are very easy to calculate from the minimized least squares cost functions).

In addition to the example demonstrated in Section 4.5.2, the RSS based model selection criterion has also been used to investigate mortality and anemotaxis in models of insect dispersal [6, 7]. These examples are summarized (focusing on the role of hypothesis testing) in [3, 9]. More recently, the RSS based model selection criterion has been used to investigate the division dependence of death rates in a model of a dividing cell population [12], to evaluate the necessity of components in viscoelastic arterial models [36] and to serve as the basis for developing a methodology to determine if a stenosis is present in a vessel based on input amplitude [4]. Thus the discussions and methods presented here have widespread applications in biological modeling as well as in more general scientific modeling efforts.

References

[1] H.T. Banks, *Modeling and Control in the Biomedical Sciences*, Lecture Notes in Biomathematics, Vol. **6**, Springer, Heidelberg, 1975.

[2] H.T. Banks, A. Choi, T. Huffman, J. Nardini, L. Poag and W.C. Thompson, Modeling CFSE label decay in flow cytometry data, *Applied Mathematical Letters*, **26** (2013), 571–577.

[3] H.T. Banks and B.G. Fitzpatrick, Statistical methods for model comparison in parameter estimation problems for distributed systems, *J. Math. Biol.*, **28** (1990), 501–527.

[4] H.T. Banks, S. Hu, Z.R. Kenz, C. Kruse, S. Shaw, J.R. Whiteman, M.P. Brewin, S.E. Greenwald and M.J. Birch, Material parameter estimation and hypothesis testing on a 1D viscoelastic stenosis model: Methodology, *J. Inverse and Ill-Posed Problems*, **21** (2013), 25-57.

[5] H.T. Banks and P. Kareiva, Parameter estimation techniques for transport equations with applications to population dispersal and tissue bulk flow models, *J. Math. Biol.*, **17** (1983), 253–272.

[6] H.T. Banks, P. Kareiva, and P.D. Lamm, Modeling insect dispersal and estimating parameters when mark-release techniques may cause initial disturbances, *J. Math. Biol.*, **22** (1985), 259–277.

[7] H.T. Banks, P. Kareiva and K. Murphy, Parameter estimation techniques for interaction and redistribution models: a predator-prey example, *Oecologia*, **74** (1987), 356–362.

[8] H.T. Banks, Z.R. Kenz and W.C. Thompson, An extension of RSS-

based model comparison tests for weighted least squares, *Intl. J. Pure and Appl. Math.*, **79** (2012), 155–183.

[9] H.T. Banks and K. Kunisch, *Estimation Techniques for Distributed Parameter Systems*, Birkhauser, Boston, 1989.

[10] H.T. Banks, K.L. Sutton, W.C. Thompson, G. Bocharov, D. Roose, T. Schenkel and A. Meyerhans, Estimation of cell proliferation dynamics using CFSE data, *Bull. Math. Biol.* **70** (2011), 116–150.

[11] H.T. Banks, K.L. Sutton, W.C. Thompson, G. Bocharov, M. Doumic, T. Schenkel, J. Argilaguet, S. Giest, C. Peligero and A. Meyerhans, A new model for the estimation of cell proliferation dynamics using CFSE data, *J. Immunological Methods,* **373** (2011), 143–160.

[12] H.T. Banks, K.L. Sutton, W.C. Thompson, G. Bocharov, D. Roose, T. Schenkel and A. Meyerhans, Estimation of cell proliferation dynamics using CFSE data, *Bull. Math. Biol.*, **70** (2011), 116–150.

[13] H.T. Banks and H.T. Tran, *Mathematical and Experimental Modeling of Physical and Biological Processes*, CRC Press, Boca Raton, FL, 2009.

[14] E.J. Bedrick and C.L. Tsai, Model selection for multivariate regression in small samples, *Biometrics*, **50** (1994), 226–231.

[15] H. Bozdogan, Model selection and Akaike's information criterion (AIC): The general theory and its analytical extensions, *Psychometrika*, **52** (1987), 345–370.

[16] H. Bozdogan, Akaike's information criterion and recent developments in information complexity, *Journal of Mathematical Psychology*, **44** (2000), 62–91.

[17] K.P. Burnham and D.R. Anderson, *Model Selection and Inference: A Practical Information-Theoretical Approach*, 2nd edition, Springer-Verlag, New York, 2002.

[18] A. Charnes, E.L. Frome and P.L. Yu, The equivalence of generalized least squares and maximum likelihood estimates in the exponential family, *Journal of the American Statistical Association*, **71** (1976), 169–171.

[19] A. Choi, T. Huffman, J. Nardini, L. Poag, W.C. Thompson and H.T. Banks, Modeling CFSE label decay in flow cytometry data, CRSC-TR12-20, November, 2012.

[20] M.J. Crowder, Maximum likelihood estimation for dependent observations, *Journal of the Royal Statistical Society. Series B (Methodological)*, **38** (1976), 45–53.

[21] G. de Vries, et al., *A Course in Mathematical Biology*, SIAM Series on Mathematical Modeling and Computation, Vol. **MM12**, SIAM, Philadelphia, 2006.

[22] I. Domowitz and H. White, Misspecified models with dependent observations, *Journal of Econometrics*, **20** (1982), 35–58.

[23] A.R. Gallant, *Nonlinear Statistical Models*, John Wiley & Sons, New York, 1987.

[24] A.R. Gallant and H. White, *A Unified Theory of Estimation and Inference for Nonlinear Dynamic Models*, Basil Blackwell Ltd., New York, 1988.

[25] J. Hasenauer, D. Schittler and F. Allgöwer, Analysis and simulation of division- and label-structured population models: A new tool to analyze proliferation assays, *Bull. Math. Biol.*, **74** (2012), 2692–2732.

[26] B. Hoadley, Asymptotic properties of maximum likelihood estimators for the independent not identically distributed case, *The Annals of Mathematical Statistics*, **42** (1971), 1977–1991.

[27] C.M. Hurvich and C.L. Tsai, Regression and time series model selection in small samples, *Biometrika*, **76** (1989), 297–307.

[28] M. Kot, *Elements of Mathematical Ecology*, Cambridge University Press, Cambridge, UK, 2001.

[29] S. Kullback and R.A. Leibler, On information and sufficiency, *Annals of Mathematical Statistics*, **22** (1951), 79–86.

[30] T. Luzyanina, D. Roose, T. Schenkel, M. Sester, S. Ehl, A. Meyerhans and G. Bocharov, Numerical modelling of label-structured cell population growth using CFSE distribution data, *Theoretical Biology and Medical Modelling*, **4** (2007), published online.

[31] C.W. Mallows, Some comments on C_p, *Technometrics*, **15** (1973), 661–675.

[32] P. McCullagh, Quasi-likelihood functions, *The Annals of Statistics*, **11** (1983), 59–67.

[33] S.I. Rubinow, *Introduction to Mathematical Biology*, John Wiley & Sons, New York, 1975.

[34] G. Schwarz, Estimating the dimension of a model, *The Annals of Statistics*, **6** (1978), 461–464.

[35] N. Sugiura, Further analysis of the data by Akaike's information criterion and the finite corrections, *Comm. Statist.*, **A7** (1978), 13–26.

[36] D. Valdez-Jasso, H.T. Banks, M.A. Haider, D. Bia, Y. Zocalo, R.L. Armentano and M.S. Olufsen, Viscoelastic models for passive arterial wall dynamics, *Adv. in Applied Math. and Mech.*, **1** (2009), 151–165.

[37] H. White, Maximum likelihood estimation of misspecified dynamic models, in *Misspecification Analysis* (T.K. Dijkstra, ed.), Springer-Verlag, New York, 1984, 1–19.

[38] H. White, *Estimation, Inference, and Specification Analysis*, Cambridge University Press, Cambridge, UK, 1994.

Chapter 5

Estimation of Probability Measures Using Aggregate Population Data

5.1 Motivation

In the mathematical modeling of physical and biological systems, situations often arise where some facet of the underlying dynamics (in the form of a parameter) is not constant but rather is distributed probabilistically within the system or across the population under study. While traditional inverse problems (as discussed in Chapter 3) involve the estimation, given a set of data/observations, of a fixed set of parameters contained within some finite dimensional admissible set, models with distributed parameters require the estimation of a probability measure or distribution over the set of admissible parameters. This problem is well known to both applied mathematicians and statisticians and both schools have developed their own set of tools for the estimation of a measure or distribution over a set. Not surprisingly, the methods developed by the two schools are best suited to the particular types of problems most frequently encountered in their respective fields. As such, the techniques for measure estimation (along with their theoretical justifications) are widely scattered throughout the literature in applied mathematics and statistics, often with few cross references to related ideas; a more complete review is given in [25].

In this chapter we present two approaches to these techniques, paying particular attention to the theoretical underpinnings of the various methods as well as to computational issues which result from the theory. Throughout this chapter, we will focus on the *nonparametric* estimation of a probability measure. That is, we want to establish a meaningful estimation problem while simultaneously placing a minimal set of restrictions on the underlying measure. The fully parametric case (that is, when the form of the underlying measure is assumed to be known and determined uniquely by a small number of parameters – for example, its mean and covariance matrix) can be considered as a standard parameter estimation problem over Euclidean space for which computational strategies are well known and readily available. Some of these were discussed in Chapter 3.

Not surprisingly, the applicability of various techniques for the nonparametric estimation of a probability measure is heavily dependent upon the type of data available for use with a given model in an inverse problem. In some studies (i.e., those considered to this point in this monograph), data is available at particular times for each individual in the population under study. Such might be the case, for example, in a pharmacokinetics study in which one desires to estimate drug absorption and excretion rates from blood samples taken serially in time from individuals in the population. For example, consider a dynamic pharmacokinetics model for an individual's blood concentration of the drug theophylline with solution

$$C(t, F, D, k_a, k_e, V) = \frac{k_a F D}{V(k_a - k_e)} \left(e^{-k_e t} - e^{-k_a t} \right), \qquad (5.1)$$

where V is blood volume, D is the dose received, F is the uptake fraction and k_a and k_e are rates of drug absorption and excretion, respectively. For sufficient data collected from a single individual, it is a standard problem in statistics (e.g., [36, Chapter 2]) and mathematics (e.g., [29, Chapter 3] and Chapter 3 above) to estimate, for example, the parameters F, k_a and k_e for V and D known. Of course, it is highly likely that individual parameters (say, k_a and/or k_e) might vary from one individual to the next within the sampled population. Thus, our goal in this case is to use the entire sample of individuals to estimate the probability measure describing the distribution of certain parameters in the full population.

In other investigations, a situation might arise in which one is only able to collect aggregate data. Such might be the case, for example, in catch and release experiments in which one cannot be certain of measuring the same individual multiple times. Unlike the pharmacokinetics example, one does not have individual data, but rather histograms showing the aggregate number of individuals sampled from the population at a given time and having a given size ([9] and [29, Chapter 9]). The goal is to estimate the probability distributions describing the variability of the parameters across the population.

As can be readily seen, these two situations lead to fundamentally different estimation problems. In the first case, one attempts to use a mathematical model describing a single individual along with data collected from multiple individuals (but data in which subsets of longitudinal data points could be identified with a specific individual) in order to determine population level characteristics. In the second case, while one again has an individual model, the data collected cannot be identified with individuals and is considered to be sampled longitudinally from the aggregate population. It is worth noting that special care must be taken in this latter case to identify the model, such as the model (5.2) introduced below in the case of structured population models as an individual model in the sense that it describes an individual sub-population. That is, all "individuals" (i.e., shrimp [14] or mosquitofish [9]) described by the model share a common growth rate function g. The confusing terminology is largely the result of the disparate fields in which such problems are commonly found. In pharmacokinetics, one typically discusses "individual" patients which together constitute a "population." In ecology, one often works with an "aggregate" population which itself consists of smaller "individual" subpopulations. Mathematically, here we define the "individual" in terms of the underlying parameters, using "individual" to describe the unit characterized by a single parameter set (excretion rate, growth rate, damping rate, relaxation times, etc.).

In this chapter, we first consider two generic estimation problems for which we give explicit examples below. In the first example, we consider the case, as in the structured density model below, that one has a mathematical model for individual dynamics but only aggregate data (we refer to this as the *individual dynamics/aggregate data* problem). The second possibility—that one has only an aggregate model (i.e., the dynamics depend explicitly on a distribution of parameters across the population) with aggregate data – will also be examined (we refer to this as the *aggregate dynamics/aggregate data* problem). Such examples arise in electromagnetic models with a distribution of polarization relaxation times for molecules (e.g., [20, 21]); in biology with HIV cellular models [6, 7, 8]; and in wave propagation in viscoelastic materials such as biotissue [22, 23, 27]. The measure estimation problem for such examples is sufficiently similar to the individual dynamics/aggregate data situation and accordingly we do not consider aggregate dynamics models as a separate case but only offer some explicit comments as warranted.

In a final section of this chapter, we consider a related problem, the classical Nonparametric Maximum Likelihood (NMPL) estimation problem as proposed by Mallet [52, 53]. In these related (but not equivalent) problems one again wishes to estimate a probability distribution for parameters across a population, but the standard theory (as developed by Mallet, Lindsay [50, 51] and others) assumes that one can follow and obtain data for individuals (i.e., *individual dynamics/individual data* problem).

In both generic estimation problems discussed, the underlying goal is the determination of the probability measure which describes the distribution of parameters across all members of the population. Thus, two main issues are of interest. First, a sensible framework must be established for each situation so that the estimation problem is meaningful. Thus we must decide what type of information and/or estimates are desired (e.g., mean, variance, complete distribution function, etc.) and determine how these decisions will depend on the type of data available. Second, we must examine what techniques are available for the computation of such estimates. Because the space of probability measures is an infinite dimensional space, we must make some type of finite dimensional approximations so that the estimation problem is amenable to computation. Moreover, here we are interested in frameworks for approximation and convergence which are not restricted to classic parametric estimates. We shall see that, while the individual dynamics/individual data and individual dynamics/aggregate data problems are fundamentally different, there are notable similarities in the approximation techniques (namely, use of discrete measures) common to both problems.

5.2   Type I: Individual Dynamics/Aggregate Data Inverse Problems

The first class of problems to be discussed, which we call *Type I* problems, involves individual dynamics which do not depend explicitly on the population level probability distribution P itself. In this case one has data across the population (*aggregate data*) for the expected value of the population state variables. Dynamics are available for individual subpopulation trajectories $v(t; g)$ for given (individual level) parameters $g \in \mathcal{G}$. A large class of structured population models provides a major set of examples for these inverse problems.

5.2.1   Structured Population Models

In these models, the "spatial" variable ξ is actually "weight" or "size" in place of spatial location (see [29, Chapter 9]) and such models have been effectively used to model marine populations such as shrimp [13, 14, 15] and, earlier, mosquitofish [9, 18, 19]. The early versions of these size structured population models were first proposed by Sinko and Streifer [64] in 1967. Cell population versions were proposed by Bell and Anderson [31] almost simultaneously. Other versions of these models called "age structured models," where age can be "physiological age" as well as chronological age, are discussed in [54].

The *Sinko–Streifer (SS) model* for weight-structured shrimp populations is given by

$$\frac{\delta v}{\delta t} + \frac{\delta}{\delta \xi}(gv) = -\mu v, \quad \xi_0 < \xi < \xi_1, \quad t > 0$$

$$v(0, \xi) = \Phi(\xi) \tag{5.2}$$

$$g(t, \xi_0)v(t, \xi_0) = \int_{\xi_0}^{\xi_1} K(t, \xi)v(t, \xi)d\xi.$$

Here $v(t, \xi)$ represents number density (given in numbers per unit weight) or population density of individuals with size ξ at time t. The growth rate of individual shrimp with size ξ at time t is assumed given by $g(t, \xi)$, where

$$\frac{d\xi}{dt} = g(t, \xi) \tag{5.3}$$

for each individual (all shrimp of a given weight have the same growth rate). We assume existence of a maximum weight ξ_1 so that $g(t, \xi_1) = 0$ for all t.

In the SS model $\mu(t, \xi)$ represents the *mortality rate* of shrimp with size ξ at time t, and the function Φ represents the initial weight density of the

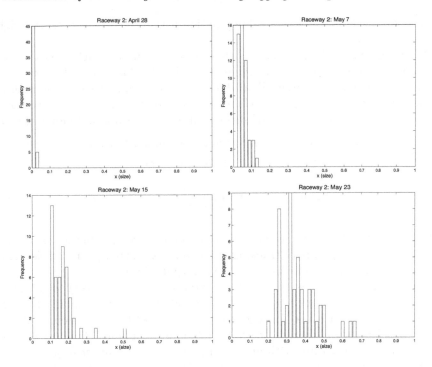

FIGURE 5.1:   Histograms for longitudinal data for a typical shrimp race-way demonstrating the need for variability in growth rates in the model.

population, while K represents the *fecundity kernel*. The boundary condition at $\xi = \xi_0$ is *recruitment*, or *birth rate*. The SS model *cannot* be used as formulated above to model the shrimp population because it *does not predict dispersion* of the population in time under biologically reasonable assumptions [11, 12, 13, 14, 15], as depicted in Figure 5.1.

In light of the dispersion in growth depicted in this figure, we will replace the growth rate g by a family $\mathcal{G}$ of growth rates and reconsider the model with a probability distribution P on this family. The population density is then given by summing "cohorts" of subpopulations where individuals belong to the same subpopulation if they have the same growth rate [9, 18, 19]. Thus, in the so-called Growth Rate Distribution (GRD) model, the population density $u(t, \xi; P)$, first discussed in [9] for mosquitofish, developed more fully in [18] and subsequently used for the shrimp models, is actually given by

$$u(t,\xi;P) = \int_{\mathcal{G}} v(t,\xi;g)dP(g) = \mathbb{E}(v(t,\xi;\cdot)|P), \qquad (5.4)$$

where $\mathcal{G}$ is a collection of admissible growth rates, P is a probability measure on $\mathcal{G}$ and $v(t,\xi;g)$ is the solution of the SS equation (5.2) with growth rate g. This model assumes the population is made up of *collections of subpopulations* with individuals in the same subpopulation if they possess the same weight dependent growth rate.

Many inverse problems, such as those discussed in the earlier part of this monograph, involve individual dynamics and individual data. For example, if one had individual longitudinal data y_{ij} corresponding to the structured population density $v(t_i, \xi_j; g)$, where v is the solution to (5.2), corresponding to an (individual) cohort of shrimp all with the same individual growth rate $\dfrac{d\xi}{dt} = g(t, \xi)$, one could then formulate a standard ordinary least squares (OLS) problem for estimation of g. This would entail finding

$$\hat{g} = \arg\min_{g \in \mathcal{G}} J(g) = \arg\min_{g \in \mathcal{G}} \sum_{i,j} |v_{ij} - v(t_i, \xi_j; g)|^2, \qquad (5.5)$$

where $\mathcal{G}$ is a family of admissible growth rates for a given population of shrimp. However, for such problems tracking of individuals is usually impossible. Thus we turn to methods where one can use aggregate data.

As outlined above, we find that the expected weight density $u(t, \xi; P) = \mathbb{E}(v(t, \xi; \cdot)|P)$ is described by a general probability distribution P where the density $v(t, \xi; g)$ satisfies, for a given g, the Sinko–Streifer system (5.2). In these problems, even though one has "individual dynamics" (the $v(t, \xi; g)$ for a given cohort of shrimp with growth rate g), one has only *aggregate* or population level longitudinal data available. This is common in marine, insect, etc., *catch and release* experiments [26] where one samples at different times from the same population but cannot be guaranteed of observing the same set of individuals at each sample time. This type of data is also typical in experiments where the organism or population member being studied is sacrificed in the process of making a single observation (e.g., certain physiologically based pharmacokinetic (PBPK) modeling [28, 56] and whole organism transport models [26]). In these cases one may still have dynamic (i.e., time course) models for individuals as in the shrimp, but no individual data is available.

Since we must use aggregate population data $\mathbf{u} = \{u_{ij}\}$ to estimate P itself in the corresponding typical inverse problems, we are therefore required to understand the qualitative properties (continuity, sensitivity, etc.) of $u(t, \xi; P)$ as a function of P. The data for parameter estimation problems are u_{ij}, which are observations for $u(t_i, \xi_j; P)$. The corresponding OLS inverse problem consists of minimizing

$$J(P; \mathbf{u}) = \sum_{ij} |u_{ij} - u(t_i, \xi_j; P)|^2 \qquad (5.6)$$

over $P \in \mathbb{P}(\mathcal{G})$, the set of probability distributions on $\mathcal{G}$, or over some suitably chosen subset of $\mathbb{P}(\mathcal{G})$.

5.3   Type II: Aggregate Dynamics/Aggregate Data Inverse Problems

The second class of problems, which we call *Type II* problems, involves dynamics which depend explicitly on the probability distribution P itself. Hence, we do not take the expected value of the population state variable. No dynamics are available for individual (e.g., particle or molecular level in materials) trajectories $v(t, \boldsymbol{\theta})$ for given (population or molecular level) parameters $\boldsymbol{\theta} \in \Omega_\theta$. Electromagnetic models in non-magnetic dielectric material (detailed below) [20, 21] as well as models in viscoelasticity [22, 23, 27] are precisely examples of this nature. Such problems also arise in biology (as discussed in Section 5.1).

5.3.1   Probability Measure Dependent Systems— Viscoelasticity

One class of our motivating problems arose in joint collaborations with scientists and engineers at MedAcoustics, Inc., [3, 27, 30] as part of the Industrial Applied Mathematics Program at North Carolina State University (see http://www.ncsu.edu/crsc/) and has continued in recent years through collaborations with researchers at Brunel University and Queen Mary Hospital in England [22, 23, 47]. We summarize briefly here the salient points of this problem, while referring the reader to the above references for more details. Turbulent blood flow due to arterial stenosis in partially occluded arteries produces normal forces on arterial walls. This results in vibrations in the surrounding body tissue which are transmitted to body surfaces in two forms: compressional waves and shear waves. The shear waves are at low frequencies ($\leq 2\ kHz$) with low propagation speed and attenuation, and hence they are the focus of the detection efforts. Devices involving multiple arrays of piezoceramic sensors were developed at MedAcoustics, with the goal of measuring shear wave propagation at the surface of the chest. The resulting signals are then processed to determine the location and extent of the stenosis. A part of this overall detection and diagnostic problem is the focus of our efforts: modeling of the shear wave propagation through a chest cavity. The cavity is composed of soft body tissues (lung, muscular and connective tissue including blood) as well as bone. For simplicity, we consider shear wave propagation through a viscoelastic heterogeneous medium. At MedAcoustics, early experiments and data collection were carried out on a cylindrical shaped mold of tissue-like synthetic gel, as depicted schematically in Figure 5.2. The cylinder of gel surrounded a tube in which source disturbances simulating disrupted flows were produced. Thus the tube with disturbances mimicked an artery with stenosis or partial blockages. Shear waves were measured with sensor arrays mounted on the gel outer surface. In [3, 22, 23], the authors developed and

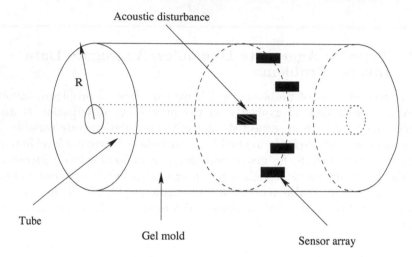

FIGURE 5.2:   A cylindrical model geometry with artificial "stenosis" and surface sensors.

used with some success a one-dimensional (axial symmetry was assumed) homogeneous medium model for computations and analysis of the gel mold. We summarize here a version of these models which allows for "molecular" level "heterogeneity" (in the sense of families of molecules with different relaxation characteristics) in the tissue (gel), while retaining the overall geometry and axial symmetry. An inner radius R_1 and outer radius R_2 for the gel is assumed, with the gel initially at rest. In one configuration, the outer surface of the gel is a free surface. The gel model, which is a reasonable first simplification to illustrate the relevant inverse problems, is far from physiologically realistic.

In [3] the following one-dimensional model (see Figure 5.2) is introduced to describe the propagation of shear waves in soft tissue:

$$\rho\frac{\partial^2 u}{\partial t^2}(t,x) - \frac{\partial}{\partial x}\sigma(t,x) = F(t,x), \quad R_1 < x < R_2, \qquad (5.7)$$

where ρ is the mass density, u is the Cartesian shear displacement, σ is the shear stress, F represents a body forcing term and x is the spatial variable. We assume that at the inner or left boundary, $(x = R_1)$, we have a pure shear force $f(t)$, while the outer or right boundary, $(x = R_2)$, is a free surface and hence

$$\sigma(t,R_1) = f(t), \quad \text{and} \quad \sigma(t,R_2) = 0. \qquad (5.8)$$

In this simplest formulation the initial conditions are

$$u(0,x) = u_0(x), \quad \text{and} \quad u_t(0,x) = u_1(x), \quad R_1 < x < R_2. \qquad (5.9)$$

To complete the model one needs to find an adequate constitutive relationship that relates the shear stress, σ, and strain, $\varepsilon(t,x) = u_x(t,x)$, in soft tissue (arteries, muscle, skin, lung, etc.). In [3] an internal strain variable model that also includes Kelvin–Voigt (rate-dependent) damping is considered in the form

$$\sigma(t,x) = C_D \dot{\varepsilon}(t,x) + \sum_{j=1}^{m} p_j^m \varepsilon_j(t,x), \tag{5.10}$$

with internal strains ε_j defined by

$$\frac{d\varepsilon_j(t,x)}{dt} = -\frac{1}{\tau_j}\varepsilon_j(t,x) + C_j \frac{d}{dt}\sigma_e(u_x(t,x)) \tag{5.11}$$

$$\varepsilon_j(0,x) = 0, \tag{5.12}$$

for $j = 1, \ldots m$, and σ_e is the elastic response function defined in Fung [41, Chapter 7] and given as

$$\sigma_e(u_x) = \gamma + \beta e^{\alpha u_x}. \tag{5.13}$$

In the above equations, C_D is the damping parameter for the Kelvin–Voigt damping term, τ_j, $j = 1, 2, \ldots, m$ denote the relaxation times, p_j^m is the proportion of the material subject to relaxation time τ_j so that $\sum_{j=1}^{m} p_j^m = 1$, and C_j, $j = 1, 2, \ldots, m$, γ, β and α are some constants. The internal strains are at the *molecular or microscopic* level and, as seen in (5.11), are dynamically driven by the elastic response function σ_e, which is a material *macroscopic* quantity. Furthermore, both σ and u are aggregate quantities (summed over the molecular structure), and hence our classification of the associated inverse problems as "aggregate dynamics/aggregate data" problems.

Fung [41] further proposes the constitutive relationship

$$\sigma(t,x) = \int_0^t G(t-s) \frac{d\sigma_e(u_x(s,x))}{ds} ds, \tag{5.14}$$

where G is the relaxation function given in the form

$$G(t) = \left\{ 1 + C \left[E_1 \left(\frac{t}{\tau_2} \right) - E_1 \left(\frac{t}{\tau_1} \right) \right] \right\} \left[1 + C \ln \left(\frac{\tau_2}{\tau_1} \right) \right]^{-1}, \tag{5.15}$$

which can be generalized to also include Kelvin–Voigt damping terms as in (5.10). Here $E_1(z) = \int_z^\infty \frac{e^{-t}}{t} dt$, C represents the degree to which viscous effects are present, and τ_1 and τ_2 represent the fast and slow viscous time phenomena. We note that the internal strain variable formulation (5.10)–(5.12) without Kelvin–Voigt damping is equivalent to the constitutive relationship proposed by Fung if one considers an approximation of the relaxation function G by a sum of exponential terms. Various internal strain variable models are investigated in [3] and a good agreement is demonstrated between the internal strain variable model ($m = 2$ so $\sigma(t,x) = \varepsilon_1(t,x) + \varepsilon_2(t,x)$) and undamped simulated data based on the Fung kernel G. We also remark that theoretical well-posedness results for the one internal strain variable model (which can be readily extended to treat finite multiple internal strain variables) are given in [30].

Fung's proposed model provides a continuous spectrum of relaxation times in contrast with the finite multiple internal strain model. In the latter, the material is assumed to have discrete relaxation times, τ_j, that correspond to a discrete hysteresis spectrum. In a more general formulation (which also includes Kelvin–Voigt damping), one may treat a continuous spectrum of relaxation times in the form

$$\sigma(t,x;P) = C_D\dot{\varepsilon}(t,x) + \int_0^t G(t-s;P)\frac{d}{ds}\sigma_e(u_x(s,x))ds, \qquad (5.16)$$

with

$$G(t;P) = \int_{\mathcal{T}} g(t;\tau)dP(\tau), \qquad (5.17)$$

where $\mathcal{T} = [\tau_1, \tau_2] \subset (0, \infty)$, P is a cumulative distribution function on $\mathcal{T}$ and g is a continuous function of relaxation times τ on $\mathcal{T}$. We note that the case $g(t;\tau) = e^{-\frac{1}{\tau}t}$ corresponds to a continuum of internal strain variable models "weighted" by the cumulative distribution function P, and includes (5.10)–(5.12) as a special case of a discrete probability measure $P_m = \sum_{j=1}^m p_j^m \Delta_{\tau_j}$, where Δ_τ is a Dirac measure with atom at τ. This can be seen readily by applying the variation of constants formula in (5.11)–(5.12) (with $C_j = 1$, $j = 1, \ldots, m$):

$$\varepsilon_j(t,x) = \int_0^t e^{-\frac{1}{\tau_j}(t-s)}\frac{d\sigma_e(u_x(s,x))}{ds}ds,$$

$$\sigma(t,x;P_m) = C_D\dot{u}_x(t,x) + \int_{\mathcal{T}}\int_0^t e^{-\frac{1}{\tau}(t-s)}\frac{d\sigma_e(u_x(s,x))}{ds}ds\,dP_m(\tau). \qquad (5.18)$$

This is an approximation to a generalized continuum model using a product measure on $\mathcal{T} \times \mathcal{C}$ to include a continuum of τ_j's and a continuum of C_j's. We remark that to guarantee well-posedness, most often one formulates the models in a weak sense (for details see [2]).

In these problems one is given measurement data $\mathbf{u} = (u_{ij})$ for $u(t_i, x_j; P)$ where u satisfies (5.7)–(5.9) with $\sigma(t,x;P)$ given by (5.16) so that u depends on P in a complex, implicit non-linear manner (note that in (5.18) σ_e is explicitly dependent on P). The corresponding OLS inverse problems consist of minimizing

$$J(P;\mathbf{u}) = \sum_{ij} |u_{ij} - u(t_i, x_j; P)|^2 \qquad (5.19)$$

over $P \in \mathbb{P}(\mathcal{T})$, the set of cumulative distribution functions on $\mathcal{T}$, or over some suitably chosen subset of $\mathbb{P}(\mathcal{T})$.

5.3.2   Probability Measure Dependent Systems—Maxwell's Equations

We turn to electromagnetics to provide a second example of these aggregate dynamics/aggregate data inverse problems. The equations and ideas underlying the theory of electricity and magnetism in materials (where these concepts are most important) are quite complex. We consider Maxwell's equations in a complex, heterogeneous material (see [10] and the extensive electromagnetics references therein including [1, 33, 39, 46, 58] for details):

$$\nabla \times \mathbf{E} = -\frac{\partial \mathbf{B}}{\partial t} \qquad \nabla \times \mathbf{H} = \frac{\partial \mathbf{D}}{\partial t} + \mathbf{J} \qquad (5.20)$$

$$\nabla \cdot \mathbf{D} = \rho \qquad \nabla \cdot \mathbf{B} = 0, \qquad (5.21)$$

where $\mathbf{E}$ is the total electric field (force), $\mathbf{H}$ is the magnetic field (force), $\mathbf{D}$ is the electric flux density (also called the electric displacement), $\mathbf{B}$ is the magnetic flux density and ρ is the density of charges in the medium.

To complete this system, we need constitutive (material dependent) relations:

$$\mathbf{D} = \epsilon_0 \mathbf{E} + \mathbf{MP}, \quad \mathbf{B} = \mu_0 \mathbf{H} + \mathbf{M}, \quad \mathbf{J} = \mathbf{J}_c + \mathbf{J}_s, \qquad (5.22)$$

where $\mathbf{MP}$ is macroscopic electric polarization, $\mathbf{M}$ is magnetization, $\mathbf{J}_c$ is the conduction current density, $\mathbf{J}_s$ is the source current density, ϵ_0 is the dielectric permittivity in free space and μ_0 is the magnetic permeability in free space.

General polarization (a constitutive law) is given by

$$\mathbf{MP}(t, \mathbf{x}) = \int_0^t \tilde{g}(t - s)\mathbf{E}(s, \mathbf{x})ds.$$

Here, $\tilde{g}$ is the polarization susceptibility kernel, or dielectric response function (DRF), and $\mathbf{x} = (x, y, z)$ is the spatial coordinate.

One widely encountered polarization model is the *Debye* model, which describes reasonably well a polar material (such as water). This model is also called dipolar or orientational polarization and the corresponding DRF is defined by

$$\tilde{g}(t - s) = g(t - s) + \epsilon_0(\epsilon_\infty - 1)\delta(t - s), \qquad (5.23)$$

where δ is the Dirac delta function, ϵ_∞ denotes the relative permittivity of the medium in the limit of high frequency and

$$g(t - s) = e^{\frac{-(t-s)}{\tau}} \left(\frac{\epsilon_0(\epsilon_s - \epsilon_\infty)}{\tau} \right). \qquad (5.24)$$

Here τ represents a molecular relaxation time and ϵ_s is the relative permittivity of the medium at zero frequency. This formulation corresponds to $\mathbf{MP} = \epsilon_0(\epsilon_\infty - 1)\mathbf{E} + \widetilde{\mathbf{MP}}$ where $\widetilde{\mathbf{MP}}$ satisfies

$$\dot{\widetilde{\mathbf{MP}}} + \frac{1}{\tau}\widetilde{\mathbf{MP}} = \epsilon_0(\epsilon_s - \epsilon_\infty)\mathbf{E}.$$

Other models for polarization include the Lorentz model (e.g., see [10] for details), which is used for electronic polarization or electron cloud materials. For complex composite materials, the standard Debye or Lorentz polarization models are not adequate, e.g., one needs multiple relaxation times τ's in some kind of distribution [5, 20, 21]. As an example, the multiple Debye model becomes

$$\mathbf{MP}(t, \mathbf{x}; P) = \int_{\mathcal{T}} \mathbf{mp}(t, \mathbf{x}; \tau) dP(\tau),$$

where $\mathcal{T}$ is a set of possible relaxation parameters τ, as in the previous example, $P \in \mathbb{P}(\mathcal{T})$ where $\mathbb{P}(\mathcal{T})$ is again the set of cumulative distribution functions on $\mathcal{T}$ and $\mathbf{mp}$ is the microscopic polarization. It is important to note that $\mathbf{MP}$ represents the *macroscopic* polarization, as opposed to the *microscopic polarization* $\mathbf{mp}$, which describes polarization at the molecular level and is given by

$$\mathbf{mp}(t, \mathbf{x}; \tau) = \int_0^t \tilde{g}(t - s; \tau)\mathbf{E}(s, \mathbf{x}) ds. \tag{5.25}$$

We observe that the macroscopic polarization $\mathbf{MP}$ is over the microscopic (at the molecular level) polarization $\mathbf{mp}$ (just as the macroscopic electric field $\mathbf{E}$ is summed over the microscopic electric field $\mathbf{e}$ in most careful derivations [39, 46, 58]). It is in this sense that we have an aggregate dynamics or macroscopic model for the $\mathbf{E}$ field as well as aggregate data (for the $\mathbf{E}$ and $\mathbf{H}$ fields, which are obtained by summing over the microscopic $\mathbf{mp}$ and $\mathbf{e}$ in the fundamental Maxwell equations derivations).

Assuming $\rho = 0$ and $\mathbf{M} = 0$ (no fixed charges, non-magnetic materials), we find this system becomes

$$\nabla \times \mathbf{E} = -\frac{\partial}{\partial t}(\mu_0 \mathbf{H})$$

$$\nabla \times (\mathbf{H}(t, \mathbf{x})) = \frac{\partial}{\partial t}\left[\epsilon_0 \mathbf{E}(t, \mathbf{x}) + \int_0^t G(t - s; P)\mathbf{E}(s, \mathbf{x}) ds\right] + \mathbf{J}(t, \mathbf{x}) \tag{5.26}$$

$$\nabla \cdot \mathbf{D} = 0$$

$$\nabla \cdot \mathbf{H} = 0,$$

where $P \in \mathbb{P}(\mathcal{T})$ and $G(t - s; P) = \int_{\mathcal{T}} \tilde{g}(t - s; \tau) dP(\tau)$.

Thus we note that the Maxwell dynamics in the scalar case (in second order form for either E or H) themselves depend implicitly on a probability measure

P, e.g.,

$$\frac{\partial^2 u}{\partial t^2} + \frac{\partial^2 u}{\partial x^2} = f(u, P),$$

where, for example, the electromagnetic field $u = E$ depends on summing with a probability distribution the effects of multiple mechanisms of polarization across all particles in the material. That is, to obtain the macroscopic polarization, we sum over all the parameters. We cannot separate dynamics for the electric and magnetic fields to obtain individual or particle dynamics, i.e., P itself depends u. Therefore we have (5.26) as an example where the dynamics for the E and H fields depend implicitly on the probability measure $P \in \mathbb{P}(\mathcal{T})$. For inverse problems, data is given in terms of field (either E or H field) measurements $\mathbf{u} = (u_{ij})$ for $u(t_i, x_j; P)$. In ordinary least squares problems, we have

$$J(P; \mathbf{u}) = \sum_{ij} |u_{ij} - u(t_i, x_j; P)|^2, \qquad (5.27)$$

to be minimized over $P \in \mathbb{P}(\mathcal{T})$ (or over some subset of $\mathbb{P}(\mathcal{T})$). We note that while (5.27) and (5.6) may appear similar, this appearance is somewhat misleading. In (5.6) we explicitly sum with the probability measure over a family of "individual" dynamics, while in (5.27), the equation for the observable depends implicitly on the probability distribution of interest.

To carry out analysis of the minimization problems using either (5.6) or (5.27), we must have a concept of regularity (in the topological or metric sense) in probability spaces. We turn to this topic in the next section.

5.4   Aggregate Data and the Prohorov Metric Framework

Note that the dynamics for all the motivating examples given above are driven by partial differential equations. However, for notational convenience, we will consider the case where the model solution is a function only of time (that is, no spatial dependency) in the subsequent discussions in this chapter. We remark that one can similarly treat the case where the model solution depends on both time and spatial coordinates by treating $\{(t_i, x_j)\}_{i=1, j=1}^{N_t, N_x}$ as $\{t_k\}_{k=1}^{N_t N_x}$.

The use of aggregate data requires a slightly different form for the statistical model as compared to the earlier models in this monograph. The individual mathematical model is given by

$$v(t) = v(t; \boldsymbol{\theta}, \psi), \qquad (5.28)$$

where it is assumed that the parameter $\boldsymbol{\theta} \in \Omega_\theta$ varies from individual to individual in the population and that some probability measure P characterizes the distribution of the parameter. The parameter $\psi \in \Omega_\psi$ is common (e.g., environment, etc.) to all individuals in the population. We will, without loss of generality, in this treatment assume ψ is known and focus only on inverse problems for the determination of the probability distribution P over the parameters $\boldsymbol{\theta}$.

For the individual dynamics/aggregate data problems, the aggregate state for the entire population can be defined in terms of an expectation

$$u(t; P) = \mathbb{E}(v(t; \cdot)|P) = \int_{\Omega_\theta} v(t; \boldsymbol{\theta}) dP(\boldsymbol{\theta}). \qquad (5.29)$$

Whether in this case or in the more complex modeling situation of Type II aggregate dynamics/aggregate data, when aggregate data $\{u_j\}_{j=1}^N$ is collected, this represents a sampling from the aggregate population, and the data collection process is assumed to be represented by the statistical model

$$U_j = u(t_j; P_0) + \mathcal{E}_j, \quad j = 1, \ldots, N. \qquad (5.30)$$

Here $\mathbb{E}(\mathcal{E}_j) = 0$ (again corresponding to the tacit assumption that the mathematical model is accurate and there exists a "true" distribution P_0). As usual, we must also specify the covariance properties of the error random variables $\mathcal{E}_j$. For notational simplicity, we will make the assumption of independent, constant variance errors so that $\text{Var}(\mathcal{E}_j) = \sigma^2$ for all j. This is not absolutely necessary, as one could similarly assume a constant coefficient of variation statistical model, or even a more general form featuring a general covariance matrix for the vector of errors. This will not pose any problem for the approximation results presented below, provided the resulting cost function J is a continuous function of P (which is the case for all examples of interest to us in this monograph). In fact, one could even use the maximum likelihood estimation method in the place of the least squares method if the errors $\mathcal{E}_j$ admit a known distribution.

With the statistical model of the data established, we turn to a discussion of estimation techniques. As we have stated, the goal is to determine the measure P describing the distribution of the parameters $\boldsymbol{\theta}$ in the population. Methods for the estimation of the measure P have traditionally focused on a least squares framework. The constant variance statistical model (5.30) allows the use of the notation for ordinary least squares, which is appealing in its simplicity. Thus we seek to determine the estimator (we drop the subscript on the estimators and estimates since we only consider OLS formulations in this section)

$$P^N = \arg \min_{\mathbb{P}(\Omega_\theta)} J_{\text{OLS}}^N(P; \boldsymbol{U}) = \arg \min_{\mathbb{P}(\Omega_\theta)} \sum_{j=1}^N (U_j - u(t_j; P))^2 \qquad (5.31)$$

and corresponding estimate

$$\hat{P}^N = \arg\min_{\mathbb{P}(\Omega_\theta)} J^N_{\text{OLS}}(P; \boldsymbol{u}) = \arg\min_{\mathbb{P}(\Omega_\theta)} \sum_{j=1}^{N} (u_j - u(t_j; P))^2 \qquad (5.32)$$

where $J^N_{\text{OLS}}(P; \boldsymbol{u})$ is the OLS cost of the model given the parameters $P \in \mathbb{P}(\Omega_\theta)$ and the data $\boldsymbol{u} = (u_1, u_2, \ldots, u_N)^T$ are realizations of the random variables $\boldsymbol{U} = (U_1, U_2, \ldots, U_N)^T$. For more general measurement errors (e.g., relative error or constant coefficient of variation) the cost function in (5.31) would need to be appropriately redefined. The minimization (5.31) must be carried out over the infinite dimensional space $\mathbb{P}(\Omega_\theta)$ of probability measures on the space Ω_θ. As such, one must establish a finite dimensional approximation and convergence theory. To that end, we introduce the *Prohorov Metric Framework* (PMF), developed over the past decade for use in the estimation of a probability measure (a summary from a functional analytic perspective along with a discussion of other topologies for probability distributions can be found in [2, 38, 45]). This framework is more common in the applied mathematics literature compared to the methods for individual data discussed previously, and is named for the Russian probabilist Y.V. Prohorov, whose metric was introduced for quite different applications [57] but now features prominently in theoretical and computational efforts [4, 6, 8, 19, 20, 21, 27] establishing convergence results for the computational methods discussed here.

We begin by defining the Prohorov metric on the set of probability measures. Consider the admissible parameter space Ω_θ with its metric $\tilde{d}$ (the assumption that Ω_θ is Hausdorff will be made throughout this chapter) and the set of probability measures (cumulative distribution functions) $\mathbb{P}(\Omega_\theta)$ on Ω_θ. For any closed set $\mathbb{F} \subset \Omega_\theta$, define

$$\mathbb{F}^\epsilon = \left\{ \boldsymbol{\theta} \in \Omega_\theta : \inf_{\tilde{\boldsymbol{\theta}} \in \mathbb{F}} \tilde{d}(\boldsymbol{\theta}, \tilde{\boldsymbol{\theta}}) < \epsilon \right\}.$$

Then we have the following definition of the Prohorov metric [32, pp. 237–238]:

Definition 5.4.1 *Let* $\mathbb{F} \subset \Omega_\theta$ *be any closed set and define* $\mathbb{F}^\epsilon$ *as above. For* $P, Q \in \mathbb{P}(\Omega_\theta)$, *the* <u>Prohorov metric</u> *is given by*

$$\rho(P, Q) =$$
$$\inf \left\{ \epsilon > 0 \,|\, Q(\mathbb{F}) \le P(\mathbb{F}^\epsilon) + \epsilon \text{ and } P(\mathbb{F}) \le Q(\mathbb{F}^\epsilon) + \epsilon, \text{ for all } \mathbb{F} \text{ closed in } \Omega_\theta \right\}.$$
$$(5.33)$$

While several alternative (and equivalent) definitions of the Prohorov metric are possible ([32, pp. 236–237],[38, pp. 393–398]), it is clear from the definition above that its meaning and application are far from intuitive. Yet one can provide several useful characterizations.

Define the space $C_B(\Omega_\theta) = \{f : \Omega_\theta \to \mathbb{R} \mid f \text{ is bounded and continuous}\}$ and let $C_B^*(\Omega_\theta)$ denote its topological dual.

Theorem 5.4.1 *Assume the parameter space Ω_θ with its metric $\tilde{d}$ is separable. Then the Prohorov metric metrizes the weak* topology of the space $C_B^*(\Omega_\theta)$ and the following are equivalent:*

(i) $\rho(P_M, P) \to 0$.

(ii) $\int_{\Omega_\theta} f(\boldsymbol{\theta}) dP_M(\boldsymbol{\theta}) \to \int_{\Omega_\theta} f(\boldsymbol{\theta}) dP(\boldsymbol{\theta})$ *for bounded, uniformly continuous functions f.*

(iii) $P_M(A) \to P(A)$ *for all Borel sets $A \subset \Omega_\theta$ with $P(\partial A) = 0$.*

We remark that the separability of the metric space $(\Omega_\theta, \tilde{d})$ is not strictly necessary. The so-called weak topology of probability measures (weak* topology) and the topology induced by the Prohorov metric are equivalent provided every P in $\mathbb{P}(\Omega_\theta)$ is a separable measure [32, Theorem 5, p. 238]. We can be guaranteed of this property in the event Ω_θ is separable [32, p. 236]. From a functional analysis perspective, if we view $\mathbb{P}(\Omega_\theta) \subset C_B^*(\Omega_\theta)$, then (see the further discussions below) convergence in the ρ topology is equivalent to weak* convergence in $\mathbb{P}(\Omega_\theta)$.

To be more precise, the Riesz Representation Theorem on the space of bounded continuous functions is useful. This theorem can be used to characterize the weak* topology on the topological dual of the space of bounded continuous functions, which provides an intuitive motivation for the weak topology on the space of probability measures. It is no surprise then that the two topologies are equivalent on the set of probability measures.

Theorem 5.4.2 (Riesz) *Assume $(\Omega_\theta, \tilde{d})$ is a compact (Hausdorff) space. For every $f^* \in C_B^*(\Omega_\theta)$, there exists a unique finite signed Borel measure μ such that*

$$f^*(f) = \int_{\Omega_\theta} f(\boldsymbol{\theta}) d\mu(\boldsymbol{\theta})$$

for all $f \in C_B(\Omega_\theta)$. Moreover, $|f| = |\mu|(\Omega_\theta)$.

For a readable proof, see [59, pp. 357–358].

Given this identification, we may write $f^* = f_\mu^*$ when convenient. We see that the set $\mathbb{P}(\Omega_\theta)$ of probability measures on $(\Omega_\theta, \tilde{d})$ can be identified with those $f_\mu^* \in C_B^*(\Omega_\theta)$ such that $f_\mu^*(f) \geq 0$ for all $f \geq 0$ and $|f_\mu^*| = \mu(\Omega_\theta) = 1$. Thus we have, in a sense, that $\mathbb{P}(\Omega_\theta) \subset C_B^*(\Omega_\theta)$. In fact, given any $f \in C_B(\Omega_\theta)$, the map from $C_B(\boldsymbol{\theta})$ into $\mathbb{R}$ given by $f_f^{**}(f^*) = f^*(f)$ defines the natural embedding [2] of $C_B(\Omega_\theta) \hookrightarrow C_B^{**}(\Omega_\theta)$. The image of f^{**} induces a topology on the space $C_B^*(\Omega_\theta)$, known to functional analysts as the weak* topology [2, pp. 49–57]. (That is, $f_n^* \xrightarrow{w^*} f^*$ if and only if $f_n^*(f) \to f^*(f)$

for all $f \in C_B(\Omega_\theta)$.) When viewed in the context of $\mathbb{P}(\Omega_\theta) \subset C_B^*(\Omega_\theta)$, this is the *weak convergence of measures* known from the theory of probability and stochastic processes. We refer the reader to Chapter 2 and our comments there on weak and weak* convergence (specifically see (2.114)), and to [2, Chapter 14]) for more detailed discussions of these two types of convergence.

The following results can be used to establish the existence of a minimizer to the least squares optimization problem. Let Ω_θ be separable when taken with its metric $\tilde{d}$. Then

(i) $(\mathbb{P}(\Omega_\theta), \rho)$ is a complete metric space if and only if $(\Omega_\theta, \tilde{d})$ is a complete metric space.

(ii) If $(\Omega_\theta, \tilde{d})$ is a compact metric space, then $(\mathbb{P}(\Omega_\theta), \rho)$ is a compact metric space.

While it is sequential compactness that will be of interest in establishing the hypotheses of the convergence theorems (Theorems 5.4.4 and 5.4.5), we note that in a metric space (i.e., when $\mathbb{P}(\Omega_\theta)$ is metrized by ρ) there is no distinction between sequential compactness and compactness. Throughout the remainder of this chapter, we will use the shortened notation $\mathbb{P}(\Omega_\theta)$ to denote $(\mathbb{P}(\Omega_\theta), \rho)$ when no confusion will result.

At this point, we have used the Prohorov metric to define a metric over the space of probability measures. As an added benefit, when Ω_θ is compact and separable (which is often the case in applications, particularly when Ω_θ is a closed bounded subset of Euclidean space), the space of probability measures (taken with the Prohorov metric) is compact. Hence, provided the least squares cost functional J_{OLS}^N is continuous in P, we are guaranteed the existence of a minimizer in (5.31) for the given $\mathbf{U}$.

We now turn our attention to strategies for the actual computation of a minimizing probability measure P (or an approximation) as defined in (5.31). In order to make the estimation problem amenable to computational solution, the infinite dimensional space $\mathbb{P}(\Omega_\theta)$ must be approximated in a meaningful way. Traditional parametric methods, which assume the optimal measure has a particular distributional form, are not preferred, as they are overly restrictive and will produce inaccurate results if the parametric form is misspecified. One particular result will be useful in establishing computational tools for the parameter estimation problems of interest.

Theorem 5.4.3 *Assume $(\Omega_\theta, \tilde{d})$ is a separable, compact metric space. Let $\Omega_{\theta D} = \{\theta_k\}_{k=1}^\infty$ be an enumeration of a countable dense subset of Ω_θ. Take $\mathbb{Q} \subset \mathbb{R}$ to be the set of all rational numbers. Define*

$$\tilde{\mathbb{P}}_D(\Omega_\theta)$$

$$= \left\{ P \subset \mathbb{P}(\Omega_\theta) \middle| P = \sum_{k=1}^M p_k \Delta_{\theta_k}, \theta_k \in \Omega_{\theta D}, M \in \mathbb{N}, p_k \in [0,1] \cap \mathbb{Q}, \sum_{k=1}^M p_k = 1 \right\},$$

where Δ_a is the Dirac measure with atom at a. (That is, $\tilde{\mathbb{P}}_D(\Omega_\theta)$ is the collection of all convex combinations of Dirac measures on Ω_θ with atoms $\theta_k \in \Omega_{\theta D}$ and rational weights.) Then $\tilde{\mathbb{P}}_D(\Omega_\theta)$ is dense in $(\mathbb{P}(\Omega_\theta), \rho)$, and thus $\mathbb{P}(\Omega_\theta)$ is separable.

A proof of the above result can be found in [4].

For each $M \in \mathbb{N}$, define

$$\mathbb{P}_M(\Omega_\theta) = \left\{ P \in \tilde{\mathbb{P}}_D(\Omega_\theta) \;\middle|\; P = \sum_{k=1}^M p_k \Delta_{\theta_k} \right\}. \tag{5.34}$$

That is, for each M, $\mathbb{P}_M(\Omega_\theta)$ is the set of all atomic probability measures with nodes placed at the first M elements in the enumeration of the countable dense subset of Ω_θ. Note that $\mathbb{P}_M(\Omega_\theta) \subset \tilde{\mathbb{P}}_D(\Omega_\theta) \subset \mathbb{P}(\Omega_\theta)$ for all M. The following theorem provides the desired convergence result.

Theorem 5.4.4 *Assume Ω_θ is compact and separable and consider $\mathbb{P}(\Omega_\theta)$ taken with the Prohorov metric. If the mapping $P \to J_{OLS}^N(P; u)$ is continuous, then there exists a sequence of minimizers $\{\hat{P}_M^N\}$ of $J_{OLS}^N(P; u)$ over $\mathbb{P}_M(\Omega_\theta)$. Moreover, this sequence has at least one convergent subsequence. The limit $\hat{P}^{N*}$ (as $M \to \infty$) of such a subsequence minimizes J_{OLS}^N over $\mathbb{P}(\Omega_\theta)$.*

This theorem follows immediately from the theoretical framework and convergence theorems of [26] and the results above. The power of this theorem is that it transforms an infinite dimensional minimization problem (over $\mathbb{P}(\Omega_\theta)$) into an M dimensional optimization problem. In practice, M nodes $\{\theta_k^M\}_{k=1}^M$ from the dense, countable subset $\Omega_{\theta D} = \{\theta_k\}_{k=1}^\infty$ are fixed in advance and then the M weights p_k^M are estimated in a minimization problem. Thus the least squares problem (5.32), i.e.,

$$\hat{P}^N = \arg\min_{\mathbb{P}(\Omega_\theta)} J_{OLS}^N(P; u) = \arg\min_{\mathbb{P}(\Omega_\theta)} \sum_{j=1}^N (u_j - u(t_j; P))^2, \tag{5.35}$$

is approximated by the family of finite-dimensional problems

$$\hat{P}_M^N = \arg\min_{\mathbb{P}_M(\Omega_\theta)} \sum_{j=1}^N (u_j - u(t_j; P))^2$$

$$= \arg\min_{\tilde{\mathbb{R}}^M} \sum_{j=1}^N \left(u_j - u\left(t_j; \sum_{k=1}^M p_k^M \Delta_{\theta_k^M} \right) \right)^2. \tag{5.36}$$

In (5.36) we are seeking optimal weights $p^M = (p_1^M, \ldots, p_M^M)^T \in \tilde{\mathbb{R}}^M = \left\{ p^M \;\middle|\; p_k^M \in \mathbb{R}^+, \sum_{k=1}^M p_k^M = 1 \right\}$, which will then characterize the optimal discrete measure (since the nodes are assumed to be fixed in advance).

In most practical problems, especially those of Type II, the model $u(t; P)$ cannot be computed exactly and must be approximated with $u_{\tilde{N}}(t; P)$ by some numerical scheme (e.g., finite difference methods, Galerkin methods, etc.). Thus, given a set of realizations $\mathbf{u}$ of the random variables $\mathbf{U}$, what one computes in practice is

$$
\hat{P}_{\tilde{N},M}^{N} = \arg \min_{P \in \mathbb{P}_M(\Omega_\theta)} J_{\tilde{N},M}^{N}(P; \mathbf{u})
$$

$$
\equiv \arg \min_{\widetilde{\mathbb{R}}^M} \sum_{j=1}^{N} \left(u_j - u_{\tilde{N}} \left(t_j; \sum_{k=1}^{M} p_k^M \Delta_{\boldsymbol{\theta}_k^M} \right) \right)^2. \qquad (5.37)
$$

This leads to a constrained optimization problem which can in many cases (especially Type I problems – see further remarks below) of interest be eminently amenable to parallel computational methods [19]. The methods based on these ideas have been successfully used in a wide range of applied problems [4, 6, 8, 11, 12, 19, 27].

One can give a convergence result for this further (state) approximation. These results are outlined in [26] and are included again here without proof.

Theorem 5.4.5 *Let $(\Omega_\theta, \tilde{d})$ be a compact, separable metric space and consider the space $(\mathbb{P}(\Omega_\theta), \rho)$ of probability measures on Ω_θ with the Prohorov metric, as before. Let $\mathbb{P}_M(\Omega_\theta)$ be as defined in (5.34). Let $u_{\tilde{N}}(t; P)$ be a family of state approximations to $u(t; P)$. Assume further*

(1) the map $P \mapsto J_{\tilde{N},M}^{N}(P; \mathbf{u})$ is continuous for all $\tilde{N}, M, N$;

(2) for any sequence of probability measures $P_k \to P$ in $\mathbb{P}(\Omega_\theta)$, $u_{\tilde{N}}(t; P_k) \to u(t; P)$ as $\tilde{N}, k \to \infty$;

(3) $u(t; P)$ is uniformly bounded for all t, P.

Then there exists minimizers $\hat{P}_{\tilde{N},M}^{N}$ satisfying (5.37). Moreover, for fixed N, there exists a subsequence (as $\tilde{N}, M \to \infty$) of the approximate estimates $\hat{P}_{\tilde{N},M}^{N}$ which converges to some (possibly non-unique) $\hat{P}^{N*}$ which minimizes $J_{OLS}^{N}(P; \mathbf{u})$ over $\mathbb{P}(\Omega_\theta)$.

Significantly, the Prohorov metric framework is computationally constructive. In practice, one does not construct a sequence of estimates for increasing values of M and $\tilde{N}$; rather, one fixes the values of M and $\tilde{N}$ to be sufficiently large to attain a desired level of accuracy. By Theorem 5.4.3, we need only to have some enumeration of the elements of $\mathbb{P}_M(\Omega_\theta)$ in order to compute an approximate estimate $\hat{P}_{\tilde{N},M}^{N}$. (We will not consider the choice of $\tilde{N}$, as this will depend upon the numerical framework by which approximate model solutions $u_{\tilde{N}}(t; P)$ are obtained – this is discussed at length in [26].) Practically, this is accomplished by selecting M nodes, $\{\boldsymbol{\theta}_k^M\}_{k=1}^{M}$, in Ω_θ. The optimization problem (5.37) is then reduced to a standard constrained estimation problem over $\widetilde{\mathbb{R}}^M$ in which one determines the values of the weights p_k^M corresponding to each node. In the *individual dynamics/aggregate dynamics case (Type I problems)* this takes a particularly appealing form given by

$$
\hat{P}_{\tilde{N},M}^N = \arg\min_{\mathbb{P}_M(\Omega_\theta)} \sum_{j=1}^N \left(u_j - u_{\tilde{N}}(t_j; P) \right)^2
$$

$$
= \arg\min_{\mathbb{P}_M(\Omega_\theta)} \sum_{j=1}^N \left(u_j - \int_{\Omega_\theta} v_{\tilde{N}}(t_j; \boldsymbol{\theta}) dP(\boldsymbol{\theta}) \right)^2
$$

$$
= \arg\min_{\widetilde{\mathbb{R}}^M} \sum_{j=1}^N \left(u_j - \left(\sum_{k=1}^M v_{\tilde{N}}(t_j; \boldsymbol{\theta}_k^M) p_k^M \right) \right)^2,
$$

where in the final line we seek the weights $\boldsymbol{p}^M = (p_1^M, \ldots, p_M^M)^T \in \widetilde{\mathbb{R}}^M$. These are sufficient to characterize the approximating discrete estimate $\hat{P}_{\tilde{N},M}^N$ since the nodes are assumed to be fixed in advance. Moreover, if we define

$$
H_{kl} = 2 \sum_{j=1}^N \left(v_{\tilde{N}}(t_j; \boldsymbol{\theta}_k^M) \right) \left(v_{\tilde{N}}(t_j; \boldsymbol{\theta}_l^M) \right)
$$

$$
f_k = -2 \sum_{j=1}^N u_j \left(v_{\tilde{N}}(t_j; \boldsymbol{\theta}_k^M) \right)
$$

$$
c = \sum_{j=1}^N (u_j)^2,
$$

then one can equivalently compute

$$
\hat{P}_{\tilde{N},M}^N = \arg\min_{\widetilde{\mathbb{R}}^M} \left(\frac{1}{2} \left(\boldsymbol{p}^M \right)^T H \boldsymbol{p}^M + f^T \boldsymbol{p}^M + c \right). \tag{5.38}
$$

From this reformulation, it is clear that the approximate problem (5.38) has a unique solution if H is positive definite – we are minimizing a convex function over a convex compact set. If the individual mathematical model (5.28) is independent of P (this is the precisely the case for Type I problems (see the motivating examples of Section 5.2.1 above), but *not* the case of the Type II examples discussed in Section 5.3), then the matrices H and f can be precomputed in advance. Then one can rapidly (and exactly) compute the gradient and Hessian of the objective function in a numerical optimization routine. As M grows large, the quadratic optimization problem (5.38) becomes poorly conditioned [11]. Thus there is a trade-off: M must be chosen sufficiently large so that the computational approximation is accurate, but not so large that ill-conditioning leads to large numerical errors. The efficient choice of M as well as the choice of the nodes $\{\boldsymbol{\theta}_k\}_{k=1}^{M}$ is an open research problem. We remark that values of M under 50 have been studied computationally in [4, 11].

It is clear from the statement of Theorems 5.4.4 and 5.4.5 that the uniqueness of the minimizer $\hat{P}^{N*}$ is not guaranteed (as there could be multiple convergent subsequences). In particular, in the implementation of the approximations on $\mathbb{P}_M(\Omega_\theta)$, for a fixed set of nodes $\{\boldsymbol{\theta}_k\}_{k=1}^{M}$, it is possible to establish first-order necessary conditions for the uniqueness of the computational problem [4, p. 102]. But these should not be confused with conditions for global uniqueness of a minimizing measure P_0. Unlike the non-parametric maximum likelihood (NPML) method for individual data discussed later in this chapter, the parameter ψ of (5.28) does not require a separate minimization process, and can be estimated simultaneously with the M weights.

Theorem 5.4.5 provides a set of conditions under which a subsequence of approximate estimates $\hat{P}_{\tilde{N},M}^{N}$ converges to an estimate $\hat{P}^{N*}$ of interest. This estimate is itself a realization (for a particular data set) of the estimator P^N, which has been shown to exist and will be shown below, under appropriate assumptions, to be consistent, so that $P^N \to P_0$ with probability one. Thus we have some reasonable assurance that a computed approximate estimate $\hat{P}_{\tilde{N},M}^{N}$ reflects the true distribution P_0. The assumptions of Theorem 5.4.5 are not restrictive. In typical problems (and, indeed, in the assumptions of other theorems appearing in this chapter) it is assumed that the parameter space Ω_θ is compact (see, e.g., Section 5.5). In such a case, Assumptions (1) and (3) above are satisfied if the individual model solutions $v(t;\boldsymbol{\theta})$ or aggregate solutions $u(t;P)$ are reasonably smooth. Assumption (2) is then simply a condition on the convergence of the numerical procedure used in obtaining model solutions.

As we have already pointed out, the uniqueness of the computational problem (i.e., when H is positive definite for a Type I problem) is not sufficient to ensure the uniqueness of the limiting estimate $\hat{P}^{N*}$ in Theorem 5.4.5 (as there could be multiple convergent subsequences). However, if $J_{\mathrm{OLS}}^{N}(P;\boldsymbol{u})$ is uniquely minimized, then every subsequence of $\hat{P}_{\tilde{N},M}^{N}$ which converges as $\tilde{N}, M$ grow large must converge to that unique minimizer.

Given realizations $\mathbf{u}$ of the random variables $\mathbf{U}$ (which we will sometimes write $\boldsymbol{u}^N$ and $\boldsymbol{U}^N$ for notational convenience), the goal of an inverse or parameter estimation problem is to produce an estimate of the hypothesized true parameter P_0. Of course, the estimated parameters should be those that best fit the data in some appropriate sense. Thus this problem first involves a choice of an approximation framework in which to work. Given that choice of framework, one must establish a set of theoretical and computation tools with which to treat the parameter estimation problem. For the results presented here, we have focused on a frequentist approach using a generalization of least squares estimation. Theoretical results for maximum likelihood estimation method (also in a frequentist framework) can be established with little difficulty from the results presented here. For the moment, we do not consider a Bayesian approach to the estimation of the unknown distribution P_0. There does seem to be some commonality between the non-parametric estimation of a probability distribution and the determination of a Bayesian posterior estimator [25] *when one has longitudinal individual data.* In this case a recent study [47] is of interest. For an extended discussion of a Bayesian approach, interested readers are referred to [62].

The remaining question of great interest is statistical. Assuming that $\hat{P}^N_{\tilde{N},M}$ approaches $\hat{P}^N$ as M and $\tilde{N}$ grow large, how does this estimate compare with P_0? Put another way, given any fixed N observations, one obtains an estimate $\hat{P}^N$ of P^N. How does this estimate improve as more data is collected (that is, as N grows large)? This is a question of the *consistency* of the least squares estimator P^N. For the least squares problem, it is most convenient to carry out the discussion in the context of *estimators* defined by (5.31).

5.5   Consistency of the PMF Estimator

We now turn our attention to characterizing the least squares estimator (5.31) and its corresponding estimate (5.35). In the present section we ignore any computational approximations and establish results concerning the theoretical consistency of the least squares estimator, regardless of our ability to compute estimates (although the method of proof does foreshadow the computational approach suggested above).

Theorem 5.4.4 demonstrates that for any fixed N the estimator P^N and the corresponding estimate $\hat{P}^N$ exist. An obvious question, then, is what the resulting measures P^N or estimates $\hat{P}^N$ represent. Since $\hat{P}^N$ is just a realization of P^N (given a specific set of data), we focus on characterization of the properties of the estimator P^N. Given the problem formulation (5.31) and the statistical model (5.30), one would certainly hope that the estimator provides some information regarding the underlying "true" distribution P_0. In particular, we would hope that $P^N \to P_0$ in some appropriate sense. (If this is the case, then the estimator is said to be *consistent.*) Of course, the estimator itself is a random element (or random variable – see Remark 2.2.1 – we shall use the terms interchangeably in this chapter), and thus this convergence must be discussed in terms of probability. To discuss consistency, we also need to consider random variables $U = (U_1, U_2, \ldots, U_\infty)^T$ with corresponding realizations $u = (u_1, u_2, \ldots, u_\infty)^T$. We then consider sequences $U^N = (U_1, U_2, \ldots, U_N)^T$ of increasing amounts of data and the corresponding problems for $J^N_{\text{OLS}}(P; U^N) = \sum_{j=1}^{N} (U_j - u(t_j; P))^2$.

With this in mind, we make the following set of assumptions.

(A1) For each fixed N, the error random variables $\mathcal{E}_j$, $j = 1, 2, \ldots, N$, are independent and identically distributed with zero mean and constant variance σ^2, and they are defined on some probability space $(\Omega, \mathcal{F}, \text{Prob})$.

(A2) $(\Omega_\theta, \tilde{d})$ is a separable, compact metric space; the space $\mathbb{P}(\Omega_\theta)$ is taken with the Prohorov metric ρ.

(A3) For all j, $1 \leq j \leq N$, $t_j \in [t_0, t_f]$ for some compact set $[t_0, t_f]$.

(A4) The model function $u \in C([t_0, t_f], C(\mathbb{P}(\Omega_\theta)))$.

(A5) There exists a measure μ on $[t_0, t_f]$ such that

$$\frac{1}{N} \sum_{j=1}^{N} h(t_j) = \int_{t_0}^{t_f} h(t) d\mu_N(t) \to \int_{t_0}^{t_f} h(t) d\mu(t)$$

for all $h \in C([t_0, t_f])$.

(A6) The functional

$$J^0(P) = \sigma^2 + \int_{t_0}^{t_f} (u(t; P_0) - u(t; P))^2 \, d\mu(t)$$

is uniquely minimized at $P_0 \in \mathbb{P}(\Omega_\theta)$.

Assumption (A1) establishes the probability space on which the error random variables $\mathcal{E}_j$ are assumed to be defined. As we will see, this probability space will permit us to make probabilistic statements regarding the consistency of the estimator P^N. These assumptions, as well as the two theorems below, follow closely the theoretical approach of [17], which establishes the consistency of the ordinary least squares estimator for a traditional non-linear least squares problem for finite dimensional vector space parameters. The key idea is to first argue that the functions $J_{\text{OLS}}^N(P; U^N)$ converge to $J^0(P)$ as N increases; then the minimizer P^N of J_{OLS}^N should converge to the unique minimizer P_0 of J^0. We note the similarity with the asymptotic theory in Chapter 3.

Because the functions J_{OLS}^N are functions of the vector U^N, which itself depends on the random variables $\mathcal{E}_j$, these functions are themselves random variables, as are the estimators P^N. It will occasionally be convenient to evaluate these functions at points in the underlying probability space. Thus we may write $P^N(\omega)$, $J_{\text{OLS}}^N(P; U^N)(\omega)$, $\mathcal{E}_j(\omega)$, etc., whenever the particular value of ω is of interest.

Theorem 5.5.1 *Under assumptions (A1)–(A6), there exists a set $\mathbb{A} \in \mathcal{F}$ with $\text{Prob}\{\mathbb{A}\} = 1$ such that for all $\omega \in \mathbb{A}$,*

$$\frac{1}{N} J_{OLS}^N(P; U^N)(\omega) \to J^0(P)$$

as $N \to \infty$ and for each $P \in \mathbb{P}(\Omega_\theta)$. Moreover, the convergence is uniform on $\mathbb{P}(\Omega_\theta)$.

Theorem 5.5.2 *Under assumptions (A1)–(A6), the estimators $P^N \to P_0$ as $N \to \infty$ with probability 1. That is,*

$$Prob\left\{\omega \middle| P^N(\omega) \to P_0\right\} = 1.$$

The ideas and calculations underlying the previous two theorems follow closely those in [17] and indeed are essentially probability measure versions of Theorem 3.1 and Corollary 3.2 of [17]. We briefly sketch the proof of consistency. Take the set $\mathbb{A}$ as in the previous theorem and fix $\omega \in \mathbb{A}$. Then by the previous theorem, $\frac{1}{N}J^N_{\mathrm{OLS}}(P; \boldsymbol{U}^N)(\omega) \to J^0(P)$ for all $P \in \mathbb{P}(\Omega_\theta)$. Let $\nu > 0$ be arbitrary and define the open neighborhood $\mathbb{O} = B_\nu(P_0)$ of P_0. Then $\mathbb{O}$ is open in $\mathbb{P}(\Omega_\theta)$ and $\mathbb{O}^c$ is compact (a closed subspace of a compact space). Since P_0 is the unique minimizer of $J^0(P)$ by assumption (A6), there exists $\epsilon > 0$ such that

$$J^0(P) - J^0(P_0) > \epsilon$$

for all $P \in \mathbb{O}^c$. By the previous theorem, there exists N_0 such that for $N \geq N_0$,

$$\left|\frac{1}{N}J^N_{\mathrm{OLS}}(P; \boldsymbol{U}^N)(\omega) - J^0(P)\right| < \frac{\epsilon}{4}$$

for all $P \in \mathbb{P}(\Omega_\theta)$. Then for $N \geq N_0$ and $P \in \mathbb{O}^c$,

$$\frac{1}{N}J^N_{\mathrm{OLS}}(P; \boldsymbol{U}^N)(\omega) - \frac{1}{N}J^N_{\mathrm{OLS}}(P_0; \boldsymbol{U}^N)(\omega)$$

$$= \frac{1}{N}J^N_{\mathrm{OLS}}(P; \boldsymbol{U}^N)(\omega) - J^0(P) + J^0(P) - J^0(P_0)$$

$$+ J^0(P_0) - \frac{1}{N}J^N_{\mathrm{OLS}}(P_0; \boldsymbol{U}^N)(\omega)$$

$$\geq -\frac{\epsilon}{4} + \epsilon - \frac{\epsilon}{4} > 0.$$

But $J^N_{\mathrm{OLS}}(P^N; \boldsymbol{U}^N)(\omega) \leq J^N_{\mathrm{OLS}}(P_0; \boldsymbol{U}^N)(\omega)$ by definition of P^N. Hence we must have $P^N \in \mathbb{O} = B_\nu(P_0)$ for all $N \geq N_0$, which implies $P^N(\omega) \to P_0$ since $\nu > 0$ was arbitrary.

Theorem 5.5.2 establishes the consistency of the estimator (5.31). Given a set of data $\boldsymbol{u}^N$, it follows that the estimate $\hat{P}^N$ corresponding to the estimator P^N will approximate the true distribution P_0 under the stated assumptions. We recall our discussions in Section 3.3 of Chapter 3 where quite similar assumptions were made in developing the asymptotic theory for estimation of unknown vector parameters $\boldsymbol{\theta}_0$. We remark that these assumptions are not overly restrictive (compare [17, 24, 42]) though some of the assumptions may be difficult to verify in practice. Assumptions (A2)–(A4) are mathematical in nature and may be verified directly for each specific problem. Assumption (A1) describes the error process which is assumed to generate the collected data. While it is unlikely (as in the case of the traditional least squares approach) that one will be able to prove a priori that the error process satisfies these assumptions, posterior analysis such as residual plots discussed in Chapter 3 can be used to investigate the appropriateness of the assumptions of the statistical model. Assumption (A5) reflects the manner in which data is sampled (recall the "sampling points fill up the space" comments of Section 3.3) and, together with Assumption (A6), constitutes an identifiability condition for the model. The limiting sampling distribution function μ may be known if the experimenter has complete control over the values t_j of the covariates or independent variables (e.g., if the t_j are measurement times) but this is not always the case.

5.6   Further Remarks

In the above sections we have defined several classes of parameter estimation problems in which one has a mathematical model describing the dynamics of an individual or aggregate biological or physical process but aggregate data which is sampled from a population. The data is described not by a single parameter but by a probability distribution (over all individuals) from which these individual parameters are sampled. Theoretical results for the non-parametric measure estimation problem are presented which establish the existence and consistency of the estimator. A previously proposed and numerically tested computational scheme is also discussed along with its convergence.

Several open problems remain. First, while the computational scheme is simple, it is not always clear how one should go about choosing the M nodes θ_k^M from the dense subset of Ω_θ which are then used to estimate the weights p_k^M of (5.37). From a theoretical perspective, the nodes need only to be added so that they "fill up" the parameter space in an appropriate way. In practice, however, rounding error and ill-conditioning can be quite problematic, particularly for a poor choice of nodes. A more complete computational algorithm would include information on how to optimally choose the M nodes θ_k^M (as well as the appropriate value of M).

5.7   Non-Parametric Maximum Likelihood Estimation

In the statistical literature there are problems of interest in which one needs a non-parametric estimate of an underlying probability distribution. Of course, the methods discussed so far have been quite general in terms of their assumptions for the statistical model. To this point, we have only needed to know the form of the statistical model for aggregate observations and in return we have developed methods which estimate a population distribution. An obvious question, then, is whether or not a stronger set of assumptions on the error process could lead to the more direct estimation of the underlying probability measure itself. This is the goal of the non-parametric maximum likelihood (NPML) method proposed by Mallet [52]. This approach assumes that some level of longitudinal data is available for each individual. Indeed, one assumes the population under study consists of i individuals, $1 \leq i \leq N_T$, and that for each individual i, there is longitudinal data $\{y_{ij}\}$, $j = 1, 2, \ldots, N_i$, available for the ith individual. The dynamics for the ith individual are assumed to be described by the mathematical model

$$\dot{x}_i(t) = g_i(t, x_i(t); \boldsymbol{\theta}_i, \psi). \tag{5.39}$$

In order to simplify notation we assume here that x_i is scalar-valued. (The results presented are readily generalized if x_i is vector-valued.) The parameters $\boldsymbol{\theta}_i$ are specific to each individual (e.g., rates of drug clearance, growth rates, etc.) and $\boldsymbol{\theta}_i \in \Omega_\theta$ for all i where Ω_θ is some admissible parameter set. The parameter $\psi \in \Omega_\psi$ is a parameter shared by all individuals in the population (e.g., environmental factors). In general, it is possible to have ψ be a function of t but we will not consider that additional complexity here. The model solution will be represented by

$$x_i(t) = f_i(t; \boldsymbol{\theta}_i, \psi). \tag{5.40}$$

We will assume $g_i = g$ for all i and hence $f_i = f$ for all i. This condition is almost always met in practice and thus the assumption is not particularly restrictive. Traditional inverse problems, focusing on a single model and data from a single subject, entail use of data to determine the parameters of the mathematical model $(\boldsymbol{\theta}_i, \psi)$ which best fit the data in an appropriate sense. Here we are less interested in the individual parameters $\boldsymbol{\theta}_i$ and more interested in the probability measure P, which describes the distribution of these parameters across the total population. This probability distribution must be estimated simultaneously with the estimation of the parameter ψ which is common to all individuals in the population. The interplay between individual data and population level characteristics makes hierarchical modeling a natural framework in which to work. Hierarchical modeling [36, 37] (see the review in [25]) relates individual and population level parameters to individual data and has been considered extensively in the statistical literature. For simplicity we here will explain the NPML theory in the case where ψ is known and is included in the dynamics g and observations f. Thus $g(t, x; \boldsymbol{\theta}_i, \psi) = g(t, x; \boldsymbol{\theta}_i)$, $f(t, x; \boldsymbol{\theta}_i, \psi) = f(t, x; \boldsymbol{\theta}_i)$ throughout this section.

5.7.1 Likelihood Formulation

The NPML method is designed to work in a likelihood setting (see the discussions in Chapter 4 above). Consider the statistical model

$$Y_{ij} = f(t_{ij}; \boldsymbol{\theta}_i) + \mathcal{E}_{ij}.$$

While least squares methods require only the specification of the first two statistical moments of the random variables $\mathcal{E}_{ij}$, we must now assume that the probability distribution form of $\mathcal{E}_{ij}$ is known. In this case, the observations y_{ij} (realizations of the random variables Y_{ij}) are characterized by a conditional density function

$$p(y_{ij}) = \varrho_{ij}(y_{ij}|t_{ij}; \boldsymbol{\theta}_i). \tag{5.41}$$

More generally, the functions ϱ_{ij} may also depend on the entire set $\{y_{ij}\}$ of observations (if multiple observations or individuals are correlated). We will not consider this case explicitly. It will be further assumed for simplicity that $\varrho_{ij} = \varrho$ for all i and j. It is quite common to assume independent, normally distributed errors with constant variance, $\mathcal{E}_{ij} \sim \mathcal{N}(0, \sigma^2)$, and we shall for the sake of ease in exposition do so here. Then the conditional density functions are

$$\varrho(y_{ij}|t_{ij}; \boldsymbol{\theta}_i) = \frac{1}{\sigma\sqrt{2\pi}}\exp\left(-\frac{(f(t_{ij}; \boldsymbol{\theta}_i) - y_{ij})^2}{2\sigma^2}\right).$$

Like the two-stage methods in traditional hierarchical methods [36], the likelihood function is formed by considering the multiple (individual and population) stages of the sampling process. Let $\vec{y}_i = \{y_{ij}\}_{j=1}^{N_i}$, $\vec{t}_i = \{t_{ij}\}_{j=1}^{N_i}$. Then we can define the individual likelihood functions

$$l_i(\vec{y}_i|\vec{t}_i; \boldsymbol{\theta}_i) = \prod_{j=1}^{N_i} \varrho(y_{ij}|t_{ij}; \boldsymbol{\theta}_i)$$

which describe the likelihood of the N_i data points taken from the ith individual. If the individual data points are not independent, then the form of the functions l_i may be more complex. Now, assume that the individual parameters $\boldsymbol{\theta}_i$ are all drawn from some (unknown) probability distribution P. Then the "mixing distribution" for each individual is defined as

$$\tilde{l}_i(\vec{y}_i|\vec{t}_i; P) = \int_{\Omega_\theta} l_i(\vec{y}_i|\vec{t}_i; \boldsymbol{\theta})dP(\boldsymbol{\theta}) \tag{5.42}$$

$$= \mathbb{E}(l_i(\vec{y}_i|\vec{t}_i; \cdot)|P).$$

The mixing distributions $\tilde{l}_i$ describe the likelihood of a given set of observations (for individual i) for a given parameter $\boldsymbol{\theta}$, weighted by the likelihood of that particular value of $\boldsymbol{\theta}$. Finally, the population likelihood (which is to be maximized) is

$$\mathcal{L}(P|\{\vec{y}_i\}_{i=1}^{N_T}, \{\vec{t}_i\}_{i=1}^{N_T}) = \prod_{i=1}^{N_T} \tilde{l}_i(\vec{y}_i|\vec{t}_i; P) = \prod_{i=1}^{N_T} \int_{\Omega_\theta} l_i(\vec{y}_i|\vec{t}_i; \boldsymbol{\theta})dP(\boldsymbol{\theta}).$$

The goal of maximum likelihood estimation is to determine the parameter (in this case, the probability measure P) which maximizes the likelihood of the given observations. The probability measures are again taken in the set of probability measures $\mathbb{P}(\Omega_\theta)$ on Ω_θ. The problem to be solved is thus to find

$$\hat{P} = \arg \max_{\mathbb{P}(\Omega_\theta)} \mathcal{L}(P|\{\vec{y}_i\}_{i=1}^{N_T}, \{\vec{t}_i\}_{i=1}^{N_T}). \qquad (5.43)$$

5.7.2 Computational Techniques

Of course, this is an optimization problem over an infinite dimensional space, so once again finite dimensional approximation techniques must be considered. To that end, consider the N_T-dimensional linear model

$$\mathbf{Z} = L(\boldsymbol{\theta})\boldsymbol{\beta} + \tilde{\boldsymbol{\mathcal{E}}}, \qquad (5.44)$$

where $\mathbb{E}(\tilde{\boldsymbol{\mathcal{E}}}) = \mathbf{0}_{N_T}$ $\mathrm{Cov}(\tilde{\boldsymbol{\mathcal{E}}}) = \mathbf{I}_{N_T}$, and $L(\boldsymbol{\theta})$ is the design matrix given by

$$L(\boldsymbol{\theta}) = \mathrm{diag}\left(\sqrt{l_i(\vec{y}_i|\vec{t}_i; \boldsymbol{\theta})} \right).$$

Here, the "data" $\mathbf{Z}$ depends linearly on "parameters" $\boldsymbol{\beta}$, and the goal is to choose the design (that is, find $\boldsymbol{\theta}$) to obtain the best possible (in an appropriate sense) estimate of $\boldsymbol{\beta}$ given observations of $\mathbf{Z}$. Optimal design theory involves maximizing some function of the Fisher information matrix and will be discussed in the next chapter. For the linear model above, the Fisher information matrix is

$$F(P) = \int_{\Omega_\theta} L(\boldsymbol{\theta})^T L(\boldsymbol{\theta}) dP(\boldsymbol{\theta}).$$

For a D-optimal design, one seeks to maximize the determinant $|F(P)|$ of the Fisher information matrix over all possible probability measures P. Now, observe

$$|F(P)| = \prod_{i=1}^{N_T} \int_{\Omega_\theta} l_i(\vec{y}_i|\vec{t}_i; \boldsymbol{\theta}) dP(\boldsymbol{\theta}) = \mathcal{L}(P|\{\vec{y}_i\}_{i=1}^{N_T}, \{\vec{t}_i\}_{i=1}^{N_T}).$$

Thus the problem of maximizing the population likelihood is equivalent to finding a D-optimal design for a finite dimensional linear companion problem (5.44). As such, theorems and computational techniques can be borrowed from the theory of D-optimal design. The following theorem is stated by Mallet [52]:

Theorem 5.7.1 *If Ω_θ is closed and L is continuous in $\boldsymbol{\theta}$, then the set of all Fisher information matrices,*

$$\{F(P) \mid P \in \mathbb{P}(\Omega_\theta)\},$$

is the convex closure of the set of Fisher information matrices corresponding to Dirac measures

$$\{F(P) \mid P = \Delta_{\boldsymbol{\theta}^*}\} = \{L(\boldsymbol{\theta}^*)^T L(\boldsymbol{\theta}^*) \mid \boldsymbol{\theta}^* \in \Omega_\theta\}.$$

Moreover, the D-optimal design belongs to the boundary of this set. Hence for any D-optimal design $\hat{P}$,

$$F(\hat{P}) = \sum_{k=1}^{M} L(\boldsymbol{\theta}_k)^T L(\boldsymbol{\theta}_k) p_k,$$

with $M \leq N_T$.

This theorem relies on the identification (via the Riesz Representation Theorem presented above in Theorem 5.4.2) of the space $\mathbb{P}(\Omega_\theta)$ of probability measures on Ω_θ as a subset of $C_B^*(\Omega_\theta)$. If one further assumes Ω_θ is compact, then one obtains the equivalence (see [50, 55]) of the two sets above, i.e.,

$$\{F(P) \mid P \in \mathbb{P}(\Omega_\theta)\} = \text{conv}\left(\{L(\boldsymbol{\theta}^*)^T L(\boldsymbol{\theta}^*) \mid \boldsymbol{\theta}^* \in \Omega_\theta\}\right),$$

and the closure is no longer needed. It is proven by Federov [40, Theorem 2.2.3] that the Fisher information matrix corresponding to an optimal design lies on the boundary of the convex closure of the set of matrices corresponding to one-point designs. The index limitation ($M \leq N_T$) then follows from the identification of an $N_T \times N_T$ diagonal matrix with a subset of $\mathbb{R}^{N_T}$ and Caratheodory's Theorem for the representation of points on the boundary of a convex set. Not only does this theorem guarantee the existence of an optimal solution for the NPML, it also specifies the finite dimensional form of an optimal solution. It can be shown [40, 50] that the optimal solution is unique in the likelihood geometry. (That is, if $\hat{P}_1$ and $\hat{P}_2$ are two optimal solutions, then $F(\hat{P}_1) = F(\hat{P}_2)$ and hence $\tilde{l}_i(\vec{y}_i|\vec{t}_i; \hat{P}_1) = \tilde{l}_i(\vec{y}_i|\vec{t}_i; \hat{P}_2)$ for all i.) Both Lindsay [50] and Mallet [52] offer conditions which can be used to assess whether the estimated measure $\hat{P}$ is unique.

In order to actually compute the estimate $\hat{P}$, Mallet provides an algorithm which is an extension of the work of Federov [40]. The algorithm attempts to find the $M \leq N_T$ nodes and corresponding weights which form the optimal measure. Interestingly, the algorithm stems from the Kiefer and Wolfowitz Equivalence Theorem [49] demonstrating the equivalence of D-optimal and G-optimal design. The algorithm itself can be found in [52, Sec. 4] and the references therein. See also [36, pp. 195–196] for more details regarding the role of the Equivalence Theorem. We provide only an overview of the algorithm:

1. Fix an initial node $\boldsymbol{\theta}_1 \in \Omega_\theta$ and set $P = \Delta_{\boldsymbol{\theta}_1}$.

2. Compute

$$\boldsymbol{\theta}_{new} = \arg\max_{\boldsymbol{\theta}} d(\boldsymbol{\theta}|P) = \arg\max_{\boldsymbol{\theta}} \sum_{i=1}^{N_T} \frac{l_i(\boldsymbol{\theta})}{\tilde{l}_i(P)},$$

where P is the current design, $l_i(\boldsymbol{\theta}) = l_i(\vec{y}_i|\vec{t}_i; \boldsymbol{\theta})$ and $\tilde{l}_i(P) = \tilde{l}_i(\vec{y}_i|\vec{t}_i; P)$.

3. Update the set of nodes. At iteration K, $\{\boldsymbol{\theta}_k\}_{k=1}^{K+1} = \{\boldsymbol{\theta}_k\}_{k=1}^{K} \cup \{\boldsymbol{\theta}_{new}\}$, and set $P = \displaystyle\sum_{k=1}^{K+1} p_k \Delta_{\boldsymbol{\theta}_k}$.

4. For the fixed set of nodes $\{\boldsymbol{\theta}_k\}_{k=1}^{K+1}$, determine the new weights p_k for each node $\boldsymbol{\theta}_k$ by maximizing the population likelihood function over the set $\{p_k\}$.

5. If this new design/distribution estimate has more than N_T nodes, perform an index limitation step (described in [52, Sec. 4.2]) to reduce the design to N_T nodes, and store this as the updated design P.

6. Repeat steps 2–5 until the algorithm converges, represented by obtaining $d(\boldsymbol{\theta}_{new}|P) = N_T$, or when a maximum number of iterations has been reached.

The above algorithm creates a sequence of designs which are guaranteed to increase the resulting likelihood function at each step [40, 52]. Schumitzky [60] proposed an expectation maximization type algorithm for computation of the index limited design. Another alternative algorithm proposed by Wynn [65] approximates the optimal measure by using an increasingly larger number M of nodes, without regard for whether $M \leq N_T$. This is a similar technique to the Prohorov metric framework discussed above in Section 5.4. In all cases, it is significant that the original infinite dimensional optimization problem of maximizing $\mathcal{L}(P)$ over all $P \in \mathbb{P}(\Omega_\theta)$ has been approximated by a finite dimensional problem which guarantees existence and convergence. The NPML method can be extended to estimate both parameter ψ and probability measure P in an iterative leap-frogging method: for a nominal value of $\hat{\psi}$, a measure $\hat{P}$ is constructed; this $\hat{P}$ is then fixed and an updated estimate of $\hat{\psi}$ is obtained, etc.

The NPML method has been shown to outperform two-stage methods (see the review [25]) in a computational study [53]. In fact, the NPML method can accurately estimate the underlying distribution of the parameters $\boldsymbol{\theta}_i$ even when only a single data point is available for each individual [52] so that the individual models are not identifiable, and, hence, two-stage methods do not apply. (This is not the same as the aggregate data problem.) Given the strong assumptions on the errors (by specifying a conditional density function), this is not entirely surprising. In fact, if the conditional density functions are correctly specified (which can be difficult to verify), then one actually obtains consistency [44, 48] of the NPML estimator. That is, as the number of individuals N_T goes to infinity, the sequence of estimates $\hat{P}_{N_T}$ converges pointwise to the true measure P_0 at all points of continuity of the latter. The NPML method does require the identification of sets of data taken from the same individual (so that the individual likelihoods $l_i(\vec{y}_i|\vec{t}_i; \boldsymbol{\theta}_i)$ can be formed). Yet, it is interesting that the $M \leq N_T$ nodes $\{\boldsymbol{\theta}_k\}$ do not necessarily correspond to the N_T underlying parameters $\boldsymbol{\theta}_i$ [63, p. 93].

Because of the assumed ability to track individuals, the NPML method is not directly applicable to aggregate data problems treated in this chapter. It is also not clear if the restrictive assumptions on the error random variables could be relaxed with least squares estimation used in the place of the maximum likelihood estimation method. It is common to establish the equivalence of least squares and maximum likelihood estimators by demonstrating that they solve the same set of estimating equations (see, e.g., [29, Chapter 3] and Chapter 4 above on model selection). However, the typical estimating equations (first-order optimality conditions) are obtained by taking the derivative of the likelihood function with respect to the unknown parameters, and it is not yet clear how this might be meaningfully done when the underlying parameter is a probability measure (see [16] and references therein for some efforts on this subject).

5.8   Final Remarks

The methods discussed above, the PMF-based method and the NPML method, attempt to estimate an unknown probability distribution directly, thus requiring a rich body of theoretical results. Significantly, these two methods apply to fundamentally different problems—the NPML when data is available for individuals, and the PMF when only aggregate data is available. A key distinction between the two situations is the role played by the probability measure to be estimated. For individual data, the unknown measure plays the role of a mixing distribution among multiple individuals each with a fixed mathematical model. For aggregate data, the unknown measure becomes a part of the mathematical model when the expectation is taken (see Equation (5.29)). Though seemingly simple, this is an inherent difference between the formulations and this difference is key when examining similarities (or lack thereof) in the theory between the different formulations based on data type.

The theoretical results for the NPML and PMF methods share the remarkable feature that the underlying population distribution (regardless of its smoothness) can be approximated by a discrete probability measure. This is all the more fascinating since these results are derived from different theoretical bases and developed to solve different problems. Of course, if one has a priori knowledge of smoothness of the underlying measure, additional techniques have been developed. For individual data, approximation by a convex combination of atomic measures in the NPML method can be replaced by a smooth density function which still maintains many of the desired non-parametric properties [34, 35, 43]. For aggregate data, the Prohorov metric framework can be restricted to the subset of distribution functions which are absolutely continuous. The underlying density function can then be estimated non-parametrically via splines [27].

For more traditional parametric and finite dimensional parameter estimation, methods for determining confidence intervals on parameters include asymptotic methods and bootstrapping (as we discussed in Chapter 3). The issue of finding confidence intervals becomes much more complicated once the object of interest is a probability distribution rather than a single constant parameter. For instance, asymptotic theory requires the computation of sensitivity matrices, which in the case of distribution estimation requires one to take a derivative with respect to a probability measure. This is not an easy task since the space of probability measures is not a linear space, nor is it finite dimensional. Previous work [16] examined the use of directional derivatives, and other work [12] has used standard asymptotic theory to put confidence intervals on each node in the finite dimensional approximation. The Portmanteau Theorem [32, Theorem 2.1], which establishes several equivalent notions of the convergence of probability measures, may provide useful techniques for the visualization of confidence intervals around a probability measure. This is an important avenue for future study in order to provide robust and trustworthy measure estimates in inverse problems.

References

[1] C.A. Balanis, *Advanced Engineering Electromagnetics*, John Wiley & Sons, New York, 1989.

[2] H.T. Banks, *A Functional Analysis Framework for Modeling, Estimation and Control in Science and Engineering*, Chapman and Hall/CRC Press, Boca Raton, FL, 2012.

[3] H.T. Banks, J.H. Barnes, A. Eberhardt, H. Tran and S. Wynne, Modeling and computation of propagating waves from coronary stenosis, *Comput. Appl. Math.*, **21** (2002), 767–788.

[4] H.T. Banks and K. Bihari, Modelling and estimating uncertainty in parameter estimation, *Inverse Problems*, **17** (2001), 95–111.

[5] H.T. Banks and V.A. Bokil, Parameter identification for dispersive dielectrics using pulsed microwave interrogating signals and acoustic wave induced reflections in two and three dimensions, CRSC-TR04-27, July, 2004; Revised version appeared as: A computational and statistical framework for multidimensional domain acoustoptic material interrogation, *Quart. Applied Math.*, **63** (2005), 156–200.

[6] H.T. Banks and D.M. Bortz, Inverse problems for a class of measure

dependent dynamical systems, *J. Inverse and Ill-posed Problems*, **13** (2005), 103–121.

[7] H.T. Banks, D.M. Bortz and S.E. Holte, Incorporation of variability into the mathematical modeling of viral delays in HIV infection dynamics, *Mathematical Biosciences*, **183** (2003), 63–91.

[8] H.T. Banks, D.M. Bortz, G.A. Pinter and L.K. Potter, Modeling and imaging techniques with potential for application in bioterrorism, CRSC-TR03-02, January 2003; Chapter 6 in *Bioterrorism: Mathematical Modeling Applications in Homeland Security* (H.T. Banks and C. Castillo-Chavez, eds.), Frontiers in Applied Math, **FR28**, SIAM, Philadelphia, 2003, 129–154.

[9] H.T. Banks, L.W. Botsford, F. Kappel and C. Wang, Modeling and estimation in size structured population models, LCDS-CCS Report 87-13, Brown University; *Proc. 2nd Course on Mathematical Ecology* (Trieste, December 8–12, 1986) World Press, 1988, Singapore, 521–541.

[10] H.T. Banks, M.W. Buksas and T. Lin, *Electromagnetic Material Interrogation Using Conductive Interfaces and Acoustic Wavefronts*, SIAM FR **21**, Philadelphia, 2002.

[11] H.T. Banks and J.L. Davis, A comparison of approximation methods for the estimation of probability distributions on parameters, CRSC-TR05-38, October, 2005; *Applied Numerical Mathematics*, **57** (2007), 753–777.

[12] H.T. Banks and J.L. Davis, Quantifying uncertainty in the estimation of probability distributions with confidence bands, CRSC-TR07-21, December, 2007; *Mathematical Biosciences and Engineering*, **5** (2008), 647–667.

[13] H.T. Banks, J.L. Davis, S.L. Ernstberger, S. Hu, E. Artimovich, A.K. Dhar and C.L. Browdy, A comparison of probabilistic and stochastic formulations in modeling growth uncertainty and variability, CRSC-TR08-03, February, 2008; *Journal of Biological Dynamics*, **3** (2009) 130–148.

[14] H.T. Banks, J.L. Davis, S.L. Ernstberger, S. Hu, E. Artimovich and A.K. Dhar, Experimental design and estimation of growth rate distributions in size-structured shrimp populations, CRSC-TR08-20, November, 2008; *Inverse Problems*, **25** (2009), 095003 (28 pp), Sept.

[15] H.T. Banks, J.L. Davis and S. Hu, A computational comparison of alternatives to including uncertainty in structured population models, CRSC-TR09-14, June, 2009; in *Three Decades of Progress in Systems and Control*, (X. Hu et al., eds.) Springer, New York, 2010, 19–33.

[16] H.T. Banks, S. Dediu and H.K. Nguyen, Sensitivity of dynamical systems to parameters in a convex subset of a topological vector space, *Math. Biosci. Eng.*, **4** (2007), 403–430.

[17] H.T. Banks and B.G. Fitzpatrick, Statistical methods for model comparison in parameter estimation problems for distributed systems, *J. Math. Biol.,* **28** (1990), 501–527.

[18] H.T. Banks and B.G. Fitzpatrick, Estimation of growth rate distributions in size structured population models, *Quarterly of Applied Mathematics,* **49** (1991), 215–235.

[19] H.T. Banks, B.G. Fitzpatrick, L.K. Potter and Y. Zhang, Estimation of probability distributions for individual parameters using aggregate population data, CRSC-TR98-06, January 1998; in *Stochastic Analysis, Control, Optimization, and Applications,* (W. McEneaney, G. Yin and Q. Zhang, eds.), Birkhauser, Boston, 1989.

[20] H.T. Banks and N.L. Gibson, Well-posedness in Maxwell systems with distributions of polarization relaxation parameters, CRSC-TR04-01, January, 2004; *Applied Math. Letters,* **18** (2005), 423–430.

[21] H.T. Banks and N.L. Gibson, Electromagnetic inverse problems involving distributions of dielectric mechanisms and parameters, CRSC-TR05-29, August, 2005; *Quarterly of Applied Mathematics,* **64** (2006), 749–795.

[22] H.T. Banks, S. Hu, Z.R. Kenz, C. Kruse, S. Shaw, J.R. Whiteman, M.P. Brewin, S.E. Greenwald and M.J. Birch, Material parameter estimation and hypothesis testing on a 1D viscoelastic stenosis model: methodology, CRSC-TR12-09, April, 2012; *J. Inverse and Ill-posed Problems,* **21** (2013), 25–57.

[23] H.T. Banks, S. Hu, Z.R. Kenz, C. Kruse, S. Shaw, J.R. Whiteman, M.P. Brewin, S.E. Greenwald and M.J. Birch, Model validation for a noninvasive arterial stenosis detection problem, CRSC-TR12-22, December, 2012; *Mathematical Biosciences and Engr.,* to appear.

[24] H.T. Banks, Z.R. Kenz and W.C. Thompson, An extension of RSS-based model comparison tests for weighted least squares, *Intl. J. Pure and Appl. Math,* **79** (2012), 155–183.

[25] H.T. Banks, Z.R. Kenz and W.C. Thompson, A review of selected techniques in inverse problem nonparametric probability distribution estimation, CRSC-TR12-13, May 2012; *J. Inverse and Ill-Posed Problems,* **20** (2012), 429–460.

[26] H.T. Banks and K. Kunisch, *Estimation Techniques for Distributed Parameter Systems,* Birkhausen, Boston, 1989.

[27] H.T. Banks and G.A. Pinter, A probabilistic multiscale approach to hysteresis in shear wave propagation in biotissue, CRSC-TR04-03, January, 2004; *SIAM J. Multiscale Modeling and Simulation,* **3** (2005), 395–412.

[28] H.T. Banks and L.K. Potter, Probabilistic methods for addressing uncertainty and variability in biological models: Application to a toxicokinetic model, CRSC-TR02-27, September, 2002; *Math. Biosci.*, **192** (2004), 193–225.

[29] H.T. Banks and H.T. Tran, *Mathematical and Experimental Modeling of Physical and Biological Processes*, CRC Press, Boca Raton, London, New York, 2009.

[30] H.T. Banks, H. Tran and S. Wynne, A well-posedness result for a shear wave propagation model, *Intl. Series Num. Math.*, **143**, Birkhauser Verlag, Basel, 2002, 25–40.

[31] G. Bell and E. Anderson, Cell growth and division. I. A mathematical model with applications to cell volume distributions in mammalian suspension cultures, *Biophysical Journal*, **7** (1967), 329–351.

[32] P. Billingsley, *Convergence of Probability Measures*, Wiley & Sons, New York, 1968.

[33] D.K. Cheng, *Field and Wave Electromagnetics*, Addison Wesley, Reading, MA, 1983.

[34] M. Davidian and A.R. Gallant, Smooth nonparametric maximum likelihood estimation for population pharmacokinetics, with application to quinidine, *J. Pharmacokinetics and Biopharmaceutics*, **20** (1992), 529–556.

[35] M. Davidian and A.R. Gallant, The nonlinear mixed effects model with a smooth random effects density, *Biometrika*, **80** (1993), 475–488.

[36] M. Davidian and D.M. Giltinan, *Nonlinear Models for Repeated Measurement Data*, Chapman and Hall, London, 2000.

[37] M. Davidian and D. Giltinan, Nonlinear models for repeated measurement data: An overview and update, *J. Agricultural, Biological, and Environmental Statistics*, **8** (2003), 387–419.

[38] R.M. Dudley, *Real Analysis and Probability*, Cambridge University Press, Cambridge, UK, 2002.

[39] R.S. Elliot, *Electromagnetics: History, Theory, and Applications*, IEEE Press, New York, 1993.

[40] V. Fedorov, *Theory of Optimal Experiments*, New York, Academic Press, 1972.

[41] Y.C. Fung, *Biomechanics: Mechanical Properties of Living Tissues*, Springer-Verlag, New York, 1993.

[42] A.R. Gallant, *Nonlinear Statistical Models*, John Wiley and Sons, New York, 1987.

[43] A.R. Gallant and D. Nychka, Semi-nonparametric maximum likelihood estimation, *Econometrica*, **55** (1987), 363–390.

[44] B. Hoadley, Asymptotic properties of maximum likelihood estimators for the independent not identically distributed case, *Ann. Math. Stat.*, **42** (1971), 1977–1991.

[45] P. Huber, *Robust Statistics,* John Wiley and Sons, New York, 1981.

[46] J.D. Jackson, *Classical Electrodynamics*, J. Wiley & Sons, New York, 1975.

[47] Z.R. Kenz, H.T. Banks and R.C. Smith, Comparison of frequentist and Bayesian confidence analysis methods on a viscoelastic stenosis model, CRSC-TR13-05, April, 2013; *SIAM/ASA Journal on Uncertainty Quantification*, 1(2013), 348–369.

[48] J. Kiefer and J. Wolfowitz, Consistency of the maximum likelihood estimator in the presence of infinitely many incidental parameters, *Ann. Math. Stat.*, **27** (1956), 887–906.

[49] J. Kiefer and J. Wolfowitz, The equivalence of two extremum problems, *Canadian J. Mathematics*, **12** (1960), 363–366.

[50] B. Lindsay, The geometry of mixture likelihoods: a general theory, *Ann. Stat.*, **11** (1983), 86–94.

[51] B. Lindsay, *Mixture Models: Theory, Geometry and Applications*, NSF-CBMS Regional Conference Series in Probability and Statistics, Volume 5, Institute of Mathematical Statistics, Hayward, CA, 1995.

[52] A. Mallet, A maximum likelihood estimation method for random coefficient regression models, *Biometrika*, **73** (1986), 645–656.

[53] A. Mallet, France Mentré, Jean-Louis Steimer and François Lokiec, Nonparametric maximum likelihood estimation for population pharmacokinetics with application to cyclosporine, *J. Pharmacokinetics and Biopharmaceutics*, **16** (1988), 311–327.

[54] J.A.J. Metz and E.O. Diekmann, *The Dynamics of Physiologically Structured Populations*, Lecture Notes in Biomathematics, Vol. **68**, Springer, Heidelberg, 1986.

[55] R. Phelps, *Lectures on Choquet's Theorem*, Van Nostram, Princeton, 1955.

[56] L.K. Potter, *Physiologically Based Pharmacokinetic Models for the Systemic Transport of Trichloroethylene*, Ph.D. Thesis, N.C. State University, August, 2001 (http://www.lib.ncsu.edu).

[57] Yu.V. Prohorov, Convergence of random processes and limit theorems in probability theory, *Theor. Prob. Appl.*, **1** (1956), 157–214.

[58] J.R. Reitz, R.W. Christy and F.J. Milford, *Foundations of Electromagnetic Theory*, Addison Wesley, Reading, MA, 1992.

[59] H.L. Royden, *Real Analysis*, 3rd ed., Prentice Hall, Upper Saddle River, NJ, 1988.

[60] A. Schumitzky, Nonparametric EM algorithms for estimating prior distributions, *Applied Mathematics and Computation*, **45** (1991), 143–157.

[61] G.A. Seber and C.J. Wild, *Nonlinear Regression*, Wiley, Hoboken, 2003.

[62] R.C. Smith, *Uncertainty Quantification: Theory, Implementation and Applications*, SIAM, Philadelphia, 2013.

[63] J.L. Steimer, A. Mallet and F. Mentre, Estimating interindividual pharmacokinetic variability, in *Variability in Drug Therapy: Description, Estimation, and Control*, M. Rowland et al., eds., Raven Press, New York, 1985.

[64] J. Sinko and W. Streifer, A new model for age-size structure of a population, *Ecology*, **48** (1967), 910–918.

[65] H.P. Wynn, The sequential generation of D-optimal experimental designs, *Ann. of Math. Stat.*, **41** (1970), 1655–1664.

Chapter 6

Optimal Design

6.1 Introduction

To this point we have discussed various aspects of uncertainty arising in inverse problem techniques. All discussions have been in the context of *a given set* or *sets* of data carried out under various assumptions on how (e.g., independent sampling, absolute measurement error, relative measurement error) the data were collected. For many years [4, 7, 16, 17, 18, 21, 22, 23] now scientists (and especially engineers) have been actively involved in designing experimental protocols to best study engineering systems, including parameters describing mechanisms. Recently, with increased involvement of scientists working in collaborative efforts with biologists and quantitative life scientists, renewed interest in the design of "best" experiments to elucidate mechanisms has been seen [4]. Thus a major question that experimentalists and inverse problem investigators alike often face is how to best collect the data to enable one to efficiently and accurately estimate model parameters. This is the well-known and widely studied *optimal design* problem. We would be remiss in this monograph if we did not provide at least a brief introduction to the ideas and issues that arise in this methodology.

Traditional optimal design methods (D-optimal, E-optimal, c-optimal) [7, 16, 17, 18] use information from the model to find the sampling distribution or mesh for the observation times (and/or locations in spatially distributed problems) that minimizes a design criterion, quite often a function of the Fisher Information Matrix (FIM). Experimental data taken on this optimal mesh are then expected to result in accurate parameter estimates. In many scientific fields where mathematical modeling is utilized, mathematical models grow increasingly complex over time, containing possibly more state variables and parameters, as the underlying governing processes of a system are better understood and refinements in mechanisms are considered. Additionally, as technology invents and improves devices to measure physical and biological phenomena, new data become available to inform mathematical modeling efforts. The world is approaching an era in which the vast amounts of information available to researchers may be overwhelming or even counterproductive to efforts. We outline a framework based on the FIM for a system of ordinary differential equations (ODEs) to determine *when an experimenter should*

195

take samples and *what variables to measure* when collecting information on a physical or biological process modeled by a dynamical system.

Inverse problem methodologies are discussed in the previous chapters in the context of dynamical system or mathematical model when a sufficient number of observations of one or more states (variables) is available. The choice of method depends on assumptions the modeler makes on the form of the error between the model and the observations (the statistical model). The most prevalent source of error is observation error, which is made when collecting data. (One can also consider model error, which originates from the differences between the model and the underlying process that the model describes. But this is often quite difficult to quantify.) Measurement error is most readily discussed in the context of statistical models. The three techniques commonly addressed are maximum likelihood estimation (MLE), used when the probability distribution form of the error is known; ordinary least squares (OLS), for error with constant variance across observations; and generalized least squares (GLS), used when the variance of the data can be expressed as a non-constant function. Uncertainty quantification is also described for optimization problems of this type, namely, in the form of observation error covariances, standard errors, residual plots and sensitivity matrices. Techniques to approximate the variance of the error are also included in these discussions. In [11], the authors develop an experimental design theory using the FIM to identify optimal sampling times for experiments on physical processes (modeled by an ODE system) in which scalar or vector data is taken. The experimental design technique developed is applied in numerical simulations to the logistic curve, a simple ODE model describing glucose regulation and a harmonic oscillator example.

In addition to when to take samples, the question of what variables to measure is also very important in designing effective experiments, especially when the number of state variables is large. Use of such a methodology to optimize what to measure would further reduce testing costs by eliminating extra experiments to measure variables neglected in previous trials (see [9]). In [6], the best set of variables for an ODE system modeling the Calvin cycle [24] is identified using two methods. The first, an ad hoc statistical method, determines which variables directly influence an output of interest at any one particular time. Such a method does not utilize the information on the underlying time-varying processes given by the dynamical system model. The second method is based on optimal design ideas. Extension of this method is developed in [12, 13]. Specifically, in [12] the authors compare the SE-optimal design introduced in [10] and [11] with the well-known methods of D-optimal and E-optimal design on a 6-compartment HIV model [3] and a 31 dimensional model of the Calvin cycle [24]. Models for which there may be a wide range of variables to possibly observe are not only ideal on which to test the proposed methodology, but also are widely encountered in applications. For example, the methods have been recently used in [14, 15] to design optimal data collection in terms of the location of sensors and the number needed for optimal design in electroencephalography (EEG) in the recording of electrical activity along the scalp. We turn to an outline of this methodology to make observations for best times and best variables.

6.2 Mathematical and Statistical Models

We return to the notation of Chapter 3. We consider inverse or parameter estimation problems in the context of a parameterized (with vector parameter $q \in \mathbb{R}^{\kappa_q}$) n-dimensional vector dynamical system or **mathematical model**

$$\frac{d\boldsymbol{x}}{dt}(t) = \boldsymbol{g}(t, \boldsymbol{x}(t), \boldsymbol{q}), \tag{6.1}$$

$$\boldsymbol{x}(t_0) = \boldsymbol{x}_0, \tag{6.2}$$

with **observation process**

$$\boldsymbol{f}(t; \boldsymbol{\theta}) = \boldsymbol{C}\boldsymbol{x}(t; \boldsymbol{\theta}), \tag{6.3}$$

where $\boldsymbol{\theta} = \text{column}\,(\mathbf{q}, \tilde{\mathbf{x}}_0) \in \mathbb{R}^{\kappa_q+\tilde{n}} = \mathbb{R}^{\kappa_\theta}, \tilde{n} \leq n$, and the observation operator $\mathcal{C}$ maps $\mathbb{R}^n$ to $\mathbb{R}^m$. In most of the discussions below we assume without loss of generality that some $\tilde{\boldsymbol{x}}_0$ of the initial values $\boldsymbol{x}_0$ are also unknown.

If we were able to observe all states, each measured by a possibly different sampling technique, then $m = n$ and $\mathcal{C} = \mathbf{I}_n$ would be a possible choice; however, this is most often not the case because of the impossibility of or the expense in measuring all state variables. In other cases (such as the HIV example below) we may be able to directly observe only combinations of the states. In the formulation below we will be interested in collections or sets of up to K (where $K \leq n$) one-dimensional observation operators or maps so that sets of $m = 1$ observation operators will be considered.

In order to discuss the amount of uncertainty in parameter estimates, we formulate a **statistical model** of the form

$$\boldsymbol{Y}(t) = \boldsymbol{f}(t; \boldsymbol{\theta}_0) + \boldsymbol{\mathcal{E}}(t), \quad t \in [t_0, t_f], \tag{6.4}$$

where $\boldsymbol{\theta}_0$ is the hypothesized true values of the unknown parameters and $\boldsymbol{\mathcal{E}}(t)$ is a random vector that represents observation error for the measured variables at time t. We make the standard assumptions:

$\mathbb{E}(\boldsymbol{\mathcal{E}}(t)) = \mathbf{0}, \quad t \in [t_0, t_f],$
$\text{Var}(\boldsymbol{\mathcal{E}}(t)) = V_0(t) = \text{diag}(\sigma_{0,1}^2(t), \sigma_{0,2}^2(t), \dots, \sigma_{0,m}^2(t)), \quad t \in [t_0, t_f],$
$\text{Cov}\{\mathcal{E}_i(t), \mathcal{E}_i(s)\} = 0, \quad s \neq t, \, s, t \in [t_0, t_f],$
$\text{Cov}\{\mathcal{E}_i(t), \mathcal{E}_j(s)\} = 0, \quad i \neq j, \, s, t \in [t_0, t_f].$

Realizations of the statistical model (6.4) are written

$$\boldsymbol{y}(t) = \boldsymbol{f}(t; \boldsymbol{\theta}_0) + \boldsymbol{\epsilon}(t), \quad t \in [t_0, t_f].$$

When collecting experimental data, it is often difficult to take continuous measurements of the observed variables. Instead, we assume that we have N observations at times t_j, $j = 1, \ldots, N$, $t_0 \le t_1 < t_2 < \cdots < t_N \le t_f$. We then write the observation process (6.3) as

$$\boldsymbol{f}(t_j; \boldsymbol{\theta}) = C\boldsymbol{x}(t_j; \boldsymbol{\theta}), \quad j = 1, 2, \ldots, N, \tag{6.5}$$

the discrete statistical model as

$$\boldsymbol{Y}_j = \boldsymbol{f}(t_j; \boldsymbol{\theta}_0) + \boldsymbol{\mathcal{E}}(t_j), \quad j = 1, 2, \ldots, N, \tag{6.6}$$

and a realization of the discrete statistical model as

$$\boldsymbol{y}_j = \boldsymbol{f}(t_j; \boldsymbol{\theta}_0) + \boldsymbol{\epsilon}(t_j), \quad j = 1, 2, \ldots, N.$$

We will use this mathematical and statistical framework to develop a methodology to identify *sampling variables* (for a fixed number K of variables or combinations of variables) and the most informative *times* (for a fixed number N) at which the samples should be taken so as to provide the most information pertinent to estimating a given set of parameters.

6.2.1   Formulation of the Optimal Design Problem

Several methods exist to solve the inverse problem. A major factor in determining which method to use is additional assumptions made about $\boldsymbol{\mathcal{E}}(t)$. It is common practice to make the assumption that realizations of $\boldsymbol{\mathcal{E}}(t)$ at particular time points are independent and identically distributed (i.i.d.). If, additionally, the distributions describing the behavior of the components of $\boldsymbol{\mathcal{E}}(t)$ are known, then a maximum likelihood estimation method may be used to find an estimate of $\boldsymbol{\theta}_0$. On the other hand, if the distributions for $\boldsymbol{\mathcal{E}}(t)$ are not known but the covariance matrix $V_0(t)$ (also unknown) is assumed to vary over time, weighted least squares methods are often used. We propose an optimal design problem formulation using a general weighted least squares criterion.

Let $\mathbb{P}_1([t_0, t_f])$ denote the set of all bounded distributions on the interval $[t_0, t_f]$. We consider the generalized weighted least squares cost functional for systems with vector output

$$J_{\text{WLS}}(\boldsymbol{\theta}; \boldsymbol{y}) = \int_{t_0}^{t_f} [\boldsymbol{y}(t) - \boldsymbol{f}(t; \boldsymbol{\theta})]^T V_0^{-1}(t) [\boldsymbol{y}(t) - \boldsymbol{f}(t; \boldsymbol{\theta})] dP_1(t), \tag{6.7}$$

where $P_1 \in \mathbb{P}_1([t_0, t_f])$ is a general measure on the interval $[t_0, t_f]$. For a given continuous data set $\boldsymbol{y}(t)$, we search for a parameter $\hat{\boldsymbol{\theta}}$ that minimizes $J_{\text{WLS}}(\boldsymbol{\theta}; \boldsymbol{y})$.

We next consider the case of observations collected at discrete times. If we choose a set of N time points $\tau = \{t_j\}_{j=1}^{N}$, where $t_0 \leq t_1 < t_2 < \cdots < t_N \leq t_f$ and take

$$P_1 = P_\tau = \sum_{j=1}^{N} \Delta_{t_j}, \tag{6.8}$$

where Δ_a represents the Dirac measure with atom at a, then the weighted least squares criterion (6.7) for a finite number of observations becomes

$$J_{\text{WLS}}^N(\boldsymbol{\theta}; \boldsymbol{y}) = \sum_{j=1}^{N} [\boldsymbol{y}(t_j) - \boldsymbol{f}(t_j; \boldsymbol{\theta})]^T V_0^{-1}(t_j)[\boldsymbol{y}(t_j) - \boldsymbol{f}(t_j; \boldsymbol{\theta})].$$

Note in this chapter we do not normalize the time "distributions" such as (6.8) by a factor of $\dfrac{1}{N}$ so that they are not the usual cumulative distribution functions but would be if we normalized each distribution by the integral of its corresponding density to obtain a true probability measure. A similar remark holds for the "variables" observation operator distributions introduced below where without loss of generality we could normalize by a factor of $\dfrac{1}{K}$ when using K 1-dimensional sampling maps.

To select a useful distribution of time points and set of observation variables, we introduce the m by κ_θ sensitivity matrices $\dfrac{\partial \boldsymbol{f}(t; \boldsymbol{\theta})}{\partial \boldsymbol{\theta}}$ and the n by κ_θ sensitivity matrices $\dfrac{\partial \boldsymbol{x}(t; \boldsymbol{\theta})}{\partial \boldsymbol{\theta}}$ that are determined using the differential operator in row vector form $(\partial_{\theta_1}, \partial_{\theta_2}, \ldots, \partial_{\theta_{\kappa_\theta}})$ represented by $\nabla_{\boldsymbol{\theta}}$ and the observation operator defined in (6.3),

$$\nabla_{\boldsymbol{\theta}} \boldsymbol{f}(t; \boldsymbol{\theta}) = \dfrac{\partial \boldsymbol{f}(t; \boldsymbol{\theta})}{\partial \boldsymbol{\theta}} = \mathcal{C} \dfrac{\partial \boldsymbol{x}(t; \boldsymbol{\theta})}{\partial \boldsymbol{\theta}} = \mathcal{C} \nabla_{\boldsymbol{\theta}} \boldsymbol{x}(t; \boldsymbol{\theta}). \tag{6.9}$$

Using the sensitivity matrix $\nabla_{\boldsymbol{\theta}} \boldsymbol{f}(t; \boldsymbol{\theta}_0)$, we may formulate the Generalized Fisher Information Matrix (GFIM). Consider the set (assumed compact) $\Omega_C \subset \mathbb{R}^{1 \times n}$ of admissible observation maps and let $\mathbb{P}_2(\Omega_C)$ represent the set of all bounded distributions P_2 on Ω_C. Then the GFIM may be written

$$F(P_1, P_2, \boldsymbol{\theta}_0) \equiv \int_{t_0}^{t_f} \int_{\Omega_C} \frac{1}{\sigma^2(t, c)} \nabla_{\boldsymbol{\theta}}^T \boldsymbol{f}(t; \boldsymbol{\theta}_0) \nabla_{\boldsymbol{\theta}} \boldsymbol{f}(t; \boldsymbol{\theta}_0) dP_2(c) dP_1(t) \tag{6.10}$$

$$= \int_{t_0}^{t_f} \int_{\Omega_C} \frac{1}{\sigma^2(t, c)} \nabla_{\boldsymbol{\theta}}^T \left(c\boldsymbol{x}(t; \boldsymbol{\theta}_0) \right) \nabla_{\boldsymbol{\theta}} \left(c\boldsymbol{x}(t; \boldsymbol{\theta}_0) \right) dP_2(c) dP_1(t).$$

In fact we shall be interested in collections of K 1-dimensional "variable" observation operators and a choice of which K variables provide the best information to estimate the desired unknown parameters in a given model.

Thus taking K different sampling maps in Ω_C, represented by the $1 \times n$-dimensional matrices $\mathcal{C}_k$, $k = 1, 2, \ldots, K$, we construct the discrete distribution on $\Omega_C^K = \bigotimes_{i=1}^{K} \Omega_C$ (the k-fold cross products of Ω_C)

$$P_{\mathcal{S}} = \sum_{k=1}^{K} \Delta_{\mathcal{C}_k}, \qquad (6.11)$$

where Δ_a represents the Dirac measure with atom at a. Using $P_{\mathcal{S}}$ in (6.10), we obtain the GFIM for multiple discrete observation methods taken continuously over $[t_0, t_f]$ given by

$$F(P_1, P_{\mathcal{S}}, \boldsymbol{\theta}_0) = \int_{t_0}^{t_f} \sum_{k=1}^{K} \frac{1}{\sigma^2(t, \mathcal{C}_k)} \nabla_{\boldsymbol{\theta}}^T \left(\mathcal{C}_k \boldsymbol{x}(t; \boldsymbol{\theta}_0) \right) \nabla_{\boldsymbol{\theta}} \left(\mathcal{C}_k \boldsymbol{x}(t; \boldsymbol{\theta}_0) \right) dP_1(t)$$

$$= \int_{t_0}^{t_f} \sum_{k=1}^{K} \frac{1}{\sigma^2(t, \mathcal{C}_k)} \nabla_{\boldsymbol{\theta}}^T \boldsymbol{x}(t; \boldsymbol{\theta}_0) \mathcal{C}_k^T \mathcal{C}_k \nabla_{\boldsymbol{\theta}} \boldsymbol{x}(t; \boldsymbol{\theta}_0) dP_1(t)$$

$$= \int_{t_0}^{t_f} \sum_{k=1}^{K} \nabla_{\boldsymbol{\theta}}^T \boldsymbol{x}(t; \boldsymbol{\theta}_0) \mathcal{C}_k^T \frac{1}{\sigma^2(t, \mathcal{C}_k)} \mathcal{C}_k \nabla_{\boldsymbol{\theta}} \boldsymbol{x}(t; \boldsymbol{\theta}_0) dP_1(t)$$

$$= \int_{t_0}^{t_f} \nabla_{\boldsymbol{\theta}}^T \boldsymbol{x}(t; \boldsymbol{\theta}_0) \sum_{k=1}^{K} \left(\mathcal{C}_k^T \frac{1}{\sigma^2(t, \mathcal{C}_k)} \mathcal{C}_k \right) \nabla_{\boldsymbol{\theta}} \boldsymbol{x}(t; \boldsymbol{\theta}_0) dP_1(t)$$

$$= \int_{t_0}^{t_f} \nabla_{\boldsymbol{\theta}}^T \boldsymbol{x}(t; \boldsymbol{\theta}_0) \left(\mathcal{S}^T V_K^{-1}(t) \mathcal{S} \right) \nabla_{\boldsymbol{\theta}} \boldsymbol{x}(t; \boldsymbol{\theta}_0) dP_1(t), \qquad (6.12)$$

where $\mathcal{S} = \text{column}(\mathcal{C}_1, \mathcal{C}_2, \ldots, \mathcal{C}_K) \in \mathbb{R}^{K \times n}$ is the set of observation operators defined above and $V_K(t) = \text{diag}(\sigma^2(t, \mathcal{C}_1), \ldots, \sigma^2(t, \mathcal{C}_K))$ is the corresponding covariance matrix for K 1-dimensional observation operators. Applying the distribution P_τ as described in (6.8) to the GFIM (6.12) for discrete observation operators measured continuously yields the discrete $\kappa_\theta \times \kappa_\theta$ Fisher Information Matrix (FIM) for discrete observation operators measured at discrete times.

$$F(\tau, \mathcal{S}, \boldsymbol{\theta}_0) \equiv F(P_\tau, P_{\mathcal{S}}, \boldsymbol{\theta}_0) =$$

$$\sum_{j=1}^{N} \nabla_{\boldsymbol{\theta}}^T \boldsymbol{x}(t_j; \boldsymbol{\theta}_0) \mathcal{S}^T V_K^{-1}(t_j) \mathcal{S} \nabla_{\boldsymbol{\theta}} \boldsymbol{x}(t_j; \boldsymbol{\theta}_0). \qquad (6.13)$$

This describes the amount of information about the κ_θ parameters of interest that is captured by the observed quantities described by the sampling maps $\mathcal{C}_k$, $k = 1, 2, \ldots, K$, defining $\mathcal{S}$, when they are measured at the time points in τ.

The questions of determining the best (in some sense) $\mathcal{S}$ and τ are the important questions in the optimal design of an experiment. Recall that the set of time points τ has an associated distribution $P_\tau \in \widetilde{\mathbb{P}}_1([t_0, t_f])$, where $\widetilde{\mathbb{P}}_1([t_0, t_f])$ is the set of all bounded discrete distributions on $[t_0, t_f]$. Similarly, the set of sampling maps $\mathcal{S}$ has an associated bounded discrete distribution $P_{\mathcal{S}} \in \widetilde{\mathbb{P}}_2(\Omega_C^K)$. Define the space of bounded discrete distributions $\widetilde{\mathbb{P}}([t_0, t_f] \times \Omega_C^K) = \widetilde{\mathbb{P}}_1([t_0, t_f]) \times \widetilde{\mathbb{P}}_2(\Omega_C^K)$ with elements $P = (P_\tau, P_{\mathcal{S}}) \in \widetilde{\mathbb{P}}$. We may, without loss of generality, assume that $\Omega_C^K \subset \mathbb{R}^{K \times n}$ is closed and bounded, and assume that there exists a functional $\mathcal{J} : \mathbb{R}^{\kappa_\theta \times \kappa_\theta} \to \mathbb{R}^+$ of the GFIM (6.12). Then the **optimal design problem** associated with $\mathcal{J}$ is selecting a discrete distribution $\hat{P} \in \widetilde{\mathbb{P}}([t_0, t_f] \times \Omega_C^K)$ such that

$$\mathcal{J}\left(F(\hat{P}, \boldsymbol{\theta}_0)\right) = \min_{P \in \widetilde{\mathbb{P}}([t_0, t_f] \times \Omega_C^K)} \mathcal{J}\left(F(P, \boldsymbol{\theta}_0)\right), \qquad (6.14)$$

where $\mathcal{J}$ depends continuously on the elements of $F(P, \boldsymbol{\theta}_0)$.

The Prohorov metric introduced earlier in Chapter 5 provides a general theoretical framework for the existence of $\hat{P}$ and approximation in $\mathbb{P}([t_0, t_f] \times \Omega_C)$ (a general theoretical framework with proofs is developed in [8, 10]). The application of the Prohorov metric to optimal design problems formulated as (6.14) is explained more fully in [10]: briefly, define the Prohorov metric ρ on the space $\mathbb{P}([t_0, t_f] \times \Omega_C^K)$, and consider the metric space $(\mathbb{P}([t_0, t_f] \times \Omega_C^K), \rho)$. Since $[t_0, t_f] \times \Omega_C^K$ is compact, $(\mathbb{P}([t_0, t_f] \times \Omega_C^K), \rho)$ is also compact. Additionally, by the properties of the Prohorov metric, $(\mathbb{P}([t_0, t_f] \times \Omega_C^K), \rho)$ is complete and separable. Therefore an optimal distribution P^* exists in $\mathbb{P}([t_0, t_f] \times \Omega_C^K)$ and may be approximated by an optimal discrete distribution $\hat{P}$ in $\widetilde{\mathbb{P}}([t_0, t_f] \times \Omega_C^K)$.

The formulation of the cost functional (6.14) may take many forms. We outline the use of traditional optimal design methods, D-optimal, E-optimal or SE-optimal design criteria, to determine the form of $\mathcal{J}$. Each of these design criteria is a function of the inverse of the FIM (assumed hereafter to be invertible) defined in (6.13).

In D-optimal design, the cost functional is written

$$\mathcal{J}_D(F) = \det\left((F(\tau, \mathcal{S}, \boldsymbol{\theta}_0))^{-1}\right) = \frac{1}{\det\left(F(\tau, \mathcal{S}, \boldsymbol{\theta}_0)\right)}.$$

By minimizing $\mathcal{J}_D$, we minimize the volume of the confidence interval ellipsoid describing the uncertainty in our parameter estimates. Since F is symmetric and positive semi-definite, $\mathcal{J}_D(F) \geq 0$. Additionally, since F is assumed invertible, $\mathcal{J}_D(F) \neq 0$; therefore, $\mathcal{J}_D : \mathbb{R}^{\kappa_\theta \times \kappa_\theta} \to (0, \infty)$.

In E-optimal design, the cost functional $\mathcal{J}_E$ is the largest eigenvalue of $(F(\tau, \mathcal{S}, \boldsymbol{\theta}_0))^{-1}$, or equivalently

$$\mathcal{J}_E(F) = \max\left\{\frac{1}{\text{eig}\left(F(\tau, \mathcal{S}, \boldsymbol{\theta}_0)\right)}\right\}.$$

To obtain a smaller standard error, we must reduce the length of the principal axis of the confidence interval ellipsoid. Since F is positive definite, all eigenvalues are therefore positive. Thus $\mathcal{J}_E : \mathbb{R}^{\kappa_\theta \times \kappa_\theta} \to (0, \infty)$.

In SE-optimal design, the cost functional $\mathcal{J}_{SE}$ is a sum of the elements on the diagonal of $(F(\tau, \mathcal{S}, \boldsymbol{\theta}_0))^{-1}$ weighted by the respective parameter values [10, 11], written

$$\mathcal{J}_{SE}(F) = \sum_{i=1}^{\kappa_\theta} \frac{(F(\tau, \mathcal{S}, \boldsymbol{\theta}_0)))^{-1}_{i,i}}{\theta_{0,i}^2}.$$

Thus in SE-optimal design, the goal is to minimize the standard deviation of the parameters, normalized by the true parameter values. As the diagonal elements of F^{-1} are all positive and all parameters are assumed non-zero in $\boldsymbol{\theta} \in \mathbb{R}^{\kappa_\theta}$, $\mathcal{J}_{SE} : \mathbb{R}^{\kappa_\theta \times \kappa_\theta} \to (0, \infty)$.

In [11], it is shown that the D-, E-, and SE-optimal design criteria select different time grids and in general yield different standard errors. As we might expect, these design cost functionals will also generally choose different observation variables (maps) [12] in order to minimize different aspects of the confidence interval ellipsoid.

6.3   Algorithmic Considerations

We complete our outline of design methods with a very brief discussion of algorithmic issues. In most choice-of-variable optimal design problems, one does not have a continuum of measurement possibilities; rather, there are $K^* < \infty$ possible variable observation maps $\mathcal{C}$. Denote this set as $\Omega_C^{K^*} \subset \mathbb{R}^{K^* \times n}$. While we may still use the Prohorov metric framework to guarantee the existence and convergence of (6.14), we have a stronger result first proposed in [6] that is useful in numerical implementation. Because for a given K, $K \leq K^*$, Ω_C^K is finite, all bounded discrete distributions made from the elements of Ω_C^K have the form

$$P_{\mathcal{S}} = \sum_{k=1}^K \Delta_{\mathcal{C}_k}.$$

Moreover, the set $\widetilde{\mathbb{P}}_2(\Omega_C^K)$ of all discrete distributions that use K sampling methods is also finite. For a fixed distribution of time points P_τ, we may compute using (6.13) the set of all possible FIM $F(\tau, \mathcal{S}, \boldsymbol{\theta})$ that could be formulated from $P_{\mathcal{S}} \in \widetilde{\mathbb{P}}_2(\Omega_C^K)$. By the properties of matrix multiplication and addition, this set is also finite. Then the functional (6.14) applied to all F in the set produces a finite set contained in $\mathbb{R}^+$. Because this set is finite, it is well-ordered by the relation $\leq$ and therefore has a minimal element. Therefore, for any distribution of time points P_τ, we may find at least one solution $\hat{P}_{\mathcal{S}} \in \widetilde{\mathbb{P}}_2(\Omega_C^K)$. Moreover, $\hat{P}_{\mathcal{S}}$ may be determined by a search over all matrices $\mathcal{S} = \text{column}(\mathcal{C}_1, \mathcal{C}_2, \dots, \mathcal{C}_K)$ formed by K elements having support in Ω_C^K.

Due to the computational demands of performing non-linear optimization for N time points and K observation maps (for a total of $N + K$ dimensions), we instead solve the coupled set of equations

$$\hat{S} = \arg \min_{\{S | P_S \in \tilde{\mathbb{P}}_2(\Omega_C^K)\}} \mathcal{J}\left(F(\hat{\tau}, S, \boldsymbol{\theta}_0)\right) \tag{6.15}$$

$$\hat{\tau} = \arg \min_{\{\tau | P_\tau \in \tilde{\mathbb{P}}_1([t_0, t_f])\}} \mathcal{J}\left(F(\tau, \hat{S}, \boldsymbol{\theta}_0)\right), \tag{6.16}$$

where $S \in \mathbb{R}^{K \times n}$ represents a set of K sampling maps and $\tau = \{t_j\}_{j=1}^N$, $t_0 \leq t_1 < t_2 < \cdots < t_N \leq t_f$, is an ordered set of N sampling times. These equations are solved iteratively as

$$\hat{S}_i = \arg \min_{\{S | P_S \in \tilde{\mathbb{P}}_2(\Omega_C^K)\}} \mathcal{J}\left(F(\hat{\tau}_{i-1}, S, \boldsymbol{\theta}_0)\right) \tag{6.17}$$

$$\hat{\tau}_i = \arg \min_{\{\tau | P_\tau \in \tilde{\mathbb{P}}_1([t_0, t_f])\}} \mathcal{J}\left(F(\tau, \hat{S}_i, \boldsymbol{\theta}_0)\right), \tag{6.18}$$

where $\mathcal{J}$ is the D-, E-, or SE-optimal design criterion. We begin by solving for $\hat{S}_1$ where $\hat{\tau}_0$ is specified by the user. The system (6.17)–(6.18) is solved until $\left| \mathcal{J}\left(F(\hat{\tau}_i, \hat{S}_i, \boldsymbol{\theta}_0)\right) - \mathcal{J}\left(F(\hat{\tau}_{i-1}, \hat{S}_{i-1}, \boldsymbol{\theta}_0)\right) \right| < \epsilon$ or until $\hat{S}_i = \hat{S}_{i-1}$. For each iteration, (6.17) is solved using a global search over all possible S. Since the sensitivity equations cannot be easily solved for in the models chosen here and in [12, 13, 14, 15] to illustrate this method, one can use a modified version of *tssolve.m* [5], which implements the *myAD* package developed in [19]. Solving (6.18) requires using a non-linear constrained optimization algorithm. While MATLAB's *fmincon* is a natural choice for such problems, as reported in [11], it does not perform well in this situation. Instead, we recommend the optimization tool *SolvOpt* developed by Kuntsevich and Kappel [20]. Once either of the convergence requirements is met and $\hat{S}$ and $\hat{\tau}$ are determined, one can compute standard errors using the asymptotic theory described in Chapter 3. We use an HIV example discussed in [12] to illustrate the application of the techniques discussed above in choosing which state variables to observe (the selection of best times has been widely discussed in the literature, including [10, 11, 12, 13]).

6.4   Example: HIV model

We examined the performance of observation operator selection algorithms on the log-scaled version of the HIV model developed in [2]. While the analytic solution of this system cannot easily be found, this model has been studied, improved, and successfully fit to and indeed validated with several sets of longitudinal data using parameter estimation techniques [1, 3]. Additionally, the sampling or observation operators used to collect data in a clinical setting as well as the relative usefulness of these sampling techniques are known [3]. The model includes uninfected and infected CD4+ T cells, called type 1 target cells (T_1 and T_1^*, respectively), uninfected and infected macrophages (subsequently determined to more likely be resting or inactive $CD4^+$ T-cells), called type 2 target cells (T_2 and T_2^*), infectious free virus V_I, and immune response E produced by CD8+ T cells. The HIV model with treatment function $u(t)$ is given by

$$\dot{T}_1 = \lambda_1 - d_1 T_1 - (1 - \epsilon_1 u(t)) k_1 V_I T_1$$
$$\dot{T}_2 = \lambda_2 - d_2 T_2 - (1 - f\epsilon_1 u(t)) k_2 V_I T_2$$
$$\dot{T}_1^* = (1 - \epsilon_1 u(t)) k_1 V_I T_1 - \delta T_1^* - m_1 E T_1^*$$
$$\dot{T}_2^* = (1 - f\epsilon_1 u(t)) k_2 V_I T_2 - \delta T_2^* - m_2 E T_2^* \tag{6.19}$$
$$\dot{V} = N_T \delta(T_1^* + T_2^*) - [c + (1 - \epsilon_2 u(t)) 10^3 k_1 T_1 + (1 - f\epsilon_1 u(t)) 10^3 k_2 T_2] V_I$$
$$\dot{E} = \lambda_E + b_E \frac{T_1^* + T_2^*}{T_1^* + T_2^* + K_b} E - d_E \frac{T_1^* + T_2^*}{T_1^* + T_2^* + K_d} E - \delta_E E.$$

In [3] this model's parameters are estimated for each of 45 different patients who were in a treatment program for HIV at Massachusetts General Hospital (MGH). These individuals experienced interrupted treatment schedules, in which the patient did not take any medication for viral load control. Seven of these patients adhered to a structured treatment interruption schedule planned by the clinician. We use the parameter values estimated to fit the data of one of these patients, Patient 4 in [3], as our "true" parameters $\boldsymbol{\theta}_0$ for this model (these are detailed in Table 1 of [12]). This patient was chosen because the patient continued treatment for an extended period of time (1919 days) and the corresponding data set contains 158 measurements of viral load V and 107 measurements of CD4+ T-cell count ($T_1 + T_1^*$) that exhibit a response to interruption in treatment, and the estimated parameter values yield a model exhibiting trends similar to that in the data. More details are given in [2, 12] on the choices of $\boldsymbol{\theta}_0$ and parameters to be estimated in using the observation operator selection ideas outlined above. Design problems were considered for problems in which the parameters and initial conditions to be estimated were $\boldsymbol{\theta} = (\lambda_1, d_1, \epsilon_1, k_1, \epsilon_2, N_T, c, b_E, T_1^0, T_2^0, V^0)^T$, where T_1^0, T_2^0 and V^0 are the initial values for T_1, T_2 and V, respectively. Based on currently available measurement methods, we allowed $K^* = 4$ possible types of observations, including (1) infectious virus V, (2) immune response E, (3) total CD4+ T cells $T_1 + T_1^*$ and (4) type 2 target cells $T_2 + T_2^*$.

To simulate the experience of a clinician gathering data as a patient regularly returns for scheduled testing, we fixed the sampling times and chose the optimal observables. We considered choices of $K = 1, 2$ or 3 sampling operators out of the $K^* = 4$ possible observables, all K of which are to be measured at $N = 51, 101, 201$ and 401 evenly spaced times over 2000 days, corresponding to measurements every 40, 20, 10 and 5 days, respectively. The K optimal sampling maps were determined for each of the three optimal design criteria,

TABLE 6.1: Number of observables, number of time points, observables selected by D-, E-, or SE-optimal cost functional, and the minimum and maximum standard error and associated parameter for the parameter subset in the HIV model (6.19).

K	N	Observables	min(ASE)	max(ASE)
		D-optimal cost		
1	51	$T_1 + T_1^*$	$\text{ASE}(\lambda_1) = 0.18$	$\text{ASE}(b_E) = 6.35$
2	51	$T_1 + T_1^*, T_2 + T_2^*$	$\text{ASE}(\lambda_1) = 0.095$	$\text{ASE}(V^0) = 2.31$
3	51	$E, T_1 + T_1^*, T_2 + T_2^*$	$\text{ASE}(b_E) = 0.045$	$\text{ASE}(V^0) = 2.26$
		E-optimal cost		
1	51	V	$\text{ASE}(d_1) = 0.27$	$\text{ASE}(b_E) = 4.27$
2	51	$V, T_2 + T_2^*$	$\text{ASE}(d_1) = 0.12$	$\text{ASE}(b_E) = 2.18$
3	51	$V, E, T_1 + T_1^*$	$\text{ASE}(b_E) = 0.045$	$\text{ASE}(V^0) = 0.77$
		SE-optimal cost		
1	51	V	$\text{ASE}(d_1) = 0.27$	$\text{ASE}(b_E) = 4.27$
2	51	$T_1 + T_1^*, T_2 + T_2^*$	$\text{ASE}(\lambda_1) = 0.095$	$\text{ASE}(V^0) = 2.31$
3	51	$E, T_1 + T_1^*, T_2 + T_2^*$	$\text{ASE}(b_E) = 0.045$	$\text{ASE}(V^0) = 2.26$

and the asymptotic standard errors (ASE) for each estimated parameter and initial condition are calculated after carrying out the corresponding inverse problem calculations. Here we just report on the results corresponding to the clinically reasonable observation operators for $N = 51$ (every 40 days for 2000 days).

In Table 6.1 we display the optimal observation operators determined by the D-, E-, and SE-optimal design cost functions, respectively, as well as the corresponding lowest and highest ASE computed using the sensitivity methodology described in Chapter 3 for the case $N = 51$. In all three optimal design criteria, there was a distinct best choice of observables (listed in more extended tables in [12]) for each pair of N and K. When only $K = 1$ observable could be measured, each design criterion consistently picked the same observable for all N; similarly, at $K = 2$, both the D-optimal and SE-optimal design criteria were consistent in their selection over all N, and E-optimal only differed at $N = 401$. Even at $K = 3$, each optimal design method specified at least two of the same observables at all N.

The observables that were rated best changed between criteria, affirming the fact that each optimal design method minimizes different aspects of the standard error ellipsoid. At $K = 1$ observable, D-optimal selects the CD4+ T-cell count, while E-optimal and SE-optimal choose the infectious virus count. As a result, the min(ASE) calculated for a parameter estimation problem using the D-optimal observables is approximately 1/3 lower than the min(ASE) of E- and SE-optimal for all tested time point distributions. Similarly, the max(ASE) calculated for E- and SE-optimal design parameter estimation problems is approximately 1/3 lower than that of D-optimal. Thus at $K = 1$, based on minimum and maximum asymptotic standard errors, there is no clear best choice of an optimal design cost function.

When selecting $K = 3$ observables, each of the three design criteria selects many of the same observables. This is to be expected as $K^* = 4$ in this simulation. For $N = 51$, 101 and 201, both total CD4+ T-cell count and immune response are selected by all design criteria. The D-optimal criterion also chooses type 2 cell count, so the lack of information on virus count as measured by V leads to its high max(ASE), ASE(V^0). E-optimal (and at larger N, SE-optimal) choose to measure infectious virus count, reducing ASE(V^0) and thus reducing the max(ASE) by more than 50%. While at low N, E-optimal has the lowest min(ASE) and max(ASE), and SE-optimal performs better at high N, so when selecting $K = 3$ observables, the number of time points N may affect which optimal design cost function performs best.

When taking samples on a uniform time grid, D-, E-, and SE-optimal design criteria all choose observation operators that yield favorable ASE's, with some criteria performing best under certain circumstances. For example, the SE-optimal observables at $N = 401$, $K = 3$ yield the smallest standard errors; however, for all other values of N at $K = 3$, E-optimal performs best. At $K = 2$, E-optimal is a slightly weaker scheme. The examples in [11] also reveal that D-, E-, and SE-optimal designs are all competitive when only selecting time points for several different models. Examples where one investigates the performance of these three criteria when selecting both an observation operator and a sampling time distribution using the algorithm described by Equations (6.17) and (6.18) can be found in detail in [12].

References

[1] B.M. Adams, *Non-parametric Parameter Estimation and Clinical Data Fitting with a Model of HIV Infection*, PhD Thesis, NC State Univ., 2005.

[2] B.M. Adams, H.T. Banks, M. Davidian, H. Kwon, H. T. Tran, S.N. Wynne and E.S. Rosenberg, HIV dynamics: Modeling, data analysis, and optimal treatment protocols, CRSC-TR04-05, February, 2004; *J. Comp. Appl. Math.*, **184** (2005), 10–49.

[3] B.M. Adams, H.T. Banks, M. Davidian and E.S. Rosenberg, Model fitting and prediction with HIV treatment interruption data, CRSC-TR05-40, October, 2005; *Bulletin of Math. Biology*, **69** (2007), 563–584.

[4] A.C. Atkinson and R.A. Bailey, One hundred years of the design of

experiments on and off the pages of *Biometrika*, *Biometrika*, **88** (2001), 53–97.

[5] A. Attarian, tssolve.m, Retrieved August 2011, from http://www4. ncsu.edu/ arattari/.

[6] M. Avery, H.T. Banks, K. Basu, Y. Cheng, E. Eager, S. Khasawinah, L. Potter and K.L. Rehm, Experimental design and inverse problems in plant biological modeling, CRSC-TR11-12, October, 2011; *J. Inverse and Ill-posed Problems*, DOI 10.1515/jiip-2012-0208.

[7] A.C. Atkinson and A.N. Donev, *Optimum Experimental Designs*, Oxford University Press, New York, 1992.

[8] H.T. Banks and K.L. Bihari, Modeling and estimating uncertainty in parameter estimation, CRSC-TR99-40, December, 1999; *Inverse Problems*, **17** (2001), 95–111.

[9] H.T. Banks, A. Cintrón-Arias and F. Kappel, Parameter selection methods in inverse problem formulation, CRSC-TR10-03, revised November 2010; in *Mathematical Model Development and Validation in Physiology: Application to the Cardiovascular and Respiratory Systems*, Lecture Notes in Mathematics, Vol. 2064, Mathematical Biosciences Subseries, Springer-Verlag, Berlin, 2013.

[10] H.T. Banks, S. Dediu, S.L. Ernstberger and F. Kappel, Generalized sensitivities and optimal experimental design, CRSC-TR08-12, September, 2008, Revised November, 2009; *J. Inverse and Ill-posed Problems*, **18** (2010), 25–83.

[11] H.T. Banks, K. Holm and F. Kappel, Comparison of optimal design methods in inverse problems, CRSC-TR10-11, July, 2010; *Inverse Problems*, **27** (2011), 075002.

[12] H.T. Banks and K.L. Rehm, Experimental design for vector output systems, CRSC-TR12-11, April, 2012; *Inverse Problems in Sci. and Engr.*, (2013), 1–34. DOI: 10.1080/17415977.2013.797973.

[13] H.T. Banks and K.L. Rehm, Experimental design for distributed parameter vector systems, CRSC-TR12-17, August, 2012; *Applied Mathematics Letters*, **26** (2013), 10–14; http://dx.doi.org/10.1016/ j.aml.2012.08.003.

[14] H.T. Banks, D. Rubio, N. Saintier and M.I. Troparevsky, Optimal design techniques for distributed parameter systems, CRSC-TR13-01, January, 2013; *Proceedings 2013 SIAM Conference on Control Theory*, SIAM, San Diego, 83–90.

[15] H.T. Banks, D. Rubio, N. Saintier and M.I. Troparevsky, Optimal electrode positions for the inverse problem of EEG in a simplified

model in 3D, *Proceedings MACI 2013: Fourth Conference on Applied, Computational and Industrial Mathematics*, May 15–17, 2013, Buenos Aires, AR, **4** (2013), 521–524.

[16] M.P.F. Berger and W.K. Wong (Editors), *Applied Optimal Designs*, John Wiley & Sons, Chichester, UK, 2005.

[17] V.V. Fedorov, *Theory of Optimal Experiments*, Academic Press, New York and London, 1972.

[18] V.V. Fedorov and P. Hackel, *Model-Oriented Design of Experiments*, Springer-Verlag, New York, 1997.

[19] M. Fink, myAD, Retrieved August 2011, from http://www.mathworks. com/matlabcentral/fileexchange/15235-automatic-differentiation-for-matlab.

[20] A. Kuntsevich and F. Kappel, SolvOpt, Retrieved December 2009, from http://www.kfunigraz.ac.at/imawww/kuntsevich/solvopt/.

[21] W. Müller and M. Stehlik, Issues in the optimal design of computer simulation experiments, *Appl. Stochastic Models in Business and Industry*, **25** (2009), 163–177.

[22] M. Patan and B. Bogacka, Optimum experimental designs for dynamic systems in the presence of correlated errors, *Computational Statistics and Data Analysis*, **51** (2007), 5644–5661.

[23] D. Ucinski and A.C. Atkinson, Experimental design for time-dependent models with correlated observations, *Studies in Nonlinear Dynamics and Econometrics*, **8(2)** (2004), Article 13: The Berkeley Electronic Press.

[24] X.-G. Zhu, E. de Sturler and S. P. Long, Optimizing the distribution of resources between enzymes of carbon metabolism can dramatically increase photosynthetic rate: A numerical simulation using an evolutionary algorithm, *Plant Physiology*, **145** (2007), 513–526.

Chapter 7

Propagation of Uncertainty in a Continuous Time Dynamical System

Uncertainty is ubiquitous in almost all branches of science and their applications including physics, chemistry, biology, engineering, environmental science and social science. Following the seminal work of [88], uncertainty is often classified into two types. One is *epistemic* (or reducible) *uncertainty*, which can possibly be reduced by improved measurements or improvements in the modeling process. The other is *aleatory* (or irreducible) *uncertainty*, which is the result of intrinsic variability/stochasticity of the system.

Uncertainty propagation through a dynamic system has enjoyed considerable research attention during the past decade due to the wide applications of mathematical models in studying the dynamical behavior of systems. In this chapter, we consider uncertainty propagation in a continuous time dynamical system through two types of differential equations, where the dynamical system is described by the system of ordinary differential equations

$$\dot{\mathbf{x}}(t) = \mathbf{g}(t, \mathbf{x}(t)), \quad \mathbf{x}(0) = \mathbf{x}_0, \tag{7.1}$$

with $\mathbf{x} = (x_1, x_2, \ldots, x_n)^T$, $\mathbf{g} = (g_1, g_2, \ldots, g_n)^T$ being n-dimensional non-random functions of t and $\mathbf{x}$, and $\mathbf{x}_0$ being an n-dimensional column vector. One type of differential equation is a stochastic differential equation (SDE), a classification reserved for differential equations driven by white noise (defined later). The other one is a random differential equation (RDE, a term popularized for some years since its early use in [104, 105]), a classification reserved for differential equations driven by other types of random inputs such as colored noise (defined later) and both colored noise and white noise.

One of the goals of this chapter is to give a short and introductory review of a number of theoretical results on SDEs and RDEs presented in different fields with the hope of making researchers in one field aware of the work in other fields and to explain the connections between them. Specifically, we focus on the equations describing the time evolution of the probability density functions of the associated stochastic processes. As we shall see below, these equations have their own applications in a wide range of fields, such as physics, chemistry, biological systems and engineering. A second goal of this chapter is to discuss the relationship between the stochastic processes resulting from SDEs and RDEs.

Unless otherwise indicated, throughout our presentation a capital letter is used to denote a random variable, a bold capital letter is for a random vector and their corresponding small letters are for their respective realizations.

7.1   Introduction to Stochastic Processes

Before addressing our main task, we shall use this section to review some background materials on continuous time stochastic processes. For convenience of presentation, we concentrate on one-dimensional continuous-time stochastic processes. It should be noted that most of the results presented in this section can be extended to the multi-dimensional case.

Let $(\Omega, \mathcal{F}, \mathrm{Prob})$ denote the probability space. Then a stochastic process $\{X(t, \cdot) : t \in \mathbb{T}\}$ is a collection of random variables on $(\Omega, \mathcal{F}, \mathrm{Prob})$ indexed by time parameter t, where $\mathbb{T} \subset \mathbb{R}$ is usually a halfline $[0, \infty)$ (this will be used in the remainder of this section unless otherwise indicated), and it may also be any bounded interval $[t_0, t_f] \subset \mathbb{R}$. We assume that these random variables are real valued; that is, $X : \mathbb{T} \times \Omega \to \mathbb{R}$. We note that for any fixed $t \in \mathbb{T}$, the function $X(t, \cdot)$ is a random variable associated with t (for this reason, a stochastic process is also referred to as a *random function* in some literature). For a fixed sample point $\omega \in \Omega$, the function $X(\cdot, \omega)$ is a *realization* (also referred to as *sample path* or *sample function*) of the stochastic process associated with ω. As is usually done and also for notational convenience, we will suppress the dependence of the stochastic process on ω in this chapter; that is, we denote $X(t, \cdot)$ by $X(t)$, and bear in mind that $X(t)$ is a random variable for any fixed t.

Similar to the equivalence between two random variables, the equivalence between two stochastic processes can also be defined in several senses. Here we give two widely used ones.

Definition 7.1.1 *Two stochastic processes $\{X(t) : t \in \mathbb{T}\}$ and $\{Y(t) : t \in \mathbb{T}\}$ defined on the same probability space are said to be a* version *(or* modification*) of each other if for any $t \in \mathbb{T}$ we have*

$$Prob\{X(t) = Y(t)\} = 1. \tag{7.2}$$

The above definition implies that if $\{X(t) : t \in \mathbb{T}\}$ is a version of $\{Y(t) : t \in \mathbb{T}\}$, then at each time t the random variables $X(t)$ and $Y(t)$ are equal almost surely.

Definition 7.1.2 *Two stochastic processes* $\{X(t) : t \in \mathbb{T}\}$ *and* $\{Y(t) : t \in \mathbb{T}\}$ *defined on the same probability space are said to be* indistinguishable *if*

$$Prob\left\{\bigcap_{t \in \mathbb{T}}\{X(t) = Y(t)\}\right\} = 1. \qquad (7.3)$$

By Definitions 7.1.1 and 7.1.2, we know that if two stochastic processes are indistinguishable, then they are versions of each other. However, the converse is usually not true. In addition, Definition 7.1.2 implies that for two indistinguishable stochastic processes almost all their sample paths agree. Hence, this is the strongest equivalence between two stochastic processes.

Definition 7.1.3 *A collection of σ-algebra subsets* $\{\mathcal{F}_t\} \equiv \{\mathcal{F}_t : t \in \mathbb{T}\}$ *is a filtration on* $(\Omega, \mathcal{F})$ *if* $\mathcal{F}_s \subset \mathcal{F}_t \subset \mathcal{F}$ *for any* $s, t \in \mathbb{T}$ *and* $s < t$.

Given a stochastic process $\{X(t) : t \in \mathbb{T}\}$, the simplest choice of a filtration is the one that is generated by the process itself; that is,

$$\mathcal{F}_t^X = \sigma(X(s); 0 \le s \le t),$$

the smallest σ algebra with respect to which $X(s)$ is measurable for every s in $[0, t]$. This filtration, $\{\mathcal{F}_t^X\}$, is called the *natural filtration* of the stochastic process $\{X(t) : t \in \mathbb{T}\}$.

Definition 7.1.4 *The stochastic process* $\{X(t) : t \in \mathbb{T}\}$ *is* adapted *(non-anticipating) to the filtration* $\{\mathcal{F}_t\}$ *if* $X(t)$ *is an* $\mathcal{F}_t$-measurable random variable *for any* $t \in \mathbb{T}$.

By the above definition, we see that a stochastic process is adapted to its natural filtration. In addition, $\{X(t) : t \in \mathbb{T}\}$ is $\{\mathcal{F}_t\}$-adapted if and only if $\mathcal{F}_t^X \subset \mathcal{F}_t$ for each $t \in \mathbb{T}$.

The concept of the independence of two stochastic processes is based on the concept of the independence of random vectors and is given as follows.

Definition 7.1.5 *Two stochastic processes* $\{X(t) : t \in \mathbb{T}\}$ *and* $\{Y(t) : t \in \mathbb{T}\}$ *are said to be* independent *if for all positive integers m and all sets of distinct times* $\{t_j\}_{j=1}^m \subset \mathbb{T}$ *($t_j < t_{j+1}$, $j = 1, 2, \ldots, m - 1$), the random vectors* $(X(t_1), X(t_2), \ldots, X(t_m))^T$ *and* $(Y(t_1), Y(t_2), \ldots, Y(t_m))^T$ *are independent.*

7.1.1 Distribution Functions of a Stochastic Process

Note that at each time t, $X(t)$ is a random variable. Hence, we have a cumulative distribution function (simply referred to as a *distribution function* in this chapter) defined by

$$P(t, x) = \text{Prob}\{X(t) \le x\}$$

to describe the stochastic process at each time t. The corresponding probability density function, if it exists, is defined by

$$p(t, x) = \frac{\partial}{\partial x} P(t, x).$$

However, we see that the distribution function P (or the corresponding probability density function p) is not enough to fully describe the stochastic process. For example, we may be interested in the joint distribution function

$$P_{X_1 X_2}(t_1, x_1, t_2, x_2) = \text{Prob}\{X_1 \leq x_1, X_2 \leq x_2\}$$

at two distinct time points t_1 and t_2 with $X_1 = X(t_1)$ and $X_2 = X(t_2)$, and its corresponding joint probability density function (if it exists)

$$p_{X_1 X_2}(t_1, x_1, t_2, x_2) = \frac{\partial^2}{\partial x_1 \partial x_2} P_{X_1 X_2}(t_1, x_1, t_2, x_2).$$

In general, we may consider the joint distribution function

$$P_{X_1,\ldots,X_m}(t_1, x_1, t_2, x_2, \ldots, t_m, x_m)$$
$$= \text{Prob}\{X_1 \leq X_1, X_2 \leq x_2, \ldots, X_m \leq x_m\}$$

for the stochastic process at m distinct times $t_1, t_2, \ldots, t_m$, and this joint distribution function is called the mth *distribution function* of the stochastic process $\{X(t) : t \in \mathbb{T}\}$, where $X_j = X(t_j)$, $j = 1, 2, \ldots, m$. The corresponding mth *probability density function* of the stochastic process $\{X(t) : t \in \mathbb{T}\}$, if it exists, is defined by

$$p_{X_1,\ldots,X_m}(t_1, x_1, t_2, x_2, \ldots, t_m, x_m)$$

$$= \frac{\partial^m}{\partial x_1 \partial x_2 \ldots \partial x_m} P_{X_1,\ldots,X_m}(t_1, x_1, t_2, x_2, \ldots, t_m, x_m).$$

Similar to random variables, a lower level probability density function can be derived from a higher level probability density function by integrating the higher level probability density function with respect to the rest of the variables. For example, the second probability density function $p_{X_1 X_2}$ can be derived from the mth $(m > 2)$ probability density function $p_{X_1,\ldots,X_m}$ by integrating $p_{X_1,\ldots,X_m}$ with respect to the rest of the variables $x_3, \ldots, x_m$; that is,

$$p_{X_1 X_2}(t_1, x_1, t_2, x_2) = \int_{\mathbb{R}^{m-2}} p_{X_1,\ldots,X_m}(t_1, x_1, t_2, x_2, \ldots, t_m, x_m) dx_3 \ldots dx_m.$$

For the remainder of this monograph, if no confusion occurs, the first probability density function of the stochastic process is simply called the *probability density function of the stochastic process*.

Recall the discussion in Section 2.2.4 that two random variables equal in distribution may not be equal almost surely. Hence, two stochastic processes having the same finite dimensional distribution functions may not be indistinguishable. However, in practice all these finite dimensional distribution functions are often assumed to be sufficient to characterize a stochastic process. Unfortunately, even for this weak characterization it is in general hard to characterize a stochastic process, as it is usually difficult, if not impossible, to find all these finite dimensional distribution functions.

7.1.2 Moments, Correlation and Covariance Functions of a Stochastic Process

In this section we assume that all the finite dimensional probability density functions exist, and hence the moments, correlation and covariance functions of a stochastic process can be defined in terms of these functions. Similar to the mean and variance of a random variable, the mean function and the variance function of the stochastic process $\{X(t) : t \in \mathbb{T}\}$ are respectively defined by

$$\mathbb{E}(X(t)) = \int_{-\infty}^{\infty} xp(t, x)dx, \quad \mathrm{Var}(X(t)) = \mathbb{E}((X(t) - \mathbb{E}(X(t))^2).$$

The mth moment and mth central-moment functions can also be defined in a manner similar to the definition for random variables, and they are respectively given by

$$\mathbb{E}(X^m(t)) = \int_{-\infty}^{\infty} x^m p(t, x)dx, \quad \mathbb{E}((X(t) - \mathbb{E}(X(t))^m).$$

Thus we see that the mean function is the first moment function, and the variance function is just the second central moment function.

Besides moment functions, we may also want to know how the process is correlated over time. The two-time auto-correlation function of the stochastic process $\{X(t) : t \in \mathbb{T}\}$ is defined by

$$\mathrm{Cor}\{X(t_1), X(t_2)\} = \mathbb{E}(X(t_1)X(t_2))$$

$$= \int_{-\infty}^{\infty} \int_{-\infty}^{\infty} x_1 x_2 p_{X_1 X_2}(t_1, x_1, t_2, x_2)dx_1 dx_2, \tag{7.4}$$

and the two-time auto-covariance function is

$$\mathrm{Cov}\{X(t_1), X(t_2)\} = \mathbb{E}((X(t_1) - \mathbb{E}(X(t_1)))(X(t_2) - \mathbb{E}(X(t_2))))$$

$$= \mathbb{E}(X(t_1)X(t_2)) - \mathbb{E}(X(t_1))\mathbb{E}(X(t_2)). \tag{7.5}$$

These two functions have some very important and useful properties. For example, by (7.4) and (7.5) we see that both functions are symmetric; that is,

$$\mathcal{R}(t, s) = \mathcal{R}(s, t), \quad \mathcal{V}(t, s) = \mathcal{V}(s, t), \tag{7.6}$$

where $\mathcal{R}(t, s) = \text{Cor}\{X(t), X(s)\}$ and $\mathcal{V}(t, s) = \text{Cov}\{X(t), X(s)\}$. In addition, both $\mathcal{R}$ and $\mathcal{V}$ are non-negative definite in the sense that

$$\sum_{j=1}^{l}\sum_{k=1}^{l} c_j \mathcal{R}(t_j, t_k) c_k \geq 0, \quad \sum_{j=1}^{l}\sum_{k=1}^{l} c_j \mathcal{V}(t_j, t_k) c_k \geq 0 \tag{7.7}$$

hold for any $t_1, t_2, \ldots, t_l \in \mathbb{T}$ and for any $\mathbf{c} = (c_1, c_2, \ldots, c_l)^T \in \mathbb{R}^l$. We note that (7.7) can be easily established through the following equalities:

$$\sum_{j=1}^{l}\sum_{k=1}^{l} c_j \mathcal{R}(t_j, t_k) c_k = \mathbf{c}^T \mathbb{E}(\mathcal{X}\mathcal{X}^T)\mathbf{c},$$

$$\sum_{j=1}^{l}\sum_{k=1}^{l} c_j \mathcal{V}(t_j, t_k) c_k = \mathbf{c}^T \mathbb{E}\{(\mathcal{X} - \mathbb{E}(\mathcal{X}))(\mathcal{X} - \mathbb{E}(\mathcal{X}))^T\}\mathbf{c},$$

where $\mathcal{X} = (X(t_1), X(t_2), \ldots X(t_l))^T$.

Similarly, the m-time auto-correlation function of the stochastic process $\{X(t) : t \in \mathbb{T}\}$ is defined by

$$\text{Cor}\{X(t_1), X(t_2), \ldots, X(t_m)\}$$

$$= \mathbb{E}\left(X(t_1)X(t_2)\cdots X(t_m)\right),$$

$$= \int_{\mathbb{R}^m} \left(\prod_{j=1}^{m} x_j\right) p_{X_1,\ldots,X_m}(t_1, x_1, t_2, x_2, \ldots, t_m, x_m) dx_1 dx_2 \ldots dx_m,$$

and the m-time auto-covariance function of the stochastic process $\{X(t) : t \in \mathbb{T}\}$ is

$$\text{Cov}\{X(t_1), X(t_2), \ldots, X(t_m)\}$$

$$= \mathbb{E}\left((X(t_1) - \mathbb{E}(X(t_1)))(X(t_2) - \mathbb{E}(X(t_2)))\cdots(X(t_m) - \mathbb{E}(X(t_m)))\right).$$

We thus see that if $t_j = t, j = 1, 2, \ldots, m$, then the m-time auto-covariance function is just the mth central-moment function, and the m-time auto-correlation function is just the mth moment function. In addition, we observe that for any positive integer m the m-time auto-correlation function and m-time auto-covariance function depend on the mth probability density function. Hence, it is in general difficult, if not impossible, to find all these auto-correlation and auto-covariance functions.

It is worth noting here that the prefix "auto" in the names "auto-covariance function" and "auto-correlation function" is intended to differentiate them from "cross-covariance function" and "cross-correlation function." The *cross-covariance function* and the *cross-correlation function* involve two different stochastic processes $\{X(t) : t \in \mathbb{T}\}$ and $\{Y(t) : t \in \mathbb{T}\}$. For example, the *two-time cross-covariance function* is defined by

$$\mathrm{Cov}\{X(t_1), Y(t_2)\} = \mathbb{E}\left((X(t_1) - \mathbb{E}(X(t_1)))(Y(t_2) - \mathbb{E}(Y(t_2)))\right),$$

and the *two-time cross-correlation function* is

$$\mathrm{Cor}\{X(t_1), Y(t_2)\} = \mathbb{E}\left(X(t_1)Y(t_2)\right).$$

If no confusion occurs, the two-time auto-covariance function and the two-time auto-correlation function are often simply referred to as the *covariance function* and the *correlation function*, respectively. These are also the names that we will use in this monograph.

7.1.3   Classification of a Stochastic Process

Stochastic processes can be classified based on different criterion. Below we give two classifications. One is based on whether or not the associated distribution functions depend on the absolute origin of time. The other is based on whether the process is Gaussian or not.

7.1.3.1   Stationary vs. Non-Stationary Stochastic Processes

Based on whether or not the associated distribution functions depend on the absolute origin of time, stochastic processes can be grouped into two classes: non-stationary stochastic processes and stationary stochastic processes. A stochastic process $\{X(t) : t \in \mathbb{T}\}$ is said to be *non-stationary* if its distribution functions depend explicitly on time. This means that for a non-stationary stochastic process, the distribution functions depend on the absolute origin of time. Most stochastic processes that we encounter in practice are non-stationary.

For a stationary stochastic process, all the distribution functions are independent of the absolute time origin. The strict definition for a stationary stochastic process is given as follows.

Definition 7.1.6 *A stochastic process $\{X(t) : t \in \mathbb{T}\}$ is said to be* stationary *if its distribution functions stay invariant under an arbitrary translation of time. In other words, for any positive integer m we have*

$$\begin{aligned}
P_{X_1,\ldots,X_m}&(t_1, x_1, t_2, x_2, \ldots, t_m, x_m) \\
&= P_{X_1,\ldots,X_m}(t_1 + s, x_1, t_2 + s, x_2, \ldots, t_m + s, x_m),
\end{aligned} \tag{7.8}$$

where s is any number such that $t_j + s \in \mathbb{T}$, $j = 1, 2, \ldots, m$.

Stationary stochastic processes have a number of important properties. For example,

- Their mean function is a constant.

- Their correlation function is a function of the difference between two time points; that is, $\mathrm{Cor}\{X(t_1), X(t_2)\}$ is a function of $|t_2 - t_1|$.

The first property can be obtained by using the fact that its first distribution function does not depend on t (as Definition 7.1.6 implies that $P(t, x) = P(0, x)$). The second property is obtained because $P_{X_1 X_2}(t_1, x_1, t_2, x_2) = P_{X_1 X_2}(0, x_1, t_2 - t_1, x_2)$ if $t_2 \geq t_1$ and $P_{X_1 X_2}(t_1, x_1, t_2, x_2) = P_{X_1 X_2}(t_1 - t_2, x_1, 0, x_2)$ if $t_2 < t_1$.

Definition 7.1.6 implies that a constant stochastic process (i.e., $X(t) = Z$ for any t with Z being a random variable) is stationary. However, in practice, it is often difficult to determine whether a stochastic process is stationary or not, as (7.8) must hold for all m. Hence, for practical reasons, one is often interested in a wide-sense stationary stochastic process. The exact definition is as follows.

Definition 7.1.7 *A stochastic process $\{X(t) : t \in \mathbb{T}\}$ is said to be* wide-sense stationary *if*

$$|\mathbb{E}(X(t))| \text{ is a finite constant,}$$

$$\mathbb{E}(X^2(t)) < \infty, \quad Cor\{X(t_1), X(t_2)\} \text{ is a function of } |t_2 - t_1|. \tag{7.9}$$

A wide-sense stationary stochastic process is sometimes called a *weakly stationary* or *covariance stationary* or *second-order stationary* stochastic process. It is obvious that a stationary stochastic process with finite second moment function is also wide-sense stationary. However, the converse is usually not true.

For a wide-sense stationary stochastic process $\{X(t) : t \in \mathbb{T}\}$, its correlation function can be simply written as

$$\Upsilon(s) = \mathrm{Cor}\{X(t+s), X(t)\}, \quad t+s, t \in \mathbb{T}. \tag{7.10}$$

This function has a number of important properties. For example, Υ is an even function, which is due to the fact that $\mathrm{Cor}\{X(t+s), X(t)\} = \mathrm{Cor}\{X(t), X(t+s)\}$. In addition, for any $s \in \mathbb{T}$ we have $|\Upsilon(s)| \leq \Upsilon(0)$, which can be obtained by using the Cauchy–Schwartz inequality (2.66). Furthermore, it can be found that if Υ is continuous at the origin, then it is uniformly continuous. Another important property for Υ is that it constitutes a Fourier transform pair with the so-called *power spectral density function*, which will be defined later in Section 7.1.7.1. Interested readers can refer to [104] for more information on the properties of Υ.

7.1.3.2   Gaussian vs. Non-Gaussian Processes

Compared to non-Gaussian processes, Gaussian processes possess many important and useful properties (inherited from the normal distribution). One of the useful properties of a Gaussian process is that the stationary property is equivalent to the wide-sense stationary property if the process has a bounded second moment function. Another useful property of a Gaussian process is that the Gaussianity is preserved under all linear transformations. In addition, a Gaussian process is one of the most important and useful stochastic processes in both practice and applications. The strict definition of a Gaussian process is given as follows.

Definition 7.1.8 *A stochastic process* $\{X(t) : t \in \mathbb{T}\}$ *is a Gaussian process if all its finite dimensional distributions are multivariate normal.*

Definition 7.1.8 means that for any given distinct time points $t_1, t_2, \ldots, t_m$ with m being a positive integer, the m-dimensional random vector $\mathcal{X} = (X(t_1), X(t_2), \ldots X(t_m))^T$ is multivariate normally distributed, and its mean is $\boldsymbol{\mu}_{\mathcal{X}} = (\mathbb{E}(X(t_1), \mathbb{E}(X(t_2), \ldots, \mathbb{E}(X(t_m))^T$ and covariance matrix is $\boldsymbol{\Sigma}_{\mathcal{X}} \in \mathbb{R}^{m \times m}$ with its (j, k)th element given by $\mathrm{Cov}\{X(t_j), X(t_k)\}$, $j, k = 1, 2, \ldots, m$. Hence, the probability density function of $\mathcal{X}$ (that is, the mth probability density function of Gaussian process $\{X(t) : t \in \mathbb{T}\}$) is

$$p_{X_1,\ldots,X_m}(t_1, x_1, t_2, x_2, \ldots, t_m, x_m)$$

$$= \frac{1}{(2\pi)^{m/2}|\boldsymbol{\Sigma}_{\mathcal{X}}|^{1/2}} \exp\left(-\frac{1}{2}(\mathbf{x} - \boldsymbol{\mu}_{\mathcal{X}})^T \boldsymbol{\Sigma}_{\mathcal{X}}^{-1}(\mathbf{x} - \boldsymbol{\mu}_{\mathcal{X}})\right),$$

where $\mathbf{x} = (x_1, x_2, \ldots, x_m)^T$. Thus we see that a Gaussian process is completely characterized by its mean and covariance functions.

7.1.4   Methods of Studying a Stochastic Process

Before we introduce some special types of stochastic processes, we give a brief introduction of two common methods, the sample function approach and the mean square calculus approach, used to study the properties of a stochastic process, and we refer interested readers to [56, 57, 104, 107] for more information on this topic.

7.1.4.1   Sample Function Approach

For the sample function approach, one considers each realization of a stochastic process. Note that each realization of a stochastic process is a deterministic function. Hence, one is able to use the ordinary calculus approach to treat each sample function. For example, in this framework, a stochastic process is said to be continuous if almost all its sample functions are continuous functions of t. This type of continuity is often called *sample continuity*. In addition, a stochastic process is said to be *sample integrable* if almost all its sample paths are Rieman integrable (or Lebesque integrable). Such defined integrals are referred to as *path by path Riemann integrals* (or *path by path*

Lebesgue integrals). Hence, we see that this approach may become difficult to use in practice, as one needs to consider every sample function of the given stochastic process. In this sense, the mean square calculus approach is much more relaxed, as we shall see below that its properties are determined only by its associated correlation functions.

7.1.4.2   Mean Square Calculus Approach

Mean square calculus is based on the concept of *mean square convergence* for a sequence of random variables, and it is exclusively used to study a second-order stochastic process, which is one that has finite second-order moment functions (that is, $\mathbb{E}(X^2(t)) < \infty$ for all $t \in \mathbb{T}$). Based on the Cauchy–Schwartz inequality (2.66), we know that a stochastic process is a second-order stochastic process if and only if its associated correlation functions exist and are finite. As we shall see below, the mean square continuous, differentiable and integral properties of a second-order stochastic process are totally determined by the corresponding properties of its associated correlation functions. This implies the mean square calculus of a second-order stochastic process follows the broad outline of ordinary calculus. For the remainder of this section, we denote $\text{Cor}\{X(t), X(s)\}$ by $\mathcal{R}(t, s)$.

Definition 7.1.9 *A second-order stochastic process $\{X(t) : t \in \mathbb{T}\}$ is continuous in mean square (or m.s. continuous) at a fixed t if*

$$\lim_{s \to 0} \mathbb{E}((X(t+s) - X(t))^2) = 0. \tag{7.11}$$

The following theorem states that the m.s. continuity property of a second-order stochastic process is determined by the continuity property of its associated correlation function.

Theorem 7.1.1 *A second-order stochastic process $\{X(t) : t \in \mathbb{T}\}$ is m.s. continuous at t if and only if $\mathcal{R}$ is continuous at (t, t), that is,*

$$\lim_{(s, \tilde{s}) \to (t, t)} \mathcal{R}(s, \tilde{s}) = \mathcal{R}(t, t).$$

A second-order stochastic process is said to be m.s. continuous in $\mathbb{T}$ if it is m.s. continuous at any time point in $\mathbb{T}$. By Theorem 7.1.1 as well as the the discussion in Section 7.1.3.1 we know that for a wide-sense stationary stochastic process, it is m.s. continuous if and only if its correlation function Υ is continuous at $s = 0$.

It is worth noting that m.s. continuity does not imply sample continuity. One such counterexample is the Poisson process that we will introduce later.

Definition 7.1.10 *A second-order stochastic process $\{X(t) : t \in \mathbb{T}\}$ has a mean square derivative (or m.s. derivative) $\dot{X}(t)$ at a fixed t if*

$$\lim_{s \to 0} \mathbb{E}\left(\left(\dot{X}(t) - \frac{X(t+s) - X(t)}{s}\right)^2\right) = 0.$$

Theorem 7.1.2 *A second-order stochastic process $\{X(t) : t \in \mathbb{T}\}$ is m.s. differentiable at t if and only if $\dfrac{\partial^2 \mathcal{R}}{\partial s \partial \tilde{s}}$ exists and is finite at (t,t).*

A second-order stochastic process is said to be m.s. differentiable in $\mathbb{T}$ if it is m.s. differentiable at any $t \in \mathbb{T}$. By the above theorem, we see that for a wide-sense stationary stochastic process, it is m.s. differentiable if and only if the first- and second-order derivatives of Υ exist and are finite at $s = 0$. In addition, by (2.119) one can establish the following equalities:

$$\begin{aligned}
\mathbb{E}(\dot{X}(t)) &= \frac{d}{dt}\mathbb{E}(X(t)), \quad \mathbb{E}(\dot{X}(t)X(s)) = \frac{\partial \mathcal{R}(t,s)}{\partial t}, \\
\mathbb{E}(X(t)\dot{X}(s)) &= \frac{\partial \mathcal{R}(t,s)}{\partial s}, \quad \mathbb{E}(\dot{X}(t)\dot{X}(s)) = \frac{\partial^2 \mathcal{R}(t,s)}{\partial t \partial s}.
\end{aligned} \quad (7.12)$$

Definition 7.1.11 *Let $\{X(t) : t \in [t_0, t_f]\}$ be a second-order stochastic process, $h : [t_0, t_f] \times \mathbb{T} \to \mathbb{R}$ be some deterministic function of t and s, and Riemann integrable for every given s. Then $h(t,s)X(t)$ is said to be mean square Riemann integrable on the interval $[t_0, t_f]$ if for any given $s \in \mathbb{T}$ there exists $Y(s)$ such that*

$$\lim_{\substack{l \to \infty \\ \Delta_l \to 0}} \mathbb{E}((Y_l(s) - Y(s))^2) = 0.$$

In the above equation, $\Delta_l = \displaystyle\max_{0 \le j \le l-1}\{t^l_{j+1} - t^l_j\}$ with $\{t^l_j\}^l_{j=0}$ being a partition of $[t_0, t_f]$, and $Y_l(s) = \displaystyle\sum_{j=1}^{l} h(\tilde{t}^l_j, s)X(\tilde{t}^l_j)(t^l_j - t^l_{j-1})$, where $\tilde{t}^l_j \in [t^l_{j-1}, t^l_j)$, $j = 1, 2, \ldots, l$. The limit $Y(s)$ is called the mean square Riemann integral of $h(t,s)X(t)$ over the interval $[t_0, t_f]$, and is denoted by

$$Y(s) = \int_{t_0}^{t_f} h(t,s)X(t)\,dt. \quad (7.13)$$

Theorem 7.1.3 *The integral in (7.13) exists if and only if the ordinary double Riemann integral $\displaystyle\int_{t_0}^{t_f}\int_{t_0}^{t_f} h(t,s)h(\tilde{t},s)\mathcal{R}(t,\tilde{t})\,dt\,d\tilde{t}$ exists and is finite.*

The m.s. Riemann integral has a number of important properties similar to those of the ordinary Riemann integral. For example, the mean square continuity of $\{X(t) : t \in [t_0, t_f]\}$ implies mean square Riemann integrability of $\{X(t) : t \in [t_0, t_f]\}$. In addition, if $\{X(t) : t \in [t_0, t_f]\}$ is mean square continuous, then the stochastic process $\{Y(t) : t \in [t_0, t_f]\}$ defined by

$Y(t) = \int_{t_0}^{t} X(s)ds$ is m.s. continuous, and it is also m.s. differentiable with

$\dot{Y}(t) = X(t)$. Furthermore, if $\{X(t) : t \in [t_0, t_f]\}$ is m.s. differentiable, then

$$X(t) - X(t_0) = \int_{t_0}^{t} \dot{X}(s)ds.$$

The following theorem states that the Gaussianity of a Gaussian process is preserved under m.s. differentiation and integration.

Theorem 7.1.4 *If the m.s. derivative $\dot{X}(t)$ of a Gaussian process $\{X(t) : t \in \mathbb{T}\}$ exists, then $\{\dot{X}(t) : t \in \mathbb{T}\}$ is a Gaussian process. In addition, if the m.s. integral $Y(t)$ of a Gaussian process $\{X(t) : t \in \mathbb{T}\}$ defined by*

$$Y(t) = \int_{0}^{t} h(t, s)X(s)ds$$

exists, then $\{Y(t) : t \in \mathbb{T}\}$ is a Gaussian process.

The mean square (m.s.) Riemann–Stieltjes integral can be defined similarly as the m.s. Riemann integral, and its existence is determined by the existence of appropriate ordinary integrals.

Theorem 7.1.5 *The m.s. Riemann–Stieltjes integral $\int_{t_0}^{t_f} h(t)dX(t)$ (h is some deterministic function) exists if and only if the ordinary double Riemann–Stieltjes integral*

$$\int_{t_0}^{t_f} \int_{t_0}^{t_f} h(t)h(s)dd\mathcal{R}(t, s) \tag{7.14}$$

exists and is finite.

By the existence theorem for the ordinary Riemann–Stieltjes integral, we know that if h is continuous on $[t_0, t_f]$ and $\mathcal{R}$ has bounded variation on $[t_0, t_f] \times [t_0, t_f]$, then the double integral (7.14) exists (e.g., see [41] for details).

7.1.5   Markov Processes

In this section, we give an elementary introduction of a special class of stochastic processes – Markov processes, which are extensively studied in the literature and widely used in practice. Interested readers can refer to [48] for more sophisticated treatment of Markov processes. Intuitively speaking, a stochastic process is a Markov process if what happens in the future does not depend on its whole past history, but only depends on the present state. This property is analogous to the property of a non-delayed deterministic dynamic system whose value at a future time is uniquely determined by its value at the current time. The strict definition for the Markov process is as follows.

Definition 7.1.12 *A stochastic process* $\{X(t) : t \geq 0\}$ *is said to be a* Markov process *with respect to the filtration* $\{\mathcal{F}_t\}$ *if for any* $t \geq 0$, $s > 0$, *the conditional distribution function of* $X(t+s)$ *given* $\mathcal{F}_t$ *is the same as the conditional distribution function* $X(t+s)$ *given* $X(t)$; *that is,*

$$Prob\{X(t+s) \leq y | \mathcal{F}_t\} = Prob\{X(t+s) \leq y | X(t)\}. \tag{7.15}$$

If $\{\mathcal{F}_t\} = \{\mathcal{F}_t^X\}$, *then we simply say that* $\{X(t) : t \geq 0\}$ *is a Markov process.*

Condition (7.15) is often referred to as the *Markov property*. Definition 7.1.12 implies that for any $0 \leq t_1 < \ldots < t_k$ with $k \geq 2$ being a positive integer we have

$$\begin{aligned}
&\text{Prob}\{X(t_k) \leq x_k \mid X(t_{k-1}) = x_{k-1}, \ldots, X(t_1) = x_1\} \\
&= \text{Prob}\{X(t_k) \leq x_k \mid X(t_{k-1}) = x_{k-1}\}.
\end{aligned} \tag{7.16}$$

Suppose that $\text{Prob}\{X(t_k) \leq x_k \mid X(t_{k-1}) = x_{k-1}, \ldots, X(t_1) = x_1\}$ is differentiable with respect to x_k, and we let

$$\rho_{X_k | X_{k-1}, \ldots, X_1}(t_k, x_k \mid t_{k-1}, x_{k-1}, \ldots, t_1, x_1)$$

$$= \frac{\partial}{\partial x_k} \text{Prob}\{X(t_k) \leq x_k \mid X(t_{k-1}) = x_{k-1}, \ldots, X(t_1) = x_1\}.$$

Then (7.16) implies that for any positive integer $k \geq 2$ we have

$$\rho_{X_k | X_{k-1}, \ldots, X_1}(t_k, x_k \mid t_{k-1}, x_{k-1}, \ldots, t_1, x_1) = \rho_{X_k | X_{k-1}}(t_k, x_k \mid t_{k-1}, x_{k-1}). \tag{7.17}$$

One of the important terminologies in studying a Markov process is the so-called *transition probability (function)*, which is defined as

$$\mathcal{P}(t, x; s, y) = \text{Prob}\{X(t) \leq x | X(s) = y\}. \tag{7.18}$$

In other words, the transition probability function $\mathcal{P}(t, x; s, y)$ is defined as the conditional probability of the process at time t given $X(s) = y$ at time $s < t$. The associated *transition probability density function* $\rho(t, x; s, y)$, if it exists, is defined by

$$\rho(t, x; s, y) = \frac{\partial}{\partial x} \mathcal{P}(t, x; s, y). \tag{7.19}$$

7.1.5.1   Characterization of a Markov Process

A Markov process $\{X(t) : t \geq 0\}$ can be completely characterized by its initial probability density function $p(0, x)$ together with its transition probability density function $\rho(t, x; s, y)$. To see that, we use the general relation (i.e., a relation that is true for any stochastic process)

$$p_{X_1, \ldots, X_m}(t_1, x_1, t_2, x_2, \ldots, t_m, x_m)$$

$$= p(t_1, x_1) \rho_{X_2 | X_1}(t_2, x_2 \mid t_1, x_1) \rho_{X_3 | X_2, X_1}(t_3, x_3 \mid t_2, x_2, t_1, x_1)$$

$$\cdots \rho_{X_m | X_{m-1}, \ldots, X_1}(t_m, x_m \mid t_{m-1}, x_{m-1}, \ldots, t_1, x_1),$$

which can be obtained by using the same arguments to obtain (2.34). By the above equality, (7.17) and (7.19) we find that for any $0 \leq t_1 < t_2 < \ldots < t_m$ we have

$$p_{X_1, \ldots, X_m}(t_1, x_1, t_2, x_2, \ldots, t_m, x_m) = p(t_1, x_1) \prod_{j=1}^{m-1} \rho(t_{j+1}, x_{j+1}; t_j, x_j).$$

$$(7.20)$$

A Markov process can also be completely specified by its first and second probability density functions. In other words, for any $0 \leq t_1 < t_2 < \ldots < t_m$ we have

$$p_{X_1, \ldots, X_m}(t_1, x_1, t_2, x_2, \ldots, t_m, x_m) = \frac{\prod_{j=1}^{m-1} p_{X_j X_{j+1}}(t_j, x_j, t_{j+1}, x_{j+1})}{\prod_{j=2}^{m-1} p(t_j, x_j)}.$$

This can be obtained by using (7.20) and the general relation

$$p_{X_j X_{j+1}}(t_j, x_j, t_{j+1}, x_{j+1}) = \rho(t_{j+1}, x_{j+1}; t_j, x_j) p(t_j, x_j), \quad j = 1, 2, \ldots, m-1.$$

7.1.5.2   The Chapman–Kolmogorov Equation

Consider the general relation

$$p_{X_1 X_3}(t_1, x_1, t_3, x_3) = \int_{-\infty}^{\infty} p_{X_1 X_2 X_3}(t_1, x_1, t_2, x_2, t_3, x_3) dx_2.$$

Then by the above equation and (7.20) we find that for any $0 \leq t_1 < t_2 < t_3$ we have

$$\rho(t_3, x_3; t_1, x_1) p(t_1, x_1) = \int_{-\infty}^{\infty} \rho(t_3, x_3; t_2, x_2) \rho(t_2, x_2; t_1, x_1) p(t_1, x_1) dx_2,$$

which implies that

$$\rho(t_3, x_3; t_1, x_1) = \int_{-\infty}^{\infty} \rho(t_3, x_3; t_2, x_2) \rho(t_2, x_2; t_1, x_1) dx_2. \quad (7.21)$$

The above equation is often called the *Chapman–Kolmogorov equation* (also referred to as the *Chapman–Kolmogrov–Smoluchowski equation*), and it plays an important role in the theory of Markov processes. By using similar arguments, we can find that for any $0 \le s \le t$ we have

$$p(t, x) = \int_{-\infty}^{\infty} p(t, x; s, y) p(s, y) dy. \tag{7.22}$$

Both (7.21) and (7.22) can be extended to a general n-dimensional stochastic process $\{\mathbf{X}(t) : t \ge 0\}$. For example, the n-dimensional version of (7.22) is given by

$$p(t, \mathbf{x}) = \int_{\mathbb{R}^n} p(t, \mathbf{x}; s, \mathbf{y}) p(s, \mathbf{y}) d\mathbf{y}. \tag{7.23}$$

7.1.5.3 An Example of a Markov Process: Wiener Process

In 1828, botanist Robert Brown observed that a pollen particle suspended in fluid moves in an irregular and random way. This phenomenon is called *Brownian motion*. In 1931, Wiener gave a rigorous treatment of Brownian motion as a stochastic process. This process is often called the *Wiener process*, which is one of the most basic and important Markov processes.

Definition 7.1.13 *The Wiener process $\{W(t) : t \ge 0\}$ is a stochastic process with the following properties.*

- $W(0) = 0$.

- *(Independent increments) $W(t) - W(s)$ is independent of the past, $t > s \ge 0$; that is, if $0 \le t_0 < t_1 < \cdots < t_m$, then $W(t_j) - W(t_{j-1})$, $j = 1, 2, \ldots, m$, are independent random variables.*

- *(Stationary normal increments) $W(t) - W(s)$ has normal distribution with mean 0 and variance $t - s$; that is,*

$$W(t) - W(s) \sim \mathcal{N}(0, t - s), \quad t > s \ge 0. \tag{7.24}$$

- *(Continuous sample paths) The sample paths of $\{W(t): t \ge 0\}$ are continuous functions of t.*

It is worth noting that the Wiener process defined here is often called the *standard Wiener process*, as the usual Wiener process is defined exactly as above except that (7.24) is changed to

$$W(t) - W(s) \sim \mathcal{N}(0, 2\mathcal{D}(t - s)), \quad t > s \ge 0,$$

where $\mathcal{D}$ is some positive constant termed the *noise intensity*. For this monograph, we mainly focus on the standard Wiener process, and hence simply refer to it as the Wiener process if no confusion occurs.

It can be shown that the Wiener process is a Gaussian process with mean and covariance functions respectively given by (e.g., see [72, Theorem 3.2])

$$\mathbb{E}(W(t)) = 0, \quad \text{Cov}\{W(t), W(s)\} = \min\{t, s\}. \tag{7.25}$$

This implies that for any fixed t, $W(t)$ is Gaussian distributed with zero mean and variance given by t; that is, $W(t) \sim \mathcal{N}(0, t)$. By Theorem 7.1.1 and (7.25) we know that a Wiener process is also m.s. continuous.

The sample path of the Wiener process has a number of interesting properties. For example, the sample path of a Wiener process is nowhere differentiable (with probability 1). The nowhere differentiability can be intuitively seen from $\mathbb{E}\left((W(t + \Delta t) - W(t))^2\right) = \Delta t$ (which suggests that the increment is roughly like $\sqrt{\Delta t}$, so $\sqrt{\Delta t}/\Delta t \to \infty$ as $\Delta t \to 0$) and Definition 7.1.10. In addition, for any fixed t almost all the sample paths of the Wiener process have quadratic variation on $[0, t]$ equal to t, where the quadratic variation of a stochastic process is defined as follows.

Definition 7.1.14 *The m-variation of a stochastic process $\{X(t) : t \geq 0\}$ on $[0, t]$ is defined by*

$$[X, X]^{(m)}([0, t]) = \lim \sum_{j=1}^{l} |X(t_j^l) - X(t_{j-1}^l)|^m, \tag{7.26}$$

where for each l, $\{t_j^l\}_{j=1}^l$ is a partition of $[0, t]$, and the limit is taken over all partitions with $\Delta_l = \max_{0 \leq j \leq l-1}(t_{j+1}^l - t_j^l) \to 0$ as $l \to \infty$ in the sense of convergence in probability. Specifically, $m = 1$ is simply called variation, and $m = 2$ is called quadratic variation.

For a Wiener process, the sums in (7.26) with $m = 2$ are found to converge to t almost surely (recall that convergence almost surely implies convergence in probability). This implies that almost all the sample paths of the Wiener process have quadratic variation on $[0, t]$ equal to t for any fixed t (as we stated earlier). Interested readers can refer to [72, Section 3.3] for more information on this.

Remark 7.1.6 *The property of non-zero quadratic variation of the sample path of the Wiener process implies that it has infinite variation on any time interval no matter how small the interval is (as the quadratic variation of any continuous function with finite variation must be zero – see [72, Theorem 1.10]). Recall that for the Riemann–Stieltjes integral the function being integrated against, the integrator, must have bounded variation if the integrand is an arbitrarily continuous function. This implies that the integrals such as $\int_{t_0}^{t_f} W(t)dW(t)$ and $\int_{t_0}^{t_f} h(t)dW(t)$ with h being an arbitrarily continuous deterministic function cannot be interpreted as path-by-path Riemann–Stieltjes*

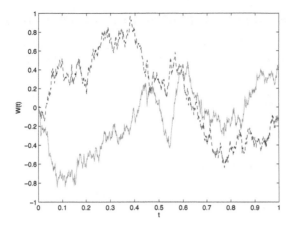

FIGURE 7.1: Two independent realizations of the Wiener process.

integrals, where path-by-path means that the integrals are defined with respect to the sample path of the Wiener process. This class of integrals with the integrator being a Wiener process is a special type of the so-called stochastic integrals, which will be explained later in Sections 7.1.6.2 and 7.2.

By the definition of the Wiener process, the transition probability is

$$\mathcal{P}(t,x;s,y) = \int_{-\infty}^{x} \frac{1}{\sqrt{2\pi(t-s)}} \exp\left(-\frac{(\xi-y)^2}{2(t-s)}\right) d\xi \qquad (7.27)$$

with its corresponding transition probability density function given by

$$\rho(t,x;s,y) = \frac{1}{\sqrt{2\pi(t-s)}} \exp\left(-\frac{(x-y)^2}{2(t-s)}\right).$$

To help to visualize the Wiener process, two independent realizations of this process on the time interval $[0,1]$ are shown in Figure 7.1. They are simulated with MATLAB based on the normal and independent increments of the Wiener process with equally spaced increment 0.001 on $[0,1]$. The exact procedure is as follows: Let $t_j = 0.001j, j = 0,1,2,\ldots,1000$. Then $W(t)$ at time point t_j is given by

$$W(t_j) = W(t_{j-1}) + \mathcal{E}_j, \quad j = 1,2,\ldots,1000,$$

where $W(0) = 0$, $\mathcal{E}_j$, $j = 1,2,\ldots,1000$, are independent and normally distributed with mean zero and variance 0.001.

Definition 7.1.15 *The l-dimensional Wiener process is defined as $\{\mathbf{W}(t) : t \geq 0\}$ with $\mathbf{W}(t) = (W_1(t), W_2(t), \ldots, W_l(t))^T$, where $\{W_j(t) : t \geq 0\}$, $j = 1,2,\ldots,l$, are independent one-dimensional Wiener processes.*

7.1.5.4   An Example of a Markov Process: Diffusion Process

A Markov process $\{X(t) : t \geq 0\}$ is said to be a *diffusion process* if the transition probability density function $\rho(t, x; s, \zeta)$ satisfies the following conditions:

$$\lim_{\Delta t \to 0+} \frac{1}{\Delta t} \int_{\mathbb{R}} (x - \zeta) \rho(t + \Delta t, x; t, \zeta) dx = g(t, \zeta),$$
$$\lim_{\Delta t \to 0+} \frac{1}{\Delta t} \int_{\mathbb{R}} (x - \zeta)^2 \rho(t + \Delta t, x; t, \zeta) dx = \sigma^2(t, \zeta),$$
(7.28)

and there exists a positive constant ϖ such that

$$\lim_{\Delta t \to 0+} \frac{1}{\Delta t} \int_{\mathbb{R}} (x - \zeta)^{2+\varpi} \rho(t + \Delta t, x; t, \zeta) dx = 0, \tag{7.29}$$

where g is referred to as the *drift coefficient* and σ^2 is called the *diffusion coefficient*. It is worth noting that (7.29) is a sufficient condition for a Markov process to have almost surely continuous sample paths. We observe that the Wiener process is a special case of a diffusion process with drift coefficient zero and diffusion coefficient one. As we shall see later in Section 7.2, the solution to a stochastic differential equation is also a diffusion process. Interested readers can refer to [62, Section 4.2] for more information on diffusion processes.

7.1.5.5   An Example of a Markov Process: Poisson Process

A Poisson process is a special example of the counting process (an integer-valued stochastic process that counts the number of events occurring in a time interval), and it is also one of the most basic and important Markov processes. The Poisson process can be used to model a series of random observations occurring in time. For example, the process could model the number of traffic accidents occurring at an intersection.

There are many ways to define a Poisson process. We give one of them below.

Definition 7.1.16 *A stochastic process $\{X(t) : t \geq 0\}$ is said to be a* Poisson process *with parameter λ (> 0) if it satisfies the following properties.*

- $X(0) = 0$.

- *(Independent increments) Numbers of events occurring in disjoint time intervals are independent random variables; that is, if $0 \leq t_0 < t_1 < \cdots < t_m$, then $X(t_j) - X(t_{j-1})$, $j = 1, 2, \ldots, m$, are independent random variables.*

- *(Stationary Poisson increment) For $t \geq 0$ and $s > 0$, $X(t + s) - X(t)$ follows a Poisson distribution with rate parameter λs; that is,*

$$Prob\{X(t + s) - X(t) = k\} = \frac{(\lambda s)^k}{k!} \exp(-\lambda s). \tag{7.30}$$

- *(Step function sample paths) The sample paths of $\{X(t)\colon t \geq 0\}$ are increasing step functions of t with jumps of size 1.*

The parameter λ is called the *rate*, and is also referred to as the *intensity*. It is used to describe the expected number of events per unit of time. The Poisson process with rate $\lambda = 1$ is called a *unit Poisson process*. In practice, the rate λ may not be a constant but rather a function of time t. For this scenario, the Poisson process is often referred to as a *non-homogeneous Poisson process* (to differentiate it from the case where λ is a constant). In this monograph, we only consider the case where λ is a constant.

It is worth noting that for a Poisson process the variation is the same as the quadratic variation, and both are finite (we remark that this is not a contradiction of Remark 7.1.6 since the Poisson process is not sample continuous). This is because for any sufficiently small interval $[t^l_{j-1}, t^l_j]$, the value of $X(t^l_j) - X(t^l_{j-1})$ is either 0 or 1. This implies that

$$(X(t^l_j) - X(t^l_{j-1}))^2 = X(t^l_j) - X(t^l_{j-1}).$$

Thus, by $X(0) = 0$, we have

$$[X, X]^{(2)}([0, t]) = \lim \sum_{j=1}^{l} |X(t^l_j) - X(t^l_{j-1})|^2$$

$$= \lim \sum_{j=1}^{l} |X(t^l_j) - X(t^l_{j-1})| = X(t).$$

By using the Taylor series expansion, we see that for any small Δt we have

$$\exp(-\lambda \Delta t) = 1 - \lambda \Delta t + o(\Delta t), \tag{7.31}$$

where the little-o notation $o(\Delta t)$ means that $\lim\limits_{\Delta t \to 0} \dfrac{o(\Delta t)}{\Delta t} = 0$. Hence, by (7.31) as well as Definition 7.1.16, we find that for any arbitrarily small Δt the transition probabilities for the Poisson process with rate λ satisfy

$$\text{Prob}\{X(t + \Delta t) - X(t) = 1 \mid X(t)\} = \text{Prob}\{X(t + \Delta t) - X(t) = 1\}$$
$$= \lambda \Delta t + o(\Delta t),$$

$$\text{Prob}\{X(t + \Delta t) - X(t) = 0 \mid X(t)\} = \text{Prob}\{X(t + \Delta t) - X(t) = 0\}$$
$$= 1 - \lambda \Delta t + o(\Delta t), \tag{7.32}$$

$$\text{Prob}\{X(t + \Delta t) - X(t) \geq 2 \mid X(t)\} = \text{Prob}\{X(t + \Delta t) - X(t) \geq 2\}$$
$$= o(\Delta t),$$

which implies that in any sufficiently small Δt, at most one event can happen.

Definition 7.1.16 also implies that for any fixed t the random variable $X(t)$ is Poisson distributed with parameter λt. Hence,

$$\mathbb{E}(X(t)) = \lambda t, \quad \text{Var}(X(t)) = \lambda t. \tag{7.33}$$

In addition, it can be easily found that the covariance function of a Poisson process with rate λ is given by

$$\text{Cov}\{X(t), X(s)\} = \lambda \min\{t, s\}. \tag{7.34}$$

Thus, by the above equality and (7.25) we see that the unit Poisson process has the same covariance function as the Wiener process. In addition, by (7.34) and Theorem 7.1.1 we know that the Poisson process is m.s. continuous. Recall that the sample paths of the Poisson process are not continuous. Hence, this is one example that m.s. continuity does not imply sample continuity.

The next theorem is about the property of *interarrival times* (the waiting time between two successive jumps) of a Poisson process and it has very important applications (as we shall see in Chapter 8).

Theorem 7.1.7 *Let* $\{X(t) : t \geq 0\}$ *be a Poisson process with rate* $\lambda > 0$ *and denote by* $0 = \mathcal{T}_0 < \mathcal{T}_1 < \mathcal{T}_2 < \ldots$ *the successive occurrence times of events (jumps). Then the interarrival times,* $S_j = \mathcal{T}_j - \mathcal{T}_{j-1}$, $j = 1, 2, \ldots$, *are independent and identically distributed (i.i.d.) random variables following an exponential distribution with mean* $\dfrac{1}{\lambda}$.

The random variable $\mathcal{T}_m$ is often called the *waiting time* to the mth jump, and the interarrival times are also called the *holding times* or *sojourn times* or *interevent times* (these names will be used interchangeably in this monograph). By Theorem 7.1.7 as well as the relationship (2.109) between the exponential and the gamma distributions, we know that for any positive integer m we have

$$\mathcal{T}_m \sim \text{Gamma}(m, \lambda).$$

The following theorem states that the converse of Theorem 7.1.7 also holds.

Theorem 7.1.8 *If the interarrival times* $\{S_j\}$ *of a counting process* $\{X(t) : t \geq 0\}$ *are i.i.d. exponential random variables with mean* $\dfrac{1}{\lambda}$, *then* $\{X(t) : t \geq 0\}$ *is a Poisson process with rate* λ.

We thus see that Theorem 7.1.8 can be used as another way to define a Poisson process. It is worth noting that if the interarrival times $\{S_j\}$ of a counting process are i.i.d. positive random variables, then the process is called the *renewal process*. Thus a Poisson process is a special case of the renewal process (i.e., a renewal process with exponentially distributed interarrival times), which can be used to model a wide variety of phenomena.

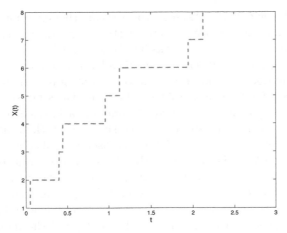

FIGURE 7.2: A realization of a unit Poisson process.

In addition, the mean function μ of a renewal process satisfies the so-called *renewal equation* given by

$$\mu(t) = P_S(t) + \int_0^t p_S(s)\mu(t-s)ds, \tag{7.35}$$

where P_S and p_S respectively denote the cumulative distribution function and probability density function of the interarrival times of the renewal process. Interested readers can refer to [92, Chapter 5] for more information on the renewal process.

To help to visualize the Poisson process, a realization of a unit Poisson process is demonstrated in Figure 7.2. This sample path is simulated with MATLAB based on Theorem 7.1.8.

7.1.5.6 Classification of a Markov Process

Based on whether or not the transition probability density function depends on the absolute origin of time, Markov processes can be grouped into two classes: homogeneous Markov processes and non-homogeneous Markov processes. Specifically, if the transition probability density functions $p(t,x;s,y)$ do not depend explicitly on s or t but only depend on the length of the time interval $t-s$, then the corresponding Markov process is said to be a *homogeneous Markov process*. Otherwise, it is called a *non-homogeneous Markov process*. It is worth noting here that the homogeneous property is different from the stationary property (recall Definition 7.1.6), as the homogeneous property does not require the distribution function P to be independent of time t. For example, by (7.27) and the fact that for any fixed t the random variable $W(t) \sim \mathcal{N}(0,t)$, we know that the Wiener process is homogeneous but not stationary (actually it is not even wide-sense stationary). In addition, by (7.32) and (7.33) we know that the Poisson process is another example of a stochastic process possessing the homogeneous property but not the stationary property.

Markov processes can also be classified based on whether or not the corresponding state spaces are discrete. Specifically, if a Markov process can only be in a finite number or countably infinitely many states, then the process is called a *Markov chain* ("chain" refers to fact that the state space is discrete). For example, a sequence of independent random variables is a special case of a Markov chain. If the time space is also discrete, then the process is called a *discrete time Markov chain* (DTMC); otherwise, it is called a *continuous time Markov chain* (CTMC). For example, the Poisson process is a continuous time Markov chain. It should be noted that in some literature the Markov chain is specifically referred to as a DTMC and a CTMC is referred to as a *Markov jump process*. In the next section, we give a brief and general introduction to the CTMC. Interested readers can refer to [6, 92] and the references therein for information on the DTMC, and also refer to [6, 7, 58, 92] for the CTMC as well as its applications in different fields.

7.1.5.7   Continuous Time Markov Chain

Without loss of generality, we assume here that the state space of a CTMC is a set of non-negative integers. The transition probabilities for a continuous time Markov chain are often written as

$$\mathcal{P}_{jk}(s,t) = \text{Prob}\{X(t) = j | X(s) = k\}, \quad 0 \le s < t \qquad (7.36)$$

for $k, j = 0, 1, 2, \cdots$. In the following, we only consider homogeneous CTMCs. Note that for a homogeneous CTMC, we can simply write the transition probabilities as

$$\mathcal{P}_{jk}(t) = \text{Prob}\{X(t+s) = j | X(s) = k\}$$

$$= \text{Prob}\{X(t) = j | X(0) = k\}, \quad t > 0, \ s \ge 0.$$

The matrix of transition probabilities $(\mathcal{P}_{jk}(t))$ is often referred to as the *transition matrix*. In most cases, these transition probabilities have the following properties:

$$\mathcal{P}_{jk}(t) \ge 0, \ k, j = 0, 1, 2, \ldots; \quad \sum_{j=0}^{\infty} \mathcal{P}_{jk}(t) = 1, \ k = 0, 1, 2, \ldots. \qquad (7.37)$$

In addition, they satisfy the *Chapman–Kolmogorov equation*, which in this case is given by

$$\mathcal{P}_{jk}(t+s) = \sum_{l=0}^{\infty} \mathcal{P}_{jl}(s)\mathcal{P}_{lk}(t), \quad k, j = 0, 1, 2, \ldots. \qquad (7.38)$$

Transition rate   Assume that the transition probabilities $\mathcal{P}_{jk}$ are continuous and differentiable for $t \geq 0$, and at $t = 0$ they are

$$\mathcal{P}_{jk}(0) = \begin{cases} 0, & \text{if } j \neq k \\ 1, & \text{if } j = k. \end{cases}$$

Then the *transition rate* q_{jk} is defined by

$$q_{jk} = \dot{\mathcal{P}}_{jk}(0) = \lim_{\Delta t \to 0+} \frac{\mathcal{P}_{jk}(\Delta t) - \mathcal{P}_{jk}(0)}{\Delta t}, \tag{7.39}$$

For example, by (7.32) and (7.39) we see that for a Poisson process with rate λ, the transition rates are given by

$$q_{jk} = \begin{cases} \lambda & \text{if } j = k + 1 \\ -\lambda & \text{if } j = k \\ 0 & \text{otherwise.} \end{cases}$$

If we assume that $\sum_{j \neq k} o(\Delta t) = o(\Delta t)$ (which is certainly true if the summation contains only a finite number of terms), then by (7.37) and (7.39) we find that $q_{kk} = -\sum_{j \neq k} q_{jk}$. Let $\lambda_k = -q_{kk}$; that is, $\lambda_k = \sum_{j \neq k} q_{jk}$. Then λ_k denotes the rate at which the process leaves state k. Hence, if λ_k is infinite, then the stochastic process will leave state k immediately; if $\lambda_k = 0$, then the stochastic process will stay in state k forever (in this case state k is called an *absorbing state*). In the following discussion, we always assume that λ_k is non-zero and finite. Then under this assumption, we have $\sum_{j=0}^{\infty} q_{jk} = 0$ and

$$\mathcal{P}_{jk}(\Delta t) = \delta_{jk} + q_{jk}\Delta t + o(\Delta t), \tag{7.40}$$

where δ_{jk} denotes the Kronecker delta, that is, $\delta_{jk} = \begin{cases} 1 & \text{if } j = k \\ 0 & \text{otherwise.} \end{cases}$

Kolmogorov's forward and backward equations   Differentiating (7.38) with respect to s and then setting $s = 0$ we have

$$\dot{\mathcal{P}}_{jk}(t) = \sum_{l=0}^{\infty} q_{jl}\mathcal{P}_{lk}(t), \quad k, j = 0, 1, 2, \ldots. \tag{7.41}$$

The above equation is often called *Kolmogorov's forward equation*, and it is also called the *master equation* or *chemical master equation* (often in the physics, chemistry and engineering literature). Using similar arguments, one can find that the transition probabilities also satisfy the equation

$$\dot{\mathcal{P}}_{jk}(t) = \sum_{l=0}^{\infty} \mathcal{P}_{jl}(t)q_{lk}, \quad k, j = 0, 1, 2, \ldots, \tag{7.42}$$

which is often called *Kolmogorov's backward equation*. Thus we see that the transition probabilities satisfy both Kolmogorov's forward and backward equations. Note that if the initial distribution function of $X(0)$ is known, then the CTMC is completely characterized by its transition probabilities. Hence, the matrix (q_{jk}) associated with Kolmogorov's forward and backward equations is often called the *infinitesimal generator* of the CTMC.

Exponentially distributed holding times   Recall from Section 7.1.5.5 that the holding times for the Poisson process are i.i.d. exponentially distributed. Here we show that the holding times for a CTMC are also exponentially distributed. However, it should be noted that these holding times may not be independent. Suppose that $X(t) = k$, and let S_k denote the amount of time that it remains in state k after entering this state (i.e., the holding time at state k). We define $R_k(t)$ as the probability that the process remains in state k for a length of time t; that is,

$$R_k(t) = \text{Prob}\{S_k > t\}. \tag{7.43}$$

Then we have

$$R_k(0) = \text{Prob}\{S_k > 0\} = 1. \tag{7.44}$$

By (7.40), we find that for any sufficiently small Δt we have

$$R_k(t + \Delta t) = R_k(t)\mathcal{P}_{kk}(\Delta t) = R_k(t)(1 - \lambda_k \Delta t + o(\Delta t)),$$

which implies that

$$\frac{R_k(t + \Delta t) - R_k(t)}{\Delta t} = -\lambda_k R_k(t) + R_k(t)\frac{o(\Delta t)}{\Delta t}.$$

Taking the limit as $\Delta t \to 0$ for the above equation we find

$$\frac{dR_k(t)}{dt} = -\lambda_k R_k(t).$$

It is easy to see that the solution to the above ordinary differential equation with initial condition (7.44) is given by

$$R_k(t) = \exp(-\lambda_k t).$$

Then by the above equation and (7.43), we know that the cumulative distribution function of S_k is given by

$$P_k(t) = 1 - \exp(-\lambda_k t), \tag{7.45}$$

which indicates that S_k is exponentially distributed with rate parameter λ_k. We summarize the above results in the following theorem.

Theorem 7.1.9 *Let* $\{X(t) : t \geq 0\}$ *be a continuous time Markov chain such that for any sufficiently small* Δt

$$Prob\{X(t + \Delta t) = j \mid X(t) = k\} = \delta_{jk} + q_{jk}\Delta t + o(\Delta t),$$

where δ_{jk} *is the Kronecker delta, and* $q_{kk} = -\sum_{j \neq k} q_{jk}$. *Let* $\lambda_k = -q_{kk}$, *and* S_k *denote the amount of time that it remains in state* k *after entering this state. Then* S_k *is exponentially distributed with rate parameter* λ_k.

Assume that $s < S_k \leq s + \Delta s$ and the chain only jumps once in $(s, s + \Delta s)$. Then by (7.40) we find that the probability that the chain jumps to state j from state k $(j \neq k)$ is given by

$$\lim_{\Delta s \to 0} \frac{\mathcal{P}_{jk}(\Delta s)}{1 - \mathcal{P}_{kk}(\Delta s)} = \frac{q_{jk}}{\sum_{j \neq k} q_{jk}} = \frac{q_{jk}}{\lambda_k}. \tag{7.46}$$

By using the inverse transform method discussed in Remark 2.5.2, one can discern which state the process will jump to by simulating a discrete random variable, whose range Ω_k is a set of non-negative integers excluding k, and the probability associated with $j \in \Omega_k$ is given by (7.40). We thus see that Theorem 7.1.9 and (7.46) provide enough information on how to simulate a CTMC, where Theorem 7.1.9 gives information on how much time the process remains in a specific state before it jumps to another state and (7.46) indicates which state it will jump to. We refer readers to Chapter 8 for a more detailed discussion on this topic.

7.1.6 Martingales

In this section, we introduce another special class of stochastic processes – martingales. The theory of martingales was initiated by J.L. Doob in the 1950s, and it plays a central role in model probability theory (e.g., the theory of Markov processes and the theory of stochastic integration) and its applications (e.g., financial mathematics). Below we give a very brief introduction to martingales, and refer interested readers to some popular books such as [45, 48, 72] for more information on this topic.

Definition 7.1.17 *A stochastic process* $\{X(t) : t \geq 0\}$ *with* $\mathbb{E}(|X(t)|) < \infty$ *for all* $t \geq 0$ *and adapted to a filtration* $\{\mathcal{F}_t\}$ *is an* $\{\mathcal{F}_t\}$-*martingale if*

$$\mathbb{E}\left(X(t + s)|\mathcal{F}_t\right) = X(t), \quad t, s \geq 0. \tag{7.47}$$

From the above definition we see that a Poisson process is not a martingale (as its sample paths are increasing step functions).

Theorem 7.1.10 *A non-constant sample-continuous martingale has a positive quadratic variation. Thus it has infinite variation on any time interval.*

The above theorem implies that if a sample-continuous martingale has finite variation, then it must be a constant. It is worth noting that there exist some martingales with finite variation. But by Theorem 7.1.10, we know that such martingales must be sample discontinuous. An example of such martingales is $\{X(t) - \lambda t : t \geq 0\}$, where $\{X(t) : t \geq 0\}$ is a Poisson process with rate λ.

7.1.6.1   Examples of Sample-Continuous Martingales

Let $\mathcal{F}_t$ be the σ-algebra generated by $\{W(s) : 0 \leq s \leq t\}$. Then by the definition of the Wiener process, it can be shown that the Wiener process is a martingale with respect to $\{\mathcal{F}_t\}$. Specifically, by Jensen's inequality (2.67) and $W(t) \sim \mathcal{N}(0, t)$ we find that for any given $t \geq 0$ we have

$$(\mathbb{E}|W(t)|)^2 \leq \mathbb{E}(|W(t)|^2) = \mathbb{E}(W^2(t)) = t. \tag{7.48}$$

The fact that $W(t)$ is $\mathcal{F}_t$-measurable implies that

$$\mathbb{E}(W(t)|\mathcal{F}_t) = W(t). \tag{7.49}$$

Note that for any $s > 0$, $W(t + s) - W(t)$ is independent of $\mathcal{F}_t$ and $W(t + s) - W(t) \sim \mathcal{N}(0, s)$. Hence, we have

$$\mathbb{E}(W(t+s) - W(t)|\mathcal{F}_t) = \mathbb{E}(W(t+s) - W(t)) = 0. \tag{7.50}$$

Thus, by (7.49) and (7.50) we find that for any $s > 0$ and $t \geq 0$ we have

$$\begin{aligned} \mathbb{E}(W(t+s)|\mathcal{F}_t) &= \mathbb{E}(W(t+s) - W(t) + W(t)|\mathcal{F}_t) \\ &= \mathbb{E}(W(t+s) - W(t)|\mathcal{F}_t) + \mathbb{E}(W(t)|\mathcal{F}_t) \\ &= W(t). \end{aligned} \tag{7.51}$$

Therefore, by (7.48) and (7.51) we know that the Wiener process is a martingale with respect to $\{\mathcal{F}_t\}$. As we shall see below, the Wiener process is one of the most basic and important examples of continuous-time martingales.

It can be found that $\{W^2(t) - t : t \geq 0\}$ is also a martingale with respect to $\{\mathcal{F}_t\}$. This provides another characterization of the Wiener process. That is, if $\{X(t) : t \geq 0\}$ is a sample-continuous martingale such that $\{X^2(t) - t : t \geq 0\}$ is also a martingale, then $\{X(t) : t \geq 0\}$ is a Wiener process. This result is due to Lèvy. Interested readers can refer to [72, Section 3.4] for more information on this.

7.1.6.2   The Role of Martingales in the Development of Stochastic Integration Theory

It is with respect to the Wiener process that stochastic integration was initially studied by K. Itô in the 1940s and then subsequently investigated extensively in the literature. As discussed in [78], the Itô theory of stochastic integration can be motivated from the viewpoint of a martingale. The goal is to define a stochastic integral $\int_{t_0}^{t_f} Y(t)dW(t)$ in such a way that the stochastic process generated by these integrals, $\left\{ \int_{t_0}^{t} Y(s)dW(s) : t_0 \leq t \leq t_f \right\}$, is a martingale. Such a stochastic integral is referred to as an *Itô integral*, and it can be defined for any integrand $\{Y(t) : t_0 \leq t \leq t_f\}$ that is adapted to the filtration generated by the Wiener process and satisfies the condition

$$\int_{t_0}^{t_f} \mathbb{E}(Y^2(t))dt < \infty.$$

It is worth noting that if the integrand is further assumed to be m.s. continuous, then the Itô integral can be defined as the limit of a sequence of Riemann–Stieltjes sums, that is,

$$\int_{t_0}^{t_f} Y(t)dW(t) = \underset{\Delta_t \to 0}{\text{l.i.m.}} \sum_{j=1}^{l} Y(t_{j-1})(W(t_j) - W(t_{j-1})),$$

where $\{t_j\}_{j=0}^{l}$ is a partition of the interval $[t_0, t_f]$ with mesh size $\Delta_t = \max_{0 \leq j \leq l-1} \{t_{j+1} - t_j\}$. From the above equation, we see that the intermediate point in the Itô integral is chosen as the left endpoint of each subinterval. Due to this choice as well as the fact that the Wiener process has non-zero quadratic variation, the Itô integral does not obey the classical chain rule and integration by parts. For example, $\int_0^t W(s)dW(s) = \frac{1}{2}W^2(t) - \frac{1}{2}t$.

However, Itô integrals have many other important properties besides the martingale property. For example, they define a linear mapping; that is, for any $c_1, c_2 \in \mathbb{R}$ we have

$$\int_{t_0}^{t_f} (c_1 Y_1(t) + c_2 Y_2(t))dW(t) = c_1 \int_{t_0}^{t_f} Y_1(t)dW(t) + c_2 \int_{t_0}^{t_f} Y_2(t)dW(t).$$

In addition, they possess the zero mean property and isometry property, which are respectively given by

$$\mathbb{E}\left(\int_{t_0}^{t_f} Y(t)dW(t) \right) = 0, \quad \mathbb{E}\left(\left(\int_{t_0}^{t_f} Y(t)dW(t) \right)^2 \right) = \int_{t_0}^{t_f} \mathbb{E}(Y^2(t))dt.$$

Furthermore, the stochastic process generated by Itô integrals is sample continuous.

It was realized by J.L. Doob in the 1950s [45] that the key property that makes stochastic integration work is the fact that the stochastic process being integrated against, the integrator, is a martingale. This observation made it possible to extend the stochastic integration theory from Wiener processes to general stochastic processes that are martingales and then to semi-martingales, which include *supmartingales* (the stochastic processes with the property $\mathbb{E}\left(X(t+s)|\mathcal{F}_t\right) \leq X(t))$, and *submartingales* (the stochastic processes satisfying $\mathbb{E}\left(X(t+s)|\mathcal{F}_t\right) \geq X(t))$ as special cases. The interested reader may consult [66, 94, 115] for excellent surveys on the history and development of stochastic integration theory and [78] for an excellent treatment of stochastic integrals.

7.1.7   White Noise vs. Colored Noise

In this section, we give a brief introduction to white noise and colored noise (these names are in analogy to white light and the effects of filtering on white light). Interested readers can refer to [47, 60, 106, 114] for some interesting discussion on white noise and colored noise as well as their applications in different fields. Before addressing our main task in this section, we first introduce an important concept, the *power spectral density function*, related to white and colored noise.

7.1.7.1   The Power Spectral Density Function

The power spectral density function, also referred to as the *power spectrum*, is defined for a wide-sense stationary stochastic process, and its definition is given as follows.

Definition 7.1.18 *The power spectral density function of a wide-sense stationary stochastic process* $\{X(t) : t \geq 0\}$ *is defined by*

$$\mathcal{S}(\omega) = \int_{-\infty}^{\infty} \Upsilon(s) \exp(-i\omega s)ds, \qquad (7.52)$$

where Υ *is the correlation function of* $\{X(t) : t \geq 0\}$ *defined by (7.10), i is the imaginary unit and* ω *is the angular frequency.*

By Definition 7.1.18 we see that the power spectral density function of a wide-sense stationary stochastic process is the Fourier transform of its correlation function. Note that Υ is an even function. Hence, $\mathcal{S}$ is also an even function with spectral height (the maximum value of $\mathcal{S}$) given by

$$\mathcal{S}(0) = \int_{-\infty}^{\infty} \Upsilon(s)ds = 2\int_{0}^{\infty} \Upsilon(s)ds.$$

It is worth noting that the unit of the power spectral density function is power per frequency. Hence, if one wants to obtain the power within a specific frequency range, then one just needs to integrate $\mathcal{S}$ within that frequency range.

There are two important parameters related to a wide-sense stationary stochastic process. One is *noise intensity* $(\mathcal{D})$, and the other is *correlation time* (τ). They are respectively given by

$$\mathcal{D} = \int_{0}^{\infty} \Upsilon(s)ds = \frac{1}{2}\mathcal{S}(0), \quad \tau = \frac{\int_{0}^{\infty} \Upsilon(s)ds}{\Upsilon(0)} = \frac{\mathcal{D}}{\Upsilon(0)}. \qquad (7.53)$$

We see that the noise intensity is defined as the area beneath the correlation function Υ, and the correlation time is a "normalized" noise intensity. Intuitively speaking, a strongly correlated stochastic process implies its correlation function decays slowly; thus there is a large area beneath the correlation function. Therefore, noise intensity and correlation time give some indication of how strongly the associated stochastic process is correlated over time. We reinforce this intuition with some examples below.

Example 7.1.1 Let $\Upsilon(s) = \exp(-|s|/\tau)$. Then by (7.52) and (7.53) one can easily find that

$$S(\omega) = \frac{2\tau}{1 + \tau^2 \omega^2},$$

and the noise intensity and the correlation time are both given by τ. In addition, we observe that for any fixed s, $\Upsilon(s)$ increases as the correlation time τ increases. This implies that a long correlation time indicates that the associated stochastic process is strongly correlated while a short correlation time indicates that it is weakly correlated. Thus the value of the correlation time determines the degree of correlation of the associated stochastic process. We also observe that the limit of the increasing correlation time is an infinite correlation time, which leads to the correlation function $\Upsilon \equiv 1$ (the maximum value of Υ), and hence the associated stochastic process is fully correlated.

Example 7.1.2 Let $\Upsilon(s) = \dfrac{D}{\tau} \exp(-|s|/\tau)$. Then by the above example we see that for this case

$$S(\omega) = \frac{2D}{1 + \tau^2 \omega^2}.$$

Based on (7.53) we observe that the noise intensity is D, and the correlation time is given by τ.

Remark 7.1.11 *The power spectral density function of a general stochastic process usually does not formally exist, as the correlation function is usually not absolutely integrable. There has been some discussion of the problem of defining the power spectral density function for a general stochastic process (e.g., see [69] and the references therein). For example, the expected sample power spectral density function is defined in [69] for a stochastic process (including a non-stationary one) in an arbitrary time frame via*

$$
\begin{aligned}
\widehat{S}(\omega) = &\int_{-t_f}^{0} \left[\frac{1}{t_f} \int_{t_0 - \frac{t}{2}}^{t_0 + t_f + \frac{t}{2}} \widehat{R}(s, t) ds \right] \exp(-i\omega t) dt \\
&+ \int_{0}^{t_f} \left[\frac{1}{t_f} \int_{t_0 + \frac{t}{2}}^{t_0 + t_f - \frac{t}{2}} \widehat{R}(s, t) ds \right] \exp(-i\omega t) dt,
\end{aligned}
\tag{7.54}
$$

where t_f is some positive constant, the constant t_0 is used to allow for an arbitrary time frame, and $\widehat{\mathcal{R}}(s,t) = Cor\left\{ X\left(s + \dfrac{t}{2}\right), X\left(s - \dfrac{t}{2}\right)\right\}$ is used to allow for non-stationary stochastic processes. By (7.54) one can find that the expected sample power spectral density function for a Wiener process is proportional to $1/\omega^2$ when $t_f \to \infty$. Interested readers can refer to [69] for more details on this topic.

7.1.7.2   White Noise

White noise is also referred to as an uncorrelated stochastic process, as the value of white noise at any time point is uncorrelated from its value at any other time point; that is, its correlation function is proportional to a Dirac delta function. Hence, the power spectral density function of white noise is a constant. This implies that white noise has infinite power in the whole frequency range. White noise is widely used in practice to account for the randomness in the inputs of systems governed by differential equations, especially in the case where one has little knowledge about the precise nature of the noise. Below we introduce two important classes of white noise, Gaussian white noise and white shot noise.

7.1.7.2.1   Gaussian white noise   Gaussian white noise is defined as the formal derivative of a Wiener process (the standard one or the usual one). By Theorem 7.1.4 as well as (7.12) and (7.25), we see that Gaussian white noise $\{\mathrm{GWN}(t) : t \geq 0\}$ is a Gaussian process with mean and covariance functions respectively given by

$$\mathbb{E}(\mathrm{GWN}(t)) = 0, \quad \mathrm{Cov}\{\mathrm{GWN}(t), \mathrm{GWN}(s)\} = 2\mathcal{D}\delta(t - s),$$

where $\mathcal{D}$ is the noise intensity ($\mathcal{D} = 1/2$ if the standard Wiener process is used), and δ denotes the Dirac delta function. Gaussian white noise has been widely used in practice, as it provides a good approximation to many real world problems. In addition, it is with this white noise that stochastic differential equations were originally studied and then subsequently investigated extensively in the literature.

To help to visualize Gaussian white noise, a realization of Gaussian white noise with $\mathcal{D} = 1/2$ on the time interval $[0,1]$ is depicted in Figure 7.3. It was simulated with MATLAB based on the normal and independent increments of a Wiener process with equally spaced increment 0.001 on $[0, 1.001]$. The exact procedure is as follows: Let $t_j = 0.001j$, $j = 0, 1, 2, \ldots, 1001$, and GWN^j be the approximation for $\mathrm{GWN}(t_j)$, $j = 0, 1, 2 \ldots, 1000$, by using the forward finite difference scheme for the derivative; that is,

$$\mathrm{GWN}^{j-1} = \frac{W(t_j) - W(t_{j-1})}{t_j - t_{j-1}}, \quad j = 1, 2, \ldots, 1001.$$

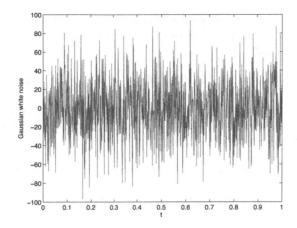

FIGURE 7.3: A realization of Gaussian white noise with $\mathcal{D} = 1/2$ on the time interval $[0, 1]$.

The properties of normal and independent increments of the Wiener process imply that GWN^j, $j = 0, 1, 2, \ldots, 1000$, are independent and normally distributed with mean zero and variance $\dfrac{1}{t_j - t_{j-1}} = 1000$. Figure 7.3 indicates that Gaussian white noise is extremely irregular and is nowhere continuous.

7.1.7.2.2 White shot noise White shot noise is another white noise that has been extensively studied in the literature. It is defined as

$$\mathrm{WSN}(t) = \sum_j \mathcal{W}_j \delta(t - \mathcal{T}_j). \tag{7.55}$$

In the above equation, $\mathcal{T}_j$ denotes the time point at which the jth pulse (or δ peak) occurs, $j = 1, 2, \ldots$, and $\{\mathcal{T}_j\}$ is assumed to form the set of successive occurrence times of events of a Poisson process $\{X(t) : t \geq 0\}$ with rate λ. In other words, $\mathcal{T}_{j+1} - \mathcal{T}_j$, $j = 1, 2, \ldots$, are independent and identically exponentially distributed random variables with mean $1/\lambda$ (that is, the average time between two successive pulses is $1/\lambda$). The corresponding weight of the jth pulse is denoted by $\mathcal{W}_j$, and these weights are either constants (i.e., $\mathcal{W}_j$ is a constant for all j) or independent and identically distributed random variables, which are assumed to be independent of the Poisson process $\{X(t) : t \geq 0\}$.

We see that the formal derivative of a Poisson process is a white shot noise. Thus, white shot noise is also known as *Poisson white noise*. If the weights $\mathcal{W}_j$, $j = 1, 2, \ldots$, in (7.55) are independent and identically distributed random variables that are independent of the Poisson process, then white shot noise is the formal derivative of the so-called *compound Poisson process* $\{\mathrm{CP}(t) : t \geq 0\}$, whose mean and covariance functions are respectively given by (e.g., see [83, p. 177] or [92, p. 130])

$$\mathbb{E}(\mathrm{CP}(t)) = \lambda t \mu_{\mathcal{W}}, \quad \mathrm{Cov}\{\mathrm{CP}(t), \mathrm{CP}(s)\} = \lambda(\mu_{\mathcal{W}}^2 + \sigma_{\mathcal{W}}^2)\min\{t, s\}. \quad (7.56)$$

Here $\mu_{\mathcal{W}}$ and $\sigma_{\mathcal{W}}^2$ respectively denote the mean and the variance of $\mathcal{W}_j$. By (7.12) and (7.56) we know that the mean and covariance functions of white shot noise are respectively given by

$$\mathbb{E}(\mathrm{WSN}(t)) = \lambda \mu_{\mathcal{W}}, \quad \mathrm{Cov}\{\mathrm{WSN}(t), \mathrm{WSN}(s)\} = \lambda(\mu_{\mathcal{W}}^2 + \sigma_{\mathcal{W}}^2)\delta(t - s).$$

We thus see that the white shot noise defined here has a positive mean function. Hence, in some literature white shot noise is defined by $\{\mathrm{WSN}(t) - \lambda \mu_{\mathcal{W}} : t \geq 0\}$ so that it has a zero mean function.

7.1.7.3   Colored Noise

Colored noise is also referred to as a correlated stochastic process, as its value at any time point is correlated with its value at any other time point. In general, colored noise will not only affect the form of the stationary probability of the system (e.g., studies of colored noise induced transitions) but also the dynamical aspects of the system. For example, Kubo showed in 1962 that a fluctuating magnetic field correlated over a long time scale (colored noise) can significantly modify the motion of spin, while one with a very short correlation time (almost white noise) typically does not (see [59] and the references therein for more information). Below we introduce several special types of colored noise that have been widely studied in the literature.

7.1.7.3.1   The $1/\omega^\alpha$ noise   The $1/\omega^\alpha$ noise is often called the $1/f^\alpha$ noise with f denoting frequency and α being a positive constant. It is a correlated stochastic process whose power spectral density function decays as $1/\omega^\alpha$. $1/\omega^\alpha$ noise is widely found in nature. For example, it has been observed in many contexts, including astronomy, electronic devices, brains, heartbeats, psychological mental states, stock markets and even in music (e.g., see [106, 114]). Due to its wide applications, this class of noise has drawn huge interest from researchers and has been intensively studied for decades with many attempts to mathematically describe the phenomenon, but there is not yet a unified explanation [114].

$1/\omega^\alpha$ noise includes a number of well-known stochastic processes. Recall from Remark 7.1.11 that the Wiener process corresponds to $\alpha = 2$. The Wiener process is often referred to as *red noise* (in analogy to red light). However, due to the name "Brownian motion," the Wiener process is also called *brown noise* or *brownian noise*. Since the power spectral density function of white noise is a constant, it corresponds to $\alpha = 0$. Thus, colored noise corresponding to $\alpha = 1$ is often referred to as *pink noise* (sometimes also called *flicker noise*), which is one of the most encountered forms of noise in practice. Similar to white noise, pink noise also has infinite power (in the whole positive frequency range). However, the sample path of pink noise behaves less irregularly than that of white noise. Interested readers can refer to [106] on how to simulate $1/\omega^\alpha$ noise, where it was also found that the solution to differential equations driven by such noise exhibits a strong dependence on α.

7.1.7.3.2 Gaussian colored noise A correlated Gaussian process is often called *Gaussian colored noise*. Note that it is often the case that environmental fluctuations (e.g., atmospheric disturbances) are the cumulative effect of weakly coupled environmental factors and the central limit theorem then implies that this fluctuation is Gaussian distributed (e.g., [60, 62, 95, 104]). Hence, Gaussian colored noise is widely used in practice to account for the randomness in the inputs of a system.

We remark that a stationary Gaussian process with zero mean function and exponentially decreasing correlation function

$$\Upsilon(s) = \exp(-|s|/\tau) \tag{7.57}$$

(where τ is the correlation time; see Example 7.1.1) is often used in the literature to model the random inputs of dynamic systems described by differential equations. This is in part due to the computational advantage it brings as for this case one can use the Karhunen–Loève expansion to decompose the Gaussian colored noise into a linear combination of a sequence of standard normal random variables that are mutually independent (e.g., see [54, Section 2.3] and [121, Section 4.2]). It is worth noting that the Karhunen–Loève expansion is one of the most popular methods of discretizing a second-order stochastic process into a sequence of random variables (and hence is heavily used in numerically solving a differential equation driven by colored noise; see the discussion in Section 7.3) and is based on the spectral decomposition of the covariance function of the given stochastic process. Specifically, the Karhunen–Loève expansion of a second-order stochastic process $\{X(t) : t \in \mathbb{T}\}$ is given by

$$X(t) = \mathbb{E}(X(t)) + \sum_{j=1}^{\infty} \sqrt{\beta_j}\phi_j(t)Z_j. \tag{7.58}$$

Here β_j, $j = 1, 2, \ldots$, are the eigenvalues of $\mathrm{Cov}\{X(t), X(s)\}$ and they are assumed to be enumerated in non-increasing order of magnitude. More precisely, $\beta_1 \geq \beta_2 \geq \beta_3 \geq \ldots$. The functions ϕ_j, $j = 1, 2, \ldots$, denote the corresponding eigenfunctions that are mutually orthonormal in $L^2(\mathbb{T})$ with

$$\int_{\mathbb{T}} \phi_j(t)\phi_k(t)dt = \begin{cases} 1 & \text{if } j = k \\ 0 & \text{otherwise.} \end{cases} \tag{7.59}$$

Thus, the covariance function has the following spectral decomposition:

$$\text{Cov}\{X(t), X(s)\} = \sum_{j=1}^{\infty} \beta_j \phi_j(t)\phi_j(s). \qquad (7.60)$$

Here β_j and ϕ_j are solutions of the following *Fredholm equation*:

$$\int_{\mathbb{T}} \text{Cov}\{X(t), X(s)\}\phi_j(s)ds = \beta_j\phi_j(t), \quad t \in \mathbb{T} \qquad (7.61)$$

for $j = 1, 2, \ldots$, where (7.61) is obtained due to (7.59) and (7.60). It is worth noting that (7.60) is obtained by using Mercer's theorem [84] (see also Wikipedia) as well as the boundness, symmetric and non-negative definite properties (i.e., (7.6) and (7.7)) of the covariance function. The $\{Z_j\}$ in (7.58) are mutually uncorrelated random variables with zero mean and unit variance, and $Z_j, j = 1, 2, \ldots$, are related to the given stochastic process by

$$\sqrt{\beta_j} Z_j = \int_{\mathbb{T}} (X(t) - \mathbb{E}(X(t))\phi_j(t)dt, \quad j = 1, 2, 3, \ldots.$$

In practice, one adopts a finite series expansion for the given stochastic process

$$X(t) \approx X^m(t) = \mathbb{E}(X(t)) + \sum_{j=1}^{m} \sqrt{\beta_j} \phi_j(t) Z_j.$$

With this approximation, we observe that the corresponding mean square error is given by

$$\int_{\mathbb{T}} \mathbb{E}\left((X(t) - X^m(t))^2\right) dt = \sum_{j=m+1}^{\infty} \beta_j.$$

Thus, for a fixed m the truncation error is completely determined by the decay rate of the eigenvalues. Theoretically, if the given stochastic process is correlated, then the eigenvalues β_j decrease as j increases. For example, for a stochastic process with correlation function given by (7.57), the decay rate of the eigenvalues depends inversely on the correlation time τ; that is, a large value of τ results in a fast decay of the eigenvalues while a small value of τ results in a slow decay of the eigenvalues. This implies that a stochastic process with a longer correlation time requires a smaller number of random variables to accurately represent it. For more information on the Karhunen–Loève expansion, interested readers are referred to [54, Section 2.3], [65], [80, Chapter XI], and the references therein.

7.1.7.3.3 Dichotomous Markovian noise

The dichotomous Markovian noise $\{\text{DMN}(t) : t \geq 0\}$ is a continuous time Markov chain with two states (a^- and a^+) and with constant transition rates (λ^- and λ^+) between these two

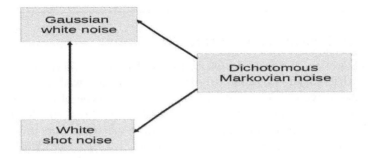

FIGURE 7.4:   A illustration of relationship among dichotomous Markovian noise, Gaussian white noise and white shot noise.

states. This was first introduced into information theory as *random telegraph noise*. Due to its simplicity and good approximation to many frequently encountered physical situations, dichotomous Markovian noise is another widely studied and used colored noise in both theory and practice (e.g., see a recent review [35] on this topic).

Dichotomous Markovian noise has a number of interesting and important properties. For example, its correlation function is an exponentially decreasing function given by

$$\text{Cor}\{\text{DMN}(t), \text{DMN}(s)\} = \frac{\mathcal{D}}{\tau} \exp\left(-\frac{|t - s|}{\tau}\right), \qquad (7.62)$$

where $\tau = \dfrac{1}{\lambda^- + \lambda^+}$, and $\mathcal{D} = -a^+ a^- \tau$. Perhaps the most interesting thing about dichotomous Markovian noise is that it has a close relationship with Gaussian white noise and white shot noise (illustrated in Figure 7.4). Specifically, symmetric dichotomous Markovian noise, the one with $a^+ = -a^- = a$ and $\lambda^- = \lambda^+ = \lambda$, reduces to a Gaussian white noise with noise intensity $\mathcal{D} = \dfrac{a^2}{2\lambda}$ in the limit $a \to \infty$ and $\lambda \to \infty$. Asymmetric dichotomous Markovian noise with $a^+/\lambda^+ + a^-/\lambda^- = 0$ reduces to a white shot noise (7.55) in the limit $a^+ \to \infty$ and $\lambda^+ \to \infty$ with constant ratio a^+/λ^+, where the average time between two successive pulses of the resulting white shot noise is $1/\lambda^-$, and the weights of pulses are independent and identically exponentially distributed random variables with mean a^+/λ^+. In addition, if $\dfrac{a^+}{\lambda^+} \to 0$ and $\lambda^- \to \infty$ with $\lambda^- \left(\dfrac{a^+}{\lambda^+}\right)^2$ remaining as a constant, then the resulting white shot noise reduces to Gaussian white noise. The interested reader can refer to [36] for details.

7.2   Stochastic Differential Equations

In this section, we consider stochastic differential equations (recall that they are differential equations driven by white noise) with particular focus on the ones driven by Gaussian white noise. This type of stochastic differential equation has been extensively studied in the literature for both theoretical and computational analysis, including existence and uniqueness of solutions, numerical methods, state estimation (often called *filtering*) techniques and stochastic control studies (e.g., see some relatively recent books [51, 52, 67, 72, 73, 89]). Interested readers can refer to [56, 93, 95] and the references therein for information on stochastic differential equations driven by Poisson white noise as well as their applications in different fields.

We consider the following form of stochastic differential equation (SDE) driven by Gaussian white noise:

$$d\mathbf{X}(t) = \mathbf{g}(t, \mathbf{X}(t))dt + \boldsymbol{\sigma}(t, \mathbf{X}(t))d\mathbf{W}(t), \quad \mathbf{X}(0) = \mathbf{X}_0, \tag{7.63}$$

where $\mathbf{X} = (X_1, X_2, \ldots, X_n)^T$, $\mathbf{g} = (g_1, g_2, \ldots, g_n)^T$ is a non-random n-dimensional function of t and $\mathbf{x}$, $\boldsymbol{\sigma}$ is a non-random $n \times l$ matrix function of t and $\mathbf{x}$, $\{\mathbf{W}(t) : t \geq 0\}$ is an l-dimensional Wiener process independent of the n-dimensional random initial vector $\mathbf{X}_0$, which is assumed to have a finite second moment. The solution to (7.63) satisfies the stochastic integral equation

$$\mathbf{X}(t) = \mathbf{X}(0) + \int_0^t \mathbf{g}(s, \mathbf{X}(s))ds + \int_0^t \boldsymbol{\sigma}(s, \mathbf{X}(s))d\mathbf{W}(s),$$

where the first integral on the right-hand side can be interpreted as a path-by-path Riemann integral, and the second one is a stochastic integral that can be interpreted as an Itô integral (as we discussed earlier). However, in physics and engineering literature, the second integral is often interpreted as a Stratonovich integral, which was introduced in 1966 by the Russian physicist Stratonovich [108], and is defined in such a way that it obeys the basic rules of classical calculus, such as the chain rule and integration by parts. It should be noted that the stochastic process generated by these integrals is not a martingale. Below we use a one-dimensional case to demonstrate the Stratonovich integral.

Definition 7.2.1 *Let t_0 and t_f be some real numbers, and the real-valued function $h : [t_0, t_f] \times \mathbb{R} \to \mathbb{R}$ be continuous in t and $\dfrac{\partial h}{\partial x}$ continuous in both t and x. Assume that the process $\{h(t, X(t)) : t \in [t_0, t_f]\}$ satisfies $\displaystyle\int_{t_0}^{t_f} \mathbb{E}(h^2(t, X(t)))dt < \infty$. Consider the partitions $\{t_0 < t_1 < \ldots t_l = t_f\}$ of* *the interval* $[t_0, t_f]$ *with mesh size* $\Delta_t = \max\limits_{0 \le j \le l-1} \{t_{j+1} - t_j\}$. *The* Stratonovich integral *is defined as*

$$\int_{t_0}^{t_f} h(t, X(t)) \circ dW(t)$$

$$= \underset{\Delta_t \to 0}{l.i.m.} \sum_{j=1}^{l} h\left(t_{j-1}, \tfrac{1}{2}(X(t_j) + X(t_{j-1}))\right) (W(t_j) - W(t_{j-1})),$$

where $\circ$ *indicates that this stochastic integral is interpreted in the* Stratonovich *sense.*

We see from the above definition that the intermediate point in the Stratonovich integral is chosen as the average of the left endpoint and right endpoint, $\tfrac{1}{2}(X(t_j) + X(t_{j-1}))$. Recall that the intermediate point in the Itô integral is chosen at the left endpoint $X(t_{j-1})$. Thus the stochastic integral is different from the Riemann–Stieltjes integral, as the choice of intermediate point has an effect on the properties of the stochastic integral: the Stratonovich integral obeys the classical chain rule and integration by parts while the Itô integral does not; the stochastic process generated by Itô integrals is a martingale while the one generated by Stratonovich integrals is not. Since both the Itô integral and the Stratonovich integral have their advantages and disadvantages, there is a long history of debate on which integral is more reasonable to use. In some situations, it is easy to determine which one to use. But there are situations where it is difficult to make the decision. Fortunately, one is able to write one integral in terms of the other one (as we shall see later). Interested readers can refer to [52, Chapter 3] and [89, Chapter 3] as well as the references therein for more information on this topic.

In this monograph, a stochastic differential equation with the stochastic integral interpreted as an Itô integral is referred to as an *Itô stochastic differential equation*, while the one with the stochastic integral interpreted as a Stratonovich integral is referred to as a *Stratonovich stochastic differential equation*. We will consider the stochastic differential equation (7.63) interpreted in both senses.

7.2.1   Itô Stochastic Differential Equations

In this section, we consider the stochastic differential equation (7.63) interpreted in the Itô sense, and we assume that the functions involved in this section are sufficiently smooth so that all the differentiations carried out are valid. Itô stochastic differential equations (introduced by Japanese mathematician K. Itô in the 1940s) are often found useful in financial mathematics, as one only takes into account information about the past (for example, in modeling stock price, the only information one has is about past events).

Itô stochastic differential equations have also been used in describing mechanisms of climate variability. For example, SDEs were used in [111] to describe the interactions between the atmosphere temperature and the ocean's surface temperature.

Existence and uniqueness of solutions   We assume that $\mathbf{g}$ and $\boldsymbol{\sigma}$ satisfy conditions guaranteeing the existence and uniqueness of solutions to the initial value problem (7.63); that is, $\mathbf{g}$ and $\boldsymbol{\sigma}$ satisfy the following two conditions (e.g., see [51, Chapter 5, Theorem 1.1] or [89, Theorem 5.2.1] for details):

(a) affine growth condition: there exists a positive constant c_l such that for all $t \in [0, t_f]$, and $\mathbf{x} \in \mathbb{R}^n$ we have

$$\|\mathbf{g}(t, \mathbf{x})\| + \|\boldsymbol{\sigma}(t, \mathbf{x})\| \leq c_l(1 + \|\mathbf{x}\|),$$

(b) global Lipschitz continuity: there exists a positive constant c_g such that for all $t \in [0, t_f]$, and $\mathbf{x}, \mathbf{y} \in \mathbb{R}^n$ we have

$$\|\mathbf{g}(t, \mathbf{x}) - \mathbf{g}(t, \mathbf{y})\| + \|\boldsymbol{\sigma}(t, \mathbf{x}) - \boldsymbol{\sigma}(t, \mathbf{y})\| \leq c_g(\|\mathbf{x} - \mathbf{y}\|).$$

Here $\|\mathbf{x}\|^2 = \sum_{j=1}^{n} x_j^2$, and $\|\boldsymbol{\sigma}\|^2 = \sum_{k=1}^{l} \sum_{j=1}^{n} \sigma_{jk}^2$, where σ_{jk} is the (j, k)th element of $\boldsymbol{\sigma}$. It is worth noting here that the affine growth condition is used to guarantee the existence of the solution on the entire interval $[0, t_f]$, and the global Lipschitz continuity condition is used to ensure the uniqueness of the solution. Here the uniqueness means *pathwise uniqueness*; that is, if both $\mathbf{X}(t)$ and $\mathbf{Y}(t)$ satisfy (7.63), then stochastic processes $\{\mathbf{X}(t) : t \in [0, t_f]\}$ and $\{\mathbf{Y}(t) : t \in [0, t_f]\}$ are indistinguishable. It should be noted that just as in the case of a system of ordinary differential equations, the global Lipschitz continuity condition can be relaxed to a local Lipschitz continuity (e.g., see [51, Chapter 5, Theorem 2.2] or [72, Theorem 6.22]):

(b′) local Lipschitz continuity: for each $\mathcal{K}$, there exists a positive constant $c_{\mathcal{K}}$ such that for all $t \in [0, t_f]$, and $\|\mathbf{x}\|, \|\mathbf{y}\| \leq \mathcal{K}$ we have

$$\|\mathbf{g}(t, \mathbf{x}) - \mathbf{g}(t, \mathbf{y})\| + \|\boldsymbol{\sigma}(t, \mathbf{x}) - \boldsymbol{\sigma}(t, \mathbf{y})\| \leq c_{\mathcal{K}}(\|\mathbf{x} - \mathbf{y}\|).$$

Solution of a stochastic differential equation as a diffusion process
It is well known that the solution to the Itô stochastic differential equation (7.63) is a diffusion process with drift coefficient $\mathbf{g}$ and diffusion coefficient $\boldsymbol{\sigma}\boldsymbol{\sigma}^T$ (e.g., see [52, Theorem 3.10] or [101, Section 4.3]). Let $\rho(t, \cdot\, ; s, \zeta)$ denote the transition probability density function of $\mathbf{X}(t)$ given $\mathbf{X}(s) = \zeta$, where

$0 \leq s < t$ and $\boldsymbol{\zeta} \in \mathbb{R}^n$. Then we have

$$\lim_{\Delta t \to 0+} \frac{1}{\Delta t} \int_{\mathbb{R}^n} (\mathbf{x} - \boldsymbol{\zeta}) \rho(t + \Delta t, \mathbf{x}; t, \boldsymbol{\zeta}) d\mathbf{x} = \mathbf{g}(t, \boldsymbol{\zeta}),$$

$$\lim_{\Delta t \to 0+} \frac{1}{\Delta t} \int_{\mathbb{R}^n} (\mathbf{x} - \boldsymbol{\zeta})(\mathbf{x} - \boldsymbol{\zeta})^T \rho(t + \Delta t, \mathbf{x}; t, \boldsymbol{\zeta}) d\mathbf{x} = \boldsymbol{\Sigma}(t, \boldsymbol{\zeta}), \qquad (7.64)$$

$$\lim_{\Delta t \to 0+} \frac{1}{\Delta t} \int_{\mathbb{R}^n} \prod_{r=1}^{\kappa} (x_{j_r} - \zeta_{j_r}) \rho(t + \Delta t, \mathbf{x}; t, \boldsymbol{\zeta}) d\mathbf{x} = 0, \ \kappa \geq 3,$$

where $\boldsymbol{\Sigma} = \boldsymbol{\sigma}\boldsymbol{\sigma}^T$, and κ is some positive integer. By the last equation in (7.64) we see that, for a diffusion process, any higher than second-order differential moments are zero.

Transition probability density function and Kolmogorov's backward equation   If $\mathbf{g}$ and $\boldsymbol{\sigma}$ are further assumed to satisfy conditions guaranteeing the existence of fundamental solutions to *Kolmogorov's backward equation* (in the setting of diffusion processes) given by

$$\frac{\partial}{\partial s} w(s, \boldsymbol{\zeta}) + \sum_{k=1}^{n} g_k(s, \boldsymbol{\zeta}) \frac{\partial w(s, \boldsymbol{\zeta})}{\partial \zeta_k} + \frac{1}{2} \sum_{k,j=1}^{n} \Sigma_{kj}(s, \boldsymbol{\zeta}) \frac{\partial^2 w(s, \boldsymbol{\zeta})}{\partial \zeta_k \zeta_j} = 0 \quad (7.65)$$

with $\Sigma_{kj}(s, \boldsymbol{\zeta})$ being the (k, j)th element of matrix $\boldsymbol{\Sigma}(s, \boldsymbol{\zeta})$, then this fundamental solution is unique and is given by the transition probability density function $\rho(t, \mathbf{x}; s, \boldsymbol{\zeta})$ (e.g., see [68, Section 5.7]). This implies that for any fixed $(t, \mathbf{x})$, the transition probability density function $\rho(t, \mathbf{x}; s, \boldsymbol{\zeta})$ satisfies Kolmogorov's backward equation with terminal condition $\rho(t, \mathbf{x}; t, \boldsymbol{\zeta}) = \delta(\boldsymbol{\zeta} - \mathbf{x})$, where $0 \leq s \leq t$, and δ denotes the Dirac delta function. In addition, for any bounded continuous function $f : \mathbb{R}^n \to \mathbb{R}$, the function w given by

$$w(s, \boldsymbol{\zeta}) = \mathbb{E}(f(\mathbf{X}(t)) \mid \mathbf{X}(s) = \boldsymbol{\zeta}) = \int_{\mathbb{R}^n} f(\mathbf{x}) \rho(t, \mathbf{x}; s, \boldsymbol{\zeta}) d\mathbf{x}$$

satisfies Kolmogorov's backward equation (7.65) with the terminal condition given by $w(t, \boldsymbol{\zeta}) = f(\boldsymbol{\zeta})$. Thus the solution of Kolmogorov's backward equation with a given terminal condition can be represented by the conditional expectation of some proper stochastic process. Conversely, the conditional expectation of a stochastic process can be obtained by solving Kolmogorov's backward equation with an appropriate terminal condition.

We remark that (7.65) is called the *backward equation* because the differentiation is with respect to the backward variable $(s, \boldsymbol{\zeta})$. Kolmogorov's backward equation is found useful in numerous applications, including option pricing theory (where it is common to hold the time of expiration constant) and genetics (where one is often interested in the fact that the system reaches a certain final state, e.g., see [62, Section 4.4]). It should be noted that there is a corresponding Kolmogorov's backward equation for the initial value problem through a change of time variable (as we shall see below in Remark 7.4.4; also see [87, Chapter 4] for further details).

Note that if the initial probability density function of $\mathbf{X}_0$ is known, then the diffusion process $\{\mathbf{X}(t) : t \geq 0\}$ is uniquely defined by its transition probability density functions (recall that this is true for any Markov process). Hence, the second-order differential operator associated with Kolmogorov's backward equation (for convenience we write it in terms of the variables $(t, \mathbf{x})$)

$$\mathscr{L}_t \phi(\mathbf{x}) = \sum_{k=1}^{n} g_k(t, \mathbf{x}) \frac{\partial \phi(\mathbf{x})}{\partial x_k} + \frac{1}{2} \sum_{k,j=1}^{n} \Sigma_{kj}(t, \mathbf{x}) \frac{\partial^2 \phi(\mathbf{x})}{\partial x_k x_j}, \tag{7.66}$$

is often called the *generator* of the diffusion process $\{\mathbf{X}(t) : t \geq 0\}$. This is also referred to as *Kolmogorov's backward operator*.

Remark 7.2.1 *Before we move on to the discussion of the evolution equation of the probability density function for the solution to the Itô SDE (7.63), we would like to comment on one important stochastic process, the Ornstein–Uhlenbeck process, which satisfies the Itô stochastic differential equation*

$$dX(t) = -aX(t)dt + \sigma_0 dW(t), \quad X(0) = X_0.$$

Here a and σ_0 are both constants, and a is assumed to be positive. In addition, X_0 is assumed to be normally distributed with mean zero and variance $\frac{\sigma_0^2}{2a}$, i.e., $\mathcal{N}\left(0, \frac{\sigma_0^2}{2a}\right)$. Then $\{X(t) : t \geq 0\}$ is a Gaussian process with zero mean and exponentially decreasing correlation function $Cor\{X(t), X(s)\}$ given by (e.g., see [68, Section 5.6])

$$Cor\{X(t), X(s)\} = \frac{\sigma_0^2}{2a} \exp\left(-a|t-s|\right). \tag{7.67}$$

By (7.52) and (7.67), it can be easily found that the corresponding spectral density function is given by

$$S(\omega) = \frac{\sigma_0^2}{\omega^2 + a^2}. \tag{7.68}$$

Hence, by (7.53) we know that the correlation time τ and the noise intensity $\mathcal{D}$ of the Ornstein–Uhlenbeck process are respectively given by $\tau = 1/a$ and $\mathcal{D} = \sigma_0^2/(2a^2)$. Thus we can rewrite (7.67) as

$$Cor\{X(t), X(s)\} = \frac{\mathcal{D}}{\tau} \exp\left(-\frac{|t-s|}{\tau}\right).$$

By the above equation and (7.62) we see that the Ornstein–Uhlenbeck process has the same form of correlation function as dichotomous Markov noise. It should be noted that the Ornstein–Uhlenbeck process is the only process that is simultaneously Gaussian, Markovian and stationary (e.g., see [72, p. 172] or [95, Section 2.5]—this follows from a result due to Doob), and it is often referred to as the Markov Gaussian colored noise.

7.2.1.1 Evolution of the Probability Density Function of $\mathbf{X}(t)$

Theorem 7.2.2 *Assume that $\mathbf{X}_0$ has probability density function p_0, and $\mathbf{X}(t)$ satisfies (7.63). Then the probability density function of $\mathbf{X}(t)$ satisfies*

$$\frac{\partial}{\partial t} p(t, \mathbf{x}) + \sum_{k=1}^{n} \frac{\partial}{\partial x_k}(g_k(t, \mathbf{x})p(t, \mathbf{x})) = \frac{1}{2} \sum_{k,j=1}^{n} \frac{\partial^2}{\partial x_k \partial x_j} [\Sigma_{kj}(t, \mathbf{x})p(t, \mathbf{x})]$$

$$(7.69)$$

with initial condition $p(0, \mathbf{x}) = p_0(\mathbf{x})$.

Equation (7.69) is often referred to as the *Fokker–Planck equation* (first used by Fokker and Planck to describe the Brownian motion of particles) or *Kolmogorov's forward equation* (rigorously developed by Kolmogorov in 1931). The Fokker–Planck equation is, of course, important in the fields of chemistry and physics. For example, the statistics of laser light and Brownian motion in potentials (important in solid-state physics, chemical physics and electric circuit theory) can both be readily treated with the Fokker–Planck equation (e.g., see [96]). In addition, the Fokker–Planck equation includes the well-known *Klein–Kramers equation* (or Kramers' equation, used to describe particle movement in position and velocity space) and *Smoluchowski's equation* (describing particle position distribution) as special cases.

There are several methods that can be used to derive the Fokker–Planck equation (7.69). One method is based on the principle of preservation of probability (see [40]) given by

$$\frac{d}{dt} \int_{\Omega_t} p(t, \mathbf{x}) d\mathbf{x} = 0,$$

$$(7.70)$$

where Ω_t denotes the region of the state space at time t. We remark that the rate of change of the integral of $p(t, \mathbf{x})$ (i.e., the left side of (7.70)) has two components: one corresponding to the rate of change of $p(t, \mathbf{x})$ in a given region Ω_t, and the other corresponding to the convective and diffusive transfer through the surface of Ω_t. Thus, the Fokker–Planck equation (7.69) can also be written in the form of a local conservation equation

$$\frac{\partial}{\partial t} p(t, \mathbf{x}) + \sum_{k=1}^{n} \frac{\partial}{\partial x_k} \left[g_k(t, \mathbf{x})p(t, \mathbf{x}) - \frac{1}{2} \sum_{j=1}^{n} \frac{\partial}{\partial x_j}(\Sigma_{kj}(t, \mathbf{x})p(t, \mathbf{x})) \right] = 0.$$

Another method involves using conditions (7.64) and employing expansion arguments as those in Moyal [86, Section 8.1] or in [90, Section 5.3.2] (for (7.69) with $n = 1$) to derive the Fokker–Planck equation (7.69). Since the arguments involved could be used to obtain more general results (which have far more widespread applications), we will present the derivation here.

PROOF   Let Δt be any positive number. Then by (7.23) we have

$$p(t + \Delta t, \mathbf{x}) = \int_{\mathbb{R}^n} p(t, \boldsymbol{\zeta}) \rho(t + \Delta t, \mathbf{x}; t, \boldsymbol{\zeta}) d\boldsymbol{\zeta}. \tag{7.71}$$

The characteristic function $\Pi(t + \Delta t, \cdot; t, \boldsymbol{\zeta})$ associated with $\rho(t + \Delta t, \cdot; t, \boldsymbol{\zeta})$ is defined by

$$\Pi(t + \Delta t, \boldsymbol{\varsigma}; t, \boldsymbol{\zeta}) = \int_{\mathbb{R}^n} \rho(t + \Delta t, \mathbf{x}; t, \boldsymbol{\zeta}) \exp(i\boldsymbol{\varsigma}^T (\mathbf{x} - \boldsymbol{\zeta})) d\mathbf{x} \tag{7.72}$$

$$\rho(t + \Delta t, \mathbf{x}; t, \boldsymbol{\zeta}) = \frac{1}{(2\pi)^n} \int_{\mathbb{R}^n} \Pi(t + \Delta t, \boldsymbol{\varsigma}; t, \boldsymbol{\zeta}) \exp(-i\boldsymbol{\varsigma}^T (\mathbf{x} - \boldsymbol{\zeta})) d\boldsymbol{\varsigma}, \tag{7.73}$$

where $\boldsymbol{\varsigma} \in \mathbb{R}^n$ and i is the imaginary unit. Expanding $\exp(i\boldsymbol{\varsigma}^T (\mathbf{x} - \boldsymbol{\zeta}))$ in terms of its argument we have

$$\exp(i\boldsymbol{\varsigma}^T(\mathbf{x}-\boldsymbol{\zeta})) = 1 + \sum_{k=1}^{\infty} \frac{i^k}{k!} \left[\sum_{j_1=1}^{n} \cdots \sum_{j_k=1}^{n} \varsigma_{j_1} \cdots \varsigma_{j_k} (x_{j_1} - \zeta_{j_1}) \cdots (x_{j_k} - \zeta_{j_k}) \right]. \tag{7.74}$$

Let $e_{j_1,\cdots,j_k}(t, \boldsymbol{\zeta}, \Delta t) = \int_{\mathbb{R}^n} (x_{j_1} - \zeta_{j_1}) \cdots (x_{j_k} - \zeta_{j_k}) \rho(t + \Delta t, \mathbf{x}; t, \boldsymbol{\zeta}) d\mathbf{x}$, where $j_1, j_2, \ldots, j_k$ are positive integers. Then substitution of (7.74) into (7.72) provides

$$\Pi(t + \Delta t, \boldsymbol{\varsigma}; t, \boldsymbol{\zeta}) = 1 + \sum_{k=1}^{\infty} \frac{i^k}{k!} \left[\sum_{j_1=1}^{n} \cdots \sum_{j_k=1}^{n} \varsigma_{j_1} \cdots \varsigma_{j_k} e_{j_1,\cdots,j_k}(t, \boldsymbol{\zeta}, \Delta t) \right]. \tag{7.75}$$

Note that

$$\frac{1}{(2\pi)^n} \int_{\mathbb{R}^n} i^k \varsigma_{j_1} \cdots \varsigma_{j_k} \exp(-i\boldsymbol{\varsigma}^T (\mathbf{x} - \boldsymbol{\zeta})) d\boldsymbol{\varsigma} = (-1)^k \frac{\partial^k}{\partial x_{j_1} \cdots \partial x_{j_k}} \delta(\mathbf{x} - \boldsymbol{\zeta}),$$

where δ is the Dirac delta function. Hence, by substituting (7.75) into Equation (7.73), we obtain

$$\rho(t + \Delta t, \mathbf{x}; t, \boldsymbol{\zeta})$$

$$= \delta(\mathbf{x} - \boldsymbol{\zeta}) + \sum_{k=1}^{\infty} (-1)^k \frac{1}{k!} \left[\sum_{j_1=1}^{n} \cdots \sum_{j_k=1}^{n} \frac{\partial^k \delta(\mathbf{x} - \boldsymbol{\zeta})}{\partial x_{j_1} \cdots \partial x_{j_k}} e_{j_1,\cdots,j_k}(t, \boldsymbol{\zeta}, \Delta t) \right].$$

Substituting the above equation into (7.71), we find

$$p(t + \Delta t, \mathbf{x}) - p(t, \mathbf{x})$$

$$= \sum_{k=1}^{\infty} (-1)^k \frac{1}{k!} \left[\sum_{j_1=1}^{n} \cdots \sum_{j_k=1}^{n} \frac{\partial^k}{\partial x_{j_1} \cdots \partial x_{j_k}} \left(p(t, \mathbf{x}) e_{j_1, \cdots, j_k}(t, \mathbf{x}, \Delta t) \right) \right].$$

Dividing both sides by Δt and letting $\Delta t \to 0$, by (7.64) we obtain (7.69).
□

Remark 7.2.3 *We remark that if* **g**, *σ and the transition probability density function are further assumed to be sufficiently smooth, then for any fixed $(s, \boldsymbol{\zeta})$ the transition probability density function $\rho(t, \mathbf{x}; s, \boldsymbol{\zeta})$ satisfies the Fokker– Planck equation (e.g., see [68, Section 5.7]). Note that the probability density function p of* $\mathbf{X}(t)$ *can be written as*

$$p(t, \mathbf{x}) = \int_{\mathbb{R}^n} p_0(\boldsymbol{\zeta}) \rho(t, \mathbf{x}; 0, \boldsymbol{\zeta}) d\boldsymbol{\zeta}.$$

Hence, p satisfies the Fokker–Planck equation with initial condition $p(0, \mathbf{x}) = p_0(\mathbf{x})$. This is the method used in [53, Section 5.2] to derive the Fokker– Planck equation (7.69). In addition, we observe that the differential operator associated with the Fokker–Planck equation

$$\mathscr{L}_t^* \psi(\mathbf{x}) = -\sum_{k=1}^{n} \frac{\partial}{\partial x_k} (g_k(t, \mathbf{x}) \psi(\mathbf{x})) + \frac{1}{2} \sum_{k,j=1}^{n} \frac{\partial^2}{\partial x_k \partial x_j} [\Sigma_{kj}(t, \mathbf{x}) \psi(\mathbf{x})] \quad (7.76)$$

is the formal adjoint of Kolmogorov's backward operator $\mathscr{L}_t$ defined in (7.66); this operator will be referred to as Kolmogorov's forward operator in this note. Furthermore, we see that both Kolmogorov's backward operator and Kolmogorov's forward operator are associated with the transition probability density function of the diffusion process $\{\mathbf{X}(t) : t \geq 0\}$.

Remark 7.2.4 *Let $\{\mathbf{X}(t) : t \geq 0\}$ be an n-dimensional Markov process with transition probability density function $\rho(t, \mathbf{x}; s, \boldsymbol{\zeta})$. Assume that for any positive integer k,*

$$\lim_{\Delta t \to 0+} \frac{1}{\Delta t} \int_{\mathbb{R}^n} (x_{j_1} - \zeta_{j_1}) \cdots (x_{j_k} - \zeta_{j_k}) \rho(t + \Delta t, \mathbf{x}; t, \boldsymbol{\zeta}) d\mathbf{x}$$

exists, where $j_1, j_2, \ldots, j_k$ are positive integers. Then, by using the same expansion arguments as those in the proof for Theorem 7.2.2, we have

$$\frac{\partial}{\partial t} p(t, \mathbf{x}) = \sum_{k=1}^{\infty} (-1)^k \frac{1}{k!} \left[\sum_{j_1=1}^{n} \cdots \sum_{j_k=1}^{n} \frac{\partial^k}{\partial x_{j_1} \cdots \partial x_{j_k}} (\alpha_{j_1, \cdots, j_k}(t, \mathbf{x}) p(t, \mathbf{x})) \right],$$

$$(7.77)$$

where $p(t, \cdot)$ denotes the probability density function of $\mathbf{X}(t)$ at time t, and $\alpha_{j_1, \cdots, j_k}$ is given by

$$\alpha_{j_1, \cdots, j_k}(t, \boldsymbol{\zeta}) = \lim_{\Delta t \to 0+} \frac{1}{\Delta t} \int_{\mathbb{R}^n} (x_{j_1} - \zeta_{j_1}) \cdots (x_{j_k} - \zeta_{j_k}) \rho(t + \Delta t, \mathbf{x}; t, \boldsymbol{\zeta}) d\mathbf{x}$$

for $k = 1, 2, 3, \ldots$. The resulting equation (7.77) is often referred to as the Kramer–Moyal expansion and is an important tool in statistical physics. The corresponding coefficients $\alpha_{j_1, \cdots, j_k}$ are called Kramer–Moyal coefficients or the kth order differential moment. It is worth noting that there is a corresponding Kramer–Moyal backward expansion (e.g., see [96, Section 4.2]). In addition, the transition probability density function $\rho(t, \mathbf{x}; s, \boldsymbol{\zeta})$ satisfies the Kramer–Moyal expansion for fixed $(s, \boldsymbol{\zeta})$, and it also satisfies the Kramer–Moyal backward expansion for fixed $(t, \mathbf{x})$ (similar to the transition probability density function for the diffusion process). We observe that the Fokker–Planck equation is a special case of the Kramer–Moyal expansion with all the Kramer–Moyal coefficients vanishing except the ones with one index (i.e., $\alpha_1, \alpha_2, \ldots, \alpha_n$) and those with two indices (i.e., $\alpha_{11}, \alpha_{12}, \ldots, \alpha_{nn}$). It is tempting to generalize the Fokker–Planck equation with some finite-order differential moments higher than two. However, based on the so-called Pawula's theorem (e.g., see [62, Section 4.5] or [96, Section 4.3]), this generalized Fokker–Planck equation leads to a contradiction of the positivity of the distribution function. In other words, the expansion term in the right-hand side of (7.77) may stop after $k > 1$ (which, as we shall see below, leads to Liouville's equation), or after $k > 2$ (which leads to the Fokker–Planck equation), or may never stop (otherwise the distribution function has negative values). We remark that the Kramer–Moyal expansion truncated at $k \geq 3$ has been found useful in certain applications (e.g., see [96, Section 4.6] for details) even though the distribution function has negative values.

7.2.1.2   Applications of the Fokker–Plank Equation in Population Dynamics

In an early biological application [71], a scalar Fokker–Planck equation (i.e., (7.69) with $n = 1$) was employed to study the random fluctuation of gene frequencies in a natural population (where x denotes gene frequency). The Fokker–Planck equation has also been effectively used in the literature (e.g., see [33] and the references therein) to model the dispersal behavior of populations such as female cabbage root fly movement in the presence of Brassica odors, and the movement of flea beetles in cultivated collard patches.

The Fokker–Planck equation was also used to describe the *population density* (numbers per unit size) in a spatially homogeneous population where individuals are characterized by their size and the growth process. Movement from one size class to another is assumed to be a diffusion process

$\{X(t) : t \geq 0\}$ satisfying the Itô stochastic differential equation

$$dX(t) = g(t, X(t))dt + \sigma(t, X(t))dW(t). \tag{7.78}$$

Here $X(t)$ represents the size of an individual at time t, $g(t, x)$ is the average or mean growth rate of individuals with size x at time t, and $\sigma(t, x)$ represents the variability in the growth rate of individuals. As is commonly done in the literature, g and σ are assumed to be non-negative. This formulation, referred to as a *stochastic formulation*, is motivated by the assumption that environmental or emotional fluctuations can be the primary influencing factor on individual growth. For example, it is known that the growth rates of shrimp are affected by several environmental factors, such as temperature, dissolved oxygen level and salinity. Thus, for this formulation the growth uncertainty is introduced into the population by the stochastic growth of each individual. In addition, individuals with the same size at the same time have the same uncertainty in growth, and individuals also have the possibility to reduce their size during their growth period. This assumption on the growth process leads to the Fokker–Planck equation for the population density u in the case of absence of sink and source (no mortality or reproduction involved; no individuals flowing into or out of the system, that is, the system is closed). The equation with appropriate boundary and initial conditions is given by

$$\frac{\partial}{\partial t}u(t, x) + \frac{\partial}{\partial x}(g(t, x)u(t, x)) = \frac{1}{2}\frac{\partial^2}{\partial x^2}(\sigma^2(t, x)u(t, x)),$$

$$g(t, \underline{x})u(t, \underline{x}) - \frac{1}{2}\frac{\partial}{\partial x}(\sigma^2(t, x)u(t, x))|_{x=\underline{x}} = 0,$$

$$g(t, \bar{x})u(t, \bar{x}) - \frac{1}{2}\frac{\partial}{\partial x}(\sigma^2(t, x)u(t, x))|_{x=\bar{x}} = 0, \tag{7.79}$$

$$u(0, x) = u_0(x),$$

where $\underline{x}$ represents the minimum size of individuals, $\bar{x}$ denotes the maximum attainable size of individuals in a lifetime, and u_0 is the initial population density. We observe that the zero-flux boundary conditions at $\underline{x}$ and $\bar{x}$ in (7.79) provide a conservation law for this model. The zero-flux boundary condition at $\underline{x}$ is due to the fact that the system is closed and there is also no reproduction involved. The zero-flux boundary condition at $\bar{x}$ is because the system is closed and $\bar{x}$ is the maximum attainable size of individuals (and hence there is no individual movement across $\bar{x}$). Since there is no source or sink involved, the total number of individuals in the population is a constant and is given by $\int_{\underline{x}}^{\bar{x}} u_0(x)dx$. In a more general case where reproduction is involved, if we assume that individuals are born with the same size $\underline{x}$, then a proper boundary condition at $\underline{x}$ is

$$g(t, \underline{x})u(t, \underline{x}) - \frac{1}{2}\frac{\partial}{\partial x}(\sigma^2(t, x)u(t, x))|_{x=\underline{x}} = \int_{\underline{x}}^{\bar{x}} \beta(t, x)u(t, x)dx,$$

where $\beta(t, x)$ is the reproduction (birth) rate of individuals with size x at time t. We observe that for this case we have a non-local boundary condition at $\underline{x}$.

Next we consider the equation for a population density in which mortality is present. We observe that the density of a population with size x at time $t + \Delta t$ can be calculated by multiplying the probability that individuals with size ζ at time t will move to x in a time increment Δt by the density of a population with size ζ at time t and summing over all the values of ζ. That is,

$$u(t + \Delta t, x) = \int_{\mathbb{R}} u(t, \zeta)(1 - \Delta t d(t, \zeta))\rho(t + \Delta t, x; t, \zeta)d\zeta, \qquad (7.80)$$

where $\rho(t + \Delta t, x; t, \zeta)$ denotes the transition probability density function for the diffusion process $\{X(t) : t \geq 0\}$, and $d(t, x)$ is the mortality rate of individuals with size x at time t. Hence, if we assume that the mortality rate is uniformly bounded, then we can use similar expansion arguments to those in the proof for Theorem 7.2.2 to obtain

$$\frac{\partial u}{\partial t}(t, x) + \frac{\partial}{\partial x}(g(t, x)u(t, x)) + d(t, x)u(t, x) = \frac{1}{2}\frac{\partial^2}{\partial x^2}(\sigma^2(t, x)u(t, x)).$$
$$(7.81)$$

It is also of interest to derive (7.81) from a purely stochastic point of view. Let Z be an exponentially distributed random variable with mean 1 that is independent of the diffusion process $\{X(t) : t \geq 0\}$. Then we define the *killing time* (also referred to as *lifetime*) by (e.g., see [68, Section 5.7])

$$K_t = \inf\left\{t \geq 0 \;\middle|\; \int_0^t d(s, X(s))ds \geq Z\right\},$$

and the killed diffusion process $\tilde{X}(t)$ by

$$\tilde{X}(t) = \begin{cases} X(t), & \text{if } t < K_t \\ \Delta, & \text{if } t \geq K_t, \end{cases} \qquad (7.82)$$

where Δ is the "coffin" (or "cemetery") state which is not in $\mathbb{R}$. Then $\{\tilde{X}(t) : t \geq 0\}$ is also a Markov process, and the associated second-order differential operator is given by

$$\mathscr{L}_t\phi(x) = \frac{1}{2}\sigma^2(t, x)\frac{\partial^2\phi(x)}{\partial x^2} + g(t, x)\frac{\partial\phi(x)}{\partial x} - d(t, x)\phi(x).$$

It is worth noting that the above differential operator is related to the famous Feynman–Kac formula (e.g., see [87, Section 4.1] for details), which is a stochastic representation of the solution of some partial differential equation (actually, a generalized Kolmogorov's backward equation with killing involved). Note that the formal adjoint $\mathscr{L}_t^*$ of $\mathscr{L}_t$ is given by

$$\mathscr{L}_t^*\varphi(x) = \frac{1}{2}\frac{\partial^2}{\partial x^2}(\sigma^2(t, x)\varphi(x)) - \frac{\partial}{\partial x}(g(t, x)\varphi(x)) - d(t, x)\varphi(x).$$

Hence, the population density function u satisfies

$$\frac{\partial u}{\partial t}(t, x) = \mathscr{L}_t^* u(t, x),$$

which is just (7.81). Thus, Equation (7.81) results from an assumed killed diffusion process.

We can generalize (7.81) to describe the dynamics of a population which is structured by multiple structure variables $\mathbf{x} \in \mathbb{R}^n$ with a growth process satisfying the Itô stochastic differential equation (7.63). Let $u(t, \mathbf{x})$ denote the population density of individuals with structure levels $\mathbf{x}$ at time t. Then u satisfies

$$\frac{\partial}{\partial t} u(t, \mathbf{x}) + \sum_{k=1}^{n} \frac{\partial}{\partial x_k} (g_k(t, \mathbf{x}) u(t, \mathbf{x})) + d(t, \mathbf{x}) u(t, \mathbf{x})$$
$$= \frac{1}{2} \sum_{k,j=1}^{n} \frac{\partial^2}{\partial x_k \partial x_j} [\Sigma_{kj}(t, \mathbf{x}) u(t, \mathbf{x})]. \tag{7.83}$$

In this monograph Equation (7.83) with appropriate boundary and initial conditions along with its special cases with non-zero diffusion coefficients will be referred to as a *Fokker–Planck physiologically structured (FPPS) population model*. In light of the above discussion, we see that FPPS population models can be associated with a stochastic process resulting from the stochasticity of growth rates of each individual along with the variability in the initial structure level of individuals across the population.

Fokker–Planck physiologically structured population models have been investigated by a number of researchers. For example, a scalar FPPS population model with a non-linear and non-local left boundary condition was studied in [42] to understand how a constant diffusion coefficient influences the stability and the Hopf bifurcation of the positive equilibrium of the system. In another example inverse problems were considered in [33] for the estimation of temporally and spatially varying coefficients in a scalar FPPS population model with a non-local left boundary condition.

7.2.2   Stratonovich Stochastic Differential Equations

We next summarize some results for the *Stratonovich stochastic differential equation*

$$d\mathbf{X}(t) = \tilde{\mathbf{g}}(t, \mathbf{X}(t))dt + \boldsymbol{\sigma}(t, \mathbf{X}(t)) \circ d\mathbf{W}(t), \quad \mathbf{X}(0) = \mathbf{X}_0, \tag{7.84}$$

where $\mathbf{X} = (X_1, X_2, \ldots, X_n)^T$, $\tilde{\mathbf{g}} = (\tilde{g}_1, \tilde{g}_2, \ldots, \tilde{g}_n)^T$ is a non-random n-dimensional vector function of t and $\mathbf{x}$, $\boldsymbol{\sigma}$ is a non-random $n \times l$ matrix function of t and $\mathbf{x}$, $\{\mathbf{W}(t) : t \geq 0\}$ is an l-dimensional Wiener process independent of the random initial vector $\mathbf{X}_0$, and $\circ$ indicates that (7.84) is interpreted in the *Stratonovich* sense. Stratonovich stochastic differential equations provide a natural model for many engineering problems, as their solutions can be obtained as the limit of the solutions of differential equations driven by Gaussian colored noises of decreasing correlation time under appropriate scaling (e.g., see [89, Section 3.3], [109, 119, 120]). This is part of the reason why Stratonovich integrals are often employed in the physics and engineering literature.

Even though the Stratonovich integral is different from the Itô integral, it can be written in terms of the Itô integral. In other words, Stratonovich stochastic differential equations can be converted to Itô stochastic differential equations. Specifically, the condition for the Itô stochastic differential equation (7.63) to be equivalent to the Stratonovich stochastic differential equation (7.84) is given by (see [52, Section 3.4] or [108])

$$\tilde{g}_k(t, \mathbf{x}) = g_k(t, \mathbf{x}) - \frac{1}{2} \sum_{r=1}^{l} \sum_{j=1}^{n} \sigma_{jr}(t, \mathbf{x}) \frac{\partial \sigma_{kr}(t, \mathbf{x})}{\partial x_j}. \tag{7.85}$$

This is often referred to as the *Wong–Zakai correction* (or the *Wong–Zakai theorem*). Thus the diffusion process obtained by the Stratonovich stochastic differential equation has drift coefficient with the kth component defined by $\tilde{g}_k(t, \mathbf{x}) + \frac{1}{2} \sum_{r=1}^{l} \sum_{j=1}^{n} \sigma_{jr}(t, \mathbf{x}) \frac{\partial \sigma_{kr}(t, \mathbf{x})}{\partial x_j}$ instead of $\tilde{g}_k(t, \mathbf{x})$, but it still has diffusion coefficient $\mathbf{\Sigma}(t, \mathbf{x}) = \boldsymbol{\sigma}(t, \mathbf{x}) \boldsymbol{\sigma}^T(t, \mathbf{x})$. In addition, (7.85) implies that the difference between these two interpretations (Stratonovich vs. Itô) disappears if $\boldsymbol{\sigma}$ is independent of the space coordinate $\mathbf{x}$.

By Theorem 7.2.2 we obtain the following results on the evolution of the probability density function of the solution to the Stratonovich stochastic differential equation (7.84).

Theorem 7.2.5 *Assume that $\mathbf{X}_0$ has probability density function p_0, and $\mathbf{X}(t)$ satisfies (7.84). Then the probability density function of $\mathbf{X}(t)$ satisfies*

$$\frac{\partial}{\partial t} p(t, \mathbf{x}) + \sum_{k=1}^{n} \frac{\partial}{\partial x_k} \left[\left(\tilde{g}_k(t, \mathbf{x}) - \frac{1}{2} \sum_{r=1}^{l} \sum_{j=1}^{n} \sigma_{kr}(t, \mathbf{x}) \frac{\partial \sigma_{jr}(t, \mathbf{x})}{\partial x_j} \right) p(t, \mathbf{x}) \right]$$

$$= \frac{1}{2} \sum_{k=1}^{n} \frac{\partial}{\partial x_k} \left[\sum_{j=1}^{n} \Sigma_{kj}(t, \mathbf{x}) \frac{\partial}{\partial x_j} p(t, \mathbf{x}) \right]$$

$$\tag{7.86}$$

with initial condition $p(0, \mathbf{x}) = p_0(\mathbf{x})$.

7.3   Random Differential Equations

In this section, we consider random differential equations (recall that they are differential equations driven by colored noise or differential equations driven by both colored noise and white noise) with particular focus on the ones only driven by colored noise.

As in studying a stochastic process, there are two common ways that can be used to study random differential equations. One is using the mean square calculus approach, and the other is by the sample function approach. To illustrate these two approaches, we consider the following general random differential equation:

$$\dot{\mathbf{X}}(t) = \mathbf{g}(t, \mathbf{X}(t)), \quad \mathbf{X}(0) = \mathbf{X}_0. \tag{7.87}$$

Here the driven colored noise is suppressed for notational convenience, $\mathbf{X} = (X_1, X_2, \ldots, X_n)^T$, $\mathbf{g} = (g_1, g_2, \ldots, g_n)^T$ is a non-random n-dimensional function of t and $\mathbf{x}$, and $\mathbf{X}_0$ is an n-dimensional column random vector. In the framework of mean square calculus, $\mathbf{X}(t)$ is said to be a *mean square solution* of (7.87) if and only if for any $t \geq 0$ we have

$$\mathbf{X}(t) = \mathbf{X}_0 + \int_0^t \mathbf{g}(s, \mathbf{X}(s)) dt$$

where the integral on the right side of the above equation is understood to be an m.s. integral. If $\{\mathbf{g}(t, \mathbf{X}) : t \in \mathbb{T}\}$ is a second-order stochastic process and it satisfies the mean square Lipschitz condition given by

$$\max_{1 \leq j \leq n} \left[\mathbb{E}\left(g_j(t, \mathbf{X}) - g_j(t, \mathbf{Y})\right)^2 \right]^{1/2} \leq c(t) \max_{1 \leq j \leq n} \left[\mathbb{E}((X_j - Y_j)^2) \right]^{1/2}$$

with c a deterministic function of t that is integrable in the given time interval, then the initial value problem (7.87) has a unique mean square solution for any given $\mathbf{X}_0$ with finite second moment (see [104, Theorem 5.1.2] for details). However, as remarked in [104], the condition for the existence and uniqueness of the mean square solution is difficult to verify in practice, as in general it is difficult, if not impossible, to determine whether or not a function of a second-order stochastic process is still a second-order process. In addition, the Lipschitz condition is very restrictive.

Compared to the mean square calculus approach, the sample function approach is more intuitive. Specifically, for this approach we consider each realization of a random differential equation. Note that each realization of a random differential equation is just a deterministic differential equation (referred to as a *sample deterministic differential equation*), which is assumed to have a unique solution. Hence, the solution to a random differential equation is obtained as a collection of solution trajectories to the sample deterministic differential equations, and this collection of trajectories can be investigated with the tools of probability theory. In this framework, the obtained solution to the RDE is called the *stochastic sample solution* (see [104, Appendix A] for details). Hence, we see that this approach also has its limitation in practice, as one needs to consider all the sample deterministic differential equations.

It is worth noting that if (7.87) has a unique mean square solution and a unique stochastic sample solution, then these two solutions are the same stochastic process uniquely defined by the given RDE (see [104] for details). Overall, the theory for random differential equations is much less advanced than that for stochastic differential equations. For this note, we just assume that the random differential equation has a unique solution and focus on discussion of the equation for the probability density function of the solution.

Due in part to their wide applicability, random differential equations have enjoyed considerable research efforts on computational methods in the past decade. Widely used approaches include Monte Carlo methods, stochastic Galerkin methods and probabilistic collocation methods (also called stochastic collocation methods). Specifically, both Monte Carlo methods and probabilistic collocation methods seek to solve deterministic realizations of the given random differential equations (and thus both methods were developed in the spirit of the sample function approach). The difference between these two methods is the manner in which one chooses the "sampling" points. Monte Carlo methods are based on large sampling of the distribution of random input variables, while probabilistic collocation methods are based on quadrature rules (or sparse quadrature rules in high-dimensional space). Stochastic Galerkin methods are based on (generalized) polynomial chaos expansions, which express the unknown stochastic process by a convergent series of (global) orthogonal polynomials in terms of random input parameters. Interested readers can refer to [121] and the references therein for details.

It should be noted that for all the above mentioned computational methods for the RDE, the necessary first step is to decompose and truncate the colored noise by a finite linear combination of random variables (e.g., using a Karhunen–Loève expansion, as discussed in Section 7.1.7.3.2). Hence, in this section we consider systems governed by random differential equations in three cases. One is differential equations with random initial conditions (in Section 7.3.1). The second is differential equations with both random initial conditions and random model parameters characterized by a finite dimensional random vector (in Section 7.3.2). Finally, we consider systems driven by a correlated stochastic process (in Section 7.3.3). As we shall see below, the first two cases have wide applications.

7.3.1   Differential Equations with Random Initial Conditions

We first consider the evolution of the probability density function for the solution to the system of ordinary differential equations with random initial conditions

$$\dot{\mathbf{x}} = \mathbf{g}(t, \mathbf{x}), \quad \mathbf{x}(0) = \mathbf{X}_0, \tag{7.88}$$

where $\mathbf{x} = (x_1, x_2, \ldots, x_n)^T$, $\mathbf{g} = (g_1, g_2, \ldots, g_n)^T$ is an n-dimensional non-random vector function of t and $\mathbf{x}$, and $\mathbf{X}_0$ is an n-dimensional random vector.

Equation (7.88) is often referred to as a *crypto-deterministic* formulation (e.g., see [52, 86, 104]) and has proved useful in a wide number of applications, including classical statistical mechanics, statistical thermodynamics, kinetic theory and biosciences (e.g., see [104, Chapter 6] for a more detailed discussion). Depending on the application, the uncertainty in initial conditions could be classified as either epistemic uncertainty (e.g., the uncertainty is due to the measurement error) or aleatory uncertainty (e.g., individuals have different initial size, as will be discussed below).

7.3.1.1 Evolution of the Probability Density Function of x(t; X₀)

Theorem 7.3.1 *Assume that* $\mathbf{X}_0$ *has probability density function* p_0, *and* (7.88) *has a mean square solution* $\mathbf{x}(t; \mathbf{X}_0)$. *Then the probability density function of the solution* $\mathbf{x}(t; \mathbf{X}_0)$ *satisfies*

$$\frac{\partial}{\partial t} p(t, \mathbf{x}) + \sum_{k=1}^{n} \frac{\partial}{\partial x_k} (g_k(t, \mathbf{x}) p(t, \mathbf{x})) = 0 \tag{7.89}$$

with initial condition $p(0, \mathbf{x}) = p_0(\mathbf{x})$.

The above result is often referred to as *Liouville's "theorem,"* which is a key result in statistical mechanics and statistical thermodynamics. The resulting equation (7.89) is sometimes called *Liouville's equation*. Here we use the terminology "Liouville's theorem" and "Liouville's equation" interchangeably to refer to this *conservation law*. We remark that Liouville's equation (7.89) includes the equation for describing the evolution of the probability density function for a conservative Hamiltonian system (e.g., see [86, Section 1.3]) as a special case. We also observe that Liouville's equation with $n = 3$ is the *equation of continuity* in classical continuum mechanics, in which $p(t, \mathbf{x})$ denotes the density of the material at location $\mathbf{x}$ at time t, and $\mathbf{g}(t, \mathbf{x})$ denotes the velocity of a particle at location $\mathbf{x}$ at time t.

There are several methods that can be used to derive Liouville's equation. One is from a probabilistic point of view using the concept of characteristic functions (e.g., see [104, p. 147]) following the derivation of Kozin [77] from 1961, which we next give.

PROOF   For any fixed t, the characteristic function of $\mathbf{X}(t) \equiv \mathbf{x}(t; \mathbf{X}_0)$ is given by

$$\Pi(t, \varsigma) = \mathbb{E}\left(\exp\left(i\left(\varsigma^T \mathbf{X}(t)\right)\right)\right), \tag{7.90}$$

where $\varsigma = (\varsigma_1, \varsigma_2, \ldots, \varsigma_n)^T \in \mathbb{R}^n$, and i is the imaginary unit. Differentiating both sides of (7.90) with respect to t we obtain

$$\frac{\partial \Pi}{\partial t}(t, \varsigma) = \mathbb{E}\left(i\varsigma^T \dot{\mathbf{X}}(t) \exp\left(i\varsigma^T \mathbf{X}(t)\right)\right)$$

$$= i \sum_{k=1}^{n} \varsigma_k \mathbb{E}\left(g_k(t, \mathbf{X}(t)) \exp\left(i\varsigma^T \mathbf{X}(t)\right)\right).$$

Taking the inverse Fourier transform of both sides of the above equation we find

$$\frac{\partial}{\partial t} p(t, \mathbf{x}) = -\sum_{k=1}^{n} \frac{\partial}{\partial x_k}(g_k(t, \mathbf{x})p(t, \mathbf{x})),$$

which yields the desired result.   □

Similar to the Fokker–Planck equation, Liouville's equation can also be derived based on the principle of preservation of probability (e.g., see [100, pp. 363–364]) which posits

$$\frac{d}{dt} \int_{\Omega_t} p(t, \mathbf{x}) d\mathbf{x} = 0, \tag{7.91}$$

where Ω_t denotes the region of the state space at time t. (This method is called the *phase-space method* or *state-space method* in statistical physics.) In this case, the rate of the change of the integral of $p(t, \mathbf{x})$ (i.e., the left side of (7.91)) also has two components: one corresponding to the rate of change of $p(t, \mathbf{x})$ in a given region Ω_t, and the other corresponding to the convective transfer through the surface of Ω_t (recall that in the case of the Fokker–Planck equation the second component corresponds to both the convective and diffusive transfer through the surface of Ω_t).

We see that (7.88) is a special case of the Itô stochastic differential equation (7.63) with $\boldsymbol{\sigma} \equiv 0$. Hence, if we assume that $\mathbf{g}$ satisfies the conditions for the existence and uniqueness of solutions of (7.63), then we know that $\{\mathbf{X}(t) : t \geq 0\}$ with $\mathbf{X}(t) = \mathbf{x}(t; \mathbf{X}_0)$ is a diffusion process with drift coefficient $\mathbf{g}$ and the diffusion coefficient being the zero matrix. Let $\rho(t + \Delta t, \mathbf{x}; t, \boldsymbol{\zeta})$ denote the transition probability density function. Then we have

$$\lim_{\Delta t \to 0+} \frac{1}{\Delta t} \int_{\mathbb{R}^n} (\mathbf{x} - \boldsymbol{\zeta}) \rho(t + \Delta t, \mathbf{x}; t, \boldsymbol{\zeta}) d\mathbf{x} = \mathbf{g}(t, \boldsymbol{\zeta}),$$

$$\lim_{\Delta t \to 0+} \frac{1}{\Delta t} \int_{\mathbb{R}^n} \prod_{r=1}^{\kappa} (x_{j_r} - \zeta_{j_r}) \rho(t + \Delta t, \mathbf{x}; t, \boldsymbol{\zeta}) d\mathbf{x} = 0, \ \kappa \geq 2,$$

$$\tag{7.92}$$

where κ is a positive integer. By using the above conditions as well as the expansion arguments similar to those in the proof for Theorem 7.2.2 we can also derive Liouville's equation, which is a special case of the Kramer–Moyal expansion (7.77) with all the Kramer–Moyal coefficients vanishing except the ones with only one index (i.e., $\alpha_1, \alpha_2, \ldots, \alpha_n$).

7.3.1.2 Applications of Liouville's Equation in Population Dynamics

We observe that Liouville's equation (7.89) with $g_1 \equiv 1$ is a special case of a physiologically structured population model without mortality $(d \equiv 0)$ given by Oster and Takahashi in 1974 [91]:[1]

$$\frac{\partial}{\partial t}u(t,\mathbf{x}) + \frac{\partial}{\partial x_1}u(t,\mathbf{x}) + \sum_{j=2}^{n}\frac{\partial}{\partial x_j}(g_j(t,\mathbf{x})u(t,\mathbf{x})) + d(t,\mathbf{x})u(t,\mathbf{x}) = 0.$$

(7.93)

This model is used to describe the population density u in a spatially homogeneous population where individuals are characterized by states $\mathbf{x} = (x_1, x_2, \cdots, x_n)^T$ with x_1 denoting the chronological age and $x_2, \cdots, x_n$ representing some physiological variables such as mass, volume, chemical composition, and other quantities having an influence on individuals' growth and mortality rates. Here $(g_2(t,\mathbf{x}), g_3(t,\mathbf{x}), \ldots, g_n(t,\mathbf{x}))^T$ is the vector for the growth rates of individuals with states $\mathbf{x}$ at time t, and $d(t,\mathbf{x})$ is the mortality rate of individuals with states $\mathbf{x}$ at time t.

Equation (7.93) includes a number of well-known structured population models as special cases. Specifically, the model with only chronological age involved, i.e.,

$$\frac{\partial}{\partial t}u(t,x) + \frac{\partial}{\partial x}u(t,x) + d(t,x)u(t,x) = 0 \qquad (7.94)$$

is the *age-structured population model* given by McKendrick in 1926 [82] and Von Foerster in 1959 [112]. (Equation (7.94) is also referred to as the *McKendrick equation*, *the Lotka–McKendrick equation*, *the Lotka–Von Foerster model*, *the von Forester equation*, or the *McKendrick–Von Forester equation*. Interested readers are referred to [70] for an interesting survey about the priority issues.) Equation (7.93) with $n = 2$,

$$\frac{\partial}{\partial t}u(t,\mathbf{x}) + \frac{\partial}{\partial x_1}u(t,\mathbf{x}) + \frac{\partial}{\partial x_2}(g_2(t,\mathbf{x})u(t,\mathbf{x})) + d(t,\mathbf{x})u(t,\mathbf{x}) = 0, \quad (7.95)$$

is the age-size structured population model developed in 1967 by Sinko and Streifer [103] (with x_2 being some physiological variable such as size). It is also the model given at the same time by Bell and Anderson [34] for cell populations (with x_2 denoting the volume of the cell). Equation (7.95) without the second term, that is,

$$\frac{\partial}{\partial t}u(t,x) + \frac{\partial}{\partial x}(g(t,x)u(t,x)) + d(t,x)u(t,x) = 0, \qquad (7.96)$$

is often referred to as the *Sinko–Streifer model* or classical *size-structured population model*. Here x is the structure variable, which may represent weight,

---

[1]The connection between Liouville's equation and (7.93) was recognized by Oster and Takahashi in the case of zero mortality and constant growth rates g_i, $i = 2, \ldots, n$.

length, volume, chronological age (in this case, Equation (7.96) becomes the age-structured population model (7.94)), caloric content, maturity, label intensity, etc., depending on the specific applications (e.g., see [85]).

These physiologically structured population models with appropriate boundary and initial conditions along with their corresponding non-linear versions have been widely studied in the literature for both computational and theoretical analysis, including well-posedness studies, asymptotic analysis, sensitivity analysis, parameter estimation and control studies (e.g., [2, 3, 8, 17, 26, 27, 37, 43, 49, 64, 85, 116] and the references therein). In addition, they have been applied to many biological and medical applications, ranging from tumor models to cell level immunology to viral spread in populations to recent applications in cell proliferation (e.g., [9, 28, 98] and the references therein).

The derivation of the age-size structured population model (7.95) in [103] is essentially by conservation of total number of individuals in the population with population sinks included, i.e.,

$$\frac{d}{dt} \int_{\Omega_t} u(t, \mathbf{x}) d\mathbf{x} = - \int_{\Omega_t} d(t, \mathbf{x}) u(t, \mathbf{x}) d\mathbf{x}, \qquad (7.97)$$

where $\mathbf{x} = (x_1, x_2)^T$ and Ω_t denotes all possible states occupied by individuals at time t. It is also interesting to derive this model from a stochastic point of view. For simplicity, we take the size-structured population model (7.96) as an example. Note that the growth of each individual in the size-structured population model (7.96) follows the same deterministic law

$$\dot{x} = g(t, x), \qquad (7.98)$$

but individuals may start from different sizes at time $t = 0$. Hence, the size of an individual at time t can be viewed as a realization to the solution $x(t; X_0)$ of the following differential equation with a random initial condition:

$$\dot{x} = g(t, x), \quad x(0) = X_0. \qquad (7.99)$$

Thus, the solution $\{X(t) : t \geq 0\}$ with $X(t) = x(t; X_0)$ to (7.99) is a diffusion process with diffusion coefficient zero. Therefore we can use arguments similar to those in the proof for deriving Equation (7.81) to obtain (7.96).

Based on the above discussion, we see that all the linear physiologically structured population models presented in this section can be associated with some stochastic process, which results from the variability in the initial sizes of individuals in the population.

7.3.2   Differential Equations with Random Model Parameters and Random Initial Conditions

In this section we consider the system of random ordinary differential equations

$$\dot{\mathbf{x}} = \mathbf{g}(t, \mathbf{x}; \mathbf{Z}), \quad \mathbf{x}(0) = \mathbf{X}_0, \qquad (7.100)$$

where $\mathbf{x} = (x_1, x_2, \ldots, x_n)^T$, $\mathbf{g} = (g_1, g_2, \ldots, g_n)^T$ is an n-dimensional non-random vector function of t and $\mathbf{x}$, $\mathbf{Z}$ is an m-dimensional random vector, and $\mathbf{X}_0$ is an n-dimensional random vector.

Equation (7.100) is often found useful in the biosciences to account for the variation between individuals. For example, a system of ordinary differential equations is often used in the literature to model the HIV dynamics for any specific individual, but the model parameters vary across the population as the clinical data reveals a great deal of variability among HIV patients (e.g., see [4, 11, 21, 97]). In addition, (7.100) has wide applications in physics to account for heterogeneity of complex materials. For example, oscillator models such as the Debye model and the Lorentz model (discussed in Section 5.3.2) are often used to describe the polarization in a dielectric material, but there are now incontrovertible experimental arguments for distributions of relaxation times (parameters in the oscillator models) for complex materials (e.g., see [20] and the references therein). We remark that the uncertainty discussed in all the above cases is aleatory. It should be noted that Equation (7.100) can also be used to describe cases where uncertainty is epistemic. For example, as discussed in earlier chapters of this monograph, uncertainty arises from the parameter estimation error due to the noise in the measurements.

7.3.2.1   Evolution of the Joint Probability Density Function for $(\mathbf{x}(t; \mathbf{X}_0, \mathbf{Z}), \mathbf{Z})^T$

We consider the time evolution of the joint probability density function of $(\mathbf{x}(t; \mathbf{X}_0, \mathbf{Z}), \mathbf{Z})^T$. Let $\tilde{\mathbf{x}} = (\mathbf{x}, \mathbf{z})^T$. Then (7.100) can be rewritten as

$$\dot{\tilde{\mathbf{x}}} = \tilde{\mathbf{g}}(t, \tilde{\mathbf{x}}), \quad \tilde{\mathbf{x}}(0) = (\mathbf{X}_0, \mathbf{Z})^T. \tag{7.101}$$

Here $\tilde{\mathbf{g}}(t, \tilde{\mathbf{x}}) = \begin{bmatrix} \mathbf{g}(t, \mathbf{x}; \mathbf{z}) \\ \mathbf{0}_m \end{bmatrix}$, where $\mathbf{0}_m$ is an m-dimensional column vector with all the elements being zeros. Note that (7.101) is crypto-deterministic. Hence, Theorem 7.3.1 can be applied to (7.101) to obtain the following result on the evolution of the joint probability density function for the solution to (7.101).

Theorem 7.3.2 *Assume that (7.100) has a mean square solution* $\mathbf{x}(t; \mathbf{X}_0, \mathbf{Z})$. *Then the joint probability density function* $\tilde{\varphi}_{\mathbf{X}, \mathbf{Z}}$ *of* $\mathbf{x}(t; \mathbf{X}_0, \mathbf{Z})$ *and random vector* $\mathbf{Z}$ *satisfies*

$$\frac{\partial}{\partial t} \tilde{\varphi}_{\mathbf{X}, \mathbf{Z}}(t, \mathbf{x}, \mathbf{z}) + \sum_{k=1}^{n} \frac{\partial}{\partial x_k} (g_k(t, \mathbf{x}; \mathbf{z}) \tilde{\varphi}_{\mathbf{X}, \mathbf{Z}}(t, \mathbf{x}, \mathbf{z})) = 0 \tag{7.102}$$

with initial condition $\tilde{\varphi}_{\mathbf{X}, \mathbf{Z}}(0, \mathbf{x}, \mathbf{z}) = \tilde{\varphi}_{\mathbf{X}, \mathbf{Z}}^0(\mathbf{x}, \mathbf{z})$, *where* $\tilde{\varphi}_{\mathbf{X}, \mathbf{Z}}^0$ *is the joint probability density function of* $\mathbf{X}_0$ *and* $\mathbf{Z}$.

Equation (7.102) is sometimes referred to as the *Dostupov–Pugachev equation* (e.g., see [40]), as it seems to first appear in [46] by Dostupov and Pugachev in 1957. For the derivation of (7.102), Dostupov and Pugachev first observed that for a given realization $\mathbf{z}$ of $\mathbf{Z}$ the system (7.100) is basically the crypto-deterministic system (7.88), and they then derived Equation (7.102) through instantaneous transformation of random vectors.

It is of interest to note that we can employ arguments similar to those in the proof for Theorem 7.3.1 to (7.100) to obtain Equation (7.102). This proof was originally given in 1961 by Kozin [77] and is repeated here for completeness.

PROOF   For any fixed t, the joint characteristic function of $\mathbf{X}(t) \equiv \mathbf{x}(t; \mathbf{X}_0, \mathbf{Z})$ and $\mathbf{Z}$ is given by

$$\Pi(t, \varsigma_{\mathbf{x}}, \varsigma_{\mathbf{z}}) = \mathbb{E}\left(\exp\left(i\left(\varsigma_{\mathbf{x}}^T \mathbf{X}(t) + \varsigma_{\mathbf{z}}^T \mathbf{Z}\right)\right)\right), \qquad (7.103)$$

where $\varsigma_{\mathbf{x}} = (\varsigma_{\mathbf{x}1}, \varsigma_{\mathbf{x}2}, \ldots, \varsigma_{\mathbf{x}n})^T \in \mathbb{R}^n$, $\varsigma_{\mathbf{z}} \in \mathbb{R}^m$, and i is the imaginary unit. Differentiating both sides of (7.103) with respect to t we find

$$\frac{\partial \Pi}{\partial t}(t, \varsigma_{\mathbf{x}}, \varsigma_{\mathbf{z}}) = \mathbb{E}\left(i\varsigma_{\mathbf{x}}^T \dot{\mathbf{X}}(t) \exp\left(i\left(\varsigma_{\mathbf{x}}^T \mathbf{X}(t) + \varsigma_{\mathbf{z}}^T \mathbf{Z}\right)\right)\right)$$

$$= i \sum_{k=1}^{n} \varsigma_{\mathbf{x}k} \mathbb{E}\left(g_k(t, \mathbf{X}(t); \mathbf{Z}) \exp\left(i\left(\varsigma_{\mathbf{x}}^T \mathbf{X}(t) + \varsigma_{\mathbf{z}}^T \mathbf{Z}\right)\right)\right).$$

Taking the inverse Fourier transform of both sides of the above equation we obtain

$$\frac{\partial}{\partial t}\tilde{\varphi}_{\mathbf{X},\mathbf{Z}}(t, \mathbf{x}, \mathbf{z}) = -\sum_{k=1}^{n}\frac{\partial}{\partial x_k}(g_k(t, \mathbf{x}; \mathbf{z})\tilde{\varphi}_{\mathbf{X},\mathbf{Z}}(t, \mathbf{x}, \mathbf{z})),$$

which yields the desired result.   ☐

Since (7.101) is a crypto-deterministic formulation, it is a special case of a stochastic differential equation with all diffusion coefficients being zero. Thus one can also use the expansion arguments given above in deriving the Fokker–Planck equation (7.69) to obtain the Dostupov–Pugachev equation (7.102), which, of course, is a special case of the Kramer–Moyal expansion (7.77) with all the Kramer–Moyal coefficients vanishing except $\alpha_1, \alpha_2, \ldots, \alpha_n$.

Given the joint probability density function $\tilde{\varphi}_{\mathbf{X},\mathbf{Z}}$, we can obtain the probability density function for $\mathbf{x}(t; \mathbf{X}_0, \mathbf{Z})$ given by

$$p(t, \mathbf{x}) = \int_{\Omega_z} \tilde{\varphi}_{\mathbf{X},\mathbf{Z}}(t, \mathbf{x}, \mathbf{z})d\mathbf{z}, \qquad (7.104)$$

where Ω_z denotes the set of all possible values for $\mathbf{z}$.

7.3.2.2   Evolution of Conditional Probability Density Function of $\mathbf{x}(t; \mathbf{X}_0, \mathbf{Z})$ Given the Realization z of Z

Finally, we derive the conditional probability density function $\varphi(t, \cdot; \mathbf{z})$ of $\mathbf{x}(t; \mathbf{X}_0, \mathbf{Z})$ given the realization $\mathbf{z}$ of $\mathbf{Z}$ (as we shall see momentarily, this has applications in population dynamics). First we observe that

$$\varphi(t, \mathbf{x}; \mathbf{z}) = \frac{\tilde{\varphi}_{\mathbf{X}, \mathbf{Z}}(t, \mathbf{x}, \mathbf{z})}{\tilde{\varphi}_{\mathbf{Z}}(\mathbf{z})}, \tag{7.105}$$

where $\tilde{\varphi}_{\mathbf{Z}}$ denotes the probability density function of $\mathbf{Z}$. Hence by (7.102) and (7.105) we find φ satisfies

$$\frac{\partial}{\partial t}\varphi(t, \mathbf{x}; \mathbf{z}) + \sum_{k=1}^{n} \frac{\partial}{\partial x_k}(g_k(t, \mathbf{x}; \mathbf{z})\varphi(t, \mathbf{x}; \mathbf{z})) = 0, \tag{7.106}$$

with initial condition $\varphi(0, \mathbf{x}; \mathbf{z}) = \varphi_0(\mathbf{x}; \mathbf{z})$, where $\varphi_0(\cdot; \mathbf{z})$ is the probability density function of initial condition $\mathbf{X}_0$ given $\mathbf{Z} = \mathbf{z}$.

Note that for any given realization $\mathbf{z}$ of $\mathbf{Z}$ the system (7.100) is crypto-deterministic. Hence, by using Liouville's equation (7.89) the probability density function of solution $\mathbf{x}(t; \mathbf{X}_0, \mathbf{z})$ to (7.100) with given value $\mathbf{z}$ of $\mathbf{Z}$ satisfies (7.106). Thus, we can derive the equation for the time evolution of the conditional probability density function directly from Liouville's equation.

Given the conditional probability density function φ, the probability density function of $\mathbf{x}(t; \mathbf{X}_0, \mathbf{Z})$ is given by

$$p(t, \mathbf{x}) = \int_{\Omega_z} \varphi(t, \mathbf{x}; \mathbf{z})\tilde{\varphi}_{\mathbf{Z}}(\mathbf{z})d\mathbf{z}. \tag{7.107}$$

For the case where $\mathbf{Z}$ is a discrete random vector that takes l different values $\{\mathbf{z}_j\}_{j=1}^{l}$ with the probability associated with $\mathbf{z}_j$ given by ϖ_j, $j = 1, 2, \ldots, l$, by (2.15) we know that the probability density function $\tilde{\varphi}_{\mathbf{Z}}$ of $\mathbf{Z}$ can be written as

$$\tilde{\varphi}_{\mathbf{Z}}(\mathbf{z}) = \sum_{j=1}^{l} \varpi_j \delta(\mathbf{z} - \mathbf{z}_j). \tag{7.108}$$

Thus, for this case (7.107) can be simply written as

$$p(t, \mathbf{x}) = \sum_{j=1}^{l} \varpi_j \varphi(t, \mathbf{x}; \mathbf{z}_j).$$

7.3.2.3   Applications in Population Dynamics

Equation (7.106) for $n = 1$ with appropriate boundary and initial conditions along with (7.107) was used in [10, 12, 13, 14, 15, 18, 19] to describe the population density in a size-structured population, where the total population is partitioned into (possibly a continuum of) subpopulations with individuals in each subpopulation having the same time-size-dependent growth rate $g(\cdot, \cdot; \mathbf{z})$ (g is assumed to be non-negative for any given $\mathbf{z}$, as is commonly done in the literature). In other words, the growth process of individuals is characterized by the following random differential equation:

$$\dot{x} = g(t, x; \mathbf{Z}), \quad x(0) = X_0, \tag{7.109}$$

where x denotes the size of individuals. Hence, we can see that for this formulation the growth uncertainty is introduced into the entire population by the variability of growth rates among subpopulations, and each individual grows according to a deterministic growth model, but different individuals (even of the same size) may have different time-size-dependent growth rates. This formulation, referred to as a *probabilistic formulation*, is motivated by the fact that genetic differences or non-lethal infections of some chronic disease can have an effect on individual growth rates. For example, for many marine species such as mosquitofish, females grow faster than males, which means that individuals even with the same size may have different growth rates. In addition, it was reported in [38] that non-lethal infection of *Penaeus vannamei* postlarvae by IHHNV (infectious hypodermal and hematopoietic necrosis virus) may reduce growth rates and increase size variability.

Let $v(t, x; \mathbf{z})$ be the population density of individuals with size x at time t having growth rate $g(t, x; \mathbf{z})$. Then we see that the equation for $v(\cdot, \cdot; \mathbf{z})$ with proper boundary and initial conditions in the case of absence of sink and source is given by

$$\frac{\partial}{\partial t}v(t, x; \mathbf{z}) + \frac{\partial}{\partial x}(g(t, x; \mathbf{z})v(t, x; \mathbf{z})) = 0, \quad x \in (\underline{x}, \bar{x}), \quad t > 0$$

$$g(t, \underline{x}; \mathbf{z})v(t, \underline{x}; \mathbf{z}) = 0, \qquad\qquad\qquad\qquad (7.110)$$

$$v(0, x; \mathbf{z}) = v_0(x; \mathbf{z}),$$

where $\underline{x}$ denotes the minimum size of individuals (it is assumed to be the same among the subpopulations). Similar to the discussion in Section 7.2.1.2, the zero-flux boundary condition is due to the fact that the subpopulation system is closed and there is also no reproduction involved. We observe that one needs to set $g(t, \bar{x}; \mathbf{z})v(t, \bar{x}; \mathbf{z}) = 0$ in order to provide a conservation law for the above model. Specifically, if $\bar{x}$ denotes the maximum attainable size of individuals in a lifetime, then it is reasonable to set $g(t, \bar{x}; \mathbf{z}) = 0$ (as is commonly done in the literature). However, if we just consider the model in a short time period where the growth rate for all the individuals is still positive (that is, no individuals achieve the maximum attainable size yet), then one may choose $\bar{x}$ sufficiently large so that $v(t, \bar{x}; \mathbf{z})$ is negligible or zero if possible for any $\mathbf{z}$ and any time point in the given time interval. In a more general case where reproduction is involved, if we assume that individuals are born with the same size $\underline{x}$ and that individuals in a subpopulation can only give birth to a newborn that belongs to the same subpopulation, then a proper boundary condition is

$$g(t, \underline{x}; \mathbf{z})v(t, \underline{x}; \mathbf{z}) = \int_{\underline{x}}^{\bar{x}} \beta(t, x)v(t, x; \mathbf{z})dx.$$

Here $\bar{x}$ denotes the maximum attainable size of individuals in a lifetime, and $\beta(t, x)$ is again the reproduction (birth) rate of individuals with size x at time t.

Given the subpopulation density $v(t, x; \mathbf{z})$, the population density $u(t, x)$ for the total population with size x at time t is given by

$$u(t, x) = \int_{\Omega_z} v(t, x; \mathbf{z}) \tilde{\varphi}_{\mathbf{Z}}(\mathbf{z}) d\mathbf{z} = \int_{\Omega_z} v(t, x; \mathbf{z}) d\tilde{\Psi}_{\mathbf{Z}}(\mathbf{z}), \qquad (7.111)$$

where $\tilde{\Psi}_{\mathbf{Z}}$ denotes the cumulative distribution function of $\mathbf{Z}$. We thus see that (7.111) along with (7.110) is the motivating example given in Section 5.2, with $\tilde{\Psi}_{\mathbf{Z}}$ being the fundamental "parameter" to be determined from the aggregate data for the population.

For the case where $\mathbf{Z}$ is a discrete random vector with an associated probability density function described by (7.108), this model, (7.111) along with (7.110), is composed of l classical size-structured population models described by (7.110), with each model describing the dynamics of one subpopulation. This is the case that is often investigated in the literature (e.g., [1, 29, 30, 31]) to account for the heterogeneity of the population. For example, to account for the distinct cell proliferation properties among cells, the authors in [29, 30] divide the cell population into a finite number of subpopulations characterized by the number of cell divisions undergone by each subpopulation.

The model, (7.111) along with (7.110), was originally developed and studied in [10, 18] in connection with mosquitofish populations in rice fields (the data exhibits both bimodality and dispersion in size as time increases even though the population begins with a unimodal density (no reproduction involved)). This model was called the *growth rate distribution model* in [10, 18] and it was not formulated directly from a stochastic process. However, from the above discussion, it can be readily seen that this model in fact is associated with the stochastic process $\{x(t; X_0, \mathbf{Z}) : t \geq 0\}$, which is obtained through the variability $\mathbf{Z}$ in the individuals' growth rates along with the variability X_0 in the initial size of individuals in the population. This model has been successfully used to model the early growth of shrimp populations [14, 15], which exhibit a great deal of variability in size as time evolves even though the shrimp begin with approximately similar size (e.g., see Figure 5.1).

Now we consider this model in the presence of mortality. Note that the size of an individual at time t in a subpopulation having growth rate $g(t, x; \mathbf{z})$ can be viewed as a realization to the solution $X_{\mathbf{z}}(t) = x(t; X_{0\mathbf{z}}, \mathbf{z})$ of the following differential equation with random initial condition

$$\dot{x} = g(t, x; \mathbf{z}), \quad x(0) = X_{0\mathbf{z}}, \qquad (7.112)$$

where $X_{0\mathbf{z}}$ denotes the initial size of individuals in this subpopulation having growth rate $g(t, x; \mathbf{z})$. Hence, $\{X_{\mathbf{z}}(t) : t \geq 0\}$ is a diffusion process with the diffusion coefficient being zero. Thus we can use similar arguments as those in the proof for deriving Equation (7.81) to obtain the equation for the population density $v(t, x; \mathbf{z})$ in a subpopulation with growth rate $g(t, x; \mathbf{z})$ (see, e.g., [32]):

$$\frac{\partial}{\partial t} v(t, x; \mathbf{z}) + \frac{\partial}{\partial x}(g(t, x; \mathbf{z}) v(t, x; \mathbf{z})) + d(t, x) v(t, x; \mathbf{z}) = 0,$$

where $d(t, x)$ denotes the death rate of individuals with size x at time t. Given the subpopulation density $v(t, x; \mathbf{z})$, the population density $u(t, x)$ for the total population with size x at time t is given by

$$u(t, x) = \int_{\Omega_z} v(t, x; \mathbf{z}) \tilde{\varphi}_{\mathbf{z}}(\mathbf{z}) d\mathbf{z}.$$

We can generalize the above results to a population which is structured by multiple structure variables $\mathbf{x} \in \mathbb{R}^n$. Let $v(t, \mathbf{x}; \mathbf{z})$ denote the population density of a subpopulation with growth rate $\mathbf{g}(t, \mathbf{x}; \mathbf{z})$. Then v satisfies

$$\frac{\partial}{\partial t} v(t, \mathbf{x}; \mathbf{z}) + \sum_{j=1}^{n} \frac{\partial}{\partial x_j} (g_j(t, \mathbf{x}; \mathbf{z}) v(t, \mathbf{x}; \mathbf{z})) + d(t, \mathbf{x}) v(t, \mathbf{x}; \mathbf{z}) = 0. \quad (7.113)$$

Given the subpopulation density $v(t, \mathbf{x}; \mathbf{z})$, the population density $u(t, \mathbf{x})$ for the total population with structure variables $\mathbf{x}$ at time t is given by

$$u(t, \mathbf{x}) = \int_{\Omega_z} v(t, \mathbf{x}; \mathbf{z}) \tilde{\varphi}_{\mathbf{z}}(\mathbf{z}) d\mathbf{z}. \quad (7.114)$$

In this presentation Equation (7.114) along with (7.113) (with appropriate boundary and initial conditions) as well as its special cases (e.g., the model with zero mortality) are referred to as *growth rate distributed physiologically structured* (GRDPS) population models.

Based on the above discussion, we see that GRDPS population models can be associated with certain stochastic processes, which are obtained due to the variability in the individual's growth rate and also the variability in the initial structure level of individuals in the population. Recall the discussion in Section 7.2.1.2 that FPPS population models are also associated with some stochastic processes resulting from the uncertainties in both growth rate and initial structure level of individuals in the population. Hence, a natural question to ask is how the GRDPS model is related to the FPPS model. And how are the stochastic processes resulting from these two types of models related? These questions will be answered in Section 7.4.2.

7.3.3   Differential Equations Driven by Correlated Stochastic Processes

We turn next to differential equations driven by correlated stochastic processes and consider in this context the system of random ordinary differential equations

$$\dot{\mathbf{x}}(t; \mathbf{X}_0, \mathbf{Y}(t)) = \mathbf{g}(t, \mathbf{x}(t; \mathbf{X}_0, \mathbf{Y}(t)); \mathbf{Y}(t)), \quad \mathbf{x}(0) = \mathbf{X}_0. \quad (7.115)$$

Here $\mathbf{x} = (x_1, x_2, \dots, x_n)^T$, $\mathbf{g} = (g_1, g_2, \dots, g_n)^T$ is an n-dimensional non-random vector function of t and $\mathbf{x}$, and the *driving process* $\{\mathbf{Y}(t) : t \geq 0\}$

with $\mathbf{Y}(t) = (Y_1(t), Y_2(t), \ldots, Y_\nu(t))^T$ is a ν-dimensional correlated stochastic process independent of random initial vector $\mathbf{X}_0$. Moreover we allow that $\mathbf{Y}(t)$ may be dependent on the states; that is, there exists a feedback between the states of the system (7.115) and the driving process. Our formulation is partly motivated by applications in optimal control.

By Section 7.1.7.3 we know that (7.115) with the driving process $\{\mathbf{Y}(t) : t \geq 0\}$ taken as a Gaussian colored noise is often found useful in a wide range of applications. In addition, Equation (7.115) with the driving process chosen as a continuous time Markov chain has been used in engineering for investigating the dynamic reliability of a complex system. For example, it was used in [74] to study the reliability of the Ignalina nuclear power plant accident localization system (ALS), which was designed to protect the environment from a radioactive release after the piping rupture in the RBMK-1500 primary reactor cooling system. In this study, the states of the system consist of two components, with one representing pressure and the other describing temperature, and $\mathbf{Y}(t) = (Y_1(t), Y_2(t))^T$ with $Y_1(t)$ and $Y_2(t)$ respectively denoting the number of failures of pumps and heat exchangers at time t. Equation (7.115) with a continuous time Markov chain driving process has also been found useful in a wide range of other fields, including the environmental sciences (e.g., [95]) and physiology (e.g., in studies of electrically excitable nerve membranes in [62, Section 9.4]).

7.3.3.1 Joint Probability Density Function of the Coupled Stochastic Process

We first consider the evolution of the joint probability density function of the coupled stochastic process $\{\tilde{\mathbf{X}}(t) : t \geq 0\}$, where $\tilde{\mathbf{X}}(t) = (\mathbf{X}(t), \mathbf{Y}(t))^T$, with $\mathbf{X}(t) = \mathbf{x}(t; \mathbf{X}_0, \mathbf{Y}(t))$ being the solution to (7.115). Specifically, two cases are discussed. One is with the driving process taken as a diffusion process, and the other is with the driving process given by a continuous time Markov chain with finite state space.

7.3.3.1.1 A diffusion driving process First we consider the case where $\{\mathbf{Y}(t) : t \geq 0\}$ is a diffusion process satisfying the Itô stochastic differential equations

$$d\mathbf{Y}(t) = \boldsymbol{\mu}_{\mathbf{Y}}(t, \mathbf{X}(t), \mathbf{Y}(t))dt + \boldsymbol{\sigma}_{\mathbf{Y}}(t, \mathbf{X}(t), \mathbf{Y}(t))d\mathbf{W}(t), \quad \mathbf{Y}(0) = \mathbf{Y}_0,$$
(7.116)

where $\boldsymbol{\mu}_{\mathbf{Y}}(t, \mathbf{x}, \mathbf{y}) = (\mu_{\mathbf{Y},1}(t, \mathbf{x}, \mathbf{y}), \mu_{\mathbf{Y},2}(t, \mathbf{x}, \mathbf{y}), \ldots, \mu_{\mathbf{Y},\nu}(t, \mathbf{x}, \mathbf{y}))^T$ is the drift coefficient, $\boldsymbol{\Sigma}_{\mathbf{Y}}(t, \mathbf{x}, \mathbf{y}) = \boldsymbol{\sigma}_{\mathbf{Y}}(t, \mathbf{x}, \mathbf{y}) (\boldsymbol{\sigma}_{\mathbf{Y}}(t, \mathbf{x}, \mathbf{y}))^T$ is the diffusion coefficient, with $\boldsymbol{\sigma}_{\mathbf{Y}}$ being a non-random $\nu \times l$ matrix function of t, $\mathbf{x}$ and $\mathbf{y}$, and $\{\mathbf{W}(t) : t \geq 0\}$ is an l-dimensional Wiener process. Then by (7.115) and (7.116) we know that $\tilde{\mathbf{X}}(t)$ satisfies the Itô stochastic differential equation

$$d\tilde{\mathbf{X}}(t) = \boldsymbol{\mu}_{\tilde{\mathbf{X}}}(t, \tilde{\mathbf{X}}(t))dt + \boldsymbol{\sigma}_{\tilde{\mathbf{X}}}(t, \tilde{\mathbf{X}}(t))d\mathbf{W}(t), \quad \tilde{\mathbf{X}}(0) = \tilde{\mathbf{X}}_0, \quad (7.117)$$

where

$$\boldsymbol{\mu}_{\tilde{\mathbf{X}}}(t, \tilde{\mathbf{X}}(t)) = \begin{bmatrix} \mathbf{g}(t, \mathbf{x}(t; \mathbf{X}_0, \mathbf{Y}(t)); \mathbf{Y}(t)) \\ \boldsymbol{\mu}_{\mathbf{Y}}(t, \mathbf{X}(t), \mathbf{Y}(t)) \end{bmatrix}$$

and

$$\boldsymbol{\sigma}_{\tilde{\mathbf{X}}}(t, \tilde{\mathbf{X}}(t)) = \begin{bmatrix} \mathbf{0}_{n \times l} \\ \boldsymbol{\sigma}_{\mathbf{Y}}(t, \mathbf{X}(t), \mathbf{Y}(t)) \end{bmatrix}$$

with $\mathbf{0}_{n \times l}$ being an $n \times l$ zero matrix, and $\tilde{\mathbf{X}}_0 = (\mathbf{X}_0, \mathbf{Y}_0)^T$. Thus, if we assume that $\tilde{\mathbf{X}}_0$ is independent of the Wiener process, and $\boldsymbol{\mu}_{\tilde{\mathbf{X}}}$ and $\boldsymbol{\sigma}_{\tilde{\mathbf{X}}}$ satisfy conditions guaranteeing the existence and uniqueness of solutions to the initial value problem (7.117), then by Theorem 7.2.2 we obtain the following results on the evolution of the joint probability density function of $\tilde{\mathbf{X}}(t)$.

Theorem 7.3.3 *Assume that* $\{\mathbf{Y}(t) : t \geq 0\}$ *is a diffusion process satisfying the Itô stochastic differential equations (7.116). Then the joint probability density function of* $\tilde{\mathbf{X}}(t)$ *satisfies*

$$\frac{\partial}{\partial t} \tilde{\varphi}_{\mathbf{X}, \mathbf{Y}}(t, \mathbf{x}, \mathbf{y}) + \sum_{k=1}^{n} \frac{\partial}{\partial x_k} (g_k(t, \mathbf{x}; \mathbf{y}) \tilde{\varphi}_{\mathbf{X}, \mathbf{Y}}(t, \mathbf{x}, \mathbf{y}))$$

$$+ \sum_{k=1}^{\nu} \frac{\partial}{\partial y_k} (\mu_{\mathbf{Y}, k}(t, \mathbf{x}, \mathbf{y}) \tilde{\varphi}_{\mathbf{X}, \mathbf{Y}}(t, \mathbf{x}, \mathbf{y})) \qquad (7.118)$$

$$= \frac{1}{2} \sum_{k,j=1}^{\nu} \frac{\partial^2}{\partial y_k \partial y_j} [\Sigma_{\mathbf{Y}, kj}(t, \mathbf{x}, \mathbf{y}) \tilde{\varphi}_{\mathbf{X}, \mathbf{Y}}(t, \mathbf{x}, \mathbf{y})]$$

with initial condition $\tilde{\varphi}_{\mathbf{X}, \mathbf{Y}}(0, \mathbf{x}, \mathbf{y}) = \tilde{\varphi}^0_{\mathbf{X}, \mathbf{Y}}(\mathbf{x}, \mathbf{y})$, *where* $\Sigma_{\mathbf{Y}, kj}(t, \mathbf{x}, \mathbf{y})$ *is the* (k, j)*th element of* $\Sigma_{\mathbf{Y}}(t, \mathbf{x}, \mathbf{y})$, *and* $\tilde{\varphi}^0_{\mathbf{X}, \mathbf{Y}}$ *is the joint probability density function of* $\mathbf{X}_0$ *and* $\mathbf{Y}_0$.

It should be noted that the results presented in the above theorem also apply to the case where $\mathbf{Y}(t)$ is not affected by the states of the system (that is, $\boldsymbol{\mu}_{\mathbf{Y}}$ and $\boldsymbol{\sigma}_{\mathbf{Y}}$ are independent of $\mathbf{X}$). In addition, we observe that the resulting equation (7.118) is a special form of the Fokker–Planck equation (7.69), and this equation with $n = 1$, $g_1 \equiv 1$ and $\Sigma_{\mathbf{Y}}(t, x, \mathbf{y})$ being a diagonal matrix is a special case of another model given in 1974 by Oster and Takahashi [91]

$$\frac{\partial}{\partial t} u(t, x, \mathbf{y}) + \frac{\partial}{\partial x} u(t, x, \mathbf{y}) + \sum_{k=1}^{\nu} \frac{\partial}{\partial y_k} (\mu_{\mathbf{Y}, k}(t, x, \mathbf{y}) u(t, x, \mathbf{y}))$$

$$+ d(t, x, \mathbf{y}) u(t, x, \mathbf{y}) = \frac{1}{2} \sum_{k=1}^{\nu} \frac{\partial^2}{\partial y_k^2} [\Sigma_{\mathbf{Y}, kk}(t, x, \mathbf{y}) u(t, x, \mathbf{y})] \qquad (7.119)$$

without mortality ($d \equiv 0$). Equation (7.119) is used to describe the population density in a spatially homogeneous population, where individuals are characterized by the chronological age x and the physiological variables $\mathbf{y}$ whose growth rates are affected by various stochastic factors. The functions $\boldsymbol{\mu}_\mathbf{Y}(t, x, \mathbf{y})$, $\boldsymbol{\Sigma}_\mathbf{Y}(t, x, \mathbf{y})$ and $d(t, x, \mathbf{y})$ in this model denote the mean growth rates, diffusion coefficient for the growth rates, and mortality rates of individuals with chronological age x and physiological variables $\mathbf{y}$ at time t, respectively.

Observe that (7.119) includes the physiologically structured population model (7.93) as a special case (specifically, when $\boldsymbol{\Sigma}_\mathbf{Y}$ is a zero matrix). In addition, (7.119) with $\nu = 1$

$$
\frac{\partial}{\partial t}u(t, x, y) + \frac{\partial}{\partial x}u(t, x, y) + \frac{\partial}{\partial y}(\mu_Y(t, x, y)u(t, x, y))
$$

$$
+d(t, x, y)u(t, x, y) = \frac{1}{2}\frac{\partial^2}{\partial y^2}\left[\Sigma_Y(t, x, y)u(t, x, y)\right]
$$

(7.120)

is a model given in 1968 by Weiss [117] for describing the growth of cell populations.

By the discussion in Section 7.2.1.2 we know that model (7.119) can be associated with a killed diffusion process, and the arguments presented there can be applied to obtain (7.119).

7.3.3.1.2 A continuous time Markov chain driving process For simplicity, we consider the case where the driving process is a one-dimensional (i.e., $\nu = 1$) CTMC with finite state space, and it is assumed to be homogeneous and unaffected by the states of the system. In addition, we assume that $\mathbf{g}$ is time independent; that is, system (7.115) becomes

$$
\dot{\mathbf{x}}(t; \mathbf{X}_0, Y(t)) = \mathbf{g}(\mathbf{x}(t; \mathbf{X}_0, Y(t)); Y(t)), \quad \mathbf{x}(0) = \mathbf{X}_0.
$$

(7.121)

Without loss of generality, the homogeneous CTMC $\{Y(t) : t \geq 0\}$ is assumed to have state space $\{1, 2, 3, \ldots, l\}$. Note that system (7.121) describes the movement of particles (or individuals) under the action of l dynamical systems corresponding to the equations $\dot{\mathbf{x}} = \mathbf{g}(\mathbf{x}; j)$, $j = 1, 2, \ldots, l$, and the driving process determines which dynamical system acts at time t. Hence, under these conditions, system (7.121) is sometimes referred to as a *randomly controlled dynamical system* (see [99]).

We remark that (7.121) is a special case of *random evolutions*, an abstract mathematical model for describing evolution systems in a general (either Markovian or non-Markovian) random medium, introduced in 1969 by Griego and Hersh [55], where certain abstract Cauchy problems were found to be related to the generators of Markov processes. There is a wide range of theoretical results on random evolutions, including representation theory, control theory, stability analysis, limit theorems in averaging and diffusion approximations with respect to the properties (such as ergodicity and reducibility) of the random medium. Investigations of random evolutions are often carried out by one of two approaches, one using asymptotic perturbations theory, and the other using a martingale approach. Interested readers can refer to [63, 75, 76, 110] and the references therein for further details.

Let q_{jk} denote the transition rate of $\{Y(t) : t \geq 0\}$ from state k to state j; that is,

$$\text{Prob}\{Y(t + \Delta t) = j \mid Y(t) = k\} = q_{jk}\Delta t + o(\Delta t), \quad j \neq k.$$

Define the functions $\tilde{\varphi}_{\mathbf{X},Y}(t, \mathbf{x}, j)$ and $\tilde{\varphi}^0_{\mathbf{X},Y}(\mathbf{x}, j)$ by

$$\text{Prob}\{\tilde{\mathbf{X}}(t) \in \mathbb{B} \times \{j\}\} = \int_{\mathbb{B}} \tilde{\varphi}_{\mathbf{X},Y}(t, \mathbf{x}, j)d\mathbf{x}, \quad j = 1, 2, \ldots, l,$$

and

$$\text{Prob}\{\tilde{\mathbf{X}}(0) \in \mathbb{B} \times \{j\}\} = \int_{\mathbb{B}} \tilde{\varphi}^0_{\mathbf{X},Y}(\mathbf{x}, j)d\mathbf{x}, \quad j = 1, 2, \ldots, l,$$

respectively, where $\tilde{\mathbf{X}}(t) = (\mathbf{X}(t), Y(t))^T$. Let $\mathcal{Q} = (q_{jk})_{l \times l}$, and

$$\tilde{\boldsymbol{\varphi}}_{\mathbf{X},Y}(t, \mathbf{x}) = (\tilde{\varphi}_{\mathbf{X},Y}(t, \mathbf{x}, 1), \tilde{\varphi}_{\mathbf{X},Y}(t, \mathbf{x}, 2), \ldots, \tilde{\varphi}_{\mathbf{X},Y}(t, \mathbf{x}, l))^T,$$

$$\tilde{\boldsymbol{\varphi}}^0_{\mathbf{X},Y}(\mathbf{x}) = (\tilde{\varphi}^0_{\mathbf{X},Y}(\mathbf{x}, 1), \tilde{\varphi}^0_{\mathbf{X},Y}(\mathbf{x}, 2), \ldots, \tilde{\varphi}^0_{\mathbf{X},Y}(\mathbf{x}, l))^T.$$

Then the coupled stochastic process $\left\{\tilde{\mathbf{X}}(t) : t \geq 0\right\}$ is a Markov process on $\mathbb{R}^n \times \{1, 2, \ldots, l\}$, and $\tilde{\boldsymbol{\varphi}}_{\mathbf{X},Y}$ satisfies the following system of partial differential equations (e.g., see [99]):

$$\frac{\partial}{\partial t} \tilde{\boldsymbol{\varphi}}_{\mathbf{X},Y}(t, \mathbf{x}) = \mathscr{A} \tilde{\boldsymbol{\varphi}}_{\mathbf{X},Y}(t, \mathbf{x}) + \mathcal{Q} \tilde{\boldsymbol{\varphi}}_{\mathbf{X},Y}(t, \mathbf{x}) \tag{7.122}$$

with initial condition $\tilde{\boldsymbol{\varphi}}_{\mathbf{X},\mathbf{Y}}(0, \mathbf{x}) = \tilde{\boldsymbol{\varphi}}^0_{\mathbf{X},\mathbf{Y}}(\mathbf{x})$. Here

$$\mathscr{A} \tilde{\boldsymbol{\varphi}}_{\mathbf{X},Y}(t, \mathbf{x}) = \left(\mathscr{A}_1 \tilde{\varphi}_{\mathbf{X},Y}(t, \mathbf{x}), \mathscr{A}_2 \tilde{\varphi}_{\mathbf{X},Y}(t, \mathbf{x}), \ldots, \mathscr{A}_l \tilde{\varphi}_{\mathbf{X},Y}(t, \mathbf{x})\right)^T,$$

where $\mathscr{A}_j \tilde{\varphi}_{\mathbf{X},Y}(t, \mathbf{x}) = -\sum_{k=1}^{n} \frac{\partial}{\partial x_k}(g_k(\mathbf{x}; j)\tilde{\varphi}_{\mathbf{X},Y}(t, \mathbf{x}, j)), \; j = 1, 2, \ldots, l.$

Remark 7.3.4 *We remark that the results presented in the above two cases can be easily applied to the scenario where differential equations are driven by both colored noise and white noise. We take the random differential equation*

$$\dot{\mathbf{X}}(t) = \mathbf{g}(\mathbf{X}(t); Y(t)) + \boldsymbol{\sigma}(\mathbf{X}(t); Y(t))d\mathbf{W}(t)$$
$$\mathbf{X}(0) = \mathbf{X}_0 \tag{7.123}$$

as an example to illustrate this. Here $\{Y(t) : t \geq 0\}$ is a one-dimensional homogeneous CTMC with state space $\{1, 2, \ldots, l\}$, which is independent of the

Wiener process and the initial condition $\mathbf{X}_0$. *We see that (7.123) describes the movement of particles (or individuals) under the action of l stochastic systems corresponding to* $\dot{\mathbf{X}}(t) = \mathbf{g}(\mathbf{X}(t); j) + \boldsymbol{\sigma}(\mathbf{X}(t); j)d\mathbf{W}(t)$, $j = 1, 2, \ldots, l$, *and* $\{Y(t) : t \geq 0\}$ *determines which stochastic system acts at time t. Hence, this type of differential equation is referred to as a switching diffusion system, stochastic differential equation with Markov switching, hybrid switching diffusion, or randomly flashing diffusion. Due to its applications in a wide range of fields, such as financial mathematics, engineering, biological systems and manufacturing systems (e.g., see [122] and the references therein for details), this type of differential equation has received considerable research attention in recent years. Interested readers can refer to the recent texts [81, 123] and the references therein for information on the recent advance of this subject.*

Let $\tilde{\mathbf{X}}(t) = (\mathbf{X}(t), Y(t))^T$, *with* $\mathbf{X}(t)$ *being the solution to (7.123). Then the coupled stochastic process* $\left\{\tilde{\mathbf{X}}(t) : t \geq 0\right\}$ *is still a Markov process on* $\mathbb{R}^n \times \{1, 2, \ldots, l\}$, *and* $\tilde{\varphi}_{\mathbf{X},Y}$ *satisfies the following system of partial differential equations (again see [99]):*

$$\frac{\partial}{\partial t}\tilde{\varphi}_{\mathbf{X},Y}(t, \mathbf{x}) = \mathscr{A}\tilde{\varphi}_{\mathbf{X},Y}(t, \mathbf{x}) + \mathcal{Q}\varphi_{\mathbf{X},Y}(t, \mathbf{x}). \qquad (7.124)$$

Here $\mathscr{A}\tilde{\varphi}_{\mathbf{X},Y}(t, \mathbf{x}) = \left(\mathscr{A}_1\tilde{\varphi}_{\mathbf{X},Y}(t, \mathbf{x}), \mathscr{A}_2\tilde{\varphi}_{\mathbf{X},Y}(t, \mathbf{x}), \ldots, \mathscr{A}_l\tilde{\varphi}_{\mathbf{X},Y}(t, \mathbf{x})\right)^T$ *with the differential operators* $\mathscr{A}_r$, $r = 1, 2, \ldots, l$ *defined by*

$$\mathscr{A}_r\tilde{\varphi}_{\mathbf{X},Y}(t, \mathbf{x}) = -\sum_{k=1}^{n} \frac{\partial}{\partial x_k}(g_k(\mathbf{x}; r)\tilde{\varphi}_{\mathbf{X},Y}(t, \mathbf{x}, r))$$

$$+\frac{1}{2}\sum_{k,j=1}^{n} \frac{\partial^2}{\partial x_k \partial x_j}\left[\Sigma_{kj}(\mathbf{x}; r)\tilde{\varphi}_{\mathbf{X},Y}(t, \mathbf{x}, r)\right],$$

where $\Sigma_{kj}(\mathbf{x}; r)$ *denotes the (k, j)th element of matrix* $\boldsymbol{\sigma}(\mathbf{x}; r)(\boldsymbol{\sigma}(\mathbf{x}; r))^T$.

Remark 7.3.5 *It should be noted that for all the above cases, even though the coupled stochastic process* $\left\{\tilde{\mathbf{X}}(t) : t \geq 0\right\}$ *is Markovian, the primary process* $\{\mathbf{X}(t) : t \geq 0\}$ *is not since the evolution of the states of the system depends on the past history due to the autocorrelation (in time) of the driving process. This means that one can enforce the Markovian property by including driving processes as separate states. The advantage for doing so is obvious in terms of mathematical analysis. However, if the dimension of the driving process is very large, this will dramatically increase the dimension of the state space of the system, and hence may become impractical for computation.*

7.3.3.2 The Probability Density Function of $\mathbf{X}(t)$

In this section, we consider the probability density function of $\mathbf{X}(t)$ based on the two cases that we discussed in Section 7.3.3.1, that is, the case with the driving process being a diffusion process, and the case with the driving process being a continuous time Markov chain.

A diffusion driving process   For the given joint probability density function $\tilde{\varphi}_{\mathbf{X},\mathbf{Y}}$, we can obtain the probability density function of $\mathbf{X}(t)$

$$p(t, \mathbf{x}) = \int_{\Omega_y} \tilde{\varphi}_{\mathbf{X},\mathbf{Y}}(t, \mathbf{x}, \mathbf{y}) d\mathbf{y}, \tag{7.125}$$

where Ω_y denotes the set of all possible values for $\mathbf{y}$. Equation (7.125) combined with Equation (7.118) for $\tilde{\varphi}_{\mathbf{X},\mathbf{Y}}$ (with appropriate boundary and initial conditions) can be used to describe the population density in a population ecology system where the growth rates $\mathbf{g}$ of the species are affected by a driving diffusion process. This could be due to the climate variability, which is often described by Itô SDEs (recall the discussion in Section 7.2.1).

A continuous time Markov chain driving process   Based on the discussion in Section 7.3.3.1, we see that under this case the probability density function of $\mathbf{X}(t)$ is given by

$$p(t, \mathbf{x}) = \sum_{j=1}^{l} \tilde{\varphi}_{\mathbf{X},Y}(t, \mathbf{x}, j). \tag{7.126}$$

Equation (7.126) combined with Equation (7.122) for $\tilde{\varphi}_{\mathbf{X},Y}$ (with appropriate boundary and initial conditions) can be used to describe the population density in a population where the growth rates $\mathbf{g}$ of the species are affected by a continuous time Markov chain driving process. For example, for the case $l = 2$ it could be used to describe a population in an environment with random switching between two opposite conditions, such as drought-non-drought and stressed-unstressed conditions.

Remarks on the equation for the probability density function of $\mathbf{X}(t)$
In general, one is unable to obtain a closed form for the equation of the probability density function of the solution to (7.115) even in the one-dimensional case. Two scenarios of the following scalar random differential equation

$$\dot{x}(t; X_0, Y(t)) = f(x(t; X_0, Y(t))) + \sigma(x(t; X_0, Y(t)))Y(t), \quad x(0) = X_0 \tag{7.127}$$

were considered in [60], where the closed form can be derived for the equation of the probability density function of the resulting process.

One case is when f is an affine function of x (i.e., $f(x) = a_0 - a_1 x$), $\sigma \equiv 1$, and $\{Y(t) : t \geq 0\}$ is a general stationary Gaussian colored noise (being either Markovian or non-Markovian) with zero-mean and correlation function $\text{Cor}\{Y(t), Y(s)\} = \Upsilon(|t - s|)$. The evolution of the probability density function of $X(t)$ is described by the Fokker–Planck-like equation

$$\frac{\partial}{\partial t}p(t,x) + \frac{\partial}{\partial x}\left[(a_0 - a_1 x)p(t,x)\right]$$

$$= \left[\int_0^t \Upsilon(t-s)\exp(-a_1(t-s))ds\right]\frac{\partial^2}{\partial x^2}p(t,x) \qquad (7.128)$$

$$p(0,x) = p_0(x),$$

where p_0 is the probability density function of X_0. We observe that even with a time-independent function f and stationary Gaussian noise, the effective diffusion in (7.128) is time dependent and may even become negative in the cases where the correlation function is an oscillatory-like function. In addition, it was found that if $p_0(x) = \delta(x - x_0)$ (i.e., X_0 is a fixed constant x_0), then $\{x(t; x_0, Y(t)) : t \geq 0\}$ is a non-Markovian Gaussian process with mean and variance functions respectively given by

$$\mu(t) = x_0 \exp(-a_1 t) + \frac{a_0}{a_1}\left[1 - \exp(-a_1 t)\right]$$

and

$$\nu(t) = 2\int_0^t \exp(-2a_1(t-s))\left[\int_0^s \Upsilon(s-r)\exp(-a_1(s-r))dr\right]ds.$$

Hence, the solution of (7.128) for this case is given by

$$p(t,x) = \frac{1}{\sqrt{2\pi\nu(t)}}\exp\left(-\frac{(x-\mu(t))^2}{2\nu(t)}\right). \qquad (7.129)$$

To help understand how the correlation function of the driving process affects the probability density function of the primary process $\{X(t) : t \geq 0\}$, we plot $p(t, \cdot)$ at $t = 0.5$ and $t = 1$ in Figure 7.5, where the results are obtained with $a_0 = 1$, $a_1 = 3$, $x_0 = 10$ and $\Upsilon(|t - s|) = \exp\left(-\frac{|t-s|}{\tau}\right)$ for two different values of τ ($\tau = 0.1$ and $\tau = 1$). Note that τ is the correlation time for this particular driving process (recall Example 7.1.1 in Section 7.1.7). Hence, we observe from Figure 7.5 that for every fixed t the probability density function disperses more as the correlation time of the driving process increases.

The other case that permits a closed form equation for the probability density function is when $\{Y(t) : t \geq 0\}$ is a symmetric dichotomous Markovian noise with transition rates between a and $-a$ given by λ. The evolution of the probability density function of $X(t)$ is described by the equation

$$\frac{\partial}{\partial t}p(t,x) + \frac{\partial}{\partial x}(f(x)p(t,x))$$

$$= a^2\frac{\partial}{\partial x}\left[\sigma(x)\int_0^t \exp\left(-(2\lambda + f'(x))(t-s)\right)\left(\frac{\partial}{\partial x}(\sigma(x)p(s,x))\right)ds\right] \qquad (7.130)$$

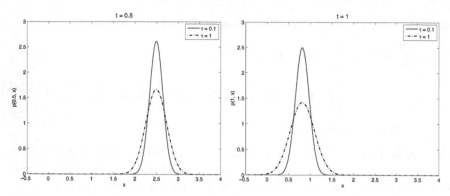

FIGURE 7.5: Probability density function $p(t, \cdot)$ at $t = 0.5$ and $t = 1$ obtained with $a_0 = 1$, $a_1 = 3$, $x_0 = 10$ and $\Upsilon(|t - s|) = \exp\left(-\dfrac{|t - s|}{\tau}\right)$ for two different values of τ.

with initial condition $p(0, x) = p_0(x)$, where f' denotes the derivative of f with respect to x. We observe from (7.130) that the evolution of the probability density function p depends on its past history, which clearly reveals that the stochastic process $\{X(t) : t \geq 0\}$ is non-Markovian. Moreover, if $f \equiv 0$ and $\sigma \equiv 1$, then by (7.130) it is easy to see that p satisfies *telegrapher's equation*

$$\frac{\partial^2}{\partial t^2} p(t, x) + 2\lambda \frac{\partial}{\partial t} p(t, x) = a^2 \frac{\partial^2}{\partial x^2} p(t, x).$$

The method for the derivation of (7.128) and (7.130) in [60] is based on the expression for $p(t, x)$ given by

$$p(t, x) = \mathbb{E}(\delta(X(t) - x)).$$

One then differentiates the above equation with respect to t, and uses the statistics of the driving process $\{Y(t) : t \geq 0\}$ as well as some functional derivative techniques to obtain the desired equation.

7.4   Relationships between Random and Stochastic Differential Equations

As noted in Section 7.2.2, the solution of a Stratonovich stochastic differential equation can be obtained as the limit of the solutions of some random differential equations (driven by Gaussian colored noise). In this section, we will give some other relationships between stochastic and random differential equations. Specifically, the probability density functions of the solutions to stochastic differential equations and crypto-deterministic differential equations can be associated with Markov operators, as detailed in the next section. In addition, there is a class of Itô stochastic differential equations with solutions having the same probability density function at each time t as those for the solution of their corresponding random differential equations. These are summarized in Section 7.4.2.

7.4.1 Markov Operators and Markov Semigroups

Markov operators and Markov semigroups have been widely used to investigate dynamical systems and dynamical systems with stochastic perturbations, as they have been proven to be suitable to study the evolution of probability density functions of the resulting processes. Hence, in this section we focus on Markov operators and Markov semigroups acting on the set of probability density functions.

Let $(\Omega, \mathcal{F}, \mu)$ be a σ-finite measure space, and $L^1(\Omega, \mathcal{F}, \mu)$ be the Banach space of all possible real-valued measurable functions $f : \Omega \to \mathbb{R}$ satisfying $\int_\Omega |f(\mathbf{x})| \mu(d\mathbf{x}) < \infty$ with the norm defined by

$$\|f\|_{L^1(\Omega, \mathcal{F}, \mu)} = \int_\Omega |f(\mathbf{x})| \mu(d\mathbf{x}).$$

In addition, $L^\infty(\Omega, \mathcal{F}, \mu)$ denotes the Banach space of all possible real-valued measurable functions that are bounded almost everywhere with the the the norm defined by

$$\|h\|_{L^\infty(\Omega, \mathcal{F}, \mu)} = \operatorname*{ess\,sup}_{\mathbf{x} \in \Omega} |h(\mathbf{x})|.$$

With the above notation, Markov operators are defined in the usual manner (e.g., see [79, Definition 3.1.1] or [44, 99]).

Definition 7.4.1 *Let $(\Omega, \mathcal{F}, \mu)$ be a σ-finite measure space, and $\mathbb{D}$ be the subset of Banach space $L^1(\Omega, \mathcal{F}, \mu)$ functions which contains all possible densities*

$$\mathbb{D} = \left\{ f \in L^1(\Omega, \mathcal{F}, \mu) \mid f \geq 0, \|f\|_{L^1(\Omega, \mathcal{F}, \mu)} = 1 \right\}.$$

A linear mapping $\mathscr{P} : L^1(\Omega, \mathcal{F}, \mu) \to L^1(\Omega, \mathcal{F}, \mu)$ is called a Markov operator if $\mathscr{P}(\mathbb{D}) \subset \mathbb{D}$.

By the above definition, we see that a Markov operator is a linear operator that maps any probability density function to a probability density function. Markov operators have a number of useful properties. For example, the Markov operator $\mathscr{P}$ is monotonic; that is, if $f_1, f_2 \in L^1(\Omega, \mathcal{F}, \mu)$ and $f_2 \geq f_1$, then $\mathscr{P}f_2 \geq \mathscr{P}f_1$. In addition, it is a contraction; that is, for any $f \in L^1(\Omega, \mathcal{F}, \mu)$ we have $\|\mathscr{P}f\|_{L^1(\Omega, \mathcal{F}, \mu)} \leq \|f\|_{L^1(\Omega, \mathcal{F}, \mu)}$. Furthermore, if Ω is a Polish space (i.e., a complete separable metric space), $\mathcal{F}$ is a Borel algebra on Ω, and μ is a Borel probability measure on Ω, then every Markov operator on $L^1(\Omega, \mathcal{F}, \mu)$ can be defined by means of a transition probability function (e.g., see [99] and the references therein). For more information on Markov operators and their applications, interested readers can refer to [79, 99] and the references therein.

Definition 7.4.2 *A family* $\{\mathscr{P}(t)\}_{t\geq 0}$ *of Markov operators which satisfies conditions*

- $\mathscr{P}(0) = \mathscr{I}$

- $\mathscr{P}(t + s) = \mathscr{P}(t)\mathscr{P}(s)$

- *for each* $f \in L^1(\Omega, \mathcal{F}, \mu)$ *the function* $t \to \mathscr{P}(t)f$ *is continuous*

is called a Markov semigroup. Here $\mathscr{I}$ *denotes the unity operator.*

An example of Markov operators: The Frobenius–Perron operator
Frobenius–Perron operators are a special case of Markov operators and are very important in the ergodic theory of chaotic dynamical systems (see [44] for more information). Before we give the precise definition, we need the following definition on non-singular transformations.

Definition 7.4.3 *Let* $(\Omega, \mathcal{F}, \mu)$ *be a* σ*-finite measure space. A transformation* $\eta : \Omega \to \Omega$ *is measurable if*

$$\eta^{-1}(\mathbb{O}) \in \mathcal{F}, \text{ for all } \mathbb{O} \in \mathcal{F}.$$

In addition, if $\mu(\eta^{-1}(\mathbb{O})) = 0$ *for all* $\mathbb{O} \in \mathcal{F}$ *such that* $\mu(\mathbb{O}) = 0$, *then* η *is non-singular.*

With the above definition, the Frobenius–Perron operator is defined as follows.

Definition 7.4.4 *Let* $(\Omega, \mathcal{F}, \mu)$ *be a* σ*-finite measure space. If* $\eta : \Omega \to \Omega$ *is a non-singular transformation, then the unique operator* $\mathscr{P} : L^1(\Omega, \mathcal{F}, \mu) \to L^1(\Omega, \mathcal{F}, \mu)$ *defined by*

$$\int_{\mathbb{O}} \mathscr{P}f(\mathbf{x})\mu(dx) = \int_{\eta^{-1}(\mathbb{O})} f(\mathbf{x})\mu(dx), \quad \mathbb{O} \in \mathcal{F} \qquad (7.131)$$

is called the Frobenius–Perron operator corresponding to η.

It is important to note that in some special cases where the transformation η is differentiable and invertible, the explicit form for $\mathscr{P}f$ is available. This is given in the following lemma (e.g., see [79, Corollary 3.2.1]).

Lemma 7.4.1 *Let $(\Omega, \mathcal{F}, \mu)$ be a σ-finite measure space, and $\boldsymbol{\eta} : \Omega \to \Omega$ be an invertible non-singular transformation (i.e., $\boldsymbol{\eta}^{-1}$ non-singular) and $\mathscr{P}$ be the associated Frobenius–Perron operator. Then for every $f \in L^1(\Omega, \mathcal{F}, \mu)$ we have*

$$\mathscr{P}(f(\mathbf{x})) = f(\boldsymbol{\eta}^{-1}(\mathbf{x}))\mathcal{J}^{-1}(\mathbf{x}), \tag{7.132}$$

where $\mathcal{J}(\mathbf{x})$ is the determinant of the Jacobian matrix $\dfrac{\partial \boldsymbol{\eta}(\mathbf{x})}{\partial \mathbf{x}}$.

Koopman operator The *Koopman operator* is widely used in applications, as it can be used to describe the evolution of the observables of a system on the phase space. Its definition is given as follows.

Definition 7.4.5 *Let $(\Omega, \mathcal{F}, \mu)$ be a σ-finite measure space. If $\boldsymbol{\eta} : \Omega \to \Omega$ is a non-singular transformation and $h \in L^\infty(\Omega, \mathcal{F}, \mu)$. Then the operator $\mathscr{U} : L^\infty(\Omega, \mathcal{F}, \mu) \to L^\infty(\Omega, \mathcal{F}, \mu)$ defined by*

$$\mathscr{U}(h(\mathbf{x})) = h(\boldsymbol{\eta}(\mathbf{x})) \tag{7.133}$$

is called the Koopman operator with respect to $\boldsymbol{\eta}$.

The Koopman operator has a number of important properties. For example, it is linear, and it is a contraction (that is, $\|\mathscr{U}(h)\|_{L^\infty(\Omega,\mathcal{F},\mu)} \leq \|h\|_{L^\infty(\Omega,\mathcal{F},\mu)}$ for every $h \in L^\infty(\Omega, \mathcal{F}, \mu)$). In addition, the Koopman operator $\mathscr{U}$ is the adjoint of the Frobenius–Perron operator $\mathscr{P}$; that is,

$$\langle \mathscr{P}f, h \rangle = \langle f, \mathscr{U}h \rangle \tag{7.134}$$

holds for every $f \in L^1(\Omega, \mathcal{F}, \mu)$ and $h \in L^\infty(\Omega, \mathcal{F}, \mu)$. Interested readers can refer to [79] for more information on Frobenius–Perron operators and Koopman operators.

7.4.1.1 Random Differential Equations

Assume that the system of crypto-deterministic formulations given by the differential equations in (7.88)

$$\dot{\mathbf{x}} = \mathbf{g}(t, \mathbf{x}), \quad \mathbf{x}(0) = \mathbf{X}_0$$

has an explicit mean square solution $\mathbf{X}(t)$. Then the solution takes the general form

$$\mathbf{X}(t) = \boldsymbol{\eta}(t, \mathbf{X}_0), \tag{7.135}$$

which essentially represents an algebraic transformation of a random vector into another. In addition, we observe that in continuum mechanics the realization of (7.135) (i.e., $\mathbf{x}(t) = \boldsymbol{\eta}(t, \mathbf{x}_0)$) provides a description of particles moving in space as time progresses; that is, it provides a Lagrangian description of the motion of the particles.

The following theorem indicates that the probability density function p of $\mathbf{X}(t)$ at any given time t can be expressed in terms of the probability density function p_0 of the initial condition $\mathbf{X}_0$.

Theorem 7.4.2 *(See [104, Theorem 6.2.1]) Suppose that the transformation $\boldsymbol{\eta}$ is continuous in $\mathbf{x}_0$ and has a continuous partial derivative with respect to $\mathbf{x}_0$, and defines a one-to-one mapping. Let the inverse transform be written by*

$$\mathbf{X}_0 = \boldsymbol{\eta}^{-1}(t, \mathbf{X}(t)). \tag{7.136}$$

Then the probability density function p of $\mathbf{X}(t)$ is given by

$$p(t, \mathbf{x}) = p_0(\boldsymbol{\eta}^{-1}(t, \mathbf{x}))|\mathcal{J}^{-1}|, \tag{7.137}$$

where $\mathcal{J}$ is the determinant of the Jacobian matrix $\dfrac{\partial \mathbf{x}}{\partial \mathbf{x}_0}$.

Observe that in continuum mechanics the realization of (7.136) (i.e., $\mathbf{x}_0 = \boldsymbol{\eta}^{-1}(t, \mathbf{x}(t))$) provides a tracing of the particle which now occupies the position $\mathbf{x}$ in the current configuration from its original position $\mathbf{x}_0$ in the initial configuration, and thus it provides an Eulerian description of the motion of particles. Hence, (7.137) gives an explicit formula for the density of material $p(t, \mathbf{x})$ in the current configuration in terms of the density of material $p_0(\mathbf{x})$ in the original configuration, where $\mathcal{J}$ in this case denotes the determinant of the *configuration gradient* (often, in something of a misnomer, referred to as the *deformation gradient* in the literature – see [25]).

By Lemma 7.4.1 and Theorem 7.4.2 we know that for any given t the probability density function $p(t, \cdot)$ can be expressed in terms of the Frobenius–Perron operator $\mathscr{P}(t)$ (corresponding to the map $\boldsymbol{\eta}(t, \cdot)$); that is,

$$p(t, \mathbf{x}) = \mathscr{P}(t)p_0(\mathbf{x}),$$

where the operator $\mathscr{P}(t)$ is given by

$$\mathscr{P}(t)f(\mathbf{x}) = f(\boldsymbol{\eta}^{-1}(t, \mathbf{x}))|\mathcal{J}^{-1}|.$$

Moreover, if $\mathbf{g}$ is time independent, then $\{\mathscr{P}(t)\}_{t\geq 0}$ forms a Frobenius–Perron semigroup with an *infinitesimal generator* given by (e.g., see [79, Chapter 7])

$$\mathscr{A}\psi(\mathbf{x}) = -\sum_{k=1}^{n} \frac{\partial}{\partial x_k}\left(g_k(\mathbf{x})\psi(\mathbf{x})\right),$$

which implies that p satisfies Liouville's equation

$$\frac{\partial}{\partial t}p(t, \mathbf{x}) = \mathscr{A}p(t, \mathbf{x}).$$

Remark 7.4.3 *Consider the time-homogeneous case of (7.88) (i.e., $\mathbf{g}$ is time independent) with mean-square solution $\boldsymbol{\eta}(t, \mathbf{X}_0)$. Then by the definition of the Koopman operator we have $\mathscr{U}(t)\phi(\mathbf{x}) = \phi(\boldsymbol{\eta}(t, \mathbf{x}))$. It can be shown that $\{\mathscr{U}(t)\}_{t \geq 0}$ forms a semigroup (if we restrict ourselves to continuous functions with compact support) with an infinitesimal generator given by (e.g., see [79, Section 7.6])*

$$\mathscr{L}\phi(\mathbf{x}) = \sum_{k=1}^{n} g_k(\mathbf{x})\frac{\partial\phi(\mathbf{x})}{\partial x_k},$$

which is the adjoint to the infinitesimal generator $\mathscr{A}$ of the Frobenius–Perron semigroup. Let $w(t, \mathbf{x}) = \mathscr{U}(t)w_0(\mathbf{x})$. Then w satisfies

$$\frac{\partial}{\partial t}w(t, \mathbf{x}) = \mathscr{L}w(t, \mathbf{x})$$

with initial condition $w(0, \mathbf{x}) = w_0(\mathbf{x})$.

7.4.1.2 Stochastic Differential Equations

We recall that any Stratonovich stochastic differential equation can be converted to its corresponding Itô stochastic differential equation. Hence, in this section we focus on the Itô stochastic differential equation (7.63) given by

$$d\mathbf{X}(t) = \mathbf{g}(t, \mathbf{X}(t))dt + \boldsymbol{\sigma}(t, \mathbf{X}(t))d\mathbf{W}(t), \quad \mathbf{X}(0) = \mathbf{X}_0.$$

By Remark 7.2.3 we know that if $\mathbf{g}$ and $\boldsymbol{\sigma}$ are assumed to satisfy conditions guaranteeing the existence of the fundamental solution to the Fokker–Planck equation (7.69), then the solution of (7.69) can be written as

$$p(t, \mathbf{x}) = \mathscr{P}(t)p_0(\mathbf{x}),$$

where the operator $\mathscr{P}(t)$ is defined by

$$\mathscr{P}(t)\psi(\mathbf{x}) = \int_{\mathbb{R}^n} \rho(t, \mathbf{x}; 0, \boldsymbol{\zeta})\psi(\boldsymbol{\zeta})d\boldsymbol{\zeta}, \quad t \geq 0.$$

We see that the operators $\mathscr{P}(t), t \geq 0$, are Markov operators. In addition, if $\mathbf{g}$ and $\boldsymbol{\sigma}$ are time independent, then the operators $\{\mathscr{P}(t)\}_{t \geq 0}$ form a Markov semigroup with an *infinitesimal generator* (e.g., see [79, Section 11.8])

$$\mathscr{A}\psi(\mathbf{x}) = -\sum_{k=1}^{n} \frac{\partial}{\partial x_k}(g_k(\mathbf{x})\psi(\mathbf{x})) + \frac{1}{2}\sum_{k,j=1}^{n} \frac{\partial^2}{\partial x_k \partial x_j}[\Sigma_{kj}(\mathbf{x})\psi(\mathbf{x})],$$

which is just Kolmogorov's forward operator (for the time-homogeneous case). Hence, under this case we can write the Fokker–Planck equation concisely as

$$\frac{\partial}{\partial t}p(t, \mathbf{x}) = \mathscr{A}p(t, \mathbf{x}).$$

Remark 7.4.4 *Note that for the time-homogeneous case of (7.63) (that is, the case where* **g** *and* $\boldsymbol{\sigma}$ *are time independent), the transition probability density function* $\rho(t, \boldsymbol{\zeta}; s, \mathbf{x})$ *of the resulting diffusion process* $\{\mathbf{X}(t) : t \geq 0\}$ *depends only on the difference of the arguments* $t - s$. *Hence, in this case we denote the transition probability density function by* $\tilde{\rho}(t - s, \boldsymbol{\zeta}, \mathbf{x})$*; that is,*

$$\tilde{\rho}(t - s, \boldsymbol{\zeta}, \mathbf{x}) = \rho(t, \boldsymbol{\zeta}; s, \mathbf{x}).$$

We define a family of operators $\{\mathscr{T}(t)\}_{t \geq 0}$ *by*

$$\mathscr{T}(t)\phi(\mathbf{x}) = \mathbb{E}\left(\phi(\mathbf{X}(t)) \mid \mathbf{X}(0) = \mathbf{x}\right) = \int_{\mathbb{R}^n} \tilde{\rho}(t, \boldsymbol{\zeta}, \mathbf{x})\phi(\boldsymbol{\zeta})d\boldsymbol{\zeta}, \quad t \geq 0.$$

Then $\{\mathscr{T}(t)\}_{t \geq 0}$ *forms a semigroup with the infinitesimal generator* $\mathscr{L}$ *given by (e.g., see [89, Theorem 7.3.3])*

$$\mathscr{L}\phi(\mathbf{x}) = \sum_{k=1}^{n} g_k(\mathbf{x})\frac{\partial\phi(\mathbf{x})}{\partial x_k} + \frac{1}{2}\sum_{k,j=1}^{n} \Sigma_{kj}(\mathbf{x})\frac{\partial^2\phi(\mathbf{x})}{\partial x_k x_j},$$

which is just Kolmogorov's backward operator (for the time-homogeneous case). The operator $\mathscr{L}$ *is often termed the infinitesimal generator of the (time-homogeneous) diffusion process* $\{\mathbf{X}(t) : t \geq 0\}$. *Let* w_0 *be a twice continuously differentiable function with compact support, and define*

$$w(t, \mathbf{x}) = \mathscr{T}(t)w_0(\mathbf{x}).$$

Then w *satisfies the equation (e.g., see [89, Theorem 8.1.1])*

$$\frac{\partial}{\partial t}w(t, \mathbf{x}) = \mathscr{L}w(t, \mathbf{x}) \tag{7.138}$$

with initial condition $w(0, \mathbf{x}) = w_0(\mathbf{x})$. *Equation (7.138) is also referred to as Kolmogorov's backward equation (for the initial value problem), as it can be converted to the form with respect to the backward variable (i.e., the form similar to (7.65)) by using a transformation on the time variable (as we discussed in Section 7.2.1).*

7.4.2   Pointwise Equivalence Results between Stochastic Differential Equations and Random Differential Equations

Both stochastic differential equations and random differential equations lead to some underlying stochastic processes and a natural question is how these two processes might differ. Somewhat surprisingly, however, it was shown in [22] that there are classes of Itô stochastic differential equations for which their solutions have the same probability density functions at each time t as those for the solutions of their corresponding random differential equations, wherein RDEs are treated by using the sample function approach.

Definition 7.4.6 *If the probability density function of the solution of an SDE is the same as that of the solution of an RDE at each time t, then the SDE and the RDE are said to be* pointwise equivalent.

The original motivation for such research was to provide an alternative way to numerically solve the Fokker–Planck equation since it is well known that the Fokker–Planck equation is difficult to solve when the drift dominates the diffusion (the case of primary interest in many situations).

In establishing the pointwise equivalence results in [22], the relationship between the normal distribution and the log-normal distribution as well as (2.86) are heavily used. In addition, the following basic result on the stochastic process generated by Itô integrals is another main tool used in establishing the equivalence results in [22].

Lemma 7.4.5 *(See [72, Sec 4.3, Thm 4.11]) Let $f \in L^2(0, t_f)$ be a non-random function, where t_f is some positive constant. Then the stochastic process $\{Q(t) : 0 \leq t \leq t_f\}$ generated by Itô integrals*

$$Q(t) = \int_0^t f(s)dW(s), \quad 0 \leq t \leq t_f$$

is a Gaussian process with the zero-mean function and the covariance function given by

$$Cov(Q(t), Q(t + \varpi)) = \int_0^t f^2(s)ds$$

for all $\varpi \geq 0$.

We next summarize the pointwise equivalence results from [22], where it was shown how to transform from a given Itô SDE to the corresponding pointwise equivalent RDE, and from a given RDE to the corresponding pointwise equivalent Itô SDE. It is worth noting that one can easily apply these pointwise equivalence results to obtain the mapping between the Stratonovich SDEs and their corresponding RDEs based on the equivalent condition (7.85) between Itô SDEs and Stratonovich SDEs.

7.4.2.1 Scalar Affine Differential Equations (Class 1)

In the first class of systems, the random differential equation considered has the following form:

$$\frac{dx(t; X_0, \mathbf{Z})}{dt} = \alpha(t)x(t; X_0, \mathbf{Z}) + \gamma(t) + \mathbf{Z} \cdot \varrho(t),$$

$$x(0; X_0, \mathbf{Z}) = X_0.$$

(7.139)

In the above equation, $\mathbf{Z} = (Z_0, Z_1, \ldots, Z_{m-1})^T$, where $Z_j \sim \mathcal{N}(\mu_j, \sigma_j^2)$, $j = 0, 1, 2, \ldots, m-1$, are mutually independent, with $\mu_j, \sigma_j, j = 0, 1, 2, \ldots, m-1$, being some positive constants. In addition, α, γ and $\boldsymbol{\varrho} = (\varrho_0, \varrho_1, \ldots, \varrho_{m-1})^T$ are non-random functions of t, and $\mathbf{Z} \cdot \boldsymbol{\varrho}(t)$ denotes the dot product of $\mathbf{Z}$ and $\varrho(t)$ (that is, $\mathbf{Z} \cdot \boldsymbol{\varrho}(t) = \mathbf{Z}^T \boldsymbol{\varrho}(t) = \sum_{j=0}^{m-1} Z_j \varrho_j(t)$). The corresponding stochastic differential equation takes the form

$$dX(t) = [\alpha(t)X(t) + \xi(t)]dt + \eta(t)dW(t), \quad X(0) = X_0, \tag{7.140}$$

where ξ and η are some non-random functions of t, and $\{W(t) : t \geq 0\}$ is a Wiener process.

The conditions for the pointwise equivalence between the RDE (7.139) and the SDE (7.140) are stated in the following theorem.

Theorem 7.4.6 *If functions ξ, η and γ, and functions ϱ_j and constants μ_j and σ_j, $j = 0, 1, 2, \ldots, m-1$ satisfy the two equalities*

$$\int_0^t \xi(s) \exp\left(\int_s^t \alpha(\tau)d\tau\right) ds = \int_0^t [\gamma(s) + \boldsymbol{\mu} \cdot \boldsymbol{\varrho}(s)] \exp\left(\int_s^t \alpha(\tau)d\tau\right) ds, \tag{7.141}$$

and

$$\int_0^t \left[\eta(s)\exp\left(\int_s^t \alpha(\tau)d\tau\right)\right]^2 ds = \sum_{j=0}^{m-1} \sigma_j^2 \left[\int_0^t \varrho_j(s)\exp\left(\int_s^t \alpha(\tau)d\tau\right) ds\right]^2, \tag{7.142}$$

then RDE (7.139) and SDE (7.140) yield stochastic processes that are pointwise equivalent. Here $\boldsymbol{\mu} = (\mu_0, \mu_1, \ldots, \mu_{m-1})^T$.

Conversion of the RDE to the SDE Based on equivalent conditions (7.141) and (7.142), the specific form of the corresponding pointwise equivalent SDE can be given in terms of the known RDE. To do that, we assume that ϱ has the property that the function

$$h(t) = \sum_{j=0}^{m-1} \sigma_j^2 \varrho_j(t) \int_0^t \varrho_j(s) \exp\left(\int_s^t \alpha(\tau)d\tau\right) ds$$

is non-negative for any $t \geq 0$. By the equivalent conditions (7.141) and (7.142), the corresponding pointwise equivalent SDE for RDE (7.139) is then given by

$$dX(t) = [\alpha(t)X(t) + \gamma(t) + \boldsymbol{\mu} \cdot \boldsymbol{\varrho}(t)]dt + \sqrt{2h(t)}dW(t), \quad X(0) = X_0.$$

Conversion of the SDE to the RDE We consider the conversion from the SDE to the equivalent RDE. The equivalence conditions (7.141) and (7.142) imply that there are numerous different choices for the RDE (through choosing different values for m). One of the simple choices of the corresponding pointwise equivalent RDE for the SDE (7.140) is given by

$$\frac{dx(t; X_0, Z)}{dt} = \alpha(t)x(t; X_0, Z) + \xi(t) + Z\varrho_0(t), \quad x(0; X_0, Z) = X_0,$$

where the random variable $Z \sim \mathcal{N}(0, 1)$, and the function ϱ_0 is defined by

$$\varrho_0(t) = \frac{d}{dt}\left[\left(\int_0^t \eta^2(s)\exp\left(2\int_s^t \alpha(\tau)d\tau\right)ds\right)^{\frac{1}{2}}\right]$$

$$-\alpha(t)\left[\int_0^t \eta^2(s)\exp\left(2\int_s^t \alpha(\tau)d\tau\right)\right]^{\frac{1}{2}}.$$

The interested reader can refer to [22] for other possible choices of a pointwise equivalent RDE.

7.4.2.2 Scalar Affine Differential Equations (Class 2)

In this case, the random differential equation considered has the following form:

$$\frac{dx(t; X_0, \mathbf{Z})}{dt} = (\mathbf{Z} \cdot \boldsymbol{\varrho}(t) + \gamma(t))(x(t; X_0, \mathbf{Z}) + c),$$

$$\quad (7.143)$$

$$x(0; X_0, \mathbf{Z}) = X_0.$$

In the above equation, $\mathbf{Z} = (Z_0, Z_1, \ldots, Z_{m-1})^T$, where $Z_j \sim \mathcal{N}(\mu_j, \sigma_j^2)$, $j = 0, 1, 2, \ldots, m-1$, are mutually independent, with $\mu_j, \sigma_j, j = 0, 1, 2, \ldots, m-1$, being some positive constants. In addition, $\boldsymbol{\varrho} = (\varrho_0, \varrho_1, \ldots, \varrho_{m-1})^T$ is a non-random vector function of t, γ is a non-random function of t, and c is a given constant. The stochastic differential equation takes the form

$$dX(t) = \xi(t)(X(t) + c)dt + \eta(t)(X(t) + c)dW(t), \quad X(0) = X_0, \quad (7.144)$$

where ξ and η are some non-random functions of t.

The conditions for the pointwise equivalence between the RDE (7.143) and the SDE (7.144) are stated in the following theorem.

Theorem 7.4.7 *If the functions ξ, η and γ, and the function ϱ_j and the constants μ_j and σ_j, $j = 0, 1, 2, \ldots, m-1$, satisfy the two equalities*

$$\int_0^t (\xi(s) - \tfrac{1}{2}\eta^2(s))ds = \int_0^t (\boldsymbol{\mu} \cdot \boldsymbol{\varrho}(s) + \gamma(s))\,ds \quad (7.145)$$

and

$$\int_0^t \eta^2(s)ds = \sum_{j=0}^{m-1} \sigma_j^2 \left(\int_0^t \varrho_j(s)ds\right)^2, \quad (7.146)$$

then the RDE (7.143) and the SDE (7.144) yield stochastic processes that are pointwise equivalent. Here $\boldsymbol{\mu} = (\mu_0, \mu_1, \ldots, \mu_{m-1})^T$.

Conversion of the RDE to the SDE   Based on equivalent conditions (7.145) and (7.146), the specific form of the corresponding pointwise equivalent SDE can be given in terms of a known RDE. To do that, we assume that ϱ has the property that the function

$$\tilde{h}(t) = \sum_{j=0}^{m-1} \sigma_j^2 \varrho_j(t) \left(\int_0^t \varrho_j(s) ds \right)$$

is non-negative for any $t \geq 0$. Then the corresponding pointwise equivalent SDE for the RDE (7.143) is given by

$$dX(t) = \left(\boldsymbol{\mu} \cdot \boldsymbol{\varrho}(t) + \gamma(t) + \tilde{h}(t) \right) (X(t) + c) dt + \sqrt{2\tilde{h}(t)} (X(t) + c) dW(t),$$

$$X(0) = X_0.$$

Conversion of the SDE to the RDE   The equivalent conditions (7.145) and (7.146) imply that there are many different choices for an equivalent RDE (through choosing different values for m). One of the simple choices for the corresponding pointwise equivalent RDE for the SDE (7.144) is as follows:

$$\frac{dx(t; X_0, Z)}{dt} = \left(Z \varrho_0(t) + \xi(t) - \tfrac{1}{2}\eta^2(t) \right) (x(t; X_0, Z) + c), \quad x(0; X_0, Z) = X_0,$$

where the random variable $Z \sim \mathcal{N}(0, 1)$, and the function ϱ_0 is defined by

$$\varrho_0(t) = \frac{d}{dt} \left[\left(\int_0^t \eta^2(s) ds \right)^{\frac{1}{2}} \right].$$

Interested readers can refer to [22] for other possible choices for the pointwise equivalent RDE.

7.4.2.3   Vector Affine Systems

In this class of systems, we consider vector random differential equations of the form

$$\frac{d\mathbf{x}(t; \mathbf{X}_0, \mathbf{Z})}{dt} = \mathcal{A}(t)\mathbf{x}(t; \mathbf{X}_0, \mathbf{Z}) + \boldsymbol{\gamma}(t) + \mathcal{H}(t)\mathbf{Z},$$

$$\mathbf{x}(0; \mathbf{X}_0, \mathbf{Z}) = \mathbf{X}_0.$$

(7.147)

Here $\mathbf{Z}$ is an m-dimensional random vector that is multivariate normal distributed with mean vector $\boldsymbol{\mu}_{\mathbf{Z}}$ and covariance matrix $\boldsymbol{\Sigma}_{\mathbf{Z}}$ (that is, $\mathbf{Z} \sim \mathcal{N}(\boldsymbol{\mu}_{\mathbf{Z}}, \boldsymbol{\Sigma}_{\mathbf{Z}})$),

$\mathcal{A}$ is a non-random $n \times n$ matrix function of t, $\boldsymbol{\gamma}$ is a non-random n-dimensional column vector function of t, and $\mathcal{H}$ is a non-random $n \times m$ matrix function of t. The corresponding system of stochastic differential equations has the form

$$d\mathbf{X}(t) = [\mathcal{A}(t)\mathbf{X}(t) + \boldsymbol{\xi}(t)]dt + \mathcal{V}(t)d\mathbf{W}(t), \quad \mathbf{X}(0) = \mathbf{X}_0, \qquad (7.148)$$

where $\boldsymbol{\xi}$ is a non-random n-dimensional column vector function of t, $\mathcal{V}$ is a non-random $n \times l$ matrix function of t, and $\{\mathbf{W}(t) : t \geq 0\}$ is an l-dimensional Wiener process.

Let $\boldsymbol{\Phi}$ be the solution of the associated deterministic initial value problem

$$\frac{d\boldsymbol{\Phi}(t)}{dt} = \mathcal{A}(t)\boldsymbol{\Phi}(t), \quad \boldsymbol{\Phi}(0) = \mathbf{I}_n, \qquad (7.149)$$

where $\mathbf{I}_n$ is the $n \times n$ identity matrix. Then the conditions for the pointwise equivalence between the system of RDEs (7.147) and the system of SDEs (7.148) are stated in the following theorem.

Theorem 7.4.8 *If functions* $\boldsymbol{\xi}$, $\mathcal{V}$, $\boldsymbol{\gamma}$, $\mathcal{H}$, $\boldsymbol{\mu}_{\mathbf{Z}}$ *and* $\boldsymbol{\Sigma}_{\mathbf{Z}}$ *satisfy the two equalities*

$$\int_0^t \boldsymbol{\Phi}^{-1}(s)\left(\boldsymbol{\gamma}(s) + \mathcal{H}(s)\boldsymbol{\mu}_{\mathbf{Z}}\right)ds = \int_0^t \boldsymbol{\Phi}^{-1}(s)\boldsymbol{\xi}(s)ds, \qquad (7.150)$$

and

$$\left[\int_0^t \boldsymbol{\Phi}^{-1}(s)\mathcal{H}(s)ds\right]\boldsymbol{\Sigma}_{\mathbf{Z}}\left[\int_0^t \boldsymbol{\Phi}^{-1}(s)\mathcal{H}(s)ds\right]^T$$
$$= \int_0^t \boldsymbol{\Phi}^{-1}(s)\mathcal{V}(s)\left(\boldsymbol{\Phi}^{-1}(s)\mathcal{V}(s)\right)^T ds, \qquad (7.151)$$

then the system of random differential equations (7.147) and the system of stochastic differential equations (7.148) are pointwise equivalent.

Conversion of Systems of RDEs to Systems of SDEs   Based on the equivalent conditions (7.150) and (7.151), the specific form of the corresponding pointwise equivalent system of SDEs can be given in terms of parameters for a given system of RDEs. To do that, we assume that the matrix function

$$\boldsymbol{\Pi}_p(t) = \left[\int_0^t \boldsymbol{\Phi}^{-1}(s)\mathcal{H}(s)ds\right]\boldsymbol{\Sigma}_{\mathbf{Z}}\left[\int_0^t \boldsymbol{\Phi}^{-1}(s)\mathcal{H}(s)ds\right]^T$$

has the property that for any $t \geq 0$ the matrix $\boldsymbol{\Phi}(t)\dot{\boldsymbol{\Pi}}_p(t)(\boldsymbol{\Phi}(t))^T$ is positive-semidefinite. Then the corresponding pointwise equivalent system of SDEs for the system of RDEs (7.147) is given by

$$d\mathbf{X}(t) = [\mathcal{A}(t)\mathbf{X}(t) + \boldsymbol{\gamma}(t) + \mathcal{H}(t)\boldsymbol{\mu}_{\mathbf{Z}}]dt + \mathcal{V}(t)d\mathbf{W}(t), \quad \mathbf{X}(0) = \mathbf{X}_0,$$

where the matrix function $\mathcal{V}$ satisfies

$$\mathcal{V}(t)\mathcal{V}^T(t) = \boldsymbol{\Phi}(t)\dot{\boldsymbol{\Pi}}_p(t)(\boldsymbol{\Phi}(t))^T.$$

Conversion of Systems of SDEs to Systems of RDEs   The equivalent conditions (7.150) and (7.151) imply that there are many different choices for the equivalent system of RDEs. One of the simple choices for the corresponding pointwise equivalent system of RDEs for the system of SDEs (7.148) is given by

$$\frac{d\mathbf{x}(t;\mathbf{X}_0,\mathbf{Z})}{dt} = \mathcal{A}(t)\mathbf{x}(t;\mathbf{X}_0,\mathbf{Z}) + \boldsymbol{\xi}(t) + \boldsymbol{\Phi}(t)\dot{\boldsymbol{\Lambda}}(t)\mathbf{Z}, \quad \mathbf{x}(0;\mathbf{X}_0,\mathbf{Z}) = \mathbf{X}_0,$$

where $\mathbf{Z}$ is an n-dimensional random vector with $\mathbf{Z} \sim \mathcal{N}(\mathbf{0}_n, \mathbf{I}_n)$, and the matrix function $\boldsymbol{\Lambda}$ satisfies the equality

$$\boldsymbol{\Lambda}(t)(\boldsymbol{\Lambda}(t))^T = \int_0^t \boldsymbol{\Phi}^{-1}(s)\mathcal{V}(s)\left(\boldsymbol{\Phi}^{-1}(s)\mathcal{V}(s)\right)^T ds.$$

7.4.2.4   Non-Linear Differential Equations

The discussion in the above three subsections implies that if a non-linear stochastic differential equation (or a system of non-linear stochastic differential equations) can be reduced to one of the forms in (7.140) or (7.144) (or the form (7.148)) by some invertible transformation, then one can find its corresponding pointwise equivalent random differential equation. The converse is true for random differential equations.

To transform a stochastic differential equation to another stochastic differential equation, we need to use the so-called *Itô formula* (the chain rule for the Itô calculus), which is stated as follows.

Theorem 7.4.9 *Let h be a function of t and x that is continuously differentiable in t and twice continuously differentiable in x, and*

$$dX(t) = g(t, X(t))dt + \sigma(t, X(t))dW(t).$$

Then we have

$$dh(t, X(t)) = \frac{\partial h(t, X(t))}{\partial t}dt + \frac{\partial h(t, X(t))}{\partial x}dX(t)$$

$$+ \frac{1}{2}\sigma^2(t, X(t))\frac{\partial^2 h(t, X(t))}{\partial x^2}dt. \tag{7.152}$$

By (7.152) we see that it does not obey the classical chain rules (as we stated earlier).

The following two theorems (given in [52, Section 4.1]) show that there is a large class of non-linear stochastic differential equations that can be reduced to linear stochastic differential equations after some invertible transformation and thus are pertinent to our discussion here.

Theorem 7.4.10 *Consider the stochastic differential equation*

$$dX(t) = g(t, X(t))dt + \sigma(t, X(t))dW(t), \qquad (7.153)$$

where g and σ are non-random functions of t and x. If the equality

$$\frac{\partial}{\partial x}\left\{\sigma(t,x)\left[\frac{1}{\sigma^2(t,x)}\frac{\partial \sigma}{\partial t}(t,x) - \frac{\partial}{\partial x}\left(\frac{g}{\sigma}\right)(t,x) + \frac{1}{2}\frac{\partial^2 \sigma}{\partial x^2}(t,x)\right]\right\} = 0$$

holds, then (7.153) can be reduced to the linear stochastic differential equation

$$dY(t) = \bar{g}(t)dt + \bar{\sigma}(t)dW(t).$$

Here $\bar{\sigma}$ is some non-random function determined from

$$\frac{d}{dt}\bar{\sigma}(t) = \bar{\sigma}(t)\sigma(t,x)\left[\frac{1}{\sigma^2(t,x)}\frac{\partial \sigma}{\partial t}(t,x) - \frac{\partial}{\partial x}\left(\frac{g}{\sigma}\right)(t,x) + \frac{1}{2}\frac{\partial^2 \sigma}{\partial x^2}(t,x)\right],$$

and $\bar{g}$ is some deterministic function given by

$$\bar{g}(t) = \frac{\partial h}{\partial t}(t,x) + \frac{\partial h}{\partial x}(t,x)g(t,x) + \frac{1}{2}\frac{\partial^2 h}{\partial x^2}(t,x)\sigma^2(t,x)$$

with h being some smooth invertible function computed from

$$\frac{\partial h}{\partial x}(t,x) = \frac{\bar{\sigma}(t)}{\sigma(t,x)}.$$

The relationship between $Y(t)$ and $X(t)$ is given by $Y(t) = h(t, X(t))$.

Theorem 7.4.11 *The autonomous stochastic differential equation*

$$dX(t) = g(X(t))dt + \sigma(X(t))dW(t)$$

can be reduced to the linear stochastic differential equation

$$dY(t) = (\lambda_0 + \lambda_1 Y(t))dt + (\nu_0 + \nu_1 Y(t))dW(t)$$

if and only if

$$\psi'(x) = 0 \ or \ \left(\frac{(\sigma\psi')'}{\psi'}\right)'(x) = 0. \qquad (7.154)$$

Here g and σ are non-random functions of x, λ_0, λ_1, ν_0 and ν_1 are some constants, $\psi(x) = \frac{g(x)}{\sigma(x)} - \frac{1}{2}\sigma'(x)$ with the prime (') denoting the $\frac{d}{dx}$ derivative, and $Y(t) = h(X(t))$ with h being some invertible transformation. If the latter part of (7.154) is satisfied, then we set $\nu_1 = -\frac{(\sigma\psi')'}{\psi'}$ and choose

$$h(x) = \begin{cases} c_1 \exp\left(\nu_1 \int_a^x \frac{1}{\sigma(\varsigma)}d\varsigma\right), & \text{if } \nu_1 \neq 0 \\[2em] \nu_0 \int_a^x \frac{1}{\sigma(\varsigma)}d\varsigma + c_0, & \text{if } \nu_1 = 0, \end{cases}$$

where a, c_0 and c_1 are some constants.

Based on the equivalence results in Sections 7.4.2.1–7.4.2.3 as well as the results presented in Theorem 7.4.10 and Theorem 7.4.11, one can derive the corresponding pointwise equivalent RDEs for a large class of non-linear SDEs. We next use several examples to illustrate this transformation method to find the corresponding pointwise equivalent SDE/RDE to the RDE/SDE.

Example 7.4.1 (*Logistic Growth Dynamics with Uncertainty*) We begin with logistic growth in the RDE form and derive its pointwise equivalent SDE form. Consider the deterministic logistic equation

$$\frac{dx}{dt} = bx\left(1 - \frac{x}{\kappa}\right), \quad x(0) = x_0, \tag{7.155}$$

where b is some positive constant representing the intrinsic growth rate, and κ is a given positive constant representing the carrying capacity. Let $y = \frac{1}{x}$. Then it is easy to find that

$$\frac{dy}{dt} = -b\left(y - \frac{1}{\kappa}\right), \quad y(0) = y_0,$$

where $y_0 = \frac{1}{x_0}$. By the discussion in Section 7.4.2.2, we know that the RDE

$$\frac{dy(t;B)}{dt} = -B\left(y(t;B) - \frac{1}{\kappa}\right), \quad y(0;B) = Y_0$$

with $B \sim \mathcal{N}(\mu_0, \sigma_0^2)$ is pointwise equivalent to the SDE given by

$$dY(t) = (-\mu_0 + \sigma_0^2 t)\left(Y(t) - \frac{1}{\kappa}\right)dt + \sqrt{2}t\sigma_0\left(Y(t) - \frac{1}{\kappa}\right)dW(t), \quad Y(0) = Y_0.$$

Let $X(t) = \frac{1}{Y(t)}$. Then by Itô's formula (i.e., Theorem 7.4.9) we find that

$$dX(t) = -\frac{1}{Y^2(t)}\left[(-\mu_0 + \sigma_0^2 t)\left(Y(t) - \frac{1}{\kappa}\right)dt + \sqrt{2}t\sigma_0\left(Y(t) - \frac{1}{\kappa}\right)dW(t)\right]$$

$$+ \frac{1}{Y^3(t)}\left[\sqrt{2}t\sigma_0\left(Y(t) - \frac{1}{\kappa}\right)\right]^2 dt$$

$$= X(t)\left[(\mu_0 - \sigma_0^2 t)\left(1 - \frac{X(t)}{\kappa}\right) + 2t\sigma_0^2\left(1 - \frac{X(t)}{\kappa}\right)^2\right]dt$$

$$- \sqrt{2}t\sigma_0 X(t)\left(1 - \frac{X(t)}{\kappa}\right)dW(t).$$

Thus the random differential equation

$$\frac{dx(t;B)}{dt} = Bx(t;B)\left(1 - \frac{x(t;B)}{\kappa}\right), \quad x(0;B) = X_0 \tag{7.156}$$

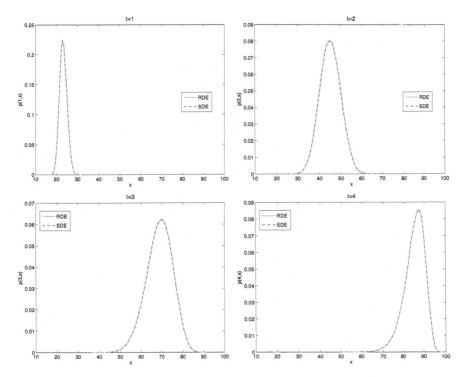

FIGURE 7.6: Probability density function $p(t, \cdot)$ obtained by simulating 10^5 sample paths for the RDE (7.156) and the SDE (7.157) at $t = 1, 2, 3$ and 4, where $\Delta t = 0.004$ is used in (7.158), and $t_f = 4$.

with $B \sim \mathcal{N}(\mu_0, \sigma_0^2)$ is pointwise equivalent to the SDE

$$
dX(t) = X(t) \left[(\mu_0 - \sigma_0^2 t) \left(1 - \frac{X(t)}{\kappa} \right) + 2t\sigma_0^2 \left(1 - \frac{X(t)}{\kappa} \right)^2 \right] dt
$$
$$
- \sqrt{2} t \sigma_0 X(t) \left(1 - \frac{X(t)}{\kappa} \right) dW(t), \tag{7.157}
$$

$$
X(0) = X_0.
$$

Figure 7.6 depicts the probability density function $p(t, \cdot)$ at different times t for the RDE (7.156) and the SDE (7.157) with $\kappa = 100$, $X_0 = 10$, $\mu_0 = 1$ and $\sigma_0 = 0.1$, where the plots are obtained by simulating 10^5 sample paths for each form and then using the kernel density estimation method (a non-parametric method to estimate the probability density function; see [102, 113] or Wikipedia for details) implemented through the MATLAB function *ksdensity* to obtain the approximation of p. Specifically, for the RDE (7.156), we analytically solve each deterministic differential equation (7.155) with b being a realization of B, which follows a normal distribution $\mathcal{N}(1, 0.01)$. This solution is given by

$$x(t; b) = \frac{x_0 \kappa \exp(bt)}{\kappa + x_0(\exp(bt) - 1)}.$$

Thus implementation of the random differential equation is extremely rapid (essentially a function evaluation). This is not true for the SDE and hence the RDE requires much less (orders of magnitude) implementation time than does the SDE. Even in examples where both methods require numerical integration, the RDE is highly preferred to the SDE in implementation of inverse problems (e.g., see [16, 19, 33]). We use an Euler explicit method to numerically approximate the sample paths for the stochastic differential equation (7.157). We remark that the scheme we used is crude but sufficient for our purposes here since our main objective is to demonstrate our theoretical equivalence results with a numerical example. Interested readers can refer to [73] for an excellent resource on the numerical methods for stochastic differential equations. Let t_f denote the final time, and m be the number of mesh point intervals. Then the mesh time points are given by $t_k = k\Delta t, k = 0, 1, 2, \ldots, m$, where $\Delta t = t_f/m$. Denote by X^k the numerical solution for $X(t_k)$; then we have the following numerical scheme:

$$X^{k+1} = X^k + X^k \left[(\mu_0 - \sigma_0^2 t_k) \left(1 - \frac{X^k}{\kappa} \right) + 2t_k \sigma_0^2 \left(1 - \frac{X^k}{\kappa} \right)^2 \right] \Delta t$$
$$- \sqrt{2t_k} \sigma_0 X^k \left(1 - \frac{X^k}{\kappa} \right) \mathcal{E}_k, \quad k = 1, 2, \ldots, m - 1$$

(7.158)

where $\mathcal{E}_k, k = 1, 2, \ldots, m-1$, are independent and identically distributed random variables following a normal distribution $\mathcal{N}(0, \Delta t)$. From Figure 7.6 we see that we obtain the same probability density function for the RDE (7.156) and the SDE (7.157), which nicely illustrates our earlier theoretical results.

Example 7.4.2 (*Gompertz Growth in the Drift*) In this example, we consider the non-linear stochastic differential equation

$$dX(t) = [a_0(t) - a_1(t) \ln(X(t))] X(t)dt + \sqrt{2d_0(t)} X(t)dW(t),$$
$$X(0) = X_0,$$

(7.159)

where a_0, a_1 and d_0 are some deterministic functions of t, and a_1 and d_0 are assumed to be positive. This equation is a stochastic version of the generalized Gompertz equation $\dot{x}(t) = [a_0(t) - a_1(t) \ln(x(t))]x(t)$, which has been extensively used in biological and medical research to describe population dynamics, such as tumor growth in humans and animals either with or without treatment (e.g., [5, 50] and the references therein) as well as in cell proliferation models (recall the discussion in Section 4.4). Next we use the transformation method to find a pointwise equivalent RDE for (7.159). Let $Y = \ln(X)$. Then by Itô's formula and (7.159) we have

$$dY(t) = \frac{1}{X(t)} \left\{ [a_0(t) - a_1(t) \ln(X(t))] X(t)dt + \sqrt{2d_0(t)} X(t)dW(t) \right\}$$

$$- \frac{1}{2X^2(t)} [2d_0(t)X^2(t)] dt$$

$$= [-a_1(t)Y(t) + (a_0(t) - d_0(t))] dt + \sqrt{2d_0(t)} dW(t).$$

By the discussion in Section 7.4.2.1 we find that a pointwise equivalent RDE for the above linear stochastic differential equation with initial condition $Y(0) = Y_0$ (where $Y_0 = \ln(X_0)$) is given by

$$\frac{dy(t; B)}{dt} = -a_1(t)y(t; B) + a_0(t) - d_0(t) + B\varrho(t), \quad y(0; B) = Y_0, \quad (7.160)$$

where $B \sim \mathcal{N}(0, 1)$, and ϱ is given by

$$\varrho(t) = \frac{d}{dt} \left(\sqrt{\varphi(t)} \right) + a_1(t)\sqrt{\varphi(t)}$$

with

$$\varphi(t) = \int_0^t 2d_0(s) \exp\left(-2 \int_s^t a_1(\varsigma)d\varsigma \right) ds.$$

Let $x = \exp(y)$. Then by (7.160) we find

$$\frac{dx(t; B)}{dt} = \exp(y(t; B)) \left[-a_1(t)y(t; B) + a_0(t) - d_0(t) + B\varrho(t) \right]$$

$$= x(t; B) \left[-a_1(t) \ln(x(t; B)) + a_0(t) - d_0(t) + B\varrho(t) \right].$$

Thus, the stochastic differential equation

$$dX(t) = [a_0(t) - a_1(t) \ln(X(t))] X(t)dt + \sqrt{2d_0(t)} X(t)dW(t),$$

$$X(0) = X_0$$

is pointwise equivalent to the random differential equation

$$\frac{dx(t; B)}{dt} = x(t; B) \left[-a_1(t) \ln(x(t; B)) + a_0(t) - d_0(t) + B\varrho(t) \right]$$

$$x(0; B) = X_0$$

with $B \sim \mathcal{N}(0, 1)$.

7.4.2.5 Remarks on the Equivalence between the SDE and the RDE

In Sections 7.4.2.1–7.4.2.4 we gave the pointwise equivalent conditions for the SDE and the RDE. Even though the stochastic processes resulting from the SDE and its pointwise equivalent RDE have the same probability density function at each time t, these two processes may <u>not</u> be the same, as they may have different mth ($m \geq 2$) probability density functions (recall the discussion in Section 7.1.1). Below we give an example to illustrate this by comparing the covariance functions of the two stochastic processes resulting from an SDE and its pointwise equivalent RDE.

Based on the results in Section 7.4.2.2 we find that the SDE

$$dX(t) = b_0(X(t) + c_0)dt + \sqrt{2t}\sigma_0(X(t) + c_0)dW(t), \quad X(0) = X_0 \quad (7.161)$$

is pointwise equivalent to the RDE

$$\frac{dx(t; X_0, B)}{dt} = (B - \sigma_0^2 t)(x(t; X_0, B) + c_0), \quad x(0; X_0, B) = X_0 \quad (7.162)$$

with $B \sim \mathcal{N}(b_0, \sigma_0^2)$. One can prove that (7.161) and (7.162) generate stochastic processes $\{X(t) : t \geq 0\}$ and $\{x(t; X_0, B) : t \geq 0\}$ with $X(t)$ and $x(t; X_0, B)$ respectively given by

$$X(t) = -c_0 + (X_0 + c_0)Y_S(t),$$

and

$$x(t; X_0, B) = -c_0 + (X_0 + c_0)Y_R(t),$$

where $Y_S(t)$ and $Y_R(t)$ are

$$Y_S(t) = \exp\left((b_0 t - \tfrac{1}{2}\sigma_0^2 t^2) + \sigma_0 \int_0^t \sqrt{2s}\,dW(s)\right), \quad (7.163)$$

and

$$Y_R(t) = \exp(Bt - \tfrac{1}{2}\sigma_0^2 t^2), \quad \text{with } B \sim \mathcal{N}(b_0, \sigma_0^2), \quad (7.164)$$

respectively.

We proceed to use (2.86) and Lemma 7.4.5 to find the covariance function of the stochastic processes $\{Y_R(t) : t \geq 0\}$ and $\{Y_S(t) : t \geq 0\}$. By (7.164) and (2.86) we find immediately

$$\mathbb{E}(Y_R(t)) = \exp(b_0 t). \quad (7.165)$$

Then, using (2.86) and (7.165), we find that the covariance function for the stochastic process $\{Y_R(t) : t \geq 0\}$ is given by

$$\begin{aligned}
\text{Cov}&\{Y_R(t), Y_R(s)\} \\
&= \mathbb{E}(Y_R(t)Y_R(s)) - \mathbb{E}(Y_R(t))\mathbb{E}(Y_R(s)) \\
&= \mathbb{E}\left(\exp\left(B(t+s) - \tfrac{1}{2}\sigma_0^2(t^2 + s^2)\right)\right) - \exp(b_0(t+s)) \quad (7.166) \\
&= \exp\left(b_0(t+s) - \tfrac{1}{2}\sigma_0^2(t^2 + s^2) + \tfrac{1}{2}\sigma_0^2(t+s)^2\right) - \exp(b_0(t+s)) \\
&= \exp(b_0(t+s))\left[\exp\left(st\sigma_0^2\right) - 1\right].
\end{aligned}$$

Let $Q(t) = \sigma_0 \int_0^t \sqrt{2\varsigma} dW(\varsigma)$. Then by Lemma 7.4.5, we have that $\{Q(t) : t \geq 0\}$ is a Gaussian process with zero mean and a covariance function given by

$$\text{Cov}\{Q(t), Q(s)\} = \sigma_0^2 \min\{t^2, s^2\}. \tag{7.167}$$

Hence, $Q(t) + Q(s)$ has a Gaussian distribution with zero mean and variance defined by

$$\begin{aligned} \text{Var}(Q(t) + Q(s)) &= \text{Var}(Q(t)) + \text{Var}(Q(s)) + 2\text{Cov}\{Q(t), Q(s)\} \\ &= \sigma_0^2 \left(t^2 + s^2 + 2\min\{t^2, s^2\}\right). \end{aligned} \tag{7.168}$$

Using (2.86), (7.163) and (7.167) we find that

$$\mathbb{E}(Y_S(t)) = \exp(b_0 t). \tag{7.169}$$

Now we use (2.86) along with Equations (7.169) and (7.168) to find the covariance function of $\{Y_S(t)\}$.

$$\begin{aligned} &\text{Cov}\{Y_S(t), Y_S(s)\} \\ &= \mathbb{E}(Y_S(t)Y_S(s)) - \mathbb{E}(Y_S(t))\mathbb{E}(Y_S(s)) \\ &= \mathbb{E}\left(\exp\left(b_0(t+s) - \tfrac{1}{2}\sigma_0^2(t^2 + s^2) + Q(t) + Q(s)\right)\right) \\ &\quad - \exp(b_0(t+s)) \\ &= \exp\left(b_0(t+s) - \tfrac{1}{2}\sigma_0^2(t^2 + s^2) + \tfrac{1}{2}\sigma_0^2\left(t^2 + s^2 + 2\min\{t^2, s^2\}\right)\right) \\ &\quad - \exp(b_0(t+s)) \\ &= \exp(b_0(t+s))\left[\exp\left(\sigma_0^2 \min\{t^2, s^2\}\right) - 1\right]. \end{aligned} \tag{7.170}$$

Thus by (7.166) and (7.170) we see that $\{Y_R(t) : t \geq 0\}$ and $\{Y_S(t) : t \geq 0\}$ have different covariance functions and hence are *not* the same stochastic process. Therefore, the stochastic process generated by (7.161) is different from the one generated by (7.162).

7.4.2.6 Relationship between the FPPS and GRDPS Population Models

To establish the pointwise equivalence between SDEs and RDEs in Sections 7.4.2.1–7.4.2.4, we assumed a normal distribution on each component of $\mathbf{Z}$ (i.e., $Z_j \sim \mathcal{N}(\mu_j, \sigma_j^2)$) for the RDE in Sections 7.4.2.1 and 7.4.2.2 (and imposed a multivariate normal distribution on $\mathbf{Z}$ in Section 7.4.2.3). However, this normal assumption is not completely reasonable in many biological applications. This is because when the variance σ_j^2 is sufficiently large relative to μ_j, the growth rate in the mathematical model may become negative for many individuals in the population, which results in the size classes of these individuals having non-negligible probability of being negative in a finite time period (here without loss of generality we assume that the minimum size of individuals is zero). A standard approach in practice to remedy this problem is to impose a *truncated normal* distribution instead of a normal distribution; that is, we restrict Z_j in some reasonable range. We also observe that the corresponding SDE also can lead to the size having non-negligible probability of being negative when the diffusion dominates the drift because $W(t) \sim \mathcal{N}(0,t)$ for any fixed t and hence decreases in size are possible. One way to remedy this situation is to set $X(t) = 0$ if $X(t) \leq 0$. Thus, if σ_j is sufficiently large relative to μ_j, then we may obtain different probability density functions for the solutions of the RDE and SDE after we have made these different modifications to each. The same anomalies (demonstrated in [16] with some computational examples) hold for the solutions of the associated FPPS and GRDPS population models themselves when we impose the zero-flux boundary condition at the minimum size zero (which is equivalent to setting $X(t) = 0$ if $X(t) \leq 0$ for the SDE in the sense that both are used to keep individuals in the system) in the FPPS population model and put constraints on each component of $\mathbf{Z}$ in the GRDPS population model.

However, for the case where the truncated normal distribution can provide a good approximation to the normal distribution (i.e., for the case where σ_j is sufficiently small relative to μ_j), the GRDPS population model and the corresponding FPPS population model can, with appropriate boundary and initial conditions as well as properly chosen parameters based on pointwise equivalent results between the SDE and the RDE, yield quite similar solutions. This is predicated on our analysis of the SDE and the RDE and their resulting FPPS and GRDPS population models, and was also demonstrated in the following example (see [16] for other computational examples).

Example 7.4.3 Model parameters in the FPPS population model (7.79) and the GRDPS population model (7.110)–(7.111) in the domain $\{(t,x) \mid (t,x) \in [0, t_f] \times [0, \bar{x}]\}$ are chosen based on the pointwise equivalent SDE (7.161) and RDE (7.162); that is, they are given by

$$\text{FPPS model: } g(x) = b_0(x + c_0), \quad \sigma(t,x) = \sqrt{2t}\sigma_0(x + c_0)$$

and

$$\text{GRDPS model: } g(t,x;b) = (b - \sigma_0^2 t)(x + c_0),$$

$$\text{where } b \in [\underline{b}, \bar{b}] \text{ with } B \sim \mathcal{N}_{[\underline{b}, \bar{b}]}(b_0, \sigma_0^2).$$

Here we choose $\underline{b} = b_0 - 3\sigma_0$ and $\bar{b} = b_0 + 3\sigma_0$. The initial condition in the FPPS model is given by $u_0(x) = 100 \exp(-100(x - 0.4)^2)$, and initial conditions in the GRDPS model are given by $v_0(x;b) = 100 \exp(-100(x - 0.4)^2)$ for $b \in [\underline{b}, \bar{b}]$. For the simulation results below we set $c_0 = 0.1$, $b_0 = 0.045$, $\sigma_0 = rb_0$ with r a positive constant, and $t_f = 10$.

Let $r_0 = \dfrac{-3 + \sqrt{4b_0 t_f + 9}}{2b_0 t_f}$ (≈ 0.3182). Then it is easy to show that if $r < r_0$, then $g(t, x; b) = (b - \sigma_0^2 t)(x + c) > 0$ in $\{(t, x) \mid (t, x) \in [0, t_f] \times [0, \bar{x}]\}$ for all $b \in [\underline{b}, \bar{b}]$. Here in this example, we just consider the case with $r < r_0$, i.e., the growth rate of each subpopulation is positive. By the discussion in Section 7.3.2.3 we know that to conserve the total population in the system, we need to choose $\bar{x}$ sufficiently large such that $v(t, \bar{x}; b)$ is negligible for any $t \in [0, t_f]$ and $b \in [\underline{b}, \bar{b}]$. In this example, we choose $\bar{x} = 3$.

We notice that with this choice of $g(t, x; b) = (b - \sigma_0^2 t)(x + c_0)$ in the GRDPS population model, we can analytically solve (7.110) by the *method of characteristics* (a general technique to solve a first-order partial differential equation; see [32, Section 9.5] and the references therein for details), and the solution is given by

$$v(t, x; b) = \begin{cases} v_0(\phi(t, x); b) \exp\left(-bt + \frac{1}{2}\sigma_0^2 t^2\right) & \text{if } \psi(t, x) \geq 0 \\ 0 & \text{if } \psi(t, x) < 0, \end{cases} \tag{7.171}$$

where $\psi(t, x) = c_0 + (x + r_0)\exp(-bt + \frac{1}{2}\sigma_0^2 t^2)$. Hence, by (2.84) and (7.111) we have

$$u(t, x) = \int_{\underline{b}}^{\bar{b}} v(t, x; b) \frac{\frac{1}{\sigma_0} \phi\left(\frac{b - b_0}{\sigma_0}\right)}{\Phi\left(\frac{\bar{b} - b_0}{\sigma_0}\right) - \Phi\left(\frac{\underline{b} - b_0}{\sigma_0}\right)} db, \tag{7.172}$$

where ϕ is the probability density function of the standard normal distribution, and Φ is its corresponding cumulative distribution function. In the simulation, the trapezoidal rule was used to calculate the integral in (7.172) with $\Delta b = (\bar{b} - \underline{b})/128$.

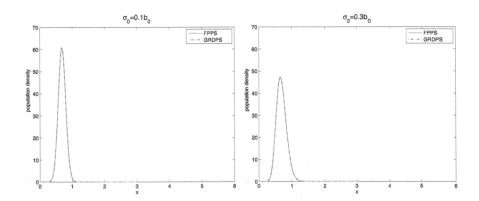

FIGURE 7.7: Numerical solutions $u(t_f, x)$ to the FPPS model and the GRDPS model with model parameters chosen as (7.4.3), where $\underline{b} = b_0 - 3\sigma_0$ and $\bar{b} = b_0 + 3\sigma_0$.

The FPPS population model (7.79) is numerically solved by the finite difference scheme developed by Chang and Cooper in [39]. This scheme provides a numerical solution which preserves some of the more important intrinsic properties of the FPPS model. In particular, the solution is non-negative, and particle conserving in the absence of sources or sinks. We refer the interested reader to [16] for details on how to use this scheme to numerically solve (7.79).

The numerical solution of the FPPS model and the solution of the GRDPS model at $t = t_f$ with $r = 0.1$ and 0.3 are shown in Figure 7.7, which indicates that we obtain quite similar population densities for these two models. This is because $\mathcal{N}_{[\underline{b}, \bar{b}]}(b_0, \sigma_0^2)$ is a good approximation of $\mathcal{N}(b_0, \sigma_0^2)$ (for this setup of $\underline{b}$ and $\bar{b}$) and σ_0 is chosen sufficiently small so that size distribution obtained in (7.161) (or (7.162)) is a good approximation of size distributions obtained by the GRDPS model and the FPPS model. (Note that the population density is just the product of the total number of individuals in the population and the probability density function of the associated stochastic process.)

Final remarks   It is well documented (e.g., see [33, 61] and the references therein) that there exist significant difficulties in numerically solving FPPS models when the drift dominates the diffusion (the case of primary interest in many situations) even for the scalar case. This is a serious drawback in forward simulations, but it is even more of a challenge in the inverse problem considered in [33]. Thus it is of great interest to explore alternative equivalent formulations that involve a much more efficient computational implementation. As mentioned above, the GRDPS models are readily solved rapidly with so-called "embarrassingly parallel" computational methods [19, 118], where the model governing each subpopulation can be solved via the method of characteristics or some other well-established numerical methods (e.g., [3, 8] and the references therein). Hence, the corresponding GRDPS model could be used as an alternative method that can be fast and efficient in numerically solving the FPPS model, especially in the case of inverse problems.

References

[1] A.S. Ackleh, H.T. Banks and K. Deng, A finite difference approximation for a coupled system of nonlinear size-structured populations, *Nonlinear Analysis*, **50** (2002), 727–748.

[2] A.S. Ackleh and K. Deng, A monotone approximation for the nonautonomous size-structured population model, *Quart. Appl. Math.*, **57** (1999), 261–267.

[3] A.S. Ackleh and K. Ito, An implicit finite-difference scheme for the

nonlinear size-structured population model, *Numer. Funct. Anal. Optim.*, **18** (1997), 865–884.

[4] B.M. Adams, H.T. Banks, M. Davidian, and E.S. Rosenberg, Model fitting and prediction with HIV treatment interruption data, *Bulletin of Math. Biology*, **69** (2007), 563–584.

[5] G. Albano, V. Giorno, P. Roman-Roman and F. Torres-Ruiz, Inferring the effect of therapy on tumors showing stochastic Gompertzian growth, *Journal of Theoretical Biology*, **276** (2011), 67–77.

[6] L.J.S. Allen, *An Introduction to Stochastic Processes with Applications to Biology*, Chapman and Hall/CRC Press, Boca Raton, FL, 2011.

[7] D.F. Anderson and T.G. Kurtz, Continuous time Markov chain models for chemical reaction networks, in *Design and Analysis of Biomolecular Circuits*, H. Koeppl, D. Densmore, G. Setti and M. di Bernardo eds. (2011), 3–42.

[8] O. Angulo and J.C. López-Marcos, Numerical integration of fully nonlinear size-structured population models, *Applied Numerical Mathematics*, **50** (2004), 291–327.

[9] H.T. Banks, V.A. Bokil, S. Hu, A.K. Dhar, R.A. Bullis, C.L. Browdy and F.C.T. Allnutt, Modeling shrimp biomass and viral infection for production of biological countermeasures, *Mathematical Biosciences and Engineering*, **3** (2006), 635–660.

[10] H.T. Banks, L.W. Botsford, F. Kappel and C. Wang, Modeling and estimation in size structured population models, LCDS-CCS Report 87-13, Brown University; Proceedings 2nd Course on Mathematical Ecology, (Trieste, December 8–12, 1986) World Press, Singapore, 1988, 521–541.

[11] H.T. Banks, M. Davidian, S. Hu, G.M. Kepler and E.S. Rosenberg, Modeling HIV immune response and validation with clinical data, *Journal of Biological Dynamics*, **2** (2008), 357–385.

[12] H.T. Banks and J.L. Davis, A comparison of approximation methods for the estimation of probability distributions on parameters, *Applied Numerical Mathematics*, **57** (2007), 753–777.

[13] H.T. Banks and J.L. Davis, Quantifying uncertainty in the estimation of probability distributions, *Math. Biosci. Engr.*, **5** (2008), 647–667.

[14] H.T. Banks, J.L. Davis, S.L. Ernstberger, S. Hu, E. Artimovich, A.K. Dhar and C.L. Browdy, A comparison of probabilistic and stochastic formulations in modeling growth uncertainty and variability, *Journal of Biological Dynamics*, **3** (2009), 130–148.

[15] H.T. Banks, J.L. Davis, S.L. Ernstberger, S. Hu, E. Artimovich and A.K. Dhar, Experimental design and estimation of growth rate

distributions in size-structured shrimp populations, *Inverse Problems*, **25** (2009), 095003 (28 pp).

[16] H.T. Banks, J.L. Davis and S. Hu, A computational comparison of alternatives to including uncertainty in structured population models, CRSC-TR09-14, June, 2009; in *Three Decades of Progress in Systems and Control*, X. Hu, U. Jonsson, B. Wahlberg and B. Ghosh (Eds.), Springer, Berlin 2010, 19–33.

[17] H.T. Banks, S.L. Ernstberger and S. Hu, Sensitivity equations for a size-structured population model, *Quarterly of Applied Mathematics*, **67** (2009), 627–660.

[18] H.T. Banks and B.G. Fitzpatrick, Estimation of growth rate distributions in size structured population models, *Quarterly of Applied Mathematics*, **49** (1991), 215–235.

[19] H.T. Banks, B.G. Fitzpatrick, L.K. Potter and Y. Zhang, Estimation of probability distributions for individual parameters using aggregate population data, CRSC-TR98-6, January, 1998; in *Stochastic Analysis, Control, Optimization and Applications*, (Edited by W. McEneaney, G. Yin and Q. Zhang), Birkhauser, Boston, 1989, 353–371.

[20] H.T. Banks and N.L. Gibson, Inverse problems involving Maxwell's equations with a distribution of dielectric parameters, *Quarterly of Applied Mathematics*, **64** (2006), 749–795.

[21] H.T. Banks, S.L. Grove, S. Hu and Y. Ma, A hierarchical Bayesian approach for parameter estimation in HIV models, *Inverse Problems*, **21** (2005), 1803–1822.

[22] H.T. Banks and S. Hu, Nonlinear stochastic Markov processes and modeling uncertainty in populations, CRSC-TR11-02, January, 2011; *Mathematical Bioscience and Engineering*, **9** (2012), 1–25.

[23] H.T. Banks and S. Hu, Propagation of growth uncertainty in a physiologically structured population, *Journal on Mathematical Modelling of Natural Phenomena*, **7** (2012), 7–23.

[24] H.T. Banks and S. Hu, Uncertainty propagation and quantification in a continuous time dynamical system, *International Journal of Pure and Applied Mathematics*, **80** (2012), 93–145.

[25] H.T. Banks, S. Hu and Z.R. Kenz, A brief review of elasticity and viscoelasticity for solids, *Advances in Applied Mathematics and Mechanics*, **3** (2011), 1–51.

[26] H.T. Banks and F. Kappel, Transformation semigroups and L^1-approximation for size-structured population models, *Semigroup Forum*, **38** (1989), 141–155.

[27] H.T. Banks, F. Kappel and C. Wang, Weak solutions and differentiability for size-structured population models, *International Series of Numerical Mathematics*, **100** (1991), 35–50.

[28] H.T. Banks, K.L. Sutton, W.C. Thompson, G. Bocharov, D. Roose, T. Schenkel and A. Meyerhans, Estimation of cell proliferation dynamics using CFSE data, *Bull. Math. Biol.*, **73** (2011), 116–150.

[29] H.T. Banks, W.C. Thompson, Mathematical models of dividing cell populations: Application to CFSE data, *Journal on Mathematical Modelling of Natural Phenomena*, **7** (2012), 24–52.

[30] H.T. Banks and W.C. Thompson, A division-dependent compartmental model with cyton and intracellular label dynamics, *Intl. J. of Pure and Applied Math.*, **77** (2012), 119–147.

[31] H.T. Banks, W.C. Thompson, C. Peligero, S. Giest, J. Argilaguet and A. Meyerhans, A compartmental model for computing cell numbers in CFSE-based lymphocyte proliferation assays, *Math. Biosci. Engr.*, **9** (2012), 699–736.

[32] H.T. Banks and H.T. Tran, *Mathematical and Experimental Modeling of Physical and Biological Processes*, CRC Press, Boca Raton, FL, 2009.

[33] H.T. Banks, H.T. Tran and D.E. Woodward, Estimation of variable coefficients in the Fokker-Planck equations using moving node finite elements, *SIAM J. Numer. Anal.*, **30** (1993), 1574–1602.

[34] G.I. Bell and E.C. Anderson, Cell growth and division. I. A mathematical model with applications to cell volume distributions in mammalian suspension cultures, *Biophysical Journal*, **7** (1967), 329–351.

[35] I. Bena, Dichotomous Markov noise: exact results for out-of-equilibrium systems, *International Journal of Modern Physics B*, **20** (2006), 2825–2888.

[36] C.V.D. Broeck, On the relation between white shot noise, Gaussian white noise, and the dichotomic Markov process, *Journal of Statistical Physics*, **31** (1983), 467–483.

[37] A. Calsina and J. Saldana, A model of physiologically structured population dynamics with a nonlinear individual growth rate, *Journal of Mathematical Biology*, **33** (1995), 335–364.

[38] F.L. Castille, T.M. Samocha, A.L. Lawrence, H. He, P. Frelier and F. Jaenike, Variability in growth and survival of early postlarval shrimp (*Penaeus vannamei Boone 1931*), *Aquaculture*, **113** (1993), 65–81.

[39] J.S. Chang and G. Cooper, A practical difference scheme for Fokker-Planck equations, *J. Comp. Phy.*, **6** (1970), 1–16.

[40] J. Chen and J. Li, A note on the principle of preservation of probability and probability density evolution equation, *Probabilistic Engineering Mechanics*, **24** (2009), 51–59.

[41] J.A. Clarkson and C.R. Adams, On functions of bounded variation for functions of two variables, *Transactions of the American Mathematical Society*, **35** (1933), 824–854.

[42] J. Chu, A. Ducrot, P. Magal and S. Ruan, Hopf bifurcation in a size-structured population dynamic model with random growth, *J. Differential Equations*, **247** (2009), 956–1000.

[43] J.M. Cushing, *An Introduction to Structured Population Dynamics*, in CMB-NSF Regional Conference Series in Applied Mathematics, SIAM, Philadelphia, PA, 1998.

[44] J. Ding, T.Y. Li and A. Zhou, Finite approximations of Markov operators, *Journal of Computational and Applied Mathematics*, **147** (2002), 137–152.

[45] J.L. Doob, *Stochastic Processes*, Wiley, New York, 1953.

[46] B.G. Dostupov and V.S. Pugachev, The equation for the integral of a system of ordinary differential equations containing random parameters, *Automation and Remote Control*, **18** (1957), 620–630.

[47] D. Ellis, About colored noise, `http://www.ee.columbia.edu/~dpwe/noise/`.

[48] S.N. Ethier and T.G. Kurtz, *Markov Processes: Characterization and Convergence*, John Wiley & Sons, New York, 1986.

[49] J.Z. Farkas and T. Hagen, Stability and regularity results for a size-structured population model, *Journal of Mathematical Analysis and Applications*, **328** (2007), 119–136.

[50] L. Ferrante, S. Bompadre, L. Possati and L. Leone, Parameter estimation in a Gompertzian stochastic model for tumor growth, *Biometrics*, **56** (2000), 1076–1081.

[51] A. Friedman, *Stochastic Differential Equations and Applications*, Dover Publications, Mineola, NY, 2006.

[52] T.C. Gard, *Introduction to Stochastic Differential Equations*, Marcel Dekker, New York, 1988.

[53] C.W. Gardiner, *Handbook of Stochastic Methods for Physics, Chemistry and the Natural Sciences*, Springer-Verlag, Berlin, 1983.

[54] R. Ghanem and P.D. Spanos, *Stochastic Finite Elements: A Spectral Approach*, revised ed., Dover Publications, Mineola, NY, 2003.

[55] R.J. Griego and R. Hersh, Random evolutions, Markov chains, and systems of partial differential equations, *Proc. of National Academy of Sciences*, **62** (1969), 305–308.

[56] M. Grigoriu, *Stochastic Calculus: Applications in Science and Engineering*, Birkhauser, Boston, 2002.

[57] M. Grigoriu, *Stochastic Systems: Uncertainty Quantification and Propagation*, Springer-Verlag, London, 2012.

[58] G.R. Grimmett and D.R. Stirzaker, *Probability and Random Process*, 2nd edition, Clarendon Press, Oxford, 1992.

[59] P. Hänggi, Colored noise in continuous dynamical systems: A functional calculus approach, in *Noise in Nonlinear Dynamical Systems*, vol. 1 (Edited by F. Moss and P. V. E. McClintock), Cambridge University Press, New York, 1989, 307–328.

[60] P. Hänggi and P. Jung, Colored noise in dynamical system, in *Advances in Chemical Physics*, Volume LXXXIX (Edited by I. Prigogine and S.A. Rice), John Wiley & Sons, Hoboken, NJ, 1995.

[61] G.W. Harrison, Numerical solution of the Fokker-Planck equation using moving finite elements, *Numerical Methods for Partial Differential Equations*, **4** (1988), 219–232.

[62] W. Horsthemke and R. Lefever, *Noise-Induced Transitions: Theory and Applications in Physics, Chemistry, and Biology*, Springer-Verlag, New York, 1984.

[63] R. Hersh, Random evolutions: A survey of results and problems, *Rocky Mountain Journal of Mathematics*, **4** (1974), 443–477.

[64] N. Hritonenko and Y. Yatsenko, Optimization of harvesting age in an integral age-dependent model of population dynamics, *Mathematical Biosciences*, **195** (2005), 154–167.

[65] S.P. Huang, S.T. Quek and K.K. Phoon, Convergence study of the truncated Karhunen–Loeve expansion for simulation of stochastic processes, *International Journal for Numerical Methods in Engineering*, **52** (2001), 1029–1043.

[66] R. Jarrow and P. Protter, A short history of stochastic integration and mathematical finance the early years, 1880–1970, *IMS Lecture Notes Monograph*, 45 (2004), 1–17.

[67] A. Jazwinski, *Stochastic Processes and Filtering Theory*, Dover Publication, New York, 2007.

[68] I. Karatzas and S.E. Shreve, *Brownian Motion and Stochastic Calculus*, 2nd edition, Springer, New York, 1991.

[69] N.J. Kasdin, Discrete simulation of colored noise and stochastic processes and $1/f^\alpha$ power law noise generation, *Proceedings of the IEEE*, **83** (1995), 802–827.

[70] B. Keyfitz and N. Keyfitz, The McKendrick partial differential equation and its uses in epidemiology and population study, *Math. Comput. Modell.*, **25** (1997), 1–9.

[71] M. Kimura, Process leading to quasi-fixation of genes in natural populations due to random fluctuation of selection intensities, *Genetics*, **39** (1954), 280–295.

[72] F. Klebaner, *Introduction to Stochastic Calculus with Applications*, 2nd edition, Imperial College Press, London, 2006.

[73] P.E. Kloeden and E. Platen, *Numerical Solution of Stochastic Differential Equations*, Springer-Verlag, New York 1992.

[74] V. Kopustinskas, J. Augutis and S. Rimkevicius, Dynamic reliability and risk assessment of the accident localization system of the Ignalina NPP RBMK-1500 reactor, *Reliability Engineering and System Safety*, **87** (2005), 77–87.

[75] V.S. Koroliuk and N. Limnios, Random evolutions toward applications, in *Encyclopedia of Statistics in Quality and Reliability*, Wiley online, 2008.

[76] V. Korolyuk and A. Swishchuk, *Evolution of Systems in Random Media*, CRC Press, Boca Ralton, FL, 1995.

[77] F. Kozin, On the probability densities of the output of some random systems, *Trans. ASME Ser. E J. Appl. Mech.*, **28** (1961), 161–164.

[78] H. Kuo, *Introduction to Stochastic Integration*, Springer, New York, 2006.

[79] A. Lasota and M.C. Mackey, *Chaos, Fractals, and Noise: Stochastic Aspects of Dynamics*, Applied Mathematical Science, **97**, Springer, New York, 1994.

[80] M. Loève, *Probability Theory*, vols. I and II, Springer, New York, 1978.

[81] X. Mao and C. Yuan, *Stochastic Differential Equations with Markovian Switching*, Imperial College Press, London, 2006.

[82] A.G. McKendrick, Applications of mathematics to medical problems, *Proceedings of the Edinburgh Mathematical Society*, **44** (1926), 98–130.

[83] J. Medhi, *Stochastic Processes*, 2nd edition, New Age International Publishers, New Delhi, 2002.

[84] J. Mercer, Functions of positive and negative type and their connection with the theory of integral equations, *Philosophical Transactions of the Royal Society A*, **209** (1909), 415–446.

[85] J.A.J. Metz and E.O. Diekmann, *The Dynamics of Physiologically Structured Populations*, Lecture Notes in Biomathematics, Vol. 68, Springer, Heidelberg, 1986.

[86] J.E. Moyal, Stochastic processes and statistical physics, *Journal of the Royal Statistical Society. Series B (Methodological)*, **11** (1949), 150–210.

[87] J. Nolen, *Partial Differential Equations and Diffusion processes*, http://math.stanford.edu/~ryzhik/STANFORD/STANF227-12/ notes227-09.pdf.

[88] W.L. Oberkampf, S.M. DeLand, B.M. Rutherford, K.V. Diegert and K.F. Alvin, Error and uncertainty in modeling and simulation, *Reliability Engineering and Systems Safety*, **75** (2002), 333–357.

[89] B. Oksendal, *Stochastic Differential Equations*, 6th edition, Springer, Berlin, 2005.

[90] A. Okubo, *Diffusion and Ecological Problems: Mathematical Models*, Biomathematics, **10** (1980), Springer-Verlag, Berlin.

[91] G. Oster and Y. Takahashi, Models for age-specific interactions in a periodic environment, *Ecological Monographs*, **44** (1974), 483–501.

[92] E. Parzen, *Stochastic Processes*, SIAM, Philadelphia, 1999.

[93] A. Pirrotta, Multiplicative cases from additive cases: Extension of Kolmogorov-Feller equation to parametric Poisson white noise processes, *Probabilistic Engineering Mechanics*, **22** (2007), 127–135.

[94] P. Protter, A book review of: *Introduction to Stochastic Integration (by K.L. Chung and R.J. Williams), Stochastic Calculus and Applications (by R.J. Elliott) and Semimartingales: A Course on Stochastic Processes (by M. Metivier)*, Technical report #85-5, Purdue University, 1985.

[95] L. Ridolfi, P. D'Odorico and F. Laio, *Noise-Induced Phenomena in the Environmental Sciences*, Cambridge University Press, New York, 2011.

[96] H. Risken, *The Fokker-Planck Equation: Methods of Solution and Applications*, Springer, New York, 1996.

[97] E.S. Rosenberg, M. Davidian and H.T. Banks, Development of structured treatment interruption strategies for HIV infection, *Drug and Alcohol Dependence*, **88S** (2007), S41–S51.

[98] R. Rudnicki, Models of population dynamics and genetics, in *From Genetics To Mathematics*, (edited by M. Lachowicz and J. Mickisz), World Scientific, Singapore, 2009, 103–148.

[99] R. Rudnicki, K. Pichór and M. Tyran-Kamińska, Markov semigroups and their applications, in *Dynamics of Dissipation*, P. Garbaczewski and R. Olkiewics (eds.), Lecture Notes in Physics, **597** (2002), Springer, Berlin, 215–238.

[100] T.L. Saaty, *Modern Nonlinear Equations*, Dover, Mineola, NY, 1981.

[101] Z. Schuss, *Theory and Applications of Stochastic Processes: An Analytical Approach*, Springer, New York, 2010.

[102] B.W. Silverman, *Density Estimation for Statistics and Data Analysis*, Chapman and Hall/CRC Press, Boca Raton, FL, 1998.

[103] J. Sinko and W. Streifer, A new model for age-size structure of a population, *Ecology*, **48** (1967), 910–918.

[104] T.T. Soong, *Random Differential Equations in Science and Engineering*, Academic Press, New York and London, 1973.

[105] T.T. Soong and S.N. Chuang, Solutions of a class of random differential equations, *SIAM J. Appl. Math.*, **24** (1973), 449–459.

[106] M. Stoyanov, M. Gunzburger and John Burkardt, Pink noise, $1/f^\alpha$ noise, and their effect on solutions of differential equations, *International Journal for Uncertainty Quantification*, **1** (2011), 257–278.

[107] J.L. Strand, Random ordinary differential equations, *Journal of Differential Equations*, **7** (1970), 538–553.

[108] R.L. Stratonovich, A new representation for stochastic integrals and equations, *J. SIAM Control*, **4** (1966), 362–371.

[109] H.J. Sussmann, On the gap between deterministic and stochastic ordinary differential equations, *Annals of Prob.*, **6** (1978), 19–41.

[110] A. Swishchuk, *Random Evolutions and Their Applications: New Trends*, Kluwer Academic Publishers, Netherlands, 2005.

[111] G.K. Vallis, Mechanisms of climate variability from years to decades, in *Stochastic Physics and Climate Modelling*, (edited by T. Palmer and P. Williams), Cambridge University Press, Cambridge, 2010, 1–34.

[112] H. Von Foerster, Some remarks on changing populations, in *The Kinetics of Cellular Proliferation*, F. Stohlman, Jr. (ed.), Grune and Stratton, New York, 1959.

[113] M.P. Wand and M.C. Jones, *Kernel Smoothing*, Chapman & Hall/CRC, Boca Raton, FL, 1995.

[114] L.M. Ward and P.E. Greenwood, $1/f$ noise, *Scholarpedia*, **2** (2007); http://www.scholarpedia.org/article/1/f_noise.

[115] S. Watanabe, The Japanese contributions to martingales, *Electronic Journal for History of Probability and Statistics*, **5** (2009); www.emis.de/journals/JEHPS/juin2009/Watanabe.pdf.

[116] G.F. Webb, *Theory of Nonlinear Age-dependent Population Dynamics*, Marcel Dekker, New York, 1985.

[117] G.H. Weiss, Equation for the age structure of growing populations, *Bull. Math. Biophy.*, **30** (1968), 427–435.

[118] A.Y. Weiße, R.H. Middleton and W. Huisinga, Quantifying uncertainty, variability and likelihood for ordinary differential equation models, *BMC Syst. Bio.*, **4**, 2010, 144 (10pp).

[119] E. Wong and M. Zakai, On the relation between ordinary and stochastic differential equations, *Int. J. Engng Sci.*, **3** (1965), 213–229.

[120] E. Wong and M. Zakai, Riemann-Stieltjes approximations of stochastic integrals, *Z. Wahr. verw. Geb.*, **12** (1969), 87–97.

[121] D. Xiu, *Numerical Methods for Stochastic Computations: A Spectral Method Approach*, Princeton University Press, Princeton and Oxford, 2010.

[122] G. Yin, X. Mao, C. Yuan and D. Cao, Approximation methods for hybrid diffusion systems with state-dependent switching processes: Numerical algorithms and existence and uniqueness of solutions, *SIAM J. Math. Anal.*, **41** (2010), 2335–2352.

[123] G. Yin and C. Zhu, *Hybrid Switching Diffusions: Properties and Applications*, Springer, New York, 2009.

Chapter 8

A Stochastic System and Its Corresponding Deterministic System

Deterministic systems involving ordinary differential equations or delay differential equations, though widely used in practice (in part because of the ease of their use in simulation studies), have proven less descriptive when applied to problems with small population size. To address this issue, stochastic systems involving continuous time Markov chain (CTMC) models or CTMC models with delays are often used when dealing with low species count.

The goals of this chapter are two-fold: one is to give an introduction on how to simulate a CTMC model or a CTMC model with delays, and the other is to demonstrate how to construct a corresponding deterministic model for a stochastic one, and how to construct a corresponding stochastic model for a deterministic one. The methodology illustrated here is highly relevant to current researchers in the biological and physical sciences. As investigations and models of disease progression become more complex and as interest in *initial phases* (i.e., HIV acute infection, initial disease outbreaks) of infections or epidemics increases, the recognition becomes widespread that many of these efforts require small population number models for which ordinary differential equations (ODEs) or delay differential equations are unsuitable. These efforts will involve CTMC models (or CTMC models with delay) that have as limits (as population size increases) ordinary differential equations (or delay differential equations). As the interest in initial stages of infection grows, so also will the need grow for efficient simulation with these small population number models. We give a careful presentation of computational issues arising in simulations with such models.

The following notation is used throughout the remainder of this chapter: t_{stop} denotes the final time, $\mathbb{Z}^n$ is the set of n-dimensional column vectors with integer components, and $\mathbf{e}_i \in \mathbb{Z}^n$ is the ith unit column vector (that is, the ith entry of $\mathbf{e}_i$ is 1 and all the other entries are zeros), $i = 1, 2, \ldots, n$.

8.1   Overview of Multivariate Continuous Time Markov Chains

We assume that $\{\mathbf{X}(t) : t \geq 0\}$ with the state space being a subset of $\mathbb{Z}^n$ is a continuous time Markov chain such that for any small time increment Δt we have

$$\text{Prob}\left\{\mathbf{X}(t + \Delta t) = \mathbf{x} + \mathbf{v}_j \mid \mathbf{X}(t) = \mathbf{x}\right\} = \lambda_j(\mathbf{x})\Delta t + o(\Delta t),$$

$$\text{Prob}\left\{\mathbf{X}(t + \Delta t) = \mathbf{x} \mid \mathbf{X}(t) = \mathbf{x}\right\} = 1 - \sum_{j=1}^{l} \lambda_j(\mathbf{x})\Delta t + o(\Delta t) \tag{8.1}$$

for $j = 1, 2, \ldots, l$ with l being the total number of transitions. Here $\mathbf{X} = (X_1, X_2, \ldots, X_n)^T$ represents the state of the system with $X_i(t)$ being the number of individuals (patients, cells, particles, etc.) in state i at time t, $\mathbf{v}_j = (v_{1j}, v_{2j}, v_{3j}, \ldots, v_{nj})^T \in \mathbb{Z}^n$ with v_{ij} denoting the change in state variable X_i caused by the jth transition, and λ_j is the transition rate (also referred to as the *transition intensity*) at the jth transition, $j = 1, 2, 3, \ldots, l$. We note that the transition is often referred to as a *reaction* in biochemistry literature, the associated transition rate is called the *propensity function*, and the matrix $(\mathbf{v}_1, \mathbf{v}_2, \ldots, \mathbf{v}_l)$ is often referred to as the *stoichiometry matrix*. In this chapter, the names "reaction" and "transition" will be used interchangeably. For simplicity, we assume that the initial condition $\mathbf{X}(0)$ is non-random throughout the rest of this presentation (the case where $\mathbf{X}(0)$ is random can be treated similarly).

8.1.1   Exponentially Distributed Holding Times

Let $S_\mathbf{x}$ denote the holding time at the given state $\mathbf{x}$ (i.e., the amount of time it remains in $\mathbf{x}$ after entering this state), and $\lambda_\mathbf{x} = \sum_{j=1}^{l} \lambda_j(\mathbf{x})$. Then, using similar arguments to those used to prove Theorem 7.1.9 and (7.46), one can show that $S_\mathbf{x}$ is exponentially distributed with rate parameter $\lambda_\mathbf{x}$, that is,

$$S_\mathbf{x} \sim \text{Exp}(\lambda_\mathbf{x}), \tag{8.2}$$

and the probability that the process jumps to state $\mathbf{x} + \mathbf{v}_j$ from state $\mathbf{x}$ is

$$\frac{\lambda_j(\mathbf{x})}{\lambda_\mathbf{x}}. \tag{8.3}$$

By using the inverse transform method discussed in Remark 2.5.2, one can know which reaction will take place after the process leaves state $\mathbf{x}$ by simulating a discrete random variable with range $\{1, 2, \ldots, l\}$ and the probability associated with $j \in \{1, 2, \ldots, l\}$ given by (8.3). Specifically, one generates a random number r from a uniform distribution on $(0, 1)$ and then chooses the reaction as follows: If $0 < r \leq \lambda_1(\mathbf{x})/\lambda_{\mathbf{x}}$, choose reaction 1; if $\lambda_1(\mathbf{x})/\lambda_{\mathbf{x}} < r \leq (\lambda_1(\mathbf{x}) + \lambda_2(\mathbf{x}))/\lambda_{\mathbf{x}}$ choose reaction 2, and so on. Thus we see that (8.2) and (8.3) provide sufficient information on how to simulate the CTMC given by (8.1). Specifically, (8.2) indicates how much time the process remains in a specific state $\mathbf{x}$ before the next reaction occurs, and (8.3) gives information on which reaction will occur after the process leaves state $\mathbf{x}$. The resulting algorithm is referred to as the *stochastic simulation algorithm* and will be discussed in detail in Section 8.2.1.

Remark 8.1.1 *To generate a realization of $S_{\mathbf{x}}$, one can either directly generate a random number from an exponential distribution with rate parameter $\lambda_{\mathbf{x}}$ (e.g., using MATLAB function* exprnd*) or use the inverse transform method discussed in Section 2.5.2. Let Z_u be a uniformly distributed random variable defined on $(0,1)$. Then by (2.92) and Theorem 2.5.1 we know that $S_{\mathbf{x}}$ satisfies the equation $1 - \exp(-S_{\mathbf{x}}\lambda_{\mathbf{x}}) = Z_u$, which implies*

$$S_{\mathbf{x}} = -\frac{1}{\lambda_{\mathbf{x}}}\ln(1 - Z_u). \tag{8.4}$$

By the property of the standard uniform distribution (2.78), we know that $1 - Z_u$ is also uniformly distributed on $(0,1)$. Hence, the holding time $S_{\mathbf{x}}$ is often written in the following form instead of (8.4):

$$S_{\mathbf{x}} = -\frac{1}{\lambda_{\mathbf{x}}}\ln(Z_u).$$

Thus, a realization $s_{\mathbf{x}}$ of $S_{\mathbf{x}}$ is given by

$$s_{\mathbf{x}} = -\frac{1}{\lambda_{\mathbf{x}}}\ln(z_u) = \frac{1}{\lambda_{\mathbf{x}}}\ln\left(\frac{1}{z_u}\right),$$

where z_u is a realization of Z_u and it can be generated by using the MATLAB command rand.

8.1.2 Random Time Change Representation

Let $R_j(t)$ denote the number of times that the jth reaction has occurred up to time t, $j = 1, 2, \ldots, l$. Then the state of CTMC (8.1) at time t can be written as

$$\mathbf{X}(t) = \mathbf{X}(0) + \sum_{j=1}^{l} \mathbf{v}_j R_j(t). \tag{8.5}$$

By the above equation and (8.1), we find that for any small Δt and for any $j \in \{1, 2, \ldots, l\}$ we have

$$\text{Prob}\{R_j(t + \Delta t) - R_j(t) = 1 \mid \mathbf{X}(s), 0 \leq s \leq t\} = \lambda_j(\mathbf{X}(t))\Delta t + o(\Delta t), \tag{8.6}$$

which implies that $\{R_j(t) : t \geq 0\}$ is a counting process with rate $\lambda_j(\mathbf{X}(t))$, $j = 1, 2, \ldots, l$. Let $\{Y_j(t) : t \geq 0\}$ be a unit Poisson process, $j = 1, 2, \ldots, l$. Then by (7.32) we know that for any small Δt we have

$$\text{Prob}\left\{ Y_j\left(\int_0^{t+\Delta t} \lambda_j(\mathbf{X}(s))ds \right) - Y_j\left(\int_0^t \lambda_j(\mathbf{X}(s))ds \right) = 1 \,\bigg|\, \mathbf{X}(s), 0 \leq s \leq t \right\}$$

$$= \int_t^{t+\Delta t} \lambda_j(\mathbf{X}(s))ds + o(\Delta t)$$

$$= \lambda_j(\mathbf{X}(t))\Delta t + o(\Delta t)$$

$$\tag{8.7}$$

for $j = 1, 2, \ldots, l$, where the last equality is obtained because the next reaction has not occurred before $t + \Delta t$ and hence $\mathbf{X}(s) = \mathbf{X}(t)$ for $s \in [t, t + \Delta t)$. By (8.6) and (8.7), we see that $R_j(t)$ can be written as

$$R_j(t) = Y_j\left(\int_0^t \lambda_j(\mathbf{X}(s))ds \right), \quad j = 1, 2, \ldots, l. \tag{8.8}$$

(We refer the interested reader to [26] for a rigorous derivation of this equation.) To ensure that only one of the reactions occurs at a time, we assume that these l unit Poisson processes $\{Y_j(t) : t \geq 0\}$, $j = 1, 2, \ldots, l$, are independent. By (8.5) and (8.8), we know that $\mathbf{X}(t)$ satisfies the stochastic equation

$$\mathbf{X}(t) = \mathbf{X}(0) + \sum_{j=1}^{l} \mathbf{v}_j Y_j\left(\int_0^t \lambda_j(\mathbf{X}(s))ds \right), \tag{8.9}$$

which indicates that the CTMC (8.1) can be characterized by unit Poisson processes. We observe that the only randomness on the right-hand side of (8.9) is due to the randomness of the unit Poisson process, and the value of $\mathbf{X}(t)$ changes only when one of these l unit Poisson processes jumps. Formulation (8.9) is often referred to as the *random time change representation*, and it has a close relationship with many algorithms that are used to simulate a CTMC (as we shall see later).

Below we give a brief introduction to the relationship between the stochastic equation (8.9) and the corresponding martingale problem as well as the relationship between the martingale problem and Kolmogorov's forward equation. The interested reader can refer to [8, 26] for more information on this subject.

8.1.2.1   Relationship between the Stochastic Equation and the Martingale Problem

Let $f : \mathbb{Z}^n \to \mathbb{R}$ be a bounded function with $f(\mathbf{x}) = 0$ for $|\mathbf{x}|$ sufficiently large, and define $f(\infty) = 0$. Then for any solution of (8.9) we have

$$U(t) = f(\mathbf{X}(t)) - f(\mathbf{X}(0)) - \int_0^t \mathscr{A} f(\mathbf{X}(s))ds \tag{8.10}$$

is a martingale. Here

$$\mathscr{A}f(\mathbf{z}) = \sum_{j=1}^{l} \lambda_j(\mathbf{z}) \left(f(\mathbf{z} + \mathbf{v}_j) - f(\mathbf{z}) \right), \tag{8.11}$$

which is often referred to as the *infinitesimal generator* of the continuous time Markov chain (8.1).

Definition 8.1.1 *A right-continuous and $\mathbb{Z}^n \cup \{\infty\}$-valued stochastic process $\{\mathbf{X}(t) : t \geq 0\}$ is said to be a solution of the martingale problem for $\mathscr{A}$ if there exists a filtration $\{\mathcal{F}_t\}$ such that for each $f : \mathbb{Z}^n \to \mathbb{R}$ with $f(\mathbf{x}) = 0$ for $|\mathbf{x}|$ sufficiently large, (8.10) is a martingale. In addition, if $\mathbf{X}(t) = \infty$ for $t \geq t_\infty$ with $t_\infty = \inf \left\{ t : \lim_{s \to t-} \mathbf{X}(s) = \infty \right\}$, then $\{\mathbf{X}(t) : t \geq 0\}$ is said to be a minimum solution.*

The next theorem states that the minimum solution of the martingale problem for $\mathscr{A}$ can be written as a solution of the stochastic equation.

Theorem 8.1.2 *If $\{\mathbf{X}(t) : t \geq 0\}$ is a minimum solution of the martingale problem for $\mathscr{A}$, then there exist independent unit Poisson processes $\{Y_j(t) : t \geq 0\}, j = 1, 2, \ldots, l$, such that*

$$\mathbf{X}(t) = \mathbf{X}(0) + \sum_{j=1}^{l} \mathbf{v}_j Y_j \left(\int_0^t \lambda_j(\mathbf{X}(s)) ds \right).$$

We thus see that the stochastic equation (8.9) and the corresponding martingale problem are equivalent.

8.1.2.2 Relationship between the Martingale Problem and Kolmogorov's Forward Equation

Let $\{\mathbf{X}(t) : t \geq 0\}$ be a solution to the stochastic equation (8.9); then by the discussion in the above section we know that $\{U(t) : t \geq 0\}$ defined by (8.10) is a martingale. Note that $U(0) = 0$. Hence, by (8.10) and the martingale property (7.47) we know that $\mathbb{E}(U(t)) = U(0) = 0$, which implies that

$$\mathbb{E}\{f(\mathbf{X}(t))\} = \mathbb{E}\{f(\mathbf{X}(0))\} + \int_0^t \mathbb{E}\{\mathscr{A}f(\mathbf{X}(s))\} ds. \tag{8.12}$$

If we choose $f(\mathbf{z}) = 1_\mathbf{x}(\mathbf{z})$ with $1_\mathbf{x}(\mathbf{z})$ defined as

$$1_\mathbf{x}(\mathbf{z}) = \begin{cases} 1, & \text{if } \mathbf{z} = \mathbf{x}, \\ 0, & \text{otherwise}, \end{cases}$$

then by (8.11) and (8.12) we have

$$\text{Prob}\{\mathbf{X}(t) = \mathbf{x}\}$$

$$= \text{Prob}\{\mathbf{X}(0) = \mathbf{x}\} + \int_0^t \sum_{j=1}^l \lambda_j(\mathbf{x} - \mathbf{v}_j)\text{Prob}\{\mathbf{X}(s) = \mathbf{x} - \mathbf{v}_j\}ds$$

$$- \int_0^t \sum_{j=1}^l \lambda_j(\mathbf{x})\text{Prob}\{\mathbf{X}(s) = \mathbf{x}\}ds.$$

Let $\Phi(t, \mathbf{x}) = \text{Prob}\{\mathbf{X}(t) = \mathbf{x}\}$. Then we have

$$\Phi(t, \mathbf{x}) = \Phi(0, \mathbf{x}) + \int_0^t \sum_{j=1}^l \lambda_j(\mathbf{x} - \mathbf{v}_j)\Phi(s, \mathbf{x} - \mathbf{v}_j)ds - \int_0^t \sum_{j=1}^l \lambda_j(\mathbf{x})\Phi(s, \mathbf{x})ds.$$

Differentiating the above equation with respect to t yields that

$$\frac{d}{dt}\Phi(t, \mathbf{x}) = \sum_{j=1}^l \lambda_j(\mathbf{x} - \mathbf{v}_j)\Phi(t, \mathbf{x} - \mathbf{v}_j) - \sum_{j=1}^l \lambda_j(\mathbf{x})\Phi(t, \mathbf{x}), \qquad (8.13)$$

which is often called *Kolmogorov's forward equation* or the *chemical master equation*.

Ideally one would like to directly solve Kolmogorov's forward equation (8.13) with given initial conditions to obtain the information for the probability distribution of the state of the system at a particular time point. However, this equation is often difficult, if not impossible, to solve. Thus, one often resorts to some simulation methods that generate exact (or approximate) sample paths and then uses these sample paths to obtain an approximation of the associated distribution function.

8.2   Simulation Algorithms for Continuous Time Markov Chain Models

In this section, we give a detailed discussion of several computational algorithms that have been widely used in the literature for simulating the CTMC model (8.1). These include the stochastic simulation algorithm, the next reaction method, and the tau-leaping method, where the first two methods generate the exact sample paths and the last method generates approximate sample paths.

8.2.1   Stochastic Simulation Algorithm

The stochastic simulation algorithm (SSA), also known as the *Gillespie algorithm* [29], is the standard method employed to simulate CTMC models.

The SSA was first introduced by Gillespie in 1976 to simulate the time evolution of the stochastic formulation of chemical kinetics. In addition to simulating chemically reacting systems, the Gillespie algorithm has become the method of choice to numerically simulate stochastic models arising in a variety of other biological applications [3, 10, 39, 43, 50, 55].

Two mathematically equivalent procedures were originally proposed by Gillespie, the *Direct method* and the *First Reaction method*. Both procedures are exact procedures rigorously based on the the same microphysical principles that underlie Kolmogorov's forward equation.

8.2.1.1   The Direct Method

The direct method can be described by the following procedure:

S1. Set the initial condition for each state at $t = 0$.

S2. For the given state $\mathbf{x}$ of the system, calculate the sum of all transition rates, $\lambda_{\mathbf{x}} = \sum_{i=1}^{l} \lambda_i(\mathbf{x})$.

S3. Generate a random number r_1 from a uniform distribution on $(0, 1)$ and set $\Delta t = \dfrac{1}{\lambda_{\mathbf{x}}} \ln \left(\dfrac{1}{r_1} \right)$.

S4. Generate a random number r_2 from a uniform distribution on $(0, 1)$ and choose the reaction as follows: If $0 < r_2 \leq \lambda_1(\mathbf{x})/\lambda_{\mathbf{x}}$, choose reaction 1; if $\lambda_1(\mathbf{x})/\lambda_{\mathbf{x}} < r_2 \leq (\lambda_1(\mathbf{x}) + \lambda_2(\mathbf{x}))/\lambda_{\mathbf{x}}$, choose reaction 2, and so on.

S5. Let reaction ι be the reaction chosen in **S4**. Update the time by setting $t = t + \Delta t$ and update the system state based on the reaction ι (i.e., set $\mathbf{x} = \mathbf{x} + \mathbf{v}_\iota$).

S6. Iterate **S2–S5** until $t \geq t_{stop}$.

Remark 8.1.1 indicates that the statement in **S3** is equivalent to generating a random number from an exponential distribution with rate parameter $\lambda_{\mathbf{x}}$. Thus we see that this algorithm is based on (8.2) and (8.3), as we stated earlier in Section 8.1. We also observe that for this method one needs to simulate two random numbers at each iteration, where one is used to find the amount of time that the process remains in a specific state $\mathbf{x}$ before the next reaction occurs, and the other one is used to determine which reaction will occur after the process leaves state $\mathbf{x}$.

8.2.1.2   The First Reaction Method

The first reaction method can be described by the following procedure:

S1. Set the initial condition for each state variable at $t = 0$.

S2. Calculate each transition rate λ_i at the given state $\mathbf{x}$ of the system and denote the resulting value by a_i (i.e., $a_i = \lambda_i(\mathbf{x})$), $i = 1, 2, \ldots, l$.

S3. Generate l independent, uniform random numbers on $(0, 1)$, $r_1, r_2, \ldots, r_l$.

S4. Set $\Delta t_i = \dfrac{1}{a_i} \ln \left(\dfrac{1}{r_i} \right)$ for $i = 1, 2, \ldots, l$.

S5. Set $\Delta t = \min\limits_{1 \leq i \leq l} \{\Delta t_i\}$, and let ι be the reaction which obtains the minimum.

S6. Update the time $t = t + \Delta t$, and update the system state based on the reaction ι (i.e., set $\mathbf{x} = \mathbf{x} + \mathbf{v}_\iota$).

S7. Iterate **S2**–**S6** until $t \geq t_{stop}$.

Remark 8.1.1 indicates that the statements in **S3** and **S4** are equivalent to generating l independent random numbers from exponential distributions with rate parameters respectively given by $a_1, a_2, \ldots, a_l$. We thus see that the equivalence between the direct method and the first reaction method can be established by the properties (2.98) and (2.100) of the exponential distribution.

We observe that for the first reaction method one needs to simulate l random numbers at each iteration. Hence, when the value of l is large, it becomes less efficient than the direct method. Thus, the direct method is the method typically implemented in practice, and it is also the method implemented in this monograph. In the rest of this chapter, whenever we talk about the SSA, it refers to the direct method.

8.2.2   The Next Reaction Method

In this section, we introduce another approach, the next reaction method, that can be used to simulate exact sample paths of a CTMC which only requires generating one random number at each subsequent iteration after the first one. This method was first proposed by Gibson and Bruck [28], and then revisited and modified by Anderson in [4].

The next reaction method is based on the random time change representation (8.9) and the property of interarrival times of the Poisson process (Theorem 7.1.7). For this method, the notion of the *internal time* of the unit Poisson process $\{Y_j(t) : t \geq 0\}$ is used, and it is defined as $T_j^I(t) = \displaystyle\int_0^t \lambda_j(\mathbf{X}(s)) ds$ for $j = 1, 2, \ldots, l$. Let $t^0 = 0$, t^k denote the time point at which the kth reaction just occurred, and $\mathbf{x}^k$ represent the value of the system state at time t^k. In addition, the following notation will be used throughout this section: $T_j^{I,k} = T_j^I(t^k)$, $j = 1, 2, \ldots, l$. This implies that for each j we have $T_j^{I,0} = 0$.

8.2.2.1 The Original Next Reaction Method

We now introduce the next reaction method proposed by Gibson and Bruck [28], and call it the *original next reaction method*. We first discuss the idea behind this method, and then outline the procedures for implementing such an algorithm.

By Theorem 7.1.7 we know that the waiting time to the first jump of a unit Poisson process is exponentially distributed with rate parameter 1. Hence, the amount of time ΔT_j^0 needed for the jth reaction to occur for the first time satisfies

$$\int_0^{\Delta T_j^0} \lambda_j(\mathbf{X}(s))ds \sim \text{Exp}(1). \tag{8.14}$$

Note that λ_j remains constant during the time interval $[t^0, t^0 + \Delta T_j^0)$ (as the next reaction has not occurred yet). Hence,

$$\int_0^{\Delta T_j^0} \lambda_j(\mathbf{X}(s))ds = \lambda_j(\mathbf{x}^0)\Delta T_j^0.$$

By (2.96), (8.14) and the above equation we find $\Delta T_j^0 \sim \text{Exp}(\lambda_j(\mathbf{x}^0))$. We see that the minimum ΔT^0 of these ΔT_j^0 is the amount of time that the process remains in a given state $\mathbf{x}^0$ before the next reaction occurs, and the index $I_{\min}$ for achieving this minimum value is the reaction that will occur after the process leaves state $\mathbf{x}^0$.

Let Δt^k be a realization of ΔT^k, and ι be a realization of $I_{\min}$. When the time is updated from t^k to $t^{k+1} = t^k + \Delta t^k$, and the system state is updated from $\mathbf{x}^k$ to $\mathbf{x}^{k+1} = \mathbf{x}^k + \mathbf{v}_\iota$, we need to find the amount of time needed for the jth reaction to occur (assuming no other reactions occur first). For $j \neq \iota$, we know that

$$\int_0^{t^{k+1}+\Delta T_j^{k+1}} \lambda_j(\mathbf{X}(s))ds$$

$$= \int_0^{t^k} \lambda_j(\mathbf{X}(s))ds + \int_{t^k}^{t^{k+1}} \lambda_j(\mathbf{X}(s))ds + \int_{t^{k+1}}^{t^{k+1}+\Delta T_j^{k+1}} \lambda_j(\mathbf{X}(s))ds$$

$$= T_j^{I,k} + \lambda_j(\mathbf{x}^k)\Delta t^k + \lambda_j(\mathbf{x}^{k+1})\Delta T_j^{k+1}.$$

Note that the internal time needed for the next jump of the unit Poisson process $\{Y_j(t) : t \geq 0\}$ is still $T_j^{I,k} + \lambda_j(\mathbf{x}^k)\Delta T_j^k$. Hence, by the above equation we have

$$T_j^{I,k} + \lambda_j(\mathbf{x}^k)\Delta t^k + \lambda_j(\mathbf{x}^{k+1})\Delta T_j^{k+1} = T_j^{I,k} + \lambda_j(\mathbf{x}^k)\Delta T_j^k,$$

which implies that

$$\Delta T_j^{k+1} = \frac{\lambda_j(\mathbf{x}^k)}{\lambda_j(\mathbf{x}^{k+1})}(\Delta T_j^k - \Delta t^k).$$

Since reaction ι just occurred, the amount of time needed for the unit Poisson process $\{Y_\iota(t) : t \geq 0\}$ to jump first again satisfies

$$\int_{t^{k+1}}^{t^{k+1}+\Delta T_\iota^{k+1}} \lambda_\iota(\mathbf{X}(s)ds) \sim \text{Exp}(1). \tag{8.15}$$

Note that λ_ι remains constant during the time interval $[t^{k+1}, t^{k+1} + \Delta T_\iota^{k+1})$ (as the next reaction has not occurred yet). Hence,

$$\int_{t^{k+1}}^{t^{k+1}+\Delta T_\iota^{k+1}} \lambda_\iota(\mathbf{X}(s)ds) = \lambda_\iota(\mathbf{x}^{k+1})\Delta T_\iota^{k+1}.$$

Then by (2.96), (8.15) and the above equation we find

$$\Delta T_\iota^{k+1} \sim \text{Exp}(\lambda_\iota(\mathbf{x}^{k+1})).$$

The minimum ΔT^{k+1} of these ΔT_j^{k+1} is the amount of time that the process remains in a given state $\mathbf{x}^{k+1}$ before the next reaction occurs, and the index $I_{\min}$ for achieving this minimum value is the reaction that will occur after the process leaves state $\mathbf{x}^{k+1}$. We update k to $k+1$, and repeat this procedure until the final time t_{stop} is reached.

Based on the above discussion, the original next reaction algorithm is outlined as follows.

S1. Set the initial condition for each state at $t = 0$.

S2. Calculate each transition rate λ_i at the given state and denote the resulting value by a_i, $i = 1, 2, \ldots, l$.

S3. Generate l independent, uniform random numbers on $(0, 1)$, $r_1, r_2, \ldots, r_l$.

S4. Set $\tilde{t}_i = \dfrac{1}{a_i} \ln\left(\dfrac{1}{r_i}\right)$ for $i = 1, 2, \ldots, l$.

S5. Set $t = \min_{1 \leq i \leq l}\{\tilde{t}_i\}$, and let ι be the reaction which obtains the minimum.

S6. Update the system state based on the reaction ι.

S7. Recalculate each transition rate λ_i at the given new state and denote the resulting value by $\bar{a}_i$, $i = 1, 2, \ldots, l$.

S8. For each $i \neq \iota$, set $\tilde{t}_i = \dfrac{a_i}{\bar{a}_i}(\tilde{t}_i - t) + t$.

S9. For transition ι, choose a uniform random number on $(0,1)$, r, and set
$$\tilde{t}_\iota = \frac{1}{\bar{a}_\iota} \ln\left(\frac{1}{r}\right) + t.$$

S10. Set $a_i = \bar{a}_i$, $i = 1, 2, \ldots, l$.

S11. Iterate **S5**–**S10** until $t \geq t_{stop}$.

Again, Remark 8.1.1 indicates that the statements in **S3** and **S4** are equivalent to generating l independent random numbers from exponential distributions with rate parameters respectively given by $a_1, a_2, \ldots, a_l$, and that the statement in **S9** is equivalent to generating a random number z_e from an exponential distribution with rate parameter $\bar{a}_\iota$ and then setting $\tilde{t}_\iota = z_e + t$. We remark that to increase the efficiency of the next reaction method the algorithm proposed in [28] uses the notion of a dependency graph and a propriety queue, where the dependency graph is used to update only those transition rates that actually change during an iteration, and the propriety queue is used to quickly determine the minimum in **S5**. We refer the interested reader to [28] for more information on this.

8.2.2.2   The Modified Next Reaction Method

The modified next reaction method was developed by Anderson in [4], and it is a modification to the original next reaction method by making more explicit use of the internal time.

Let $T_j^{J,k}$ represent the waiting time to the event (jump) occurring right after time $T_j^{I,k}$ for the jth unit Poisson process $\{Y_j(t) : t \geq 0\}$; that is,

$$T_j^{J,k} = \inf \left\{ s > T_j^{I,k} : Y_j(s) > Y_j\left(T_j^{I,k}\right) \right\}. \tag{8.16}$$

Note that $T_j^{I,0} = 0$, $j = 1, 2, \ldots, l$. Hence, $T_j^{J,0}$ is the waiting time to the first jump for the jth unit Poisson process. By Theorem 7.1.7, we know that random variables $T_j^{J,0}$, $j = 1, 2, \ldots, l$, are independent and exponentially distributed with rate parameter 1, where the independence is due to the fact that $\{Y_j(t) : t \geq 0\}, j = 1, 2, \ldots, l$, are independent unit Poisson processes.

Now we consider the process at a general time point t^k where some reaction just occurred. By Theorem 7.1.7, we know that if $T_j^{I,k}$ happens to be a jump time for the unit Poisson process $\{Y_j(t) : t \geq 0\}$, then

$$T_j^{J,k} - T_j^{I,k} \sim \text{Exp}(1). \tag{8.17}$$

Let ΔT_j^k denote the amount of time needed for the first jump of the jth unit Poisson process right after time $T_j^{I,k}$; that is,

$$\int_{t^k}^{t^k + \Delta T_j^k} \lambda_j(\mathbf{X}(s))ds = T_j^{J,k} - T_j^{I,k}. \tag{8.18}$$

Note that λ_j remains constant during the time interval $[t^k, t^k + \Delta T_j^k)$ (as the next reaction has not occurred yet). Hence,

$$\int_{t^k}^{t^k + \Delta T_j^k} \lambda_j(\mathbf{X}(s))ds = \Delta T_j^k \lambda_j(\mathbf{x}^k).$$

Thus, by (8.18) and the above equation we have

$$\Delta T_j^k = \frac{T_j^{J,k} - T_j^{I,k}}{\lambda_j(\mathbf{x}^k)}.$$

We see that the minimum ΔT^k of these ΔT^k_j is the amount of time that the process remains in a given state $\mathbf{x}^k$ before the next reaction occurs, and the index $I_{\min}$ for achieving this minimum value is the reaction that will occur after the process leaves state $\mathbf{x}^k$. Let Δt^k be a realization of ΔT^k, and ι be a realization of $I_{\min}$. When the time is updated from t^k to $t^{k+1} = t^k + \Delta t^k$, the internal time is updated as follows:

$$
\begin{aligned}
T^{I,k+1}_j &= \int_0^{t^{k+1}} \lambda_j(\mathbf{X}(s))ds \\
&= \int_0^{t^k} \lambda_j(\mathbf{X}(s))ds + \int_{t^k}^{t^k+\Delta t^k} \lambda_j(\mathbf{X}(s))ds \\
&= T^{I,k}_j + \lambda_j(\mathbf{x}^k)\Delta t^k
\end{aligned}
$$

for $j = 1, 2, \ldots, l$. Note that the ιth reaction just occurred. Hence, we have

$$
T^{I,k+1}_\iota = T^{J,k}_\iota, \quad T^{I,k+1}_j < T^{J,k}_j \text{ for } j \neq \iota.
$$

Thus, by the above equation and (8.16), we see that only the waiting time for the Poisson process $\{Y_\iota(t) : t \geq 0\}$ needs to be updated. Let $Z_e \sim \text{Exp}(1)$. Then by (8.17) we know that

$$
T^{J,k+1}_\iota = T^{J,k}_\iota + Z_e.
$$

We then update k to $k+1$, and repeat this procedure until the final time t_{stop} is reached.

Based on the above discussion, the modified next reaction algorithm is outlined as follows.

S1. Set the initial condition for each state at $t = 0$, and set $t^I_i = 0$, $i = 1, 2, \ldots, l$.

S2. Calculate each transition rate λ_i at the given state and denote the resulting value by a_i, $i = 1, 2, \ldots, l$.

S3. Generate l independent, uniform random numbers on $(0, 1)$, $r_1, r_2, \ldots, r_l$.

S4. Set $t^J_i = \ln(1/r_i)$ for $i = 1, 2, \ldots, l$.

S5. Set $\Delta t_i = (t^J_i - t^I_i)/a_i$ for $i = 1, 2, \ldots, l$.

S6. Set $\Delta t = \min_{1 \leq i \leq l} \{\Delta t_i\}$, and let ι be the reaction which obtains the minimum.

S7. Set $t = t + \Delta t$.

S8. Update the system state based on the reaction ι.

S9. Set $t^I_i = t^I_i + a_i \Delta t$ for $i = 1, 2, \ldots, l$.

S10. For transition ι, generate a uniform random number on $(0,1)$, r, and set $t^J_\iota = t^J_\iota + \ln(1/r)$.

S11. Recalculate each transition rate λ_i at the given new state and denote the resulting value by a_i, $i = 1, 2, \ldots, l$.

S12. Iterate **S5**–**S11** until $t \geq t_{stop}$.

Remark 8.1.1 indicates that the statements in **S3** and **S4** are equivalent to generating l independent random numbers from $\mathrm{Exp}(1)$. In addition, the statement in **S10** is equivalent to generating a random number z_e from $\mathrm{Exp}(1)$ and then setting $t_i^J = t_i^J + z_e$.

One of the advantages of the modified next reaction algorithm is that it can be easily extended to simulate stochastic systems with time dependent transition rates, whereas the original next reaction method cannot (see [4] for more information). In addition, this algorithm can be altered to simulate stochastic systems with fixed delays, and further altered to simulate stochastic systems with random delays (details will be discussed later).

8.2.3   Tau-Leaping Methods

All the methods mentioned so far are exact, and they keep track of each transition. Note that the holding time is an exponentially distributed random variable with rate parameter $\sum_{j=1}^{l} \lambda_j(\mathbf{x})$. Hence, if the value of this rate parameter is really large (due to the fact that the population in some/all states is really large or the number of transitions l is really large or both), then the time step size will be really small. Thus, the computational time required by these methods to generate even a single exact sample path can be prohibitive for certain applications (as we shall see later in Section 8.5). As a result, Gillepsie proposed an approximate procedure, the *tau-leaping method*, which accelerates the computational time while only sustaining a small loss in accuracy [30] (the "tau-leaping method" gets its name because τ is traditionally used to denote the time step instead of Δt in the simulation algorithms for CTMC models). Instead of keeping track of each transition as in the above mentioned methods, the tau-leaping method *leaps* from one subinterval to the next, approximating how many transitions take place during a given subinterval. It is assumed that the value of the leap, Δt, is sufficiently small that there is no significant change in the value of the transition rates along the subinterval $[t, t + \Delta t]$. This condition is known as the *leap condition*. The tau-leaping method thus has the advantage of simulating many transitions in one *leap* while not losing significant accuracy, resulting in a speed up in computational time. In this section, we consider two tau-leaping methods: an explicit and an implicit version.

8.2.3.1   An Explicit Tau-Leaping Method

The basic explicit tau-leaping method approximates the number of times a transition j is expected to occur within the time interval $[t, t + \Delta t]$, by a Poisson random variable Z_j with rate parameter $\lambda_j(\mathbf{x})\Delta t$; that is,

$$Z_j \sim \mathrm{Poisson}(\lambda_j(\mathbf{x})\Delta t) \qquad (8.19)$$

with $\mathbf{X}(t) = \mathbf{x}$. Once the number of transitions is estimated, $\mathbf{X}(t + \Delta t)$ can be approximated as

$$\mathbf{X}(t + \Delta t) \approx \mathbf{x} + \sum_{j=1}^{l} \mathbf{v}_j Z_j, \tag{8.20}$$

where Z_j, $j = 1, 2, \ldots, l$, are independent. Thus we see that (8.20) is just the explicit Euler approximation of the random change representation (8.9) on the time interval $[t, t + \Delta t]$, and the independence of these l Poisson random variables, $Z_1, Z_2, \ldots, Z_l$, is due to the independence of unit Poisson processes $\{Y_j(t) : t \geq 0\}, j = 1, 2, \ldots, l$. Therefore, this method is sometimes called *Euler tau-leaping*. It is worth noting that one can also use other approximation methods, such as midpoint or trapezoidal type methods, to approximate the random change representation (8.9). We refer the interested reader to [7] and the references therein for more information.

As mentioned previously, the procedure for selecting Δt is critical in the tau-leaping method. If Δt is chosen too small, then the tau-leaping method requires similar computational time to the exact methods; on the other hand, if the value of Δt is too large, the leap condition may not be satisfied, possibly causing significant inaccuracies in the simulation. In this section, we use a tau-selection procedure based on the algorithm in [23]. For alternative procedures for selecting Δt, we refer the reader to references [5, 23, 30, 31].

Let $\Delta X_i = X_i(t + \Delta t) - x_i$ with x_i being the ith component of $\mathbf{x}$, $i = 1, 2, \ldots n$, and ϵ be an error control parameter with $0 < \epsilon \ll 1$. In the given tau-selection procedure, Δt is chosen such that

$$\Delta X_i \leq \max\left\{\frac{\epsilon}{\varpi_i} x_i, 1\right\}, \ i = 1, \ldots, n, \tag{8.21}$$

which obviously requires the relative change in X_i to be bounded by $\dfrac{\epsilon}{\varpi_i}$ except that X_i will never be required to change by an amount less than 1 (as the values of state variables $X_i, i = 1, 2, \ldots, n$ must be integers). The value of ϖ_i in (8.21) is chosen such that the relative changes in all the transition rates will be bounded by ϵ at least to the first-order approximation (which will be explained below). To do that, we need to find the value of ϖ_i^j, which is the condition for the relative change of state variable X_i such that the relative change in transition rate λ_j is bounded by ϵ at least to the first-order approximation. Once we find all these values of ϖ_i^j, $j = 1, 2, \ldots, l$, the value of ϖ_i is given by

$$\varpi_i = \max_{1 \leq j \leq l}\{\varpi_i^j\}.$$

For example, if the transition rate λ_j has the form $\lambda_j(\mathbf{x}) = c_j x_i$ with c_j being a constant, then the reaction j is said to be *first order* and the absolute change in $\lambda_j(\mathbf{x})$ is given by

$$\Delta \lambda_j(\mathbf{x}) = \lambda_j(\mathbf{x} + \Delta \mathbf{x}) - \lambda_j(\mathbf{x}) = c_j(x_i + \Delta x_i) - c_j x_i = c_j \Delta x_i.$$

Hence, the relative change in $\lambda_j(\mathbf{x})$ is related to the relative change in the X_i by

$$\frac{\Delta\lambda_j(\mathbf{x})}{\lambda_j(\mathbf{x})} = \frac{\Delta x_i}{x_i},$$

which implies that if we set the relative change in X_i by ϵ (i.e., $\varpi_i^j = 1$), then the relative change in λ_j is bounded by ϵ. If the transition rate λ_j has the form $\lambda_j(\mathbf{x}) = c_j x_i x_r$ with c_j being a constant, then the reaction j is said to be *second order* and the absolute change in $\lambda_j(\mathbf{x})$ is given by

$$\Delta\lambda_j(\mathbf{x}) = c_j(x_i + \Delta x_i)(x_r + \Delta x_r) - c_j x_i x_r = c_j x_r \Delta x_i + c_j x_i \Delta x_r + c_j \Delta x_i \Delta x_r.$$

Hence, the relative change in $\lambda_j(\mathbf{x})$ is related to the relative changes in X_i and X_r by

$$\frac{\Delta\lambda_j(\mathbf{x})}{\lambda_j(\mathbf{x})} = \frac{\Delta x_i}{x_i} + \frac{\Delta x_r}{x_r} + \frac{\Delta x_i}{x_i}\frac{\Delta x_r}{x_r},$$

which implies that if we set the relative change in X_i by $\dfrac{\epsilon}{2}$ and the relative change in X_r by $\dfrac{\epsilon}{2}$ (i.e., $\varpi_i^j = 2$, $\varpi_r^j = 2$), then the relative change in λ_j is said to be bounded by ϵ to the *first-order approximation* (as we ignore the product of the relative change in X_i and the relative change in X_r).

The tau-leaping method employed here also includes modifications developed by Cao et al. [22] to avoid the possibility of negative populations. This could occur in some instances for which the number of individuals is small at the beginning of the leap interval and the state change involved is negative, and hence after numerous transitions it may result in a negative population. To avoid this situation, Cao et al. [22, 23] introduced another control parameter, n_c, a positive integer (normally set between 2 and 20) which is used to separate transitions into two classes, critical transitions or non-critical transitions. To do that, an estimate for the maximum number of times L_j, $j \in \{1, 2, \ldots, l\}$, that transition j is allowed to occur in the leap interval (that is, before one of the state variables involved in the transition becomes less than or equal to zero) is calculated by

$$L_j = \min_{\{1 \leq i \leq n; v_{ij} < 0\}} \left\lfloor \frac{x_i}{|v_{ij}|} \right\rfloor$$

with $\lfloor \cdot \rfloor$ indicating the floor function (that is, $\lfloor x \rfloor = k$ if $k \leq x < k+1$, where k is an integer). If $L_j \leq n_c$, then the reaction is deemed critical; otherwise, the reaction is considered to be non-critical. The algorithm for the modified explicit tau-leaping method is given below.

S1. Set the initial condition for each state at $t = 0$.

S2. For the given state $\mathbf{x}$ of the system, identify all critical transitions as described above (in our calculations below, we set $n_c = 10$), and denote the set of critical transitions by

$$\mathbb{J}_{cr} = \{j \in \{1, 2, \ldots, l\} \mid \text{the } j\text{th transition is a critical one}\},$$

and the set of non-critical ones by

$$\mathbb{J}_{ncr} = \{j \in \{1, 2, \ldots, l\} \mid \text{the } j\text{th transition is a non-critical one}\}.$$

S3. Compute Δt_1 according to the following formula:

$$\Delta t_1 = \min_{\{1 \le i \le n\}} \left\{ \frac{\max\{\epsilon x_i / \varpi_i, 1\}}{|\hat{\mu}_i(\mathbf{x})|}, \frac{\max\{\epsilon x_i / \varpi_i, 1\}^2}{\hat{\sigma}_i^2(\mathbf{x})} \right\}, \quad (8.22)$$

where ϵ is the error control parameter (in our calculations below, we set $\epsilon = 0.03$), ϖ_i is chosen such that the relative changes in all the transition rates will be bounded by ϵ at least to the first-order approximation (as described in the text), and

$$\hat{\mu}_i(\mathbf{x}) = \sum_{j \in \mathbb{J}_{ncr}} v_{ij} \lambda_j(\mathbf{x}), \quad \hat{\sigma}_i^2(\mathbf{x}) = \sum_{j \in \mathbb{J}_{ncr}} v_{ij}^2 \lambda_j(\mathbf{x}), \quad i = 1, 2, \ldots, n.$$

S4. Determine whether tau-leaping is appropriate by comparing Δt_1 to $1/\lambda_{\mathbf{x}}$, where $\lambda_{\mathbf{x}} = \sum_{j=1}^{l} \lambda_j(\mathbf{x})$. If Δt_1 is less than some multiple of $1/\lambda_{\mathbf{x}}$ (chosen to be 10 in our calculations), then abandon tau-leaping and execute a set number of SSA steps (chosen to be 100 in our calculations) and return to **S3**. Otherwise proceed to the next step.

S5. Compute the sum of all critical transition rates

$$\lambda_{\mathbf{x}}^c = \sum_{j \in \mathbb{J}_{cr}} \lambda_j(\mathbf{x}).$$

Generate a *second candidate* time leap, Δt_2, as a realization of the exponential random variable with mean $1/\lambda_{\mathbf{x}}^c$.

S6. Let $\Delta t = \min\{\Delta t_1, \Delta t_2\}$. Approximate the number of transitions within the time interval, z_j, as a realization of the Poisson random variable with mean $\lambda_j(\mathbf{x}) \Delta t$ (i.e., z_j is a realization of Z_j defined by (8.19)) for all $j \in \mathbb{J}_{ncr}$. For all critical reactions, define z_j as follows:

    S6.1 If $\Delta t = \Delta t_1$, set $z_j = 0$ for all $j \in \mathbb{J}_{cr}$ (that is, no critical transitions occur).

    S6.2 If $\Delta t = \Delta t_2$, let j_c be a realization of a discrete random variable with range $\mathbb{J}_{cr}$ and the probability associated with $j \in \mathbb{J}_{cr}$ being $\lambda_j(\mathbf{x}) / \lambda_{\mathbf{x}}^c$. Set $z_{j_c} = 1$ and $z_j = 0$ for $j \in \mathbb{J}_{cr}, j \ne j_c$ (that is, only the critical transition, j_c, occurs).

S7. If there is a negative component in $\mathbf{x} + \sum_{j=1}^{l} z_j \mathbf{v}_j$, reduce Δt_1 by half and return to **S4.** Otherwise, set $t = t + \Delta t$, and update the system state by setting

$$\mathbf{x} = \mathbf{x} + \sum_{j=1}^{l} z_j \mathbf{v}_j.$$

S8. Iterate **S2-S7** until $t \geq t_{stop}$.

It is worth noting that the inverse transform method discussed in Remark 2.5.2 can be used to generate a realization of a discrete random variable in **S6.2**.

Remark 8.2.1 *The numerical analysis of the basic explicit tau-leaping method has been studied by several researchers in recent years (e.g., see [37, 49]). For example, it was shown in [37] that with some mild conditions on the transition rates and state variables the basic explicit tau-leaping method has a strong convergence of order $1/2$ in the sense that*

$$\sup_{0 \leq j \leq r} \mathbb{E}\|\mathbf{X}^{(ex)}(t_j) - \mathbf{X}(t_j)\|^2 \leq c\Delta t,$$

and it has a weak convergence of order 1 in the sense that

$$\left| \mathbb{E}\{h(\mathbf{X}^{(ex)}(t_{stop}))\} - \mathbb{E}\{h(\mathbf{X}(t_{stop}))\} \right| \leq c\Delta t$$

for any continuous function h satisfying the exponential growth condition (i.e., there exist positive constants c_h and c_b such that $|h(\mathbf{x})| \leq c_h c_b^{\|\mathbf{x}\|}$). Here $\mathbf{X}^{(ex)}(t)$ denotes the approximation of $\mathbf{X}(t)$, $\Delta t = \max_{1 \leq j \leq r}\{\Delta t_j\}$ with Δt_j, $j = 1, 2, \ldots, r$, being the time steps, $t_j = \sum_{i=1}^{j} \Delta t_i$, $j = 1, 2, \ldots, r$, with $t_0 = 0$ and $t_r = t_{stop}$. In practice, one often switches from the explicit tau-leaping to the SSA if the condition $\lambda_{\mathbf{x}} \Delta t \gg 1$ is not satisfied (as we did in the above algorithm). To naturally enforce this condition, the numerical analysis for the basic explicit tau-leaping method was revisited recently by incorporating a scaling parameter of the system into the analysis (i.e., the error is dependent on the scaling parameter), where the scaling parameter is related to the size of the population in the system. We refer the interested reader to [6, 7, 32] for more information on this.

8.2.3.2   An Implicit Tau-Leaping Method

In many applications, such as the HIV model explained in Section 8.5, problems of "stiffness" may arise. Rathinam et al. [48] explored the nature of stiffness in discrete stochastic systems and demonstrated that an implicit tau-leaping method (similar to implicit Euler methods for ordinary differential equations) is capable of taking large time steps for stiff, discrete systems, producing accurate results for such systems while significantly reducing the computational time when compared to explicit tau-leaping methods. The implicit tau-leaping method replaces the explicit update formula given in Equation (8.20) by an implicit tau-leaping formula given by

$$\mathbf{X}(t + \Delta t) \approx \mathbf{x} + \sum_{j=1}^{l} (Z_j - \lambda_j(\mathbf{x})\Delta t + \lambda_j(\mathbf{X}(t + \Delta t))\Delta t)\, \mathbf{v}_j,$$

where Z_j is defined by (8.19). Note that the above formula typically gives a non-integer vector for $\mathbf{X}(t + \Delta t)$. To overcome this difficulty, Rathinam et al. [48] proposed a two-stage process given by

$$\widetilde{\mathbf{X}} = \mathbf{x} + \sum_{j=1}^{l} \left(Z_j - \lambda_j(\mathbf{x})\Delta t + \lambda_j(\widetilde{\mathbf{X}})\Delta t \right) \mathbf{v}_j, \qquad (8.23)$$

and

$$\mathbf{X}(t + \Delta t) = \mathbf{x} + \sum_{j=1}^{l} \left[\!\left[Z_j - \lambda_j(\mathbf{x})\Delta t + \lambda_j(\widetilde{\mathbf{X}})\Delta t \right]\!\right] \mathbf{v}_j, \qquad (8.24)$$

where $[\![a]\!]$ denotes the nearest non-negative integer corresponding to a real number a.

The implicit tau-leaping method does not have a stability limitation as the explicit tau-leaping does (i.e., the relative changes in all the transition rates are bounded by ϵ) due to the implicitness of the scheme. In [24], the step size for the stiff system is chosen to bound the relative changes of those transition rates resulting from the non-equilibrium reactions by ϵ; thus a larger time step size is allowed. However, as remarked by the authors in [24], it is generally difficult to determine whether or not a reaction is in partial equilibrium, and the partial equilibrium condition is only formulated in [24] for reversible reaction pairs for some biochemical systems.

Remark 8.2.2 *We give a brief comment on how to numerically solve the system of non-linear equations required in the implicit tau leaping method. By (8.23) we know that the system of non-linear equations to be solved is of the form*

$$\tilde{\mathbf{x}} = \mathbf{x} + \sum_{j=1}^{l} [z_j - \lambda_j(\mathbf{x})\Delta t + \lambda_j(\tilde{\mathbf{x}})\Delta t]\, \mathbf{v}_j, \qquad (8.25)$$

where z_j is a realization of Z_j. To solve (8.25) for $\tilde{\mathbf{x}}$, we first convert the non-linear equation (8.25) into the form

$$\mathbf{h}(\mathbf{y}) = 0, \qquad (8.26)$$

where $\mathbf{y} = (y_1, y_2, \ldots y_n)^T$ *with* $\tilde{\mathbf{x}} = (y_1^2, y_2^2, \ldots, y_n^2)^T$, *and*

$$\mathbf{h}(\mathbf{y}) = (y_1^2, y_2^2, \ldots, y_n^2)^T - \mathbf{x} - \sum_{j=1}^{l} \left[z_j - \lambda_j(\mathbf{x})\Delta t + \lambda_j(y_1^2, y_2^2, \ldots, y_n^2)\Delta t\right] \mathbf{v}_j.$$

Then we can use the MATLAB function fsolve *or other solver to obtain the numerical solution of* (8.26); *then the numerical solution for* (8.25) *is obtained by taking* $\tilde{\mathbf{x}} = (y_1^2, y_2^2, \ldots, y_n^2)^T$. *The reason for converting* (8.25) *into* (8.26) *is to avoid possible negative values of state variables.*

8.3   Density Dependent Continuous Time Markov Chains and Kurtz's Limit Theorem

Based on the discussion in the above section, we know that it is in general quite expensive to simulate a stochastic system with very large populations. Hence, in such cases one often wishes to know whether or not the stochastic system can be approximated by a deterministic one when the population size is sufficiently large. In this section, we introduce such a result, known as Kurtz's limit theorem, which gives a way to construct a deterministic system to approximate density dependent continuous time Markov chains (defined below) as the population size is sufficiently large.

Suppose for each positive number M, $\{\mathbf{X}^M(t) : t \geq 0\}$ is a continuous time Markov chain such that for any small time step Δt we have

$$\text{Prob}\left\{\mathbf{X}^M(t + \Delta t) = \mathbf{x}^M + \mathbf{v}_j \mid \mathbf{X}^M(t) = \mathbf{x}^M\right\} = \lambda_j(\mathbf{x}^M)\Delta t + o(\Delta t)$$

$$\text{Prob}\left\{\mathbf{X}^M(t + \Delta t) = \mathbf{x}^M \mid \mathbf{X}^M(t) = \mathbf{x}^M\right\} = 1 - \sum_{j=1}^{l} \lambda_j(\mathbf{x}^M)\Delta t + o(\Delta t)$$

$$(8.27)$$

for $j = 1, 2, \ldots, l$ with l being the total number of transitions. Here $\mathbf{X}^M(t) = (X_1^M(t), X_2^M(t), \ldots, X_n^M(t))^T \in \mathbb{Z}^n$, $\mathbf{v}_j = (v_{1j}, v_{2j}, v_{3j}, \ldots, v_{nj})^T \in \mathbb{Z}^n$ with v_{ij} denoting the change in state variable X_i^M caused by the jth transition, and λ_j is the transition rate associated with transition j. In addition, we assume that $\mathbf{X}^M(0) = \mathbf{x}_0^M$ is a constant vector in $\mathbb{Z}^n$.

To establish Kurtz's limit theorem, we need the notion of a *density-dependent* continuous time Markov chain, which is defined as follows.

Definition 8.3.1 *The family of continuous time Markov chains* $\{\mathbf{X}^M(t) : t \geq 0\}$ *defined by* (8.27) *is called* density dependent *if and only if there exist continuous functions* $f_j : \mathbb{R}^n \to \mathbb{R}$ *such that*

$$\lambda_j(\mathbf{x}) = M f_j(\mathbf{x}/M), \quad j = 1, 2, \ldots, l. \tag{8.28}$$

In the remainder of this section, we assume that the family of continuous time Markov chains $\{\mathbf{X}^M(t) : t \geq 0\}$ defined by (8.27) is *density dependent*.

The following fact about the Poisson process plays a significant role in establishing Kurtz's limit theorem.

Lemma 8.3.1 *Let $\{Y(t) : t \geq 0\}$ be a unit Poisson process. Then for any $t_f > 0$ we have*

$$\lim_{m \to \infty} \sup_{t \leq t_f} \left| \frac{Y(mt)}{m} - t \right| = 0 \quad a.s. \tag{8.29}$$

Furthermore, the following Gronwall's inequality (e.g., see [9] for details) is used in establishing Kurtz's limit theorem.

Lemma 8.3.2 *Let c_0 and c_1 be some positive constants, and h be a nonnegative real function such that for any $t \geq 0$ it satisfies*

$$h(t) \leq c_0 + c_1 \int_0^t h(s)ds.$$

Then for any $t \geq 0$ we have $h(t) \leq c_0 \exp(c_1 t)$.

8.3.1   Kurtz's Limit Theorem

Now we give the ideas behind Kurtz's limit theorem. Let $\{Y_j(t) : t \geq 0\}, j = 1, 2, \ldots, l$, be independent unit Poisson processes. Then $\mathbf{X}^M(t)$ can be written as

$$\mathbf{X}^M(t) = \mathbf{X}^M(0) + \sum_{j=1}^{l} \mathbf{v}_j Y_j \left(\int_0^t \lambda_j(\mathbf{X}^M(s))ds \right).$$

By (8.28), we can rewrite the above equation as

$$\mathbf{X}^M(t) = \mathbf{X}^M(0) + \sum_{j=1}^{l} \mathbf{v}_j Y_j \left(M \int_0^t f_j(\mathbf{X}^M(s)/M)ds \right). \tag{8.30}$$

Let $\mathbf{C}^M(t) = \mathbf{X}^M(t)/M$. Then we obtain another continuous time Markov chain $\{\mathbf{C}^M(t) : t \geq 0\}$. By (8.30) we find

$$\mathbf{C}^M(t) = \mathbf{C}^M(0) + \sum_{j=1}^{l} \mathbf{v}_j \left[\frac{1}{M} Y_j \left(M \int_0^t f_j(\mathbf{C}^M(s))ds \right) \right]. \tag{8.31}$$

Let $\tilde{Y}_j(t) = Y_j(t) - t$. Then we can rewrite (8.31) as

$$\mathbf{C}^M(t) = \mathbf{C}^M(0) + \sum_{j=1}^{l} \mathbf{v}_j \int_0^t f_j(\mathbf{C}^M(s))ds$$

$$+ \sum_{j=1}^{l} \mathbf{v}_j \left[\frac{1}{M} \tilde{Y}_j \left(M \int_0^t f_j(\mathbf{C}^M(s))ds \right) \right]. \tag{8.32}$$

Define $\mathbf{g}(\mathbf{c}) = \sum_{j=1}^{l} \mathbf{v}_j f_j(\mathbf{c})$. Then by (8.32), Lemma 8.3.1 and Gronwall's inequality, one can show that with some mild conditions on $\mathbf{g}$ the processes $\{\mathbf{C}^M(t) : t \geq 0\}$ converge to a deterministic process that is the solution of the following system of ordinary differential equations:

$$\dot{\mathbf{c}}(t) = \mathbf{g}(\mathbf{c}), \quad \mathbf{c}(0) = \mathbf{c}_0. \tag{8.33}$$

This result was originally shown by Kurtz [34, 35] (and hence it is referred to as *Kurtz's limit theorem* in this monograph), and is summarized in the following theorem (e.g., see [9, 36] for details on the proof).

Theorem 8.3.3 *Suppose that* $\lim_{M \to \infty} \mathbf{C}^M(0) = \mathbf{c}_0$ *and for any compact set* $\Omega \subset \mathbb{R}^n$ *there exists a positive constant* η_Ω *such that*

$$|\mathbf{g}(\mathbf{c}) - \mathbf{g}(\tilde{\mathbf{c}})| \leq \eta_\Omega |\mathbf{c} - \tilde{\mathbf{c}}|, \quad \mathbf{c}, \tilde{\mathbf{c}} \in \Omega.$$

Then we have

$$\lim_{M \to \infty} \sup_{t \leq t_f} |\mathbf{C}^M(t) - \mathbf{c}(t)| = 0 \ a.s. \ for \ all \ t_f > 0,$$

where $\mathbf{c}$ *denotes the unique solution to* (8.33).

Theorem 8.3.3 indicates that the convergence is in the sense of convergence almost surely. It is worth noting that in Kurtz's original work [34, 35] the convergence is in the sense of convergence in probability. In addition, it should be noted that, based on the problem considered, the parameter M can be interpreted as the total number of individuals in the population (if this number is constant over time) or the population size at the initial time or the volume occupied by the population or some other scaling factor. Hence, the parameter M is often called the *population size* or *sample size* or *scaling parameter* (these names will be used interchangeably in the remainder of this chapter). We thus see that Kurtz's limit theorem does not require the total number of individuals in the population to stay constant over time.

8.3.2 Implications of Kurtz's Limit Theorem

We see from the above discussion that a prerequisite condition for Kurtz's limit theorem is that the family of CTMCs must be density dependent. Fortunately, this condition is often satisfied in practice. One of the important implications for Kurtz's limit theorem is that the stochastic effects become less important as the population size increases, and hence it provides a justification for using ODEs when the population size is sufficiently large. Another important implication is that it can be used to construct a corresponding deterministic model for the CTMC model, and it can also be used to construct a corresponding CTMC model for a deterministic model (described by ODEs).

In general, deterministic systems are much easier to analyze compared to stochastic systems. Techniques such as parameter estimation methods are well developed for deterministic systems, whereas parameter estimation is much more difficult in a stochastic framework. Thus the advantage in constructing a corresponding deterministic system for a CTMC model is that one is able to use the well-developed methods for the deterministic system to carry out analysis for the CTMC model. For example, one can use the parameter estimation techniques (ordinary least squares, weighted or generalized least squares) introduced in Chapter 3 to estimate model parameters in a CTMC model as its model, parameters are exactly the same as those in its corresponding deterministic model. This will be further discussed in Section 8.4.

For a given CTMC model $\{\mathbf{X}^M(t) : t \geq 0\}$ satisfying (8.27) and (8.28), Kurtz's limit theorem indicates that the corresponding deterministic model for the CTMC model $\{\mathbf{C}^M(t) : t \geq 0\}$ (with $\mathbf{C}^M(t) = \mathbf{X}^M(t)/M$) is given by

$$\dot{\mathbf{c}}(t) = \sum_{j=1}^{l} \mathbf{v}_j f_j(\mathbf{c}), \quad \mathbf{c}(0) = \mathbf{c}_0. \tag{8.34}$$

Let $\mathbf{x}^M(t) = M\mathbf{c}(t)$. Then by (8.28) and (8.34) we know that the corresponding deterministic model for the CTMC model $\{\mathbf{X}^M(t) : t \geq 0\}$ is

$$\dot{\mathbf{x}}^M(t) = \sum_{j=1}^{l} \mathbf{v}_j \lambda_j(\mathbf{x}^M), \quad \mathbf{x}^M(0) = \mathbf{x}_0^M. \tag{8.35}$$

Ordinary differential equations are often used in practice to describe physical and biological processes. As interest in *initial phases* of infections or epidemics becomes more prevalent, investigators increasingly recognize that many of the corresponding investigative efforts require small population number models for which ODEs are unsuitable. Hence, it is important to know how to construct a corresponding CTMC model for this system of ODEs when the population size is small (so that the constructed CTMC model has this system of ODEs as its limit). Note that we are mainly interested in the CTMC model in terms of numbers (i.e., $\mathbf{X}^M$). Hence, if the given system of ODEs is in terms of concentration (i.e., (8.34)), then we first rewrite the system of ODEs in terms of numbers (i.e., (8.35)) and then derive the corresponding stochastic model for the resulting system of ODEs. We know that to construct a corresponding CTMC model an important task is to find the transition rates and the corresponding state change vectors. These usually can be derived based on some biological or physical interpretation of the deterministic model. For example, a flow chart is usually given to describe the movement from one compartment to another compartment. This flow chart is useful not only to develop a deterministic model but also to construct its corresponding CTMC model as the rates for the movement from one compartment to another compartment have a close relationship to the transition rates. It should be noted that starting with a given deterministic model, different biological/physical interpretation could lead to different stochastic models; that is, a deterministic model may correspond to more than one stochastic model. This will be demonstrated in detail in Section 8.5.

8.4   Biological Application: Vancomycin-Resistant Enterococcus Infection in a Hospital Unit

In this section, we present a CTMC model to describe the dynamics of vancomycin-resistant enterococcus (VRE) infection in a hospital unit, and demonstrate how to construct a corresponding deterministic model based on Kurtz's limit theorem for this stochastic model. This model was originally developed in [43], wherein a detailed discussion on VRE infection can be found.

8.4.1   The Stochastic VRE Model

VRE is the group of bacterial species of the genus enterococcus that is resistant to the antibiotic vancomycin, and it can be found in sites such as digestive/gastrointestinal and urinary tracts, and in surgical incisions. The bacteria responsible for VRE can be a member of the normal, usually commensal bacterial flora that becomes pathogenic when it multiplies in normally sterile sites. VRE infection is one of the most common infections occurring in hospitals, and this type of infection is often referred to as a *nosocomial* or *hospital-acquired infection*.

We consider a simple compartmental model in which patients in a hospital unit are classified as uncolonized U (those individuals with no VRE present in the body), colonized C (those individuals with VRE present in the body) or colonized in isolation J, as depicted in Figure 8.1. We assume that patients are admitted into the hospital at a rate of Λ per day, with some fraction m of patients already VRE colonized ($0 \le m \le 1$). Uncolonized patients are discharged from the hospital at a rate of $\mu_1 U$ per day, and colonized patients and colonized patients in isolation are discharged from the hospital at rates of $\mu_2 C$ and $\mu_2 J$ per day, respectively. An uncolonized patient becomes colonized at a rate $\beta\, U[C + (1 - \gamma)J]$ per day, where the hand-hygiene policy applied to health care workers on isolated VRE colonized patients reduces infectivity by a factor of γ ($0 < \gamma < 1$). We thus see that the rate of contact is assumed to be proportional to the population size (this assumption is often referred to as *mass action incidence*, and it is reasonable for a small population size). A colonized patient is moved into isolation at rate αC per day. A summary of the description of all the model parameters is given in Table 8.1.

We also assume that the hospital remains full at all times (that is, the overall admission rate equals the overall discharge rate), and that the

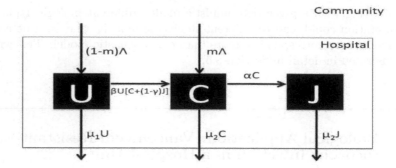

FIGURE 8.1:   A schematic of the VRE model.

TABLE 8.1:   Description of model parameters in the stochastic VRE model as well as the values of parameters used in the simulation.

Parameters	Description	Value
m	VRE colonized patient on admission rate	0.04
β	Effective contact rate	0.001 per day
γ	Hand-hygiene compliance rate	0.58
α	Patient isolation rate	0.29 per day
μ_1	Uncolonized patients discharge rate	0.16 per day
μ_2	VRE colonized patients discharge rate	0.08 per day

dynamics of VRE infection in a hospital unit can be modeled as a CTMC. Let M denote the total number of beds available in the hospital, and $\mathbf{X}^M(t) = (X_1^M(t), X_2^M(t), X_3^M(t))^T$ with $X_1^M(t), X_2^M(t), X_3^M(t)$ respectively denoting the number of uncolonized patients, VRE colonized patients, VRE colonized patients in isolation at time t in a hospital with M beds. For any positive number M, we assume that $\{\mathbf{X}^M(t) : t \geq 0\}$ is a CTMC such that for any small Δt it satisfies (8.27). In addition, we assume that the initial condition is non-random; that is, $\mathbf{X}^M(0) = \mathbf{x}_0^M$ with $\mathbf{x}_0^M \in \mathbb{Z}^3$ being non-random.

Based on the above discussion, we know that the movement from the uncolonized compartment to the colonized compartment (i.e., $x_1^M \to x_2^M$) is through either the admission (due to discharge of uncolonized patients) or the effective colonization, where the rate for the admission is given by $m\mu_1 x_1^M$ (as the discharge rate of uncolonized patients is $\mu_1 x_1^M$ with some fraction m being colonized), and the rate for the effective colonization is $\beta x_1^M (x_2^M + (1-\gamma)x_3^M)$. In addition, the movement from the colonized in isolation compartment to the colonized compartment ($x_3^M \to x_2^M$) is only through the admission (due to the discharge of colonized patients in isolation) with the rate $m\mu_2 x_3^M$. Furthermore, we see that the movement from the colonized compartment to

TABLE 8.2:   Transition rates as well as corresponding state change vectors for the stochastic VRE model.

j	Description	$\lambda_j(\mathbf{x}^M)$	$\mathbf{v}_j$
1	$x_1^M \to x_2^M$	$m\mu_1 x_1^M + \beta x_1^M(x_2^M + (1-\gamma)x_3^M)$	$-\mathbf{e}_1 + \mathbf{e}_2$
2	$x_3^M \to x_2^M$	$m\mu_2 x_3^M$	$\mathbf{e}_2 - \mathbf{e}_3$
3	$x_2^M \to x_1^M$	$(1-m)\mu_2 x_2^M$	$\mathbf{e}_1 - \mathbf{e}_2$
4	$x_3^M \to x_1^M$	$(1-m)\mu_2 x_3^M$	$\mathbf{e}_1 - \mathbf{e}_3$
5	$x_2^M \to x_3^M$	αx_2^M	$-\mathbf{e}_2 + \mathbf{e}_3$

the uncolonized compartment $(x_2^M \to x_1^M)$ is through the admission (due to the discharge of colonized patients) with rate $(1-m)\mu_2 x_2^M$, and that the movement from the colonized in isolation compartment to the uncolonized compartment $(x_3^M \to x_1^M)$ is by the admission (due to the discharge of the colonized patients in isolation) with rate $(1-m)\mu_2 x_3^M$. Finally, the movement from the colonized compartment to the colonized in isolation compartment $(x_2^M \to x_3^M)$ is through the isolation with rate αx_2^M. Based on the above discussion, we see that there are five reactions (i.e, $l = 5$) with transition rates λ_j and corresponding state change vectors $\mathbf{v}_j$, summarized in Table 8.2, where $\mathbf{e}_i \in \mathbb{Z}^3$ is the ith unit column vector, $i = 1, 2, 3$.

8.4.2   The Deterministic VRE Model

We observe that the family of CTMCs $\{\mathbf{X}^M(t) : t \geq 0\}$ with transition rates and corresponding state change vectors defined in Table 8.2 is density dependent, and the conditions in Kurtz's limit theorem are satisfied. Thus, based on Kurtz's limit theorem, the corresponding deterministic model for the stochastic VRE model is given by

$$\dot{\mathbf{x}}^M = \sum_{j=1}^{5} \mathbf{v}_j \lambda_j(\mathbf{x}^M), \quad \mathbf{x}^M(0) = \mathbf{x}_0^M.$$

Let $U^M = x_1^M$, $C^M = x_2^M$ and $J^M = x_3^M$. Then the above equation can be rewritten as

$$\dot{U}^M = -m\mu_1 U^M - \beta U^M(C^M + (1-\gamma)J^M) + (1-m)\mu_2(C^M + J^M)$$

$$\dot{C}^M = m\mu_1 U^M + \beta U^M(C^M + (1-\gamma)J^M) + m\mu_2 J^M - [(1-m)\mu_2 + \alpha]C^M$$

$$\dot{J}^M = -\mu_2 J^M + \alpha C^M$$

(8.36)

with initial condition $(U^M(0), C^M(0), J^M(0))^T = \mathbf{x}_0^M$.

Remark 8.4.1 *Using the resulting deterministic model (8.36) the authors in [43] discussed how to select a parameter subset combination that can be estimated using an ordinary least squares or generalized least squares method with given VRE surveillance data from an oncology unit. Note that model parameters in the stochastic VRE model are exactly the same as those in the deterministic one. Hence, this leads to rather obvious parameter estimation techniques for CTMC models, that is, a method for estimating parameters in CTMC models based on Kurtz's limit theorem coupled with the inverse problem methods developed for deterministic systems.*

8.4.3   Numerical Results

In this section we report on numerical results obtained for the stochastic VRE model with transition rates and corresponding state change vectors given in Table 8.2 as well as its corresponding deterministic model (8.36). Note that the number of transitions in the stochastic VRE model is very small (only 5 transitions) and the total number of patients in a realistic hospital is also relatively small. Hence, we can use an exact simulation method to simulate the stochastic VRE model. Here we choose to use the SSA due to its simplicity and efficiency.

All the simulations were run for the time period $[0, 200]$ days with parameter values given in the third column of Table 8.1 and initial conditions (adapted from Table 3 in [43]) $\mathbf{X}^M(0) = M \left(\dfrac{29}{37}, \dfrac{4}{37}, \dfrac{4}{37} \right)^T$, where the value of M is chosen to be 37, 370 or 3700. We remark that $M = 3700$ may be too large for a hospital in reality and it may also be too large for the assumption of mass action incidence to be reasonable. However, we still show the results obtained with $M = 3700$ because we want to illustrate that the solution to the deterministic VRE model indeed provides a good approximation to the solution to the corresponding stochastic VRE model when the population size is sufficiently large.

Figure 8.2 depicts five typical sample paths of each state of the stochastic VRE model obtained by the SSA in comparison to the solution for the deterministic VRE model (8.36). We observe from Figure 8.2 that all the sample paths for each state variable oscillate around their corresponding deterministic solution, and they exhibit very large differences for $M = 37$ and $M = 370$. This offers rather clear warnings for the indiscriminate use of ODEs in place of CTMC models to study dynamical systems with a small population size. We also observe from Figure 8.2 that the variation among the sample paths decreases as M increases, and that the stochastic solution becomes reasonably close to the deterministic solution when $M = 3700$. This agrees with expectations in light of Kurtz's limit theorem.

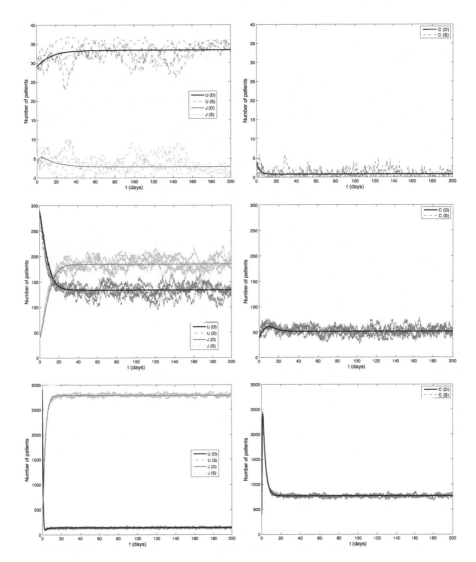

FIGURE 8.2:   Results in the left column are for uncolonized patients (U) and colonized patients in isolation (J), and the ones in the right column are for the colonized patients (C). The (D) and (S) in the legend denote the solution obtained with the deterministic VRE model and stochastic VRE model, respectively. The deterministic results are obtained using the MATLAB function *ode15s*, and the stochastic results are obtained using the SSA, where the plots in the first row are for $M = 37$, the ones in the second row are for $M = 370$ and the ones in the last row are for $M = 3700$.

8.5   Biological Application: HIV Infection within a Host

In this section, we demonstrate how to construct a corresponding continuous time Markov chain model based on Kurtz's limit theorem for a given deterministic model described by a system of ordinary differential equations. The example we use here is a deterministic HIV model proposed in [13]. Data fitting results validate this model and verify that it provides reasonable fits to all the 14 patients studied. Moreover it has impressive predictive capability when comparing model simulations (with parameters based on estimation procedures using only half of the longitudinal observations) to the corresponding full longitudinal data sets.

The resulting stochastic model is intended to be used to describe the dynamics of HIV during the early or acute stage of infection, where the uninfected cells are still at very high numbers while the infected cells are at very low numbers. Hence, we see that this model is multi-scaled in state. As we shall see later in this section, the exact methods such as the SSA for simulation of this stochastic model are impractical at the scale of an average person, while the tau-leaping methods are possible. Even though tau-leaping methods have now been widely used in the biochemistry literature, they are rarely applied to complex non-linear dynamical infectious disease models such as the HIV model that we present here (with transition rates being complicated non-linear functions of state variables rather than simple polynomial functions). Hence, we will give a detailed discussion on how to apply the tau-leaping methods to the resulting stochastic model. The step-by-step implementation recipe demonstrated here for these algorithms can be applied to a wide range of other complex biological models, specifically those in infectious disease progression and epidemics.

8.5.1   Deterministic HIV Model

HIV is a retrovirus that targets the CD4+ T cells in the immune system. Once the virus has taken control of a sufficiently large proportion of CD4+ T cells, an individual is said to have AIDS. There are a wide variety of deterministic mathematical models that have been proposed to describe the various aspects of in-host HIV infection dynamics (e.g., [1, 2, 13, 21, 25, 45, 47, 54]). The most basic of these models typically include two or three of the key dynamic compartments: virus, uninfected target cells and infected cells. These compartmental depictions lead to systems of linear or non-linear ordinary differential equations in terms of state variables representing the concentrations in each compartment and parameters describing viral production and clearance, cell infection and death rate, treatment efficacy, etc.

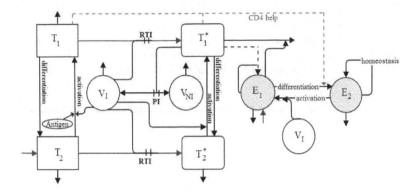

FIGURE 8.3:   A schematic of the HIV model: PI and RTI denote protease inhibitors and reverse transcriptase inhibitors, respectively, and they are the two types of drug used to treat HIV patients.

The model in [13] includes eight compartments: uninfected activated CD4+ T cells (T_1), uninfected resting CD4+ T cells (T_2), along with their corresponding infected states $(T_1^*$ and $T_2^*)$, infectious free virus (V_I), non-infectious free virus (V_{NI}), HIV-specific effector CD8+ T cells (E_1), and HIV-specific memory CD8+ T cells (E_2), as depicted in the compartmental schematic of Figure 8.3. In this model, the unit for all the cell compartments is cells/μl blood and the unit for the virus compartments is RNA copies/ml plasma. We remark that this latter unit is adopted in [13] because the viral load in the provided clinical data is reported as RNA copies/ml plasma. It is worth noting that there is no clear standard for reporting the viral load in the literature and the measurements are usually taken using either plasma, whole blood or peripheral blood leukocytes (see [33] for details).

For simplicity, we only consider this model in the case without treatment and will exclude the non-infectious free virus compartment from the model. In addition, we convert the unit for the free virus from RNA copies/ml plasma to RNA copies/μl blood to be consistent with the units of cell compartments, and modify the equation based on this change. This conversion will make the derivation of the corresponding stochastic HIV model more direct. With these changes in the model of [13], the state variables and their corresponding units for the deterministic model that we use in this section are reported in Table 8.3. The corresponding compartmental ordinary differential equation

TABLE 8.3:   Model states for the deterministic HIV model.

States	Unit	Description
T_1	cells/μl blood	uninfected activated CD4+ T cells
T_1^*	cells/μl blood	infected activated CD4+ T cells
T_2	cells/μl blood	uninfected resting CD4+ T cells
T_2^*	cells/μl blood	infected resting CD4+ T cells
V_I	RNA copies/μl blood	infectious free virus
E_1	cells/μl blood	HIV-specific effector CD8+ T cells
E_2	cells/μl blood	HIV-specific memory CD8+ T cells

model is given by

$$\dot{T}_1 = -d_{T1}T_1 - \beta_{T1}V_IT_1 - \gamma_T T_1 + n_T\left(\frac{a_T V_I}{V_I+\kappa_V} + a_A\right)T_2,$$

$$\dot{T}_1^* = \beta_{T1}V_IT_1 - \delta_V T_1^* - \delta_E E_1 T_1^* - \gamma_T T_1^* + n_T\left(\frac{a_T V_I}{V_I+\kappa_V} + a_A\right)T_2^*,$$

$$\dot{T}_2 = \zeta_T\frac{\kappa_s}{V_I+\kappa_s} + \gamma_T T_1 - d_{T2}T_2 - \beta_{T2}V_IT_2 - \left(\frac{a_T V_I}{V_I+\kappa_V} + a_A\right)T_2,$$

$$\dot{T}_2^* = \gamma_T T_1^* + \beta_{T2}V_IT_2 - d_{T2}T_2^* - \left(\frac{a_T V_I}{V_I+\kappa_V} + a_A\right)T_2^*,$$

$$\dot{V}_I = n_V\delta_V T_1^* - d_V V_I - (\beta_{T1}T_1 + \beta_{T2}T_2)V_I,$$

$$\dot{E}_1 = \zeta_E + \frac{b_{E1}T_1^*}{T_1^*+\kappa_{b1}}E_1 - \frac{d_E T_1^*}{T_1^*+\kappa_d}E_1 - d_{E1}E_1 - \gamma_E\frac{T_1+T_1^*}{T_1+T_1^*+\kappa_\gamma}E_1 + n_E\frac{a_E V_I}{V_I+\kappa_V}E_2,$$

$$\dot{E}_2 = \gamma_E\frac{T_1+T_1^*}{T_1+T_1^*+\kappa_\gamma}E_1 + \frac{b_{E2}\kappa_{b2}}{E_2+\kappa_{b2}}E_2 - d_{E2}E_2 - \frac{a_E V_I}{V_I+\kappa_V}E_2,$$

$$(8.37)$$

with an initial condition vector

$$(T_1(0), T_1^*(0), T_2(0), T_2^*(0), V_I(0), E_1(0), E_2(0))^T.$$

A summary of the description of all the model parameters in (8.37) is given in Table 8.4. Next we present a brief description of model (8.37) and focus our discussion on the interactions particularly relevant to the derivation of the corresponding stochastic HIV model. Interested readers may refer to [13] for a more detailed discussion of the rationale behind this model.

The terms $d_{T1}T_1$, $d_{T2}T_2$, $d_{T2}T_2^*$, $d_V V_I$, $d_{E1}E_1 + \frac{d_E T_1^*}{T_1^*+\kappa_d}E_1$ and $d_{E2}E_2$ denote the death (or clearance) of uninfected activated CD4+ T cells, uninfected resting CD4+ T cells, infected resting CD4+ T cells, infectious free virus, HIV-specific activated CD8+ T cells and HIV-specific memory CD8+ T cells, respectively. The term $\delta_E E_1 T_1^*$ is used to account for the elimination of infected activated CD4+ T cells T_1^* by the HIV-specific effector CD8+ T cells (that is, T_1^* is eliminated from the system at rate $\delta_E E_1$, depending on the density of HIV-specific effector CD8+ T cells).

The terms involving $\beta_{T1}V_IT_1$ represent the infection process wherein infected activated CD4+ T cells T_1^* result from encounters between uninfected activated CD4+ T cells T_1 and free virus V_I (that is, the activated CD4+

TABLE 8.4: Description of model parameters in (8.37) as well as the values of parameters used in the simulations.

	Description	Value
d_{T1}	Natural death rate of T_1	0.02 per day
β_{T1}	Infection rate of active CD4+ T cells	10^{-2} μl blood/(copies $\cdot$ day)
γ_T	Differentiation rate of activated CD4+ T cells to resting CD4+ T cells	0.005 per day
n_T	Number of daughter cells produced by one proliferating resting CD4+ T cell	2
a_T	Maximum activation rate of HIV-specific resting CD4+ T cells	0.008 per day
κ_V	Half saturation constant (for activation rate)	0.1 copies/μl blood
a_A	Activation rate of resting CD4+ T cells by non-HIV antigen	10^{-12} per day
δ_V	Death rate of T_1^*	0.7 per day
δ_E	Rate at which E_1 eliminates T_1^*	0.01 μl blood/(cells $\cdot$ day)
ζ_T	Maximum source rate of T_2	7 cells/(μl blood $\cdot$ day)
κ_s	Saturation constant (for source rate of T_2)	10^2 copies/(μl blood)
d_{T2}	Natural death rate of resting CD4+ T cells	0.005 per day
β_{T2}	Infection rate of resting CD4+ T cells	10^{-6} μl blood/(copies $\cdot$ day)
n_V	Number of copies of virions produced by one activated infected CD4+ T cell	100
d_V	Natural clearance rate of V_I	13 per day
ζ_E	Source rate of E_1	0.001 cells/(μl blood $\cdot$ day)
b_{E1}	Maximum birth rate of E_1	0.328 per day
κ_{b1}	Half saturation constant (for birth rate of E_1)	0.1 cells/μl blood
d_E	Maximum death rate of E_1 due to the impairment of a high number of infected activated CD4+ T cells	0.25 per day
κ_d	Half saturation constant (for death rate of E_1)	0.5 cells/μl blood
d_{E1}	Natural death rate of E_1	0.1 per day
a_E	Maximum activation rate of E_2 to E_1	0.1 per day
n_E	Number of daughter cells produced by one proliferating memory CD8+ T cell	3
γ_E	Maximum differentiation rate of E_1 to E_2	0.01 per day
κ_γ	Half saturation constant (for differentiation rate)	10 cells/μl blood
b_{E2}	Maximum proliferation rate of E_2	0.001 per day
κ_{b2}	Saturation constant (for proliferation rate of E_2)	100 cells/μl blood
d_{E2}	Natural death rate of E_2	0.005 per day

T cells T_1 become infected at a rate $\beta_{T1}V_I$, depending on the density of infectious virus), and $\beta_{T2}V_IT_2$ is the resulting term for the infection process wherein infected resting CD4+ T cells T_2^* result from encounters between uninfected resting CD4+ T cells T_2 and free virus V_I (that is, the resting CD4+

T cells T_2 become infected at rate $\beta_{T2}V_I$, depending on the density of infectious virus). In addition, for simplicity it is assumed that one copy of a virion is responsible for one new infection. Hence, the term $(\beta_{T1}T_1 + \beta_{T2}T_2)V_I$ in the V_I compartment is used to denote the loss of virions due to infection.

The terms involving $\gamma_T T_1$ (or $\gamma_T T_1^*$) are included in the model to account for the phenomenon of differentiation of uninfected (or infected) activated CD4+ T cells into uninfected (or infected) resting CD4+ T cells at rate γ_T. Similarly, the term $\gamma_E \frac{T_1+T_1^*}{T_1+T_1^*+\kappa_\gamma} E_1$ is used to describe the phenomenon of differentiation of HIV-specific activated CD8+ T cells into HIV-specific resting CD8+ T cells.

The terms $\left(\frac{a_T V_I}{V_I+\kappa_V} + a_A\right) T_2$ and $\left(\frac{a_T V_I}{V_I+\kappa_V} + a_A\right) T_2^*$ represent the activation of uninfected and infected resting CD4+ T cells, respectively, due to both HIV and some non-HIV antigen. We assume that each proliferating resting CD4+ T cell produces n_T daughter cells. In addition, the term $\frac{a_E V_I}{V_I+\kappa_V} E_2$ denotes the activation of HIV-specific memory CD8+ T cells, and each proliferating HIV-specific CD8+ T cell produces n_E daughter cells.

The infected activated CD4+ T cell dies at a rate δ_V and produces n_V copies of virions during its lifespan, either continuously producing virions during its lifetime (referred to as *continuous production mode*) or releasing all its virions in a single burst simultaneous with its death (called *burst production mode*). It is remarked in [46] that for HIV infection it has not yet been established whether a continuous or burst production mode is most appropriate. It should be noted that even though the viral production mode does not affect the form of this deterministic system, it does affect the form of its corresponding stochastic model (as we shall see in the next section).

The terms $\zeta_T \frac{\kappa_s}{V_I+\kappa_s}$ and $\zeta_E + \frac{b_{E1}T_1^*}{T_1^*+\kappa_{b1}} E_1$ denote the source rates for the uninfected resting CD4+ T cells and HIV-specific effector CD8+ T cells, respectively. The term $\frac{b_{E2}\kappa_{b2}}{E_2+\kappa_{b2}} E_2$ is used to account for the self-proliferation of E_2 (due to the homeostatic regulation).

We note that (8.37) is in terms of concentration. Since we are interested in deriving a corresponding stochastic model in terms of numbers, we need to rewrite (8.37). Let M denote the volume of blood (in units μl blood), and the parameter vector $\boldsymbol{\kappa} = (\kappa_V, \kappa_s, \kappa_{b1}, \kappa_d, \kappa_\gamma, \kappa_{b2})^T$ where $\kappa_V, \kappa_s, \kappa_{b1}, \kappa_d, \kappa_\gamma, \kappa_{b2}$ are the saturation constants in model (8.37). We then define $\mathbf{k} = M\boldsymbol{\kappa}$ with $\mathbf{k} = (k_V, k_s, k_{b1}, k_d, k_\gamma, k_{b2})^T$, and define

$$\mathbf{x}^M = M(T_1, T_1^*, T_2, T_2^*, V_I, E_1, E_2)^T$$

with x_j^M denoting the jth component of $\mathbf{x}^M$, $j = 1,2,\ldots,7$. We multiply both sides of (8.37) by M, and then write the resulting equations in terms of

$\mathbf{x}^M$ and obtain

$$\dot{x}_1^M = -d_{T1}x_1^M - \beta_{T1}\frac{x_5^M}{M}x_1^M - \gamma_T x_1^M + n_T\left(\frac{a_T x_5^M}{x_5^M+k_V}+a_A\right)x_3^M,$$

$$\dot{x}_2^M = \beta_{T1}\frac{x_5^M}{M}x_1^M - \delta_V x_2^M - \delta_E\frac{x_6^M}{M}x_2^M - \gamma_T x_2^M + n_T\left(\frac{a_T x_5^M}{x_5^M+k_V}+a_A\right)x_4^M,$$

$$\dot{x}_3^M = M\zeta_T\frac{k_s}{x_5^M+k_s} + \gamma_T x_1^M - d_{T2}x_3^M - \beta_{T2}\frac{x_5^M}{M}x_3^M - \left(\frac{a_T x_5^M}{x_5^M+k_V}+a_A\right)x_3^M,$$

$$\dot{x}_4^M = \gamma_T x_2^M + \beta_{T2}\frac{x_5^M}{M}x_3^M - d_{T2}x_4^M - \left(\frac{a_T x_5^M}{x_5^M+k_V}+a_A\right)x_4^M,$$

$$\dot{x}_5^M = n_V\delta_V x_2^M - d_V x_5^M - \left(\beta_{T1}x_1^M + \beta_{T2}x_3^M\right)\frac{x_5^M}{M},$$

$$\dot{x}_6^M = M\zeta_E + \frac{b_{E1}x_2^M}{x_2^M+k_{b1}}x_6^M - \frac{d_E x_2^M}{x_2^M+k_d}x_6^M - d_{E1}x_6^M - \gamma_E\frac{x_1^M+x_2^M}{x_1^M+x_2^M+k_\gamma}x_6^M$$
$$+n_E\frac{a_E x_5^M}{x_5^M+\kappa_V}x_7^M,$$

$$\dot{x}_7^M = \gamma_E\frac{x_1^M+x_2^M}{x_1^M+x_2^M+k_\gamma}x_6^M + \frac{b_{E2}k_{b2}}{x_7^M+k_{b2}}x_7^M - d_{E2}x_7^M - \frac{a_E x_5^M}{x_5^M+k_V}x_7^M.$$

$$(8.38)$$

8.5.2   Stochastic HIV Models

In this section, we derive two corresponding stochastic HIV models for the deterministic model (8.38), where one is based on the assumption of a burst production mode for the virions (originally developed in [14]), and the other one is based on the assumption of a continuous production mode. We assume that $\{\mathbf{X}^M(t) : t \geq 0\}$ is a continuous time Markov chain such that for any small Δt it satisfies (8.27), where $\mathbf{X}^M(t) = (X_1^M(t), X_2^M(t), \ldots, X_7^M(t))^T$ with $X_1^M(t), X_2^M(t), \ldots, X_7^M(t)$ respectively denoting the number of non-infected and infected activated CD4+ T cells, non-infected and infected resting CD4+ T cells, RNA copies of infectious free virus and HIV-specific effector and memory CD8+ T cells in M μl blood at time t. In addition, we assume that $\mathbf{X}^M(0) = \mathbf{x}_0^M$ with $\mathbf{x}_0^M \in \mathbb{Z}^7$ being non-random.

8.5.2.1   The Stochastic HIV Model Based on the Burst Production Mode

If we assume that a burst production mode is used to describe the production of the new virions, then based on this assumption as well as the discussion in Section 8.5.1, we can obtain the transition rate and the corresponding state change vector for each reaction (e.g., natural clearance of virus, infection of activated CD4+ T cells). The description of these reactions as well as their corresponding transition rates λ_j and state change vectors $\mathbf{v}_j$ are summarized in Table 8.5, where $\mathbf{e}_i \in \mathbb{R}^7$ is the ith unit column vector, $i = 1, 2, \ldots, 7$. From this table, we see that there are 19 reactions (i.e., $l = 19$).

TABLE 8.5:   Transition rates as well as corresponding state change vectors for the stochastic HIV model, where the burst production mode is assumed to describe the production of new virions.

j	Discriptions	$\lambda_j(\mathbf{x}^M)$	$\mathbf{v}_j$
1	Death of uninfected activated CD4+ T cells	$d_{T1}x_1^M$	$-\mathbf{e}_1$
2	Elimination of infected activated CD4+ T cells by HIV-specific effector CD8+ T cells	$\delta_E \dfrac{x_6^M}{M}x_2^M$	$-\mathbf{e}_2$
3	Death of uninfected resting CD4+ T cells	$d_{T2}x_3^M$	$-\mathbf{e}_3$
4	Death of infected resting CD4+ T cells	$d_{T2}x_4^M$	$-\mathbf{e}_4$
5	Natural clearance of virus	$d_V x_5^M$	$-\mathbf{e}_5$
6	Death of HIV-specific effector CD8+ T cells	$\left(d_{E1}+d_E\dfrac{x_2^M}{x_2^M+k_d}\right)x_6^M$	$-\mathbf{e}_6$
7	Death of HIV-specific memory CD8+ T cells	$d_{E2}x_7^M$	$-\mathbf{e}_7$
8	Infection of activated CD4+ T cells	$\beta_{T1}\dfrac{x_5^M}{M}x_1^M$	$-\mathbf{e}_1+\mathbf{e}_2-\mathbf{e}_5$
9	Infection of resting CD4+ T cells	$\beta_{T2}\dfrac{x_5^M}{M}x_3^M$	$-\mathbf{e}_3+\mathbf{e}_4-\mathbf{e}_5$
10	Differentiation from uninfected activated CD4+ T cells to uninfected resting CD4+ T cells	$\gamma_T x_1^M$	$-\mathbf{e}_1+\mathbf{e}_3$
11	Differentiation from infected activated CD4+ T cells to infected resting CD4+ T cells	$\gamma_T x_2^M$	$-\mathbf{e}_2+\mathbf{e}_4$
12	Differentiation from HIV-specific effector CD8+ T cells to HIV-specific memory CD8+ T cells	$\gamma_E \dfrac{x_1^M+x_2^M}{x_1^M+x_2^M+k_\gamma}x_6^M$	$-\mathbf{e}_6+\mathbf{e}_7$
13	Activation of uninfected resting CD4+ T cells	$\left(a_T\dfrac{x_5^M}{x_5^M+k_V}+a_A\right)x_3^M$	$n_T\mathbf{e}_1-\mathbf{e}_3$
14	Activation of infected resting CD4+ T cells	$\left(a_T\dfrac{x_5^M}{x_5^M+k_V}+a_A\right)x_4^M$	$n_T\mathbf{e}_2-\mathbf{e}_4$
15	Activation of HIV-specific memory CD8+ T cells	$a_E\dfrac{x_5^M}{x_5^M+k_V}x_7^M$	$n_E\mathbf{e}_6-\mathbf{e}_7$
16	Production of new virions simultaneous with the death of infected activated CD4+ T cells	$\delta_V x_2^M$	$-\mathbf{e}_2+n_V\mathbf{e}_5$
17	Birth of uninfected resting CD4+ T cells	$\left(\zeta_T \dfrac{k_s}{x_5^M+k_s}\right)M$	$\mathbf{e}_3$
18	Birth of HIV-specific effector CD8+ T cells	$M\zeta_E+b_{E1}\dfrac{x_2^M}{x_2^M+k_{b1}}x_6^M$	$\mathbf{e}_6$
19	Proliferation of HIV-specific memory CD8+ T cells	$b_{E2}\dfrac{k_{b2}}{x_7^M+k_{b2}}x_7^M$	$\mathbf{e}_7$

8.5.2.2   The Stochastic HIV Model Based on the Continuous Production Mode

If we assume that a continuous production mode is used to describe the production of new virions, then based on this assumption as well as the discussion in Section 8.5.1 we still obtain 19 reactions, with all the transition rates and the corresponding state change vectors being the same as the corresponding ones in Table 8.5, except for reactions 2 and 16. Under this assumption, reaction 2 describes the loss of infected activated CD4+ T cells due to either natural death or elimination by HIV-specific effector CD8+ T cells, and the transition rate λ_2 and the state change vector $\mathbf{v}_2$ are given by

$$\lambda_2(\mathbf{x}^M) = \delta_V x_2^M + \delta_E \frac{x_6^M}{M} x_2^M, \quad \mathbf{v}_2 = -\mathbf{e}_2.$$

Reaction 16 describes the production of new virions, and the transition rate λ_{16} and the state change vector $\mathbf{v}_{16}$ are given by

$$\lambda_{16}(\mathbf{x}^M) = n_V \delta_V x_2^M, \quad \mathbf{v}_{16} = \mathbf{e}_5.$$

We thus see that the production mode for the new virions does affect the form of the stochastic model.

8.5.3   Numerical Results for the Stochastic HIV Model Based on the Burst Production Mode

In this section, we report on use of the SSA, and the explicit and implicit tau-leaping algorithms on the stochastic HIV model with transition rates and corresponding state change vectors given in Table 8.5 (that is, the stochastic HIV model based on the burst production mode for the virions). We compare the computational times of the SSA, explicit and implicit tau-leaping methods with different values of M (the scale on the transition rates which effectively scales the population counts). Each of the simulations was run for the time period $[0, 100]$ days with parameter values given in the third column of Table 8.4 (adapted from Table 2 in [13]) and initial conditions

$$\mathbf{X}^M(0) = M(5, 1, 1400, 1, 10, 5, 1)^T.$$

That is, each of the simulations starts with the same initial concentrations $(5, 1, 1400, 1, 10, 5, 1)^T$.

8.5.3.1   Implementation of the Tau-Leaping Algorithms

The discussion in Section 8.2.3.1 reveals that an important step for implementation of the tau-leaping algorithms is to find ϖ_i, $i = 1, 2, \ldots, 7$. From Table 8.5 we see that all the reactions are either zero order, first order or second order. However, a lot of them have transition rates with non-constant coefficients $(\lambda_6, \lambda_{12}, \lambda_{13}, \lambda_{14}, \lambda_{15}, \lambda_{17}, \lambda_{18}, \lambda_{19})$. Hence, for this model the values of $\varpi_i, i = 1, 2, \ldots, 7$, are not just one or two. Next we will give a detailed discussion on how to obtain these values. Note that the values of $\varpi_i, i = 1, 2, \ldots, 7$, should be chosen such that $|\Delta\lambda_j(\mathbf{x}^M)| \leq \epsilon\lambda_j(\mathbf{x}^M)$ at least to the first-order approximation for all $j = 1, 2, \ldots, 19$, where $\Delta\lambda_j(\mathbf{x}^M) = \lambda_j(\mathbf{x}^M + \Delta\mathbf{x}^M) - \lambda_j(\mathbf{x}^M)$, with $\Delta\mathbf{x}^M$ being the absolute changes in the state variables, $j = 1, 2, \ldots, 19$. Hence, the important thing is to find the conditions such that the relative changes in those transition rates with non-constant coefficients are bounded by ϵ (at least to the first-order approximation); that is, find the values of $\varpi_i^j, i = 1, 2, \ldots, 7, j \in \{6, 12, 13, 14, 15, 17, 18, 19\}$. We take reaction 6 as an example to illustrate how to obtain the values of ϖ_2^6 and ϖ_6^6 (recall that x_2^M and x_6^M are the only two states related to reaction 6). Note that

$$\Delta\lambda_6(\mathbf{x}^M)$$

$$= d_{E1}\Delta x_6^M + d_E\left[\frac{(x_2^M + \Delta x_2^M)(x_6^M + \Delta x_6^M)}{x_2^M + \Delta x_2 + k_d} - \frac{x_2^M x_6^M}{x_2^M + k_d}\right]$$

$$= d_{E1}\Delta x_6^M + d_E\frac{x_2^M(x_2^M + k_d)\Delta x_6^M + k_d x_6^M \Delta x_2^M + (x_2^M + k_d)\Delta x_2^M \Delta x_6^M}{(x_2^M + \Delta x_2^M + k_d)(x_2^M + k_d)},$$

which implies that

$$\frac{\Delta\lambda_6(\mathbf{x}^M)}{\lambda_6(\mathbf{x}^M)} = \frac{d_{E1}\Delta x_6^M}{d_{E1}x_6^M + d_E\frac{x_2^M x_6^M}{x_2^M + k_d}}$$

$$+ \frac{d_E\frac{x_2^M(x_2^M + k_d)\Delta x_6^M + k_d x_6^M \Delta x_2^M + (x_2^M + k_d)\Delta x_2^M \Delta x_6^M}{(x_2^M + \Delta x_2^M + k_d)(x_2^M + k_d)}}{d_{E1}x_6^M + d_E\frac{x_2^M x_6^M}{x_2^M + k_d}}.$$

Thus, by the above equation as well as the non-negativeness of the state variables and the positiveness of the parameters we have

$$\frac{|\Delta\lambda_6(\mathbf{x}^M)|}{\lambda_6(\mathbf{x}^M)}$$

$$< \frac{|\Delta x_6^M|}{x_6^M} + \left|\frac{x_2^M(x_2^M + k_d)\Delta x_6^M + k_d x_6^M \Delta x_2^M + (x_2^M + k_d)\Delta x_2^M \Delta x_6^M}{(x_2^M + \Delta x_2^M + k_d)x_2^M x_6^M}\right|$$

$$\leq \frac{|\Delta x_6^M|}{x_6^M} + \frac{x_2^M + k_d}{|x_2^M + \Delta x_2^M + k_d|}\frac{|\Delta x_6^M|}{x_6^M} + \frac{k_d}{|x_2^M + \Delta x_2^M + k_d|}\frac{|\Delta x_2^M|}{x_2^M}$$

$$+ \frac{x_2^M + k_d}{|x_2^M + \Delta x_2^M + k_d|}\frac{|\Delta x_2^M|}{x_2^M}\frac{|\Delta x_6^M|}{x_6^M}.$$

$$(8.39)$$

If the value of ϵ is set to be $\epsilon < \frac{1}{2}$, we find $|\Delta x_i^M| \leq \epsilon_i x_i^M \leq \frac{1}{2}(x_i^M + k_d)$, which implies that

$$\frac{x_2^M + k_d}{|x_2^M + \Delta x_2^M + k_d|} \leq 2, \quad \frac{k_d}{|x_2^M + \Delta x_2^M + k_d|} \leq 1.$$

Hence, by (8.39) and the above inequality we obtain

$$\frac{|\Delta \lambda_6(\mathbf{x}^M)|}{\lambda_6(\mathbf{x}^M)} < 3\frac{|\Delta x_6^M|}{x_6^M} + \frac{|\Delta x_2^M|}{x_2^M} + 2\frac{|\Delta x_2^M|}{x_2^M}\frac{|\Delta x_6^M|}{x_6^M}.$$

Thus, from the above inequality we see that if we choose

$$|\Delta x_2^M| \leq \frac{\epsilon}{4}x_2^M, \quad |\Delta x_6^M| \leq \frac{\epsilon}{4}x_6^M, \tag{8.40}$$

then the relative change in λ_6 is bounded by ϵ (to the first-order approximation). We note that (8.40) implies

$$\varpi_2^6 = 4, \quad \varpi_6^6 = 4. \tag{8.41}$$

Similarly, to have the relative changes in the other transition rates with non-constant coefficients be bounded by ϵ (either exactly or to the first-order approximation), we need to set

Reaction 12: $|\Delta x_1^M| \leq \frac{\epsilon}{3}x_1^M, \quad |\Delta x_2^M| \leq \frac{\epsilon}{3}x_2^M, \quad |\Delta x_6^M| \leq \frac{\epsilon}{3}x_6^M,$

Reaction 13: $|\Delta x_3^M| \leq \frac{\epsilon}{4}x_3^M, \quad |\Delta x_5^M| \leq \frac{\epsilon}{4}x_5^M,$

Reaction 14: $|\Delta x_4^M| \leq \frac{\epsilon}{4}x_4^M, \quad |\Delta x_5^M| \leq \frac{\epsilon}{4}x_5^M,$

Reaction 15: $|\Delta x_5^M| \leq \frac{\epsilon}{3}x_5^M, \quad |\Delta x_7^M| \leq \frac{\epsilon}{3}x_7^M,$

Reaction 17: $|\Delta x_5^M| \leq \frac{\epsilon}{2}x_5^M,$

Reaction 18: $|\Delta x_2^M| \leq \frac{\epsilon}{3}x_2^M, \quad |\Delta x_6^M| \leq \frac{\epsilon}{3}x_6^M$

Reaction 19: $|\Delta x_7^M| \leq \frac{\epsilon}{2}x_7^M.$

This implies that

$$\varpi_1^{12} = 3, \quad \varpi_2^{12} = 3, \quad \varpi_6^{12} = 3, \quad \varpi_3^{13} = 4, \quad \varpi_5^{13} = 4,$$
$$\varpi_4^{14} = 4, \quad \varpi_5^{14} = 4, \quad \varpi_5^{15} = 3, \quad \varpi_7^{15} = 3, \quad \varpi_5^{17} = 2, \tag{8.42}$$
$$\varpi_2^{18} = 3, \quad \varpi_6^{18} = 3, \quad \varpi_7^{19} = 2.$$

For those transition rates with a constant coefficient, we can easily see that to have them be bounded by ϵ (at least to the first-order approximation) we need to set the relative changes in the state variables to be bounded by either $\frac{1}{2}\epsilon$ (for the second-order reactions: reactions 2, 8 and 9) or ϵ (for the first-order reactions: reactions 1, 3, 4, 5, 7, 10, 11 and 16); that is,

$$\varpi_2^2 = 2, \quad \varpi_6^2 = 2, \quad \varpi_1^8 = 2, \quad \varpi_5^8 = 2, \quad \varpi_3^9 = 2, \quad \varpi_5^9 = 2,$$

$$\varpi_1^1 = 1, \quad \varpi_3^3 = 1, \quad \varpi_4^4 = 1, \quad \varpi_5^5 = 1, \quad \varpi_7^7 = 1, \quad \varpi_1^{10} = 1, \qquad (8.43)$$

$$\varpi_2^{11} = 1, \quad \varpi_2^{16} = 1.$$

Thus, by (8.41), (8.42) and (8.43) we know that to have $|\Delta\lambda_j| \le \epsilon\lambda_j$ for all $j = 1, 2, \ldots, 19$, we need to set

$$\varpi_i = \max_{1 \le j \le 19}\{\varpi_i^j\} = \begin{cases} 4, \, i = 2, 3, 4, 5, 6 \\ 3, \, i = 1, 7. \end{cases}$$

As we remarked in Section 8.2.3.2, it is generally difficult to determine whether or not a reaction is in partial equilibrium. Hence, it is difficult to choose the step size based on the condition for bounding only those transition rates resulting from non-equilibrium reactions by ϵ. To overcome this difficulty, here we use (8.22) to choose Δt_1 for the implicit tau-leaping method but with a larger ϵ to allow a possibly large time step size. To avoid the possibility of negative populations (i.e., to ensure (8.23) has a non-negative solution), the algorithm for the implicit tau-leaping method is implemented in the same way as that for the explicit tau-leaping method except the update for states (i.e., replace (8.20) by (8.23) and (8.24)). For the implicit tau-leaping method, ϵ was taken as 0.12 (to allow a larger time step size). This value is chosen based on the simulation results so that the computational time is comparatively short without compromising the accuracy of the solution.

8.5.3.2   Comparison of Computational Efficiency of the SSA and the Tau-Leaping Algorithms

All three algorithms were coded in MATLAB, and all the simulation results were run on a Linux machine with a 2 GHz Intel Xeon Processor with 8 GB of RAM total. In addition, all three algorithms were implemented in the same style as well as using similar coding strategies (such as array preallocation and vectorization) to speed up computations. Thus the comparison of computational times given below should be interpreted as relative to each other for the algorithms discussed rather than in any absolute sense.

Figure 8.4 depicts the computational time of each algorithm for an average of five typical simulation runs with M varying from $10, 50, 10^2, 2 \times 10^2, 5 \times 10^2, 10^3$ for the SSA and $10, 50, 10^2, 2 \times 10^2, 5 \times 10^2, 10^3, 10^4, 10^5, 10^6, 5 \times 10^6$ for the explicit and implicit tau-leaping schemes. From this figure, we see that the computational times for the SSA increase as the value M increases. This is expected, as the mean time step size for the SSA is the inverse of the sum of all

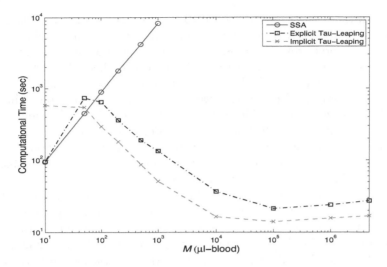

FIGURE 8.4: Comparison of computational times of different algorithms (SSA, explicit tau-leaping and implicit tau-leaping) for an average of five typical simulation runs.

transition rates, which increases as M increases (roughly proportional to M, as can be seen from the transition rates illustrated in Table 8.5). In addition, even with $M = 10^3$, it took the SSA more than 8000 seconds for one sample path (which is why we did not run any simulations for the SSA when M was greater than 10^3). Hence, it is impractical to implement the SSA if we want to run this HIV model for a normal person (generally having approximately 5×10^6 μl blood). This is expected due to the large value of uninfected resting CD4+ T cells (as can be seen from the initial condition (8.5.3)).

From Figure 8.4 we also see that the computational times for the explicit tau-leaping method increase as the value of M increases from 10 to 50 and decrease as M increases to 100. Then its computational times decrease dramatically as the value of M increases from 100 to 10^4, and stabilize somewhat for $M \geq 10^4$. This is due to the formula for Δt_1. As we can see from (8.22) and the transition rates in Table 8.5, if $\epsilon x_i / \varpi_i > 1$, then the first term inside the minimum sign of (8.22) is roughly of the same order for all the values of M while the second term is roughly proportional to the value of M. In addition, we found from the simulation results that the first term inside the minimum sign of (8.22) is much larger than the corresponding second term until $M = 10^4$ and becomes smaller than the second term when $M \geq 10^4$. Hence, Δt_1 increases as M increases when $M \leq 10^4$ and has roughly similar values for all the cases when $M \geq 10^4$. This agrees with the observation that the computational times decrease dramatically as the value of M increases from 100 to 10^4 and then stabilize there for $M \geq 10^4$. The increase of computational times as M increases from 10 to 50 is because Δt_1 is so small that a large number of SSA steps are implemented instead of tau-leaping.

We also observe from Figure 8.4 that the computational times for the implicit tau-leaping method decrease as M increases when $M \leq 10^4$ and then stabilize there for $M \geq 10^4$. This is for the same reason as that the explicit tau-leaping method. In addition, we see that the computational times for implicit tau-leaping are significantly higher than that of the SSA and explicit tau-leaping at $M = 10$, which is because under this case implicit tau-leaping is implemented many times (solving systems of non-linear equations in each implicit tau-leaping step is costly) and the time step size is not significantly larger than those of the other two methods.

Based on the above discussion, we know that for smaller values of M (less than 100) the SSA is the choice due to its simplicity, accuracy and efficiency. However, for larger values the tau-leaping methods are definitely the choice, with implicit tau-leaping performing better than explicit tau-leaping; this is expected due to the stiffness of the system (large variations in both parameter values and state variables). Overall this example illustrates how widely algorithm performances vary and demonstrates how one might perform some initial analysis and computational studies to aid in selection of appropriate algorithms.

8.5.3.3   Accuracy of the Results Obtained by Tau-Leaping Algorithms

Note that the results obtained by the SSA are exact. Hence, the differences between the histograms obtained by the SSA and the ones obtained by the tau-leaping methods provide a measure of simulation errors in tau-leaping methods when the number of simulation runs is sufficiently large. Thus, to gain some information on the accuracy of the tau-leaping methods, we plotted the histograms of state solutions to the stochastic HIV model (in terms of concentration) obtained by explicit and implicit tau-leaping methods in comparison to those obtained by the SSA. Due to long computational times to obtain these histograms, we present here the results only for the case with $M = 200$ μl blood (where computational times required by both tau-leaping methods are significantly lower than those required by the SSA, as can be seen from Figure 8.4). The histograms of state solutions at $t = 100$ days (the end time point of the simulation) are shown in Figure 8.5; they are obtained by simulating 1000 sample paths of solutions to the stochastic HIV model. We observe from this figure that the histograms obtained by tau-leaping methods are reasonably close to those obtained by the SSA (similar behavior is also observed for the histograms of state solutions at other time points). This suggests that the accuracy of results obtained by tau-leaping methods is acceptable.

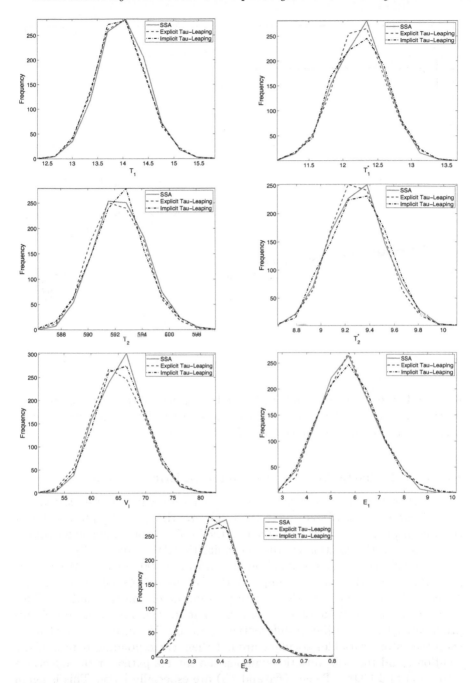

FIGURE 8.5: Histograms of state solutions to the stochastic HIV model (in terms of concentrations) with $M = 200 \ \mu l$ blood at $t = 100$ days, where histograms are obtained by simulating 1000 sample paths of solutions to the stochastic HIV model.

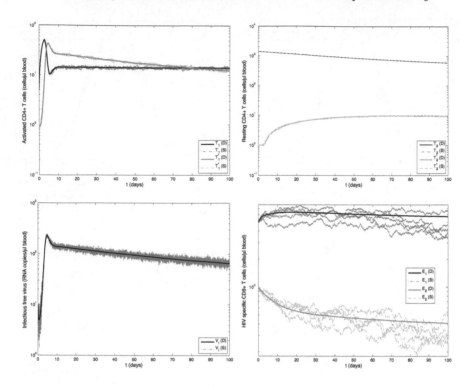

FIGURE 8.6: The (D) and (S) in the legend denote the solution obtained by the deterministic HIV model and the stochastic HIV model, respectively, where the results are obtained with $M = 100$ μl blood.

8.5.3.4   Stochastic Solution vs. Deterministic Solution

To have some idea of the dynamics of each state variable, we plotted five typical sample paths of solution to the stochastic HIV model (in terms of concentrations, i.e., $\mathbf{X}^M(t)/M$) obtained by the implicit tau-leaping algorithm in comparison to the solution for the deterministic HIV model (8.37).

Figure 8.6 depicts the results obtained with $M = 100$ μl blood. We observe from this figure that all the sample paths for the uninfected resting CD4+ T cells (T_2) are similar to their corresponding deterministic solution. This is expected, as the value of T_2 is so large that its dynamics can be well approximated by the ordinary differential equation. We also observe that all the other state variables oscillate around their corresponding deterministic solutions, and that the variations among the sample paths for the infectious virus (V_I) and CD8+ T cells (E_1 and E_2) are especially large. This is again expected, as T_2 has less effect on the dynamics of these three compartments (especially on E_1 and E_2, as observed from (8.37) and the parameter values in Table 8.4). Figure 8.7 depicts the results obtained with $M = 5 \times 10^6$ μl

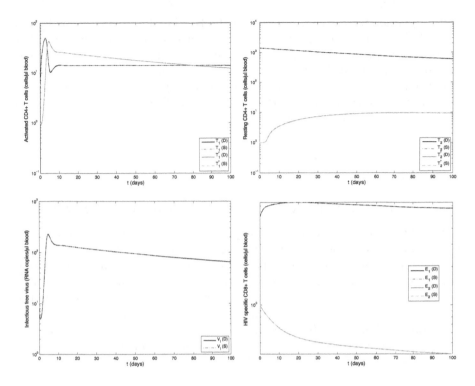

FIGURE 8.7:   The (D) and (S) in the legend denote the solutions obtained by the deterministic HIV model and the stochastic HIV model, respectively, where the results are obtained with $M = 5 \times 10^6$ μl blood.

blood. We observe from this figure that the stochastic solutions agree with the deterministic solutions. This is consistent with Kurtz's limit theorem.

8.5.3.5   Final Remark

Note that the stochastic HIV model is of interest in early or acute infection where the number of uninfected resting T cells is large (on the order of 1000 cells per μl blood). This explains why the SSA requires more than 8000 seconds to run even one sample path with $M = 1000$ μl blood. If one considers an average person having 5×10^6 μl blood, to run the SSA for even one sample path is impractical at this scale. The numerical results demonstrate that the dynamics of uninfected resting CD4+ T cells can be well approximated by ordinary differential equations even with $M = 10$ μl blood. In addition, Table 8.4 demonstrates there is a large variation between the values of the parameters. Thus this HIV model is multi-scaled in both states and time. There are some hybrid simulation methods (also referred to as multi-scale approaches) specifically designed for the multi-scaled system (the interested reader may refer to [44] for an overview of these methods). The basic idea of the hybrid methods is to partition the system into two subsystems, one containing fast reactions and the other containing slow reactions. Then the two subsystems are simulated iteratively by using numerical integration of ODEs (or SDEs) and stochastic algorithms (such as the SSA), respectively. Although these algorithms are very attractive, they are most challenging to implement and require a great deal of user intervention.

8.6 Application in Agricultural Production Networks

In addition to the infection models presented in Sections 8.4 and 8.5, similar CTMC models arise in just-in-time production networks, manufacturing and delivery, logistic/supply chains, and multi-scale (large/small) population models as well as network models in communications and security. In this section, we present a continuous time Markov chain model to describe the dynamics of a simplified pork production network. This stochastic model was originally developed in [10], where it was used to study the impact of disturbances (introduction of diseases and other disruptions in the network) on the overall performance of the pork production system.

8.6.1 The Stochastic Pork Production Network Model

In current production methods for livestock based on a "just-in-time philosophy" (the storage of unused inventory produces a waste of resources), feedstock and animals are grown in different areas and the animals are moved with precise timing from one farm to another, depending on their age. Here we consider a simplified pork production network model with four nodes: sows, nurseries, finishers and slaughterhouses. The movement of pigs from one node to the next is assumed to occur only in the forward direction (depicted in Figure 8.8). At the sow farms (node 1), new piglets are born and weaned approximately three weeks after birth. The three-week-old piglets are moved to nursery farms (node 2) to mature. They are then transferred to the finisher farms (node 3), where they grow to full market size. Once they reach market weight, the matured pigs are moved to the slaughterhouses (node 4). The

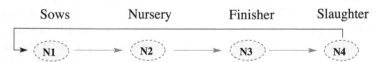

FIGURE 8.8:   A schematic of a pork production network model.

TABLE 8.6: Transition rates as well as corresponding state change vectors for the stochastic pork production network model.

j	Description	$\lambda_j(\mathbf{x}^M)$	$\mathbf{v}_j$
1	movement from node 1 to node 2	$k_1 x_1^M (L_2 - x_2^M)_+$	$-\mathbf{e}_1 + \mathbf{e}_2$
2	movement from node 2 to node 3	$k_2 x_2^M (L_3 - x_3^M)_+$	$-\mathbf{e}_2 + \mathbf{e}_3$
3	movement from node 3 to node 4	$k_3 x_3^M (L_4 - x_4^M)_+$	$-\mathbf{e}_3 + \mathbf{e}_4$
4	movement from node 4 to node 1	$k_4 \min(x_4^M, S_m)$	$-\mathbf{e}_4 + \mathbf{e}_1$

population is assumed to remain constant, and thus, the number of deaths that occur at the slaughterhouses is taken to be the same as the number of births at the sow node. It should be noted that this assumption is realistic when the network is efficient and operates at or near full capacity (i.e., when the number of animals removed from the node is immediately replaced by an equal number of new individuals).

Let $\mathbf{X}^M(t) = (X_1^M(t), X_2^M(t), X_3^M(t), X_4^M(t))^T$, with $X_i^M(t)$ denoting the number of pigs at the ith node at time t in a network where the total population size is M (i.e., $\sum_{j=1}^{4} X_j^M(t) = M$). For each M, the evolution of the system is modeled using a CTMC $\{\mathbf{X}^M(t) : t \geq 0\}$ such that for any small Δt it satisfies (8.27). In this model, it is assumed that the transition rates λ_i are proportional to $x_i^M(L_{i+1} - x_{i+1}^M)_+$, $i = 1, 2, 3$. Here L_i denotes the capacity constraint at node i, and $(z)_+ = \max(z, 0)$, which indicates that the transition rate is taken as zero if there is no capacity available. In addition, the transition rate at node 4 is assumed to be proportional to the maximum killing capacity S_m if a sufficient number of animals is available; otherwise it is assumed to be proportional to the number of animals present at the node. Based on these assumptions, the transition rates λ_j and the corresponding state change vectors $\mathbf{v}_j$ are summarized in Table 8.6, where $\mathbf{e}_i \in \mathbb{Z}^4$ is the ith unit column vector, and k_i is the service rate at node i, $i = 1, 2, 3, 4$.

8.6.2 The Deterministic Pork Production Network Model

Let $\mathbf{c}(t) = \mathbf{x}^M(t)/M$. If we re-scale the constants as follows:

$$\kappa_4 = k_4, \quad \kappa_i = Mk_i, \quad i = 1, 2, 3, \quad s_m = S_m/M, \quad l_i = L_i/M,$$

and define

$$\begin{aligned} f_1(\mathbf{c}(t)) &= \kappa_1 c_1(t)(l_2 - c_2(t)), \quad f_2(\mathbf{c}(t)) = \kappa_2 c_2(t)(l_3 - c_3(t)) \\ f_3(\mathbf{c}(t)) &= \kappa_3 c_3(t)(l_4 - c_4(t))_+, \quad f_4(\mathbf{c}(t)) = \kappa_4 \min(c_4(t), s_m), \end{aligned} \quad (8.44)$$

then we observe that the transition rates λ_j given in Table 8.6 can be written as $\lambda_j(\mathbf{x}^M) = Mf_j(\mathbf{c}(t))$, $j = 1, 2, 3, 4$, which implies that the family of

CTMCs $\{\mathbf{X}^M(t) : t \geq 0\}$ with transition rates and corresponding state change vectors defined in Table 8.6 is density dependent.

Based on Kurtz's limit theorem, we know that the corresponding deterministic model for the stochastic pork production network model with transition rates λ_j and corresponding state change vectors $\mathbf{v}_j$ given in Table 8.6 is

$$\dot{x}_1^M(t) = -k_1 x_1^M(t)(L_2 - x_2^M(t))_+ + k_4 \min(x_4^M(t), S_m)$$

$$\dot{x}_2^M(t) = -k_2 x_2^M(t)(L_3 - x_3^M(t))_+ + k_1 x_1^M(t)(L_2 - x_2^M(t))_+$$

$$\dot{x}_3^M(t) = -k_3 x_3^M(t)(L_4 - x_4^M(t))_+ + k_2 x_2^M(t)(L_3 - x_3^M(t))_+ \qquad (8.45)$$

$$\dot{x}_4^M(t) = -k_4 \min(x_4^M(t), S_m) + k_3 x_3^M(t)(L_4 - x_4^M(t))_+$$

$$\mathbf{x}^M(0) = \mathbf{x}_0^M.$$

8.6.3   Numerical Results

In this section, we report on numerical results obtained for the stochastic pork production network model with transition rates and corresponding state change vectors given in Table 8.6, as well as the results for its corresponding deterministic model (8.45). Note that the number of transitions in the stochastic pork network model is very small (only four reactions). Hence, one can use an exact simulation method to simulate the stochastic pork production network model. Here we choose to use the modified next reaction method.

All the simulations were run for the time period $[0, 50]$ days with parameter values and initial conditions summarized in Table 8.7 (taken from [10]). Here the value of M is chosen to be either 10^3, 10^4 or 10^6. Figure 8.9 depicts five typical sample paths of each state of the stochastic pork production model in comparison to the solution to the deterministic pork production model. We observe from Figure 8.9 that all the sample paths for each state variable oscillate around their corresponding deterministic solution, and that the variation among the sample paths is large for all the state variables, especially for node 2, when $M = 10^3$. In addition, as the population size M increases, the variation among the sample paths decreases. Figure 8.9 also indicates that when $M = 10^6$ the deterministic solution for node 2 provides a reasonable

TABLE 8.7:   Parameter values and initial conditions for the pork production network model.

Parameters	Values	Units	Parameters	Values	Units
l_2	$2.607 \cdot 10^{-1}$		κ_1	1.674	1/days
l_3	$7.267 \cdot 10^{-1}$		κ_2	0.323	1/days
l_4	$6.3 \cdot 10^{-3}$		κ_3	4.521	1/days
s_m	$1.39 \cdot 10^{-1}$		κ_4	1	1/days
$\mathbf{c}(0) = (2.528 \cdot 10^{-1}, 2.212 \cdot 10^{-1}, 4.739 \cdot 10^{-1}, 5.21 \cdot 10^{-2})^T$					

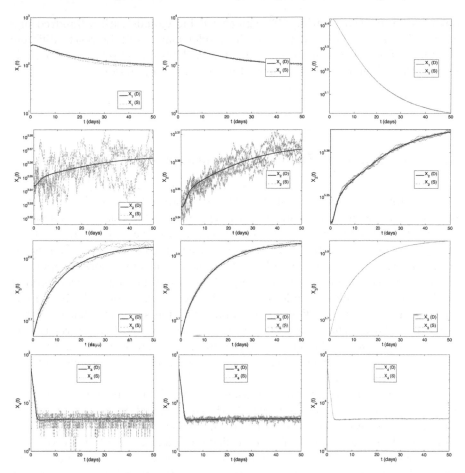

FIGURE 8.9:   Results obtained by the stochastic pork production model (S) and by the corresponding deterministic model (D) with $M = 10^3$ (left column), $M = 10^4$ (middle column) and $M = 10^6$ (right column).

approximation to its corresponding stochastic solution, and the deterministic solutions for all the other nodes agree with their corresponding stochastic solutions. Once again, this is expected in light of Kurtz's limit theorem.

We remark that an application of such a stochastic transportation model to describe the system behavior should account for the size of groups in which pigs are transported between nodes. If a hundred pigs are moved at a time, an appropriate notion of an "individual" in the context of the model might be a hundred pigs. Then treating each group of a hundred pigs as a unit would lead to a marked increase in the magnitude of stochastic fluctuations seen at the "population" level, as Figure 8.9 reveals that variation among sample paths is large when treating 10^6 pigs as 10^4 individuals with a hundred pigs as a unit. As a result, scaling in the model may result in vastly different stochastic fluctuations in model simulations. Thus one must exercise care in how data and "units" are formulated in modeling populations.

8.7   Overview of Stochastic Systems with Delays

Delays occur and are important in many physical and biological processes. For example, recent studies show that delay-induced stochastic oscillations can occur in certain elements of gene regulation networks [18]. These are similar to delay-induced phenomena recognized for many years in mechanical systems (e.g., see [40, 41, 42]). More recent applications have focused on concerns that delays might be of practical importance in general supply networks in investigations of a wide range of perturbations (either accidental or deliberate). Examples include a node being rendered inoperable due to bad weather and technical difficulties in communication systems. Hence, continuous time Markov chain models with delays incorporated (simply referred to as *stochastic models with delays*) have enjoyed considerable research attention in the past decade, especially the effort on the development of algorithms to simulate such systems (e.g., [4, 16, 18, 20, 51]).

Two types of delays are often considered in the literature. One is fixed delays, and the other is random delays. In the case of fixed delays, the amount of time needed for a delayed reaction to complete is a fixed value. In the case of random delays, the amount of time needed for a delayed reaction to complete is random; that is, each time this delayed reaction fires, the amount of time needed for this reaction to complete is different. In general, for the stochastic system with delays, the reactions can be partitioned into three groups. First are the reactions with no delay (ND). Second are the reactions where they only affect the state of the system upon completion (CD). The last are the reactions where the state of the system is affected at the initiation and the completion of the reaction (ICD).

As before, let $l \geq 1$ be the number of transitions. In addition, if the jth transition is a delayed one (CD or ICD), then we let τ_j denote the delay time between the initiation and completion for this transition. Let $\mathbf{X}(t) = \mathbf{x}$. We assume that for any small Δt the stochastic system with delays satisfies

$$\text{Prob}\{\text{reaction } j \text{ initiates in } (t, t + \Delta t] \mid \mathbf{X}(s), 0 \leq s \leq t\}$$
$$= \lambda_j(\mathbf{x})\Delta t + o(\Delta t), \quad j = 1, 2, \ldots, l. \tag{8.46}$$

Let $R_j^I(t)$ denote the number of initiations of reaction j by time t, $j = 1, 2, \ldots, l$. Then for any reaction j we have $R_j^I(t) = Y_j \left(\int_0^t \lambda_j(\mathbf{X}(s)) ds \right)$, where $\{Y_j(t) : t \geq 0\}$, $j = 1, 2, \ldots, l$, are independent unit Poisson processes.

Let $\mathbf{v}_j^L = (v_{1j}^L, v_{2j}^L, v_{3j}^L, \ldots, v_{nj}^L)^T \in \mathbb{Z}^n$, with v_{ij}^L denoting the number of X_i lost (or consumed) in the jth reaction, and $\mathbf{v}_j^C = (v_{1j}^C, v_{2j}^C, v_{3j}^C, \ldots, v_{nj}^C)^T \in \mathbb{Z}^n$, with v_{ij}^C denoting the number of X_i created (or produced) in the jth reaction. Then we see that $\mathbf{v}_j = \mathbf{v}_j^C - \mathbf{v}_j^L$. For example, for reaction 16 in Table 8.5, $\mathbf{v}_j^C = n_V \mathbf{e}_5$ and $\mathbf{v}_j^L = \mathbf{e}_2$ (as for this reaction the death of an infected activated CD4+ T cell results in n_V new virions).

Note that (8.46) is an assumption for the initiation times of reactions. Hence, we need to consider the completion times of reactions separately based on the group to which they belong. Specifically, for these three types of reactions (ND, CD and ICD), we have

- If reaction $j \in$ ND initiates at time t, then we update the system state by adding $\mathbf{v}_j$ to the current state at the time of initiation;

- If reaction $j \in$ CD initiates at time t, then we update the system state only at the time completion $t + \tau_j$, by adding $\mathbf{v}_j$ to the given state;

- If reaction $j \in$ ICD initiates at time t, then we update the system state by subtracting $\mathbf{v}_j^L$ from the current state at the time of initiation, and then update the system state again by adding $\mathbf{v}_j^C$ to the given state at the time of completion $t + \tau_j$.

If we assume that $\lambda_j(\mathbf{x}(s)) = 0$ for $s < 0$, then the stochastic system with delays can be written as follows.

$$
\begin{aligned}
\mathbf{X}(t) = \mathbf{X}(0) \;+\; &\sum_{j \in \text{ND}} \mathbf{v}_j Y_j^{\text{ND}}\left(\int_0^t \lambda_j(\mathbf{X}(s))ds\right) \\
+\; &\sum_{j \in \text{CD}} \mathbf{v}_j Y_j^{\text{CD}}\left(\int_0^t \lambda_j(\mathbf{X}(s - \tau_j))ds\right) \\
+\; &\sum_{j \in \text{ICD}} \mathbf{v}_j^C Y_j^{\text{ICD}}\left(\int_0^t \lambda_j(\mathbf{X}(s - \tau_j))ds\right) \\
-\; &\sum_{j \in \text{ICD}} \mathbf{v}_j^L Y_j^{\text{ICD}}\left(\int_0^t \lambda_j(\mathbf{X}(s))ds\right),
\end{aligned}
\tag{8.47}
$$

where $\{Y_j^{\text{ND}}(t) : t \geq 0\}$, $\{Y_j^{\text{CD}}(t) : t \geq 0\}$ and $\{Y_j^{\text{ICD}}(t) : t \geq 0\}$ are independent unit Poisson processes. It is worth noting that the solution to the stochastic system with delays (8.47) is no longer a Markov process, as the state of the system depends on its past history (due to the effect of delays).

To the best of our knowledge, it appears that there is a dearth of efforts on the convergence of solutions to stochastic systems with delays as the sample size goes to infinity. One work that we found is by Bortolussi and Hillston [17], wherein Kurtz's limit theorem was extended to the case where fixed delays are incorporated into a density dependent CTMC. Another work is by Schlicht and Winkler [53], wherein it was stated that if all the transition rates are linear, then the mean solution of the stochastic system with random delays can be described by deterministic differential equations with distributed delays. We remark that it was not explicitly stated in [53] that the approximation of a discrete sampling of delays by a continuous delay kernel has a dependence on the sample size. However, intuition would suggest that a large sample size would be needed for this approximation. This was confirmed in [12] by some numerical results. Overall, to our knowledge, there is still no analog of Kurtz's limit theorem for a general stochastic system with random delays. In the next several sections, we will outline some algorithms that can be used to simulate a stochastic system with delays (either fixed or random) and numerically explore the corresponding deterministic approximations for such systems.

8.8   Simulation Algorithms for Stochastic Systems with Fixed Delays

The discussion in the above section reveals that the method used to simulate a stochastic system with delays needs to store the completion times for those delayed reactions. For each delayed reaction j, we assign a vector $\mathbf{m}_j$ that stores the completion times of reaction j in ascending order. Based on this concept, a number of algorithms were proposed in the literature to simulate a stochastic system with fixed delays. These include the *rejection method* proposed in [16] and [18]. Specifically, the initiation times are calculated exactly like the SSA. However, if there is a stored delay reaction to complete within the computed time step, then this computed time step is discarded, and the system state is updated according to the stored delayed reaction. We thus see that lots of computations may be wasted. For this reason, Cai [20] proposed an algorithm in which no computed time steps are discarded. The difference between this method and the rejection method is that the system state and transition rates are updated due to the stored delayed reactions during the search for the next initiation time (which ensures that no computed time steps are discarded). However, as remarked in [4], the extra procedures used to compute the time steps for the delayed reactions may slow this algorithm, as compared with the rejection method. An alternative algorithm based on the modified next reaction method was then proposed by Anderson [4]. The advantage of this algorithm is that it does not discard any computation as the rejection method does. In addition, it does not need extra procedures to update the system as Cai's method does. This algorithm is outlined below.

S1 Set the initial condition for each state at $t = 0$. Set $t_i^I = 0$, and the array $\mathbf{m}_i = (\infty)$ for $i = 1, 2, \ldots, l$.

S2 Calculate each transition rate λ_i at the given state and denote the resulting value by a_i for $i = 1, 2, \ldots, l$.

S3 Generate l independent, uniform random numbers on $(0,1)$, $r_1, r_2, \ldots, r_l$.

S4 Set $t_i^J = \ln(1/r_i)$ for $i = 1, 2 \ldots, l$.

S5 Set $\Delta t_i = (t_i^J - t_i^I)/a_i$ for $i = 1, 2, \ldots, l$.

S6 Set $\Delta t = \min_{1 \le i \le l} \{\Delta t_i, \mathbf{m}_i(1) - t\}$.

S7 Set $t = t + \Delta t$.

S8 If Δt_ι obtained the minimum in **S6**, then do the following. If the ι reaction is an ND, then update the system state according to the reaction ι. If reaction ι is a CD, then store the time $t + \tau_\iota$ in the second to last position in $\mathbf{m}_\iota$. If reaction ι is an ICD, then update the system state according to the initiation of the ι reaction and store $t + \tau_\iota$ in the second to last position in $\mathbf{m}_\iota$.

S9 If $\mathbf{m}_\iota(1) - t$ obtained the minimum in **S6**, then update the system state according to the completion of reaction ι and delete the first element in $\mathbf{m}_\iota$.

S10 Set $t_i^I = t_i^I + a_i \Delta t$ for $i = 1, 2, \ldots, l$.

S11 For transition ι which initiated, choose a uniform random number on $(0,1)$, r, and set $t_\iota^J = t_\iota^J + \ln(1/r)$.

S12 Recalculate each transition rate λ_i at the given new state and denote the resulting value by a_i for $i = 1, 2, \ldots, l$.

S13 Iterate **S5**–**S12** until $t \ge t_{stop}$.

We thus see that the only change to the modified next reaction method is to keep track of the delayed reactions and store their completion times. To guarantee that the array $\mathbf{m}_j$ was sorted in ascending order, the completion time for the delay reaction j was stored in $\mathbf{m}_j$ in the second to last position at each time this reaction occurred (recall that the last element in $\mathbf{m}_j$ was initialized to infinity).

8.9   Application in the Pork Production Network with a Fixed Delay

We see from Section 8.6 that the assumption made on the pork production network model is that the movement from one node to the next is instantaneous. However, incorporating delays would give a more realistic model due to the possible long distance between the nodes, bad weather or some other disruptions/interruptions. In this section, we present a pork production network model with fixed delays. For ease of exposition, we assume that all movements occur instantaneously except for the movement from node 1 to node 2. That is, the pigs leave node 1 immediately, but the time of arrival at node 2 is delayed. Delays at other nodes can be readily treated in a similar manner.

8.9.1   The Stochastic Pork Production Network Model with a Fixed Delay

Let $\mathbf{X}^M(t) = (X_1^M(t), X_2^M(t), X_3^M(t), X_4^M(t))^T$, with $X_i^M(t)$ denoting the number of pigs at the ith node at time t in a network with a scaling parameter M (e.g., M can be interpreted as the total number of individuals in the initial population). We remark that for this case the total size of the population is no longer a constant (at time $t > 0$ some individuals may be in transition from one node to another), and hence the meaning of M is different from that in Section 8.6.

If we assume that the process starts from $t = 0$ (that is, $\mathbf{X}^M(t) = 0$ for $t < 0$), then this results in a stochastic model with a fixed delay given by

$$
\begin{aligned}
X_1^M(t) &= X_1^M(0) - Y_1\left(\int_0^t \lambda_1(\mathbf{X}^M(s))ds\right) + Y_4\left(\int_0^t \lambda_4(\mathbf{X}^M(s))ds\right) \\
X_2^M(t) &= X_2^M(0) - Y_2\left(\int_0^t \lambda_2(\mathbf{X}^M(s))ds\right) + Y_1\left(\int_0^t \lambda_1(\mathbf{X}^M(s-\tau))ds\right) \\
X_3^M(t) &= X_3^M(0) - Y_3\left(\int_0^t \lambda_3(\mathbf{X}^M(s))ds\right) + Y_2\left(\int_0^t \lambda_2(\mathbf{X}^M(s))ds\right) \\
X_4^M(t) &= X_4^M(0) - Y_4\left(\int_0^t \lambda_4(\mathbf{X}^M(s))ds\right) + Y_3\left(\int_0^t \lambda_3(\mathbf{X}^M(s))ds\right),
\end{aligned}
$$

$$(8.48)$$

where λ_j, $j = 1, 2, 3, 4$, are given as those in Table 8.6, and τ is the delayed time of arrival at node 2. Note that λ_1 only depends on the state. Hence, the assumption of $\mathbf{X}^M(t) = 0$ for $t < 0$ leads to $\lambda_1(\mathbf{X}^M(t)) = 0$ for $t < 0$. The interpretation of this stochastic model is as follows. When any of the reactions 2, 3 or 4 fires at time t, the system is updated accordingly at time t. When the reaction 1 fires at time t, one unit is subtracted from the first node. Since the completion of the transition is delayed, at time $t + \tau$ one unit is added to node 2. Thus, in this example the delayed reaction is an ICD.

8.9.2   The Deterministic Pork Production Network Model with a Fixed Delay

In [17], Bortolussi and Hillston extended Kurtz's limit theorem to a scenario with fixed delays incorporated into a density dependent CTMC, where the convergence is in the sense of convergence in probability. We will use our pork production model (8.48) to illustrate this theorem (referred to as the BH limit theorem). Using the same rescaling procedure as before ($\mathbf{C}^M(t) = \mathbf{X}^M(t)/M$ with $\mathbf{X}^M(t)$ described by (8.48)), an approximating deterministic system can be constructed based on the BH limit theorem for the scaled stochastic system with a fixed delay. This approximating deterministic system is given by

$$
\begin{aligned}
\dot{c}_1(t) &= -\kappa_1 c_1(t)(l_2 - c_2(t))_+ + \kappa_4 \min(c_4(t), s_m) \\
\dot{c}_2(t) &= -\kappa_2 c_2(t)(l_3 - c_3(t))_+ + \kappa_1 c_1(t - \tau)(l_2 - c_2(t - \tau))_+ \\
\dot{c}_3(t) &= -\kappa_3 c_3(t)(l_4 - c_4(t))_+ + \kappa_2 c_2(t)(l_3 - c_3(t))_+ \\
\dot{c}_4(t) &= -\kappa_4 \min(c_4(t), s_m) + \kappa_3 c_3(t)(l_4 - c_4(t))_+.
\end{aligned}
\tag{8.49}
$$

We note that this approximating deterministic system is no longer a system of ordinary differential equations, but rather a system of delay differential equations (DDEs) with a fixed delay. The delay differential equation is a direct result of the delay term present in (8.48). Since there is a delay term, the system is dependent on the previous states, and for this reason it is necessary to have some past history functions as initial conditions. It should be noted that past history functions *should not* be chosen in an arbitrary fashion as they should capture the limit dynamics of the scaled stochastic system with a fixed delay.

Next we illustrate how to construct the initial conditions for the delay differential equation (8.49). We observe that in the interval $[0, \tau]$ the delay term has no effect on the system (as we assumed that $\mathbf{X}^M(t) = 0$ for $t < 0$); thus we can ignore the delay term in this interval. This yields a stochastic system with no delays. Thus it can be approximated by a system of ODEs as was done previously. This deterministic system is given as

$$
\begin{aligned}
\dot{c}_1(t) &= -\kappa_1 c_1(t)(l_2 - c_2(t))_+ + \kappa_4 \min(c_4(t), s_m) \\
\dot{c}_2(t) &= -\kappa_2 c_2(t)(l_3 - c_3(t))_+ \\
\dot{c}_3(t) &= -\kappa_3 c_3(t)(l_4 - c_4(t))_+ + \kappa_2 c_2(t)(l_3 - c_3(t))_+ \\
\dot{c}_4(t) &= -\kappa_4 \min(c_4(t), s_m) + \kappa_3 c_3(t)(l_4 - c_4(t))_+ \\
\mathbf{c}(0) &= \mathbf{c}_0
\end{aligned}
\tag{8.50}
$$

for $t \in [0, \tau]$. We let $\boldsymbol{\Psi}$ denote the solution to (8.50) and thus we have that $\mathbf{C}^M(t)$ converges to $\boldsymbol{\Psi}(t)$ as $M \to \infty$ on the interval $[0, \tau]$.

In the interval $[\tau, t_{stop}]$, the delay has an affect, so we approximate with the DDE system (8.49), and the solution $\boldsymbol{\Psi}$ to the ODE (8.50) on the interval $[0, \tau]$ serves as the initial function. Explicitly the system can be written as

$$
\begin{aligned}
\dot{c}_1(t) &= -\kappa_1 c_1(t)(l_2 - c_2(t))_+ + \kappa_4 \min(c_4(t), s_m) \\
\dot{c}_2(t) &= -\kappa_2 c_2(t)(l_3 - c_3(t))_+ + \kappa_1 c_1(t - \tau)(l_2 - c_2(t - \tau))_+ \\
\dot{c}_3(t) &= -\kappa_3 c_3(t)(l_4 - c_4(t))_+ + \kappa_2 c_2(t)(l_3 - c_3(t))_+ \\
\dot{c}_4(t) &= -\kappa_4 \min(c_4(t), s_m) + \kappa_3 c_3(t)(l_4 - c_4(t))_+ \\
\mathbf{c}(s) &= \boldsymbol{\Psi}(s), \quad s \in [0, \tau].
\end{aligned}
\tag{8.51}
$$

The BH limit theorem indicates that $\mathbf{C}^M(t)$ converges in probability to the solution of (8.51) as $M \to \infty$. The delayed deterministic system (8.51) can be readily solved numerically using a linear spline approximation method as first developed in [15] and further explained in [11].

Remark 8.9.1 *We remark that in the literature one often sees an arbitrarily chosen past history function for the delay differential equation. However, this does not make sense, as the history function must depend on the dynamics of the given delay differential equation. If one knows when this process starts and the value it starts with, then the method illustrated above provides a reasonable way to construct the past history function. Otherwise, one needs to collect experimental data and resort to parameter estimation methods to estimate this past history function.*

8.9.3   Comparison of the Stochastic Model with a Fixed Delay and Its Corresponding Deterministic System

In this section we compare the results of the stochastic system with a fixed delay (8.48) to its corresponding deterministic system (in terms of number of pigs, i.e., $M\mathbf{c}(t)$ with $\mathbf{c}(t)$ being the solution to (8.51)). The stochastic system was simulated using the algorithm presented in Section 8.8 with $M = 100$, $M = 1000$ and $M = 10,000$. For all the simulations presented below, the parameter values and initial conditions are chosen as those in Table 8.7, and the value of the delay was set to be $\tau = 5$.

In Figure 8.10, the deterministic approximation is compared to five typical sample paths of the solution to the stochastic system (8.48). It is clear from this figure that the trajectories of the stochastic simulations follow the solution of its corresponding deterministic system, and that the variation among sample paths of the stochastic solution decreases as the sample size increases.

Figure 8.11 depicts the mean of the solution for the stochastic system (8.48) in comparison to the solution for the deterministic approximation, where the mean of the solution to the stochastic system was calculated by averaging 10,000 sample paths. We observe from Figure 8.11 that as the sample size increases the mean of the solution to the stochastic system becomes closer to the solution of its corresponding deterministic system. In addition, we see that the mean of the solution to the stochastic system is reasonably close to the deterministic solution even when $M = 1000$, in which the variation among sample paths of the stochastic solution is still large (see the middle column of Figure 8.10).

8.10   Simulation Algorithms for Stochastic Systems with Random Delays

Roussel and Zhu proposed a method in [51] for simulating a stochastic system with random delays. Specifically, the algorithm for a stochastic system with random delays can be implemented in a way that is a slight modification of the algorithms in Section 8.8. The only difference is that each time a

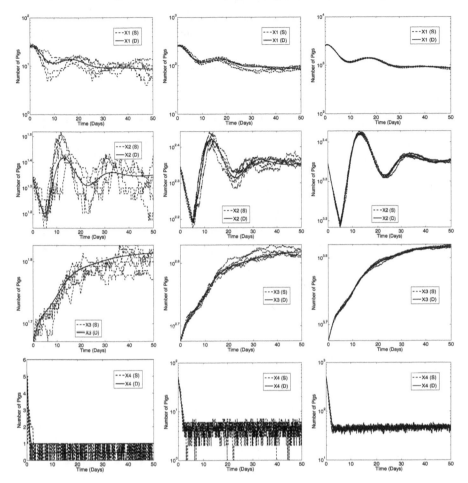

FIGURE 8.10: Results obtained by the stochastic system with a fixed delay (S) and by the corresponding deterministic system (D) with $M = 100$ (left column), $M = 1000$ (middle column) and $M = 10,000$ (right column).

delayed transition is fired, a new value of the delay is drawn from the given distribution. This implies that storing the completion time for the delayed reaction j in the array $\mathbf{m}_j$ in the second to last position does not guarantee that $\mathbf{m}_j$ was sorted in ascending order. To do that, one can store the completion time in any position in $\mathbf{m}_j$ and then sort $\mathbf{m}_j$ in ascending order. Based on this concept, one can extend the algorithms in Section 8.8 to the case where the delays are random. The full details of the algorithm for the modified reaction method applied to this case are outlined below.

S1 Set the initial condition for each state for $t = 0$. Set $t_i^I = 0$, and array

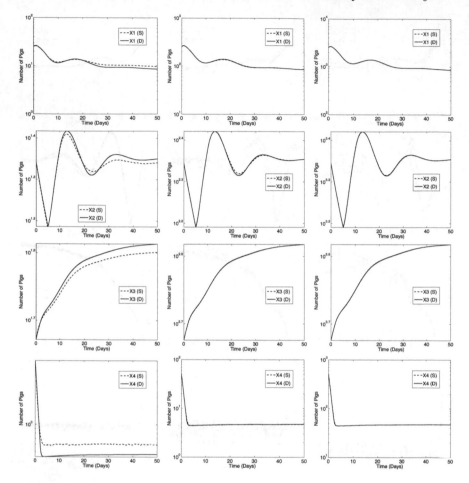

FIGURE 8.11:   The mean of the solution to the stochastic system with a fixed delay (S) and the solution to the corresponding deterministic system (D) with $M = 100$ (left column), $M = 1000$ (middle column) and $M = 10{,}000$ (right column).

$\mathbf{m}_i = (\infty)$ for $i = 1, 2, \dots, l$.

S2 Calculate each transition rate λ_i at the given state and denote the resulting value by a_i for each $i = 1, 2, \dots, l$.

S3 Generate l independent, uniform random numbers on $(0, 1)$, $r_1, r_2, \dots, r_l$.

S4 Set $t_i^J = \ln(1/r_i)$ for $i = 1, 2, \dots, l$.

S5 Set $\Delta t_i = (t_i^J - t_i^I)/a_i$ for $i = 1, 2, \dots, l$.

S6 Set $\Delta t = \min_{1 \le i \le l} \{\Delta t_i, \mathbf{m}_i(1) - t\}$.

S7 Set $t = t + \Delta t$.

S8 If Δt_ι obtained the minimum in **S6**, then do the following. If the ι reaction is an ND, then update the system state according to the reaction ι. If reaction ι is a CD, then store the time $t + \tau_\iota$ in $\mathbf{m}_\iota$ and sort $\mathbf{m}_\iota$ in ascending order, where τ_ι is sampled from a given distribution. If reaction ι is an ICD, then update the system state according to the initiation of the ι reaction and store $t + \tau_\iota$ in $\mathbf{m}_\iota$ and then sort $\mathbf{m}_\iota$ in ascending order, where τ_ι is sampled from a given distribution.

S9 If $\mathbf{m}_\iota(1) - t$ obtained the minimum in **S6**, then update the system state according to the completion of reaction ι and delete the first element in $\mathbf{m}_\iota$.

S10 Set $t_i^I = t_i^I + a_i \Delta t$ for $i = 1, 2, \ldots, l$.

S11 For the transition ι which initiated, choose a uniform random number $(0,1)$, r, and set $t_\iota^J = t_\iota^J + \ln(1/r)$.

S12 Recalculate each transition rate λ_i at the given new state and denote the resulting value by a_i for $i = 1, 2, \ldots, l$.

S13 Iterate **S5**–**S12** until $t \geq t_{stop}$.

8.11 Application in the Pork Production Network with a Random Delay

For the pork production network model in Section 8.9, we assumed that every movement from node 1 to node 2 is delayed by the same amount of time. We now want to consider the case where the amount of delayed time varies each time this movement occurs. The motivation for doing so is that in practice we would expect the amount of time it takes to travel from node 1 to node 2 to vary based on a number of conditions, e.g., weather, traffic, road construction, etc. The resulting system to describe this case will be called the *stochastic pork production network model with a random delay*. For this model, we again assume that all transitions occur instantaneously except for the arrival of pigs moving from node 1 to node 2, and that the process starts from $t = 0$ (that is, $\mathbf{X}^M(t) = 0$ for $t < 0$). In addition, for all the simulations presented below, each of the simulations was run for the time period $[0, 50]$ days with parameter values and initial conditions given in Table 8.7.

8.11.1 The Corresponding Deterministic System

As we remarked earlier in Section 8.7, Schlicht and Winkler in [53] stated that if all the transition rates are *linear*, then the mean solution of the stochastic system with a random delay can be described by a system of deterministic differential equations with a distributed delay, where the delay kernel is the probability density function of the random delay. Even though the transition rates in our pork production model are *non-linear*, we still would like to explore whether or not such a deterministic system can be used as a possible corresponding deterministic system for our stochastic system and to explore the relationship between them.

Let G be the probability density function of the random delay. Then the corresponding deterministic system for our stochastic system with a random delay is given by

$$\dot{x}_1^M(t) = -k_1 x_1^M(t)(L_2 - x_2^M(t))_+ + k_4 \min(x_4^M(t), S_m)$$

$$\dot{x}_2^M(t) = -k_2 x_2^M(t)(L_3 - x_3^M(t))_+ + \int_{-\infty}^t G(t-s)k_1 x_1^M(s)(L_2 - x_2^M(s))_+ ds$$

$$\dot{x}_3^M(t) = -k_3 x_3^M(t)(L_4 - x_4^M(t))_+ + k_2 x_2^M(t)(L_3 - x_3^M(t))_+$$

$$\dot{x}_4^M(t) = -k_4 \min(x_4^M(t), S_m) + k_3 x_3^M(t)(L_4 - x_4^M(t))_+$$

$$x_i^M(0) = x_{i0}^M, \quad i = 1, 2, 3, 4,$$

$$x_i^M(s) = 0, \quad s < 0, \quad i = 1, 2, 3, 4.$$

$$(8.52)$$

We note that the linear or more general spline approximation method as developed in [15] could be used to numerically solve a DDE with a general delay kernel G. However, here we take a different approach. It is well known that if we make additional assumptions on G, then we can transform a DDE into a system of ODEs by using the so-called *linear chain trick* [27] (e.g., see [19, 38, 52, 56] and the references therein). Specifically, a necessary and sufficient condition for the reducibility of a system with distributed delay to a system of ODEs is that the delay kernel G is a linear combination of functions

$$\exp(at), \quad t\exp(at), \quad \ldots, \quad t^m \exp(at)$$

with m being a positive integer and a being a complex number. In the following, we assume that the delay kernel has the form

$$G(t; \alpha, m) = \frac{\alpha^m t^{m-1} \exp(-\alpha t)}{(m-1)!} \tag{8.53}$$

with α being a positive number and m being a positive integer (that is, G is the probability density function of a gamma distributed random variable with mean m/α and the variance m/α^2), so that we can transform system (8.52) into a system of ODEs by way of the linear chain trick. For example, for the case $m = 1$ (that is, $G(t) = \alpha \exp(-\alpha t)$), if we let

$$x_5^M(t) = \int_{-\infty}^t \alpha \exp(-\alpha(t-s))k_1 x_1^M(s)(L_2 - x_2^M(s))_+ ds,$$

then system (8.52) can be written as

$$\dot{x}_1^M(t) = -k_1 x_1^M(t)(L_2 - x_2^M(t))_+ + k_4 \min(x_4^M(t), S_m)$$

$$\dot{x}_2^M(t) = -k_2 x_2^M(t)(L_3 - x_3^M(t))_+ + x_5^M(t)$$

$$\dot{x}_3^M(t) = -k_3 x_3^M(t)(L_4 - x_4^M(t))_+ + k_2 x_2^M(t)(L_3 - x_3^M(t))_+$$

$$\dot{x}_4^M(t) = -k_4 \min(x_4^M(t), S_m) + k_3 x_3^M(t)(L_4 - x_4^M(t))_+$$ (8.54)

$$\dot{x}_5^M(t) = \alpha k_1 x_1^M(t)(L_2 - x_2^M(t))_+ - \alpha x_5^M(t)$$

$$x_i^M(0) = x_{i0}^M, \quad i = 1, 2, 3, 4, \quad x_5^M(0) = 0.$$

The advantage of using this linear chain trick is two-fold. The resulting system of ODEs is much easier to solve compared to the system (8.52). In addition, we can use Kurtz's limit theorem in a rigorous manner to construct a corresponding stochastic system which converges to the resulting system of ODEs. This approach will be considered in Section 8.11.3.

8.11.2   Comparison of the Stochastic Model with a Random Delay and Its Corresponding Deterministic System

In this section, we consider the stochastic system with a random delay having probability density function $G(\cdot; \alpha, m)$ defined as in (8.53), where m was taken to be 1 and α to be 0.2 (which implies that the mean value of the random delay is 5 and its variance is 25). Sample sizes of $M = 100$ and $M = 10,000$ were considered for the stochastic system with a random delay. As before, the stochastic system was simulated for 10,000 trials, and the mean of the solution was computed.

Figures 8.12 and 8.13 compare the solution of deterministic system (8.54) and the results of the stochastic system with a random delay. We observe from Figure 8.12 that the trajectories of the stochastic simulations follow the solution of the deterministic system, and that the variation of the sample paths of the solution to the stochastic system with a random delay decreases as the sample size increases. It is also seen from Figure 8.13 that as the sample size increases the mean of the solution to the stochastic system with a random delay becomes closer to the solution of the deterministic system. Hence, not only are the sample paths of the solution to the stochastic system with a random delay showing less variation for larger sample sizes, but the expected value of the solution is better approximated by the solution of the deterministic system for large sample sizes. Thus the deterministic system (8.54) (or the deterministic differential equation with a distributed delay (8.52)) could be used to serve as a reasonable corresponding deterministic system for this particular stochastic system (with the given parameter values and initial conditions).

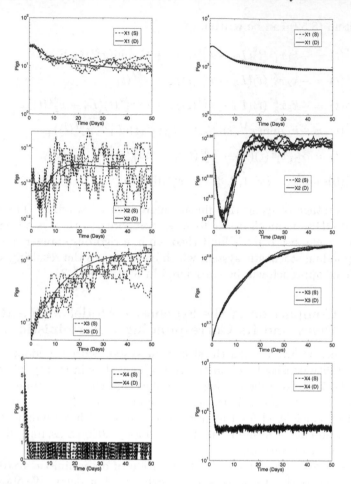

FIGURE 8.12: Results obtained by the stochastic system with a random delay (S) and by the corresponding deterministic system (8.54) (D) with $M = 100$ (left column) and $M = 10,000$ (right column).

8.11.3   The Corresponding Constructed Stochastic System

Based on Kurtz's limit theorem, one can construct a corresponding CTMC model $\{\mathbf{X}^M(t) : t \geq 0\}$ with the state space being a subset of $\mathbb{Z}^5$ to the system of ODEs given in (8.54). Its transition rate $\lambda_j(\mathbf{x}^M)$ and the corresponding

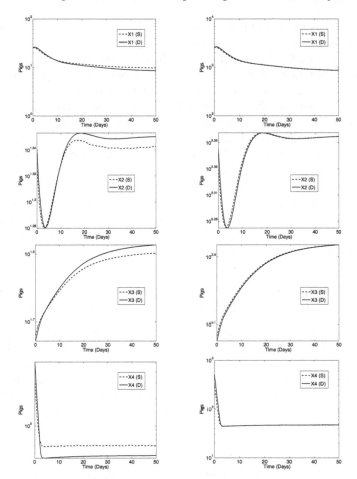

FIGURE 8.13: The mean of the solution to the stochastic system with a random delay (S) and the solution to the corresponding deterministic system (8.54) (D) with $M = 100$ (left column) and $M = 10,000$ (right column).

state change vector $\mathbf{v}_j$ at the jth transition, $j = 1, 2, \ldots, 7$, are given by

$$
\begin{aligned}
\lambda_1(\mathbf{x}^M) &= k_1 x_1^M (L_2 - x_2^M)_+, & \mathbf{v}_1 &= -\mathbf{e}_1, \\
\lambda_2(\mathbf{x}^M) &= k_2 x_2^M (L_3 - x_3^M)_+, & \mathbf{v}_2 &= -\mathbf{e}_2 + \mathbf{e}_3, \\
\lambda_3(\mathbf{x}^M) &= k_3 x_3^M (L_4 - x_4^M)_+, & \mathbf{v}_3 &= -\mathbf{e}_3 + \mathbf{e}_4, \\
\lambda_4(\mathbf{x}^M) &= k_4 \min(x_4^M, S_m), & \mathbf{v}_4 &= -\mathbf{e}_4 + \mathbf{e}_1 & (8.55) \\
\lambda_5(\mathbf{x}^M) &= x_5^M, & \mathbf{v}_5 &= \mathbf{e}_2 \\
\lambda_6(\mathbf{x}^M) &= \alpha k_1 x_1^M (L_2 - x_2^M)_+, & \mathbf{v}_6 &= \mathbf{e}_5 \\
\lambda_7(\mathbf{x}^M) &= \alpha x_5^M, & \mathbf{v}_7 &= -\mathbf{e}_5,
\end{aligned}
$$

where $\mathbf{e}_i \in \mathbb{Z}^5$ is the ith unit column vector, $i = 1, 2, \ldots, 5$. It is worth noting that even though this constructed stochastic process is a Markov process, the primary process $(X_1, X_2, X_3, X_4)^T$ in which we are interested is not (as we remarked earlier in Section 8.7, the system depends on its past history due to the effect of the delay). This means that one can enforce the Markovian property by including all the intermediate steps as separate states. However, the disadvantage of doing so is that it may dramatically increase the number of transitions and hence may greatly increase the computational time.

We remark that even though this constructed stochastic system may have no biological meaning, it will be used to serve as a comparison for the stochastic system with a random delay. The reason for doing so is that this stochastic system is constructed based on a rigorous use of Kurtz's limit theorem from the system of ODEs (8.54), which can also be used as a possible corresponding deterministic system for the stochastic system with a random delay (as we demonstrated in the previous section).

8.11.4   Comparison of the Constructed Stochastic System and Its Corresponding Deterministic System

Figure 8.14 depicts five typical sample paths of the solution to the constructed stochastic system (with transition rates and corresponding state change vectors given by (8.55)) in comparison to the solution to the corresponding deterministic system (8.54), and Figure 8.15 shows the mean of the solution to the constructed stochastic system in comparison to the solution to (8.54). Here the mean of the solution for the stochastic system was again computed by averaging 10,000 sample paths.

We observe from these two figures that, as might be expected, the variation of the sample paths of the solution to the constructed stochastic system decreases as the sample size increases, and the mean of its solution becomes closer to the solution of its corresponding deterministic system as the sample size increases. It is also noted that for each of the same sample sizes M, the variation of sample paths is much higher compared to that of the stochastic system with a random delay, especially for node 1 and node 2.

8.11.5   Comparison of the Stochastic System with a Random Delay and the Constructed Stochastic System

The numerical results in the previous section demonstrate that with the same sample size the sample paths of the solutions to the stochastic system with a random delay have less variation than those obtained for the corresponding constructed stochastic system. In this section we make a further comparison of these two stochastic systems.

For all the simulations below, the random delay is assumed to be gamma distributed with probability density function $G(\cdot; \alpha, m)$ as defined in (8.53), where the expected value of the random delay is always chosen to be 5. Each

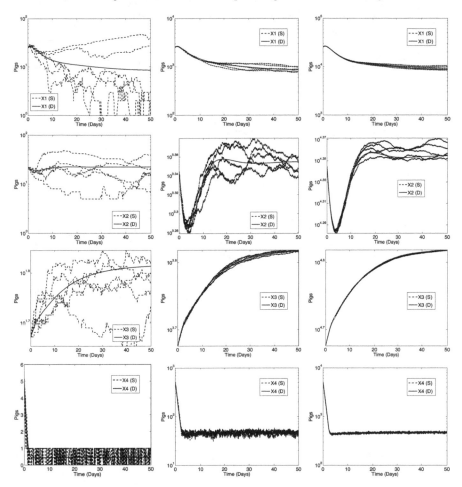

FIGURE 8.14: Results obtained for the deterministic system (8.54) (D) and the corresponding constructed stochastic system (S) with a sample size of $M = 100$ (left column), $M = 10,000$ (middle column) and $M = 100,000$ (right column).

stochastic system was simulated 10,000 times with various sample sizes. For each of the sample sizes, a histogram was constructed for $X_i^M(t)$, $i = 1, 2, 3, 4$, at $t = 50$ (the end time point of the simulation), for each stochastic system based on these 10,000 sample paths.

8.11.5.1   The Effect of Sample Size on the Comparison of These Two Stochastic Systems

In this section, we investigate how the sample size M affects the comparison of the stochastic system with a random delay and the constructed stochastic

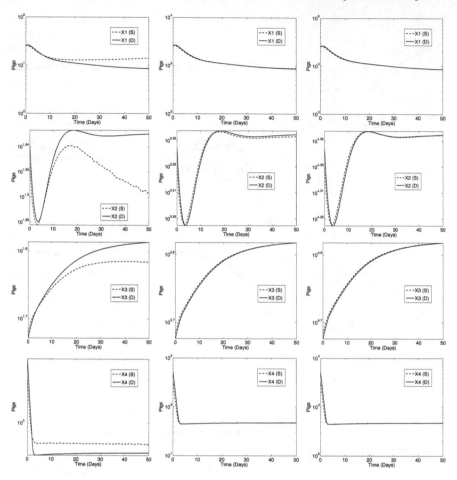

FIGURE 8.15: The solution for the deterministic system (8.54) (D) and the mean of the solution to the corresponding constructed stochastic system (S) with a sample size of $M = 100$ (left column), $M = 10,000$ (middle column) and $M = 100,000$ (right column).

system with transition rates and corresponding state change vectors given by (8.55), where $G(t; \alpha, m)$ is chosen with $m = 1$ and $\alpha = 0.2$ as before.

Using the same sample size for these two stochastic systems   Figure 8.16 depicts the histograms of each node for these two stochastic systems with $M = 100$ (left column), $M = 1000$ (middle column) and $M = 10,000$ (right column) at time point $t = 50$. We observe from this figure that the histograms for each of the stochastic systems do not match well except for node 4. This

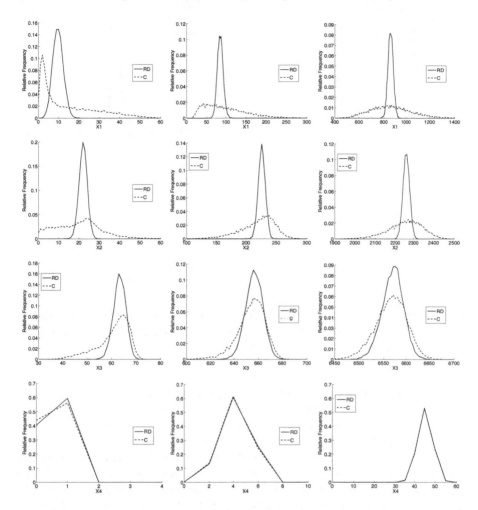

FIGURE 8.16: Histograms of the stochastic system with random delays (RD) and the constructed stochastic system (C) with $M = 100$ (left column), $M = 1000$ (middle column) and $M = 10,000$ (right column) at $t = 50$.

is probably because the delay only occurs from node 1 to node 2, and hence it has less effect on node 4 than on all the other nodes due to the movement from one node to the next occurring only in the forward direction. Specifically, it is seen that for all the sample sizes investigated, the histogram plots of nodes 1, 2 and 3 obtained for the constructed stochastic system appear more dispersed than those for the stochastic system with a random delay, and this is especially obvious for nodes 1 and 2. Recall that for the deterministic differential equation with a distributed delay, one assumes that each individual is

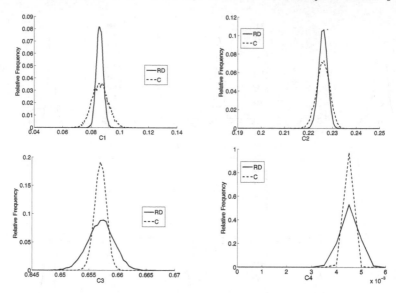

FIGURE 8.17: Histograms of the stochastic system with random delays (RD) with a sample size of $M = 10,000$ and the constructed stochastic system with a sample size of $M = 100,000$ at $t = 50$.

different and may take different times to progress from one node to another, so that for a large population one can use a distributed delay to approximate it. Hence, for the stochastic system constructed from this deterministic system, each individual may also be treated differently. However, individuals for the stochastic system with a random delay are treated the same. This means that the constructed stochastic system is more "random" than the stochastic system with a random delay, and thus it shows more variation. Figure 8.16 also reveals that the histogram plots of nodes 1, 2 and 3 obtained for the stochastic system with a random delay are symmetric for all the sample sizes investigated, while those for the constructed stochastic system are asymmetric when M is small, but become more symmetric as M increases.

Using different sample sizes for these two stochastic systems   In Figure 8.17 we compare the histogram plots obtained from the stochastic system with a random delay for a sample size of $M = 10,000$ to those obtained from the constructed stochastic system with a sample size of $M = 100,000$ and we see that the two histograms are in better agreement for nodes 1 and 2. We offer one possible explanation for this occurrence. Both stochastic systems can be approximated by the same deterministic system, yet it is possible that the two stochastic systems are "converging" to the deterministic solution at different rates; thus different values for sample size are needed to obtain the same order of approximation.

8.11.5.2   The Effect of the Variance of a Random Delay on the Comparison of These Two Stochastic Systems

What remains to be investigated is how the variance of the random delay affects the comparison of the two stochastic systems. In order to change the variance, the shape and rate parameters m and α in the probability density function $G(\cdot; \alpha, m)$ must be altered to keep the mean of the random delay the same. Specifically, the value of m determines the number of additional equations needed to reduce the deterministic system with a distributed delay into a system of ODEs. Thus, for additional variance to be considered, a new system of ODEs and the corresponding stochastic system must be derived.

For the case where $m = 10$ and $\alpha = 2$, we have a mean of 5 and a variance of 2.5 for the random delay. Now using the substitutions

$$x_{j+4}^M(t) = \int_{-\infty}^{t} \frac{\alpha^j (t-s)^{j-1}}{(j-1)!} \exp(-\alpha(t-s)) k_1 x_1^M(s)(L_2 - x_2^M(s))_+ ds$$

for $j = 1, 2, 3, \ldots, 10$, we obtain the following system of ODEs that is equivalent to (8.52) with delay kernel $G(t; \alpha, 10)$:

$$
\begin{aligned}
\dot{x}_1^M(t) &= -k_1 x_1^M(t)(L_2 - x_2^M(t))_+ + k_4 \min(x_4^M, S_m) \\
\dot{x}_2^M(t) &= -k_2 x_2^M(L_3 - x_3^M)_+ + x_{14}^M(t) \\
\dot{x}_3^M(t) &= -k_3 x_3^M(L_4 - x_4^M)_+ + k_2 x_2^M(L_3 - x_3^M)_+ \\
\dot{x}_4^M(t) &= -k_4 \min(x_4^M, S_m) + k_3 x_3^M(L_4 - x_4^M)_+ \\
\dot{x}_5^M(t) &= \alpha k_1 x_1^M(t)(L_2 - x_2^M(t))_+ - \alpha x_5^M(t) \\
\dot{x}_i^M(t) &= \alpha x_{i-1}^M(t) - \alpha x_i^M(t), \quad i = 6, 7, \ldots, 14, \\
x_i^M(0) &= x_{i0}^M, \quad i = 1, 2, 3, 4, \\
x_i^M(0) &= 0 \quad i = 6, 7, \ldots, 14.
\end{aligned}
\tag{8.56}
$$

We can construct a corresponding CTMC model $\{\mathbf{X}^M(t) : t \geq 0\}$ (with the state space being a subset in $\mathbb{Z}^{14}$) for the deterministic system (8.56), where its transition rates and corresponding state change vectors are given by

$$\lambda_1(\mathbf{x}^M) = k_1 x_1^M (L_2 - x_2^M)_+, \quad \mathbf{v}_1 = -\mathbf{e}_1,$$

$$\lambda_2(\mathbf{x}^M) = k_2 x_2^M (L_3 - x_3^M)_+, \quad \mathbf{v}_2 = -\mathbf{e}_2 + \mathbf{e}_3,$$

$$\lambda_3(\mathbf{x}^M) = k_3 x_3^M (L_4 - x_4^M)_+, \quad \mathbf{v}_3 = -\mathbf{e}_3 + \mathbf{e}_4,$$

$$\lambda_4(\mathbf{x}^M) = k_4 \min(x_4^M, S_m), \quad \mathbf{v}_4 = -\mathbf{e}_4 + \mathbf{e}_1$$

$$\lambda_5(\mathbf{x}^M) = x_{14}^M, \quad \mathbf{v}_5 = \mathbf{e}_2$$

$$\lambda_6(\mathbf{x}^M) = \alpha k_1 x_1^M (L_2 - x_2^M)_+, \quad \mathbf{v}_6 = \mathbf{e}_5$$

$$\lambda_7(\mathbf{x}^M) = \alpha x_5^M, \quad \mathbf{v}_7 = \mathbf{e}_6 - \mathbf{e}_5$$

$$\lambda_8(\mathbf{x}^M) = \alpha x_6^M, \quad \mathbf{v}_8 = \mathbf{e}_7 - \mathbf{e}_6$$

$$\lambda_9(\mathbf{x}^M) = \alpha x_7^M, \quad \mathbf{v}_9 = \mathbf{e}_8 - \mathbf{e}_7$$

$$\lambda_{10}(\mathbf{x}^M) = \alpha x_8^M, \quad \mathbf{v}_{10} = \mathbf{e}_9 - \mathbf{e}_8$$

$$\lambda_{11}(\mathbf{x}^M) = \alpha x_9^M, \quad \mathbf{v}_{11} = \mathbf{e}_{10} - \mathbf{e}_9$$

$$\lambda_{12}(\mathbf{x}^M) = \alpha x_{10}^M, \quad \mathbf{v}_{12} = \mathbf{e}_{11} - \mathbf{e}_{10}$$

$$\lambda_{13}(\mathbf{x}^M) = \alpha x_{11}^M, \quad \mathbf{v}_{13} = \mathbf{e}_{12} - \mathbf{e}_{11}$$

$$\lambda_{14}(\mathbf{x}^M) = \alpha x_{12}^M, \quad \mathbf{v}_{14} = \mathbf{e}_{13} - \mathbf{e}_{12}$$

$$\lambda_{15}(\mathbf{x}^M) = \alpha x_{13}^M, \quad \mathbf{v}_{15} = \mathbf{e}_{14} - \mathbf{e}_{13}$$

$$\lambda_{16}(\mathbf{x}^M) = \alpha x_{14}^M, \quad \mathbf{v}_{16} = -\mathbf{e}_{14}.$$

Here $\mathbf{e}_j \in \mathbb{Z}^{14}$ is the jth unit vector, $j = 1, 2, \ldots, 14$. To avoid confusion we will refer to this new stochastic system with transition rates and corresponding state change vectors given above as the constructed stochastic system with $\alpha = 2$ to distinguish it from the previously constructed system in which we had $\alpha = 0.2$.

Both the constructed stochastic system with $\alpha = 2$ and the stochastic system with a random delay having probability density function $G(\cdot; 2, 10)$ were simulated 10,000 times with a sample size of $M = 100$ and $M = 1000$. The resulting histograms are shown in Figure 8.18. From here it is seen that there is much more agreement between the histograms than the corresponding ones in Section 8.11.5.1 when the variance was 25. In addition, we observe that the histograms of the constructed system with $\alpha = 2$ are a little closer to the histograms to the stochastic system with a random delay for the larger sample size. Overall, we see that the variance of the random delay has a decided effect on the agreement between these two stochastic systems.

8.11.5.3   Summary Remarks

In Sections 8.11.5.1 and 8.11.5.2, we compared the stochastic model with a gamma distributed random delay to the stochastic system constructed based

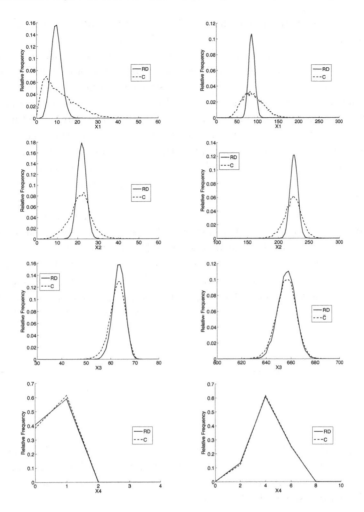

FIGURE 8.18: Histograms of the stochastic system with random delays (RD) and the constructed stochastic system with $\alpha = 2$ (C) with $M = 100$ (left column), $M = 1000$ (right column) at $t = 50$. The random delay was chosen from a gamma distribution with a mean of 5 and variance of 2.5.

on Kurtz's limit theorem from a system of deterministic differential equations with a gamma distributed delay. Even though the same system of deterministic differential equations with a gamma distributed delay can be used as the corresponding deterministic systems for these two stochastic systems, it was found that with the same sample size the histogram plots of the state solutions for the constructed stochastic system are more dispersed than the corresponding ones obtained for the stochastic model with a random delay. However, there is more agreement between the histograms of these two stochastic systems as the variance of the random delay decreases.

Note that the transition rates for the example here are non-linear. One may wonder whether the difference in the respective histograms for the stochastic system with a random delay and the constructed stochastic system is due to the non-linearity of the transition rates. However, the numerical results in [12] demonstrate that even in the case that all the transition rates are linear, there are remarkable differences in the respective histograms for these two stochastic systems. (With the same sample size, the histogram plots of the state solutions to the constructed stochastic system are also more dispersed than the corresponding ones obtained for the stochastic model with a random delay). This may suggest that the difference between the histograms is at least partially due to the different methods on which the stochastic systems are being approximated by deterministic systems. Therefore, it may be plausible that the agreement between histograms is dependent on the fact that two different methods were used to obtain a deterministic approximation (the method of Schlicht and Winkler, and the Kurtz's limit theorem) as well as being dependent on both the variance of the delay and the sample sizes used, independent of whether linear or non-linear transition rates appear in the model.

References

[1] B.M. Adams, H.T. Banks, M. Davidian, H. Kwon, H.T. Tran, S.N. Wynne and E.S. Rosenberg, HIV dynamics: Modeling, data analysis, and optimal treatment protocols, *J. Computational and Applied Mathematics*, **184** (2005), 10–49.

[2] B.M. Adams, H.T. Banks, M. Davidian and E.S. Rosenberg, Model fitting and prediction with HIV treatment interruption data, *Bulletin of Mathematical Biology*, **69** (2007), 563–584.

[3] L.J.S. Allen, *An Introduction to Stochastic Processes with Applications to Biology*, Chapman and Hall/CRC press, Boca Raton, FL, 2011.

[4] D.F. Anderson, A modified next reaction method for simulating chemical systems with time dependent propensities and delays, *The Journal of Chemical Physics*, **127** (2007), 214107 (10 p).

[5] D.F. Anderson, Incorporating postleap checks in tau-leaping, *The Journal Of Chemical Physics*, **128** (2008), 054103 (8 p).

[6] D.F. Anderson, A. Ganguly and T.G. Kurtz, Error analysis of tau-leap

simulation methods, *The Annals of Applied Probability*, **21** (2011), 2226–2262.

[7] D.F. Anderson and M. Koyama, Weak error analysis of numerical methods for stochastic models of population processes, *Multiscale Model. Simul.*, **10** (2012), 1493–1524.

[8] D.F. Anderson and T.G. Kurtz, Continuous time Markov chain models for chemical reaction networks, in *Design and Analysis of Biomolecular Circuits: Engineering Approaches to Systems and Synthetic Biology*, H. Koeppl, D. Densmore, G. Setti and M. di Bernardo eds. (2011), 3–42.

[9] H. Andersson and T. Britton, *Stochastic Epidemic Models and Their Statistical Analysis*, Springer-Verlag, New York, 2000.

[10] P. Bai, H.T. Banks, S. Dediu, A.Y. Govan, M. Last, A.L. Lloyd, H.K. Nguyen, M.S. Olufsen, G. Rempala and B.D. Slenning, Stochastic and deterministic models for agricultural production networks, *Mathematical Biosciences and Engineering*, **4** (2007), 373–402.

[11] H.T. Banks, *A Functional Analysis Framework for Modeling, Estimation and Control in Science and Engineering*, Chapman and Hall/CRC Press, Boca Raton, FL, 2012.

[12] H.T. Banks, J. Catenacci and S. Hu, A comparison of stochastic systems with different types of delays, CRSC-TR13-04, March, 2013; *Stochastic Analysis and Applications*, 31(2013), 913–955.

[13] H.T. Banks, M. Davidian, S. Hu, G. Kepler and E.S. Rosenberg, Modelling HIV immune response and validation with clinical data, *Journal of Biological Dynamics*, **2** (2008), 357–385.

[14] H.T. Banks, S. Hu, M. Joyner, A. Broido, B. Canter, K. Gayvert and K. Link, A comparison of computational efficiencies of stochastic algorithms in terms of two infection models, *Mathematical Biosciences and Engineering*, **9** (2012), 487–526.

[15] H.T. Banks and F. Kappel, Spline approximations for functional differential equations, *Journal of Differential Equations*, **34** (1979), 496–522.

[16] M. Barrio, K. Burrage, A. Leier and T. Tian, Oscillatory regulation of Hes1: discrete stochastic delay modeling and simulation, *PLoS Computational Biology*, **2** (2006), 1017–1030.

[17] L. Bortolussi and J. Hillston, Fluid approximations of CTMC with deterministic delays, *Quantitative Evaluation of Systems (QEST), 2012 Ninth International Conference*, Sept 17–20, 2012, 53–62.

[18] D. Bratsun, D. Volfson, L.S. Tsimring and J. Hasty, Delayed-induced stochastic oscillations in gene regulation, *PNAS*, **102** (2005), 14593–14598.

[19] S.N. Busenberg and C.C. Travis, On the use of reducible functional differential equations in biological models, *Journal of Mathematical Analysis and Applications*, **89** (1982), 46–66.

[20] X. Cai, Exact stochastic simulation of coupled chemical reactions with delays, *The Journal of Chemical Physics*, **126** (2007), 124108 (8 p).

[21] D.S. Callaway and A.S. Perelson, HIV-1 infection and low steady state viral loads, *Bulletin of Mathematical Biology*, **64** (2001) 29–64.

[22] Y. Cao, D.T. Gillespie and L.R. Petzold, Avoiding negative populations in explicit Poisson tau-leaping, *The Journal of Chemical Physics*, **123** (2005), 054104.

[23] Y. Cao, D.T. Gillespie and L.R. Petzold, Efficient step size selection for the tau-leaping simulation method, *The Journal of Chemical Physics*, **124** (2006), 044109.

[24] Y. Cao, D.T. Gillespie and L.R. Petzold, Adaptive explicit-implicit tau-leaping method with automatic tau selection, *The Journal of Chemical Physics*, **126** (2007), 224101.

[25] N.M. Dixit and A.S. Perelson, Complex patterns of viral load decay under antiretroviral therapy: Influence of pharmacokinetics and intracellular delay, *J. of Theoretical Biology*, **226** (2004) 95–109.

[26] S.N. Ethier and T.G. Kurtz, *Markov Processes: Characterization and Convergence*, J. Wiley & Sons, New York, 1986.

[27] D.M. Fargue, Réducibilité des systèmes héréditaires a des systèmes dynamiques, *C. R. Acad. Sci. Paris Ser. B*, **277** (1973), 471–473.

[28] M. Gibson and J. Bruck, Efficient exact stochastic simulation of chemical systems with many species and many channels, *The Journal of Physical Chemistry*, 104 (2000), 1876–1899.

[29] D.T. Gillespie, A general method for numerically simulating the stochastic time evolution of coupled chemical reactions, *The Journal of Computational Physics*, **22** (1976), 403–434.

[30] D.T. Gillespie, Approximate accelerated stochastic simulation of chemically reacting systems, *The Journal of Chemical Physics*, **115** (2001), 1716–1733.

[31] D.T. Gillespie and L.R. Petzold, Improved leap-size selection for accelerated stochastic simulation, *The Journal of Chemical Physics*, **119** (2003), 8229–8234.

[32] Y. Hu, T. Li and B. Min, The weak convergence analysis of tau-leaping methods: revisited, *Commun. Math. Sci.*, **9** (2011), 965–996.

[33] G.M. Kepler, H.T. Banks, M. Davidian and E.S. Rosenberg, A model for HCMV infection in immunosuppressed recipients, *Mathematical and Computer Modelling*, **49** (2009), 1653–1663.

[34] T.G. Kurtz, Solutions of ordinary differential equations as limits of pure jump Markov processes, *J. Appl. Prob.*, **7** (1970), 49–58.

[35] T.G. Kurtz, Limit theorems for sequences of pure jump Markov processes approximating ordinary differential processes, *J. Appl. Prob.*, **8** (1971), 344–356.

[36] T.G. Kurtz, *Approximation of Population Processes*, SIAM, Philadelphia, 1981.

[37] T. Li, Analysis of explicit tau-leaping schemes for simulating chemically reacting systems, *Multiscale Model. Simul.*, **6** (2007), 417–436.

[38] N. MacDonald, *Time Lags in Biological Models*, Lecture Notes in Biomathematics Vol. **27**, Springer-Verlag, Berlin, 1978.

[39] H.H. McAdams and A. Arkin, Stochastic mechanisms in gene expression, *PNAS*, **94** (1997), 814–819.

[40] N. Minorsky, Self-excited oscillations in dynamical systems possessing retarded actions, *Journal of Applied Mechanics*, **9** (1942), A65–A71.

[41] N. Minorsky, On non-linear phenomenon of self-rolling, *Proceedings of the National Academy of Sciences*, **31** (1945), 346–349.

[42] N. Minorsky, *Nonlinear Oscillations*, Van Nostrand, New York, 1962.

[43] A.R. Ortiz, H.T. Banks, C. Castillo-Chavez, G. Chowell and X. Wang, A deterministic methodology for estimation of parameters in dynamic Markov chain models, *Journal of Biological Systems*, **19** (2011), 71–100.

[44] J. Pahle, Biochemical simulations: Stochastic, approximate stochastic and hybrid approaches, *Brief Bioinform*, **10** (2009), 53–64.

[45] A.S. Perelson, P. Essunger, Y.Z. Cao, M. Vesanen, A. Hurley, K. Saksela, M. Markowitz and D.D. Ho, Decay characteristics of HIV-1-infected compartments during combination therapy, *Nature*, **387** (1997), 187–191.

[46] J.E. Pearson and P. Krapivsky and A.S. Perelson, Stochastic theory of early viral infection: Continuous versus burst production of virions, *PLoS Comput. Biol.*, **7** (2011), e1001058.

[47] A.S. Perelson and P.W. Nelson, Mathematical analysis of HIV-1 dynamics in vivo, *SIAM Review*, **41** (1999), 3–44.

[48] M. Rathinam, L.R. Petzold, Y. Cao and D.T. Gillespie, Stiffness in stochastic chemically reacting systems: The implicit tau-leaping method, *The Journal of Chemical Physics*, **119** (2003), 12784–12794.

[49] M. Rathinam, L.R. Petzold, Y. Cao and D.T. Gillespie, Consistency and stability of tau-leaping schemes for chemical reaction systems, *Multiscale Model. Simul.*, **4** (2005), 867–895.

[50] E. Renshaw, *Modelling Biological Populations in Space and Time*, Cambridge Univ. Press, Cambridge, 1991.

[51] M. Roussel and R. Zhu, Validation of an algorithm for delay stochastic simulation of transcription and translation in prokaryotic gene expression, *Physical Biology*, **3** (2006), 274–284.

[52] S. Ruan, Delay differential equations in single species dynamics, in *Delay Differential Equations and Applications*, O. Arino et al. (eds.), Springer, Berlin, 2006, 477–517.

[53] R. Schlicht and G. Winkler, A delay stochastic process with applications in molecular biology, *Journal of Mathematical Biology*, **57** (2008), 613–648.

[54] K. Wendelsdorf, G. Dean, S. Hu, S. Nordone and H.T. Banks, Host immune responses that promote initial HIV spread, *Journal of Theoretical Biology*, **289** (2011), 17–35.

[55] D.J. Wilkinson, Stochastic modelling for quantitative description of heterogeneous biological systems, *Nature Reviews Genetics*, **10** (2009), 122–133.

[56] A. Wörz-Busekros, Global stability in ecological systems with continuous time delay, *SIAM J. Appl. Math.*, **35** (1978), 123–134.

Frequently Used Notations and Abbreviations

Frequently Used Notations

- $\mathbb{N}$: the set of natural numbers

- $\mathbb{R}^n$: n-dimensional Euclidean space; $\mathbb{R}^1 = \mathbb{R}$

- $\mathbb{R}^+$: the set of non-negative real numbers

- $\mathbb{R}^{n \times m}$: the set of $n \times m$ matrices with real entries; $\mathbb{R}^{n \times 1} = \mathbb{R}^n$

- $\mathbb{Z}^n$: the set of n-dimensional column vectors with integer components

- $\partial \mathbb{A}$: the boundary of the set $\mathbb{A}$

- $\mathbb{A}^c$: the complement of the set $\mathbb{A}$

- Prob$\{\cdot\}$: a probability function (measure) defined on a given measurable space $(\Omega, \mathcal{F})$

- δ: Dirac delta function; that is, $\delta(x - x_j) = \begin{cases} 0 & \text{if } x \neq x_j \\ \infty & \text{if } x = x_j \end{cases}$ with the property that $\displaystyle\int_{-\infty}^{\infty} h(x)\delta(x - x_j)dx = h(x_j)$

- δ_{jk}: the Kronecker delta; that is, $\delta_{jk} = \begin{cases} 1 & \text{if } j = k \\ 0 & \text{otherwise} \end{cases}$

- $\exp(\cdot)$: exponential function

- $\ln(\cdot)$: natural logarithm function (i.e., $\log_e(\cdot)$)

- $\mathbb{E}(\cdot)$: expectation/mean of a random variable/vector

- Var$(\cdot)$: variance of a random variable/vector

- Cor$\{\cdot, \cdot\}$: correlation of two random variables or correlation function of a stochastic process

- $\mathrm{Cov}\{\cdot,\cdot\}$: covariance of two random variables/vectors or covariance function of a stochastic process

- $\mathcal{N}(\cdot,\cdot)$: normal or multivariate normal distribution

- $\mathcal{N}_{[\underline{x},\bar{x}]}(\cdot,\cdot)$: truncated normal distribution with range $[\underline{x},\bar{x}]$

- $\mathrm{Exp}(\cdot)$: exponential distribution

- $\mathrm{Gamma}(\cdot,\cdot)$: gamma distribution

- χ_k^2: chi-square distribution with k degrees of freedom

- Superscript T: transpose of a vector/matrix (e.g., $\mathbf{x}^T$ denotes the transpose of $\mathbf{x}$)

- $\approx$: approximately equal to

- $\ll$: much smaller than

- $\gg$: much larger than

- $\sim$: distributed as

- $\xrightarrow{\mathrm{d}}$: convergence in distribution

- l.i.m.: mean square convergence

- $\mathbf{I}_n$: an $n \times n$ identity matrix

- $\mathbf{0}_{n \times m}$: an $n \times m$ matrix with all the entries being zeros; $\mathbf{0}_{n \times 1} = \mathbf{0}_n$

- $\mathrm{diag}(a_1, a_2, \ldots, a_m)$: an $m \times m$ diagonal matrix with the main diagonal entries being $a_1, a_2, \ldots, a_m$

- $(a_{ij})_{m \times l}$: an $m \times l$ matrix with the (i,j)th entry being a_{ij}

- $\mathbf{a} \cdot \mathbf{b}$: the dot product of two vectors $\mathbf{a} = (a_1, a_2, \ldots, a_m)^T$ and $\mathbf{b} = (b_1, b_2, \ldots, b_m)^T$; that is, $\mathbf{a} \cdot \mathbf{b} = \mathbf{a}^T \mathbf{b} = \displaystyle\sum_{j=1}^{m} a_j b_j$

- $\mathrm{column}(\mathbf{a}, \mathbf{b})$: defined as $\begin{pmatrix} \mathbf{a} \\ \mathbf{b} \end{pmatrix}$

- $o(\Delta t)$: the little-o notation with the meaning $\displaystyle\lim_{\Delta t \to 0} \frac{o(\Delta t)}{\Delta t} = 0$

Abbreviations

- AIC: Akaike information criterion
- a.s.: almost surely
- ASE: asymptotic standard error
- CI: confidence interval
- CTMC: continuous time Markov chain
- CV: constant variance
- DDE: delay differential equation
- DRF: dielectric response function
- FIM: Fisher information matrix
- FPPS model: Fokker–Planck physiologically structured model
- GLS: generalized least squares
- GRDPS model: growth rate distributed physiologically structured model
- i.i.d.: independent and identically distributed
- K–L information: Kullback–Leibler information
- MLE: maximum likelihood estimation
- m.s.: mean square
- NCV: non-constant variance
- NPML: non-parametric maximum likelihood
- ODE: ordinary differential equation
- OLS: ordinary least squares
- PMF: Prohorov metric framework
- RDE: random differential equation
- RSS: residual sum of squares
- SDE: stochastic differential equation
- SE: standard error

- SSA: stochastic simulation algorithm
- TIC: Takeuchi's information criterion
- WLS: weighted least squares

Index